COASTAL AND MARINE GEO-INFORMATION SYSTEMS

Coastal Systems and Continental Margins

VOLUME 4

Series Editor

Bilal U. Haq

Editorial Advisory Board

M. Collins, *Dept. of Oceanography, University of Southampton, U.K.*
D. Eisma, *Emeritus Professor, Utrecht University and Netherlands Institute for Sea Research, Texel, The Netherlands*
K.E. Louden, *Dept. of Oceanography, Dalhousie University, Halifax, NS, Canada*
J.D. Milliman, *School of Marine Science, The College of William & Mary, Gloucester Point, VA, U.S.A.*
H.W. Posamentier, *Anadarko Canada Corporation, Calgary, AB, Canada*
A. Watts, *Dept. of Earth Sciences, University of Oxford, U.K.*

Coastal and Marine Geo-Information Systems

Applying the Technology to the Environment

Edited by

David R. Green

*Centre for Marine and Coastal Zone Management (CMCZM),
Department of Geopgraphy and Enrivonment,
University of Aberdeen, Scotland, U.K.*

and

Stephen D. King

*Centre for Marine and Coastal Zone Management (CMCZM),
Department of Geopgraphy and Enrivonment,
University of Aberdeen, Scotland, U.K.*

KLUWER ACADEMIC PUBLISHERS
DORDRECHT / BOSTON / LONDON

A C.I.P. Catalogue record for this book is available from the Library of Congress.

ISBN 0-7923-5686-1

Published by Kluwer Academic Publishers,
P.O. Box 17, 3300 AA Dordrecht, The Netherlands.

Sold and distributed in North, Central and South America
by Kluwer Academic Publishers,
101 Philip Drive, Norwell, MA 02061, U.S.A.

In all other countries, sold and distributed
by Kluwer Academic Publishers,
P.O. Box 322, 3300 AH Dordrecht, The Netherlands.

Printed on acid-free paper

All Rights Reserved
© 2003 Kluwer Academic Publishers and copyright holders as specified
on appropriate pages within.
No part of this work may be reproduced, stored in a retrieval system, or transmitted
in any form or by any means, electronic, mechanical, photocopying, microfilming, recording
or otherwise, without written permission from the Publisher, with the exception
of any material supplied specifically for the purpose of being entered
and executed on a computer system, for exclusive use by the purchaser of the work.

Printed in the Netherlands.

Dedicated to our Parents

Table of contents

PART I: THE SETTING

1 The Coastal Zone Environment: A Place to Work, Rest, Play and to Manage
S.D. King

Introduction	1
The Need for Better Coastal Management	2
Managing the Coastal Landscape in the UK	3
Some Problems with the SMP	4
A Different Approach	5
Legislation	6
Sea Defence (Flooding)	6
Coast Protection (Erosion)	7
New Strategies for Defence	8
Sediment Cells	10
Voluntary Coastal Defence Groups	10
Shoreline Management Plans (SMPs)	11
Coastal Processes	11
Coastal Defence	11
Coastal Defence Schemes	12
Offshore Breakwaters	13
Beach Nourishment	13
Artificial Seaweed	15
Summary	16

2 The North East Coastline of Scotland
J.S. Smith

The Geological Skeleton	21
Glaciations, Deglaciation and Changing Land-Sea Relationships	23
Shoreline Regularisation	24
The Building of Dune Systems and Human Impacts	24
Offshore Topography and Deposits - The Coastal Sediment Bank	26
Sustainability of the Beach/Dune Resources	29
Coastal Zone Land Uses	30
Conclusion	32

PART II: THE COASTAL ZONE

3 Coastal/Marine GI/GIS - A Pan-European Perspective
R. A. Longhorn

Introduction	35
Diverse Coastal Zone Activities Require Diverse Programmes of Research	36
Initiatives from EU Institutions	38
DG XII Programmes with Marine and Coastal Zone Related Actions	41
DG Joint Research Centre (JRC), Ispra, Italy	43
European Environment Agency (EEA)	45
Initiatives from Non-EU Institutions	46
Initiatives and Actions for the Mediterranean and/or Black Sea	48
Initiatives and Actions for the Baltic Sea	49
International Initiatives/Programmes Active in Europe	51
Potential Impact of the EU's Fifth Framework Programme for RTD (1998-2002)	53
GI2000: Towards a European Policy Framework for Geographic Information	54
GISDATA and AGILE	55
Conclusion and Recommendations	56

4 Plans for the Coastal Zone
P.A.G. Watts

Introduction	61
A Map-Oriented Appraisal for the Coastal Zone	62
Mapping (Geographic Information) and the Coastal Zone	63
Mapping (Geographic Information) for the Coastal Zone	64
The GIS Dimension	66
Mapping the Changes	66
Matching Mapping (Geospatial Information) to Applications	67
Shoreline Management Planning	67
Coastal Communications	69
Contingency Planning	69
Flood Risk and Hazard Analysis	70
Why GIS for Best Practice?	70
The National Geospatial Data Framework and the Coastal Zone	71
Coastal Map Data Availability and Accessibility	72
Summary	72

5 Hydrographic Data and Geographical Information Systems
P. Wright

Introduction	75
Stage 1 of the HO/OS Coastal Zone Mapping Project	76

	Stage 2 of the HO/OS Coastal Zone Mapping Project	76
	Applications	76
	Data Compatibility	81
	Data Suitability	83
	Way Ahead	83

PART III: EXAMPLE APPLICATIONS

6   GIS for Sustainable Coastal Zone Management in the Pacific - A Strategy
B. Crawley and J. Aston

Introduction	85
International and Regional Initiatives	86
Data Requirements for Coastal Management	90
Constraints to Coastal Management in the Pacific	91
Information Constraints and Sources	94
Conclusion	94

7   Managing Marine Resources: The Role of GIS in EEZ Management
S. Fletcher

Introduction	97
United Nations Convention on the Law of the Sea	98
Rights and Responsibilities in the EEZ	99
EEZ Management	99
Co-operation Amongst Coastal Nations	100
Information Requirements	101
The Role of GIS in EEZ Management	102
Conclusion	103

8   The Management Plan of the Wadden Sea and its Visualisation
M.A. Damoiseaux

Introduction	105
The Management Plan of the Wadden Sea	106
Use of GIS	106
The Sector Notes	107
The Management Plan	109
Information Management for the benefit of the Wadden Sea	109
New Initiatives	111
Acknowledgements	111

9 **Using GIS For Siting Artificial Reefs - Data Issues, Problems And Solutions: 'Real World' To 'Real World'**
   D.R. Green and S.T. Ray

Introduction	114
Artificial Reef Siting	115
Moray Firth Project	116
Datasets	124
Discussion	127
Solutions and Recommendations	128
Summary and Conclusions	128
Additional Reading on Artificial Reefs	131

10 **Collating the Past for Assessing the Future: Analysis of the Subtidal and Intertidal Data Records Within GIS**
   C.I.S. Pater

Introduction	133
Research Objectives	134
Shoreline Management Plans	136
The Selection of Datum	137
Digital Representation of Subtidal and Intertidal Topography	138
Boston Port Authority Surveys	138
Environment Agency Intertidal Levelling	140
Remote Sensing and Aerial Imagery	141
Critique of Methodology	141
Historical and Modern Inconsistencies	143
Suitability of Analysis within a GIS	144
The Value of GIS to Shoreline Management Plans	144
Conclusions	146
Acknowledgements	146

11 **Identifying Sites for Flood Protection - A Case Study from the River Clyde**
   G. Jones

Introduction	149
Main Reasons for Attempting to Predict Flood Risk	150
Predictions for Sea Level Change	150
Computer Model of the Clyde Estuary	151
Interpreting the Results	154
Analysis of Flood Scenario	156
Conclusion	159

12	**Arctic Coastal and Marine Environmental Monitoring**	
	H. Goodwin and R. Palerud	
	Background	163
	Data Collection	164
	Software Development	165
	Data Input to the Environmental Database	166
	GIS Interface	166
	Modelling with GIS	167
	The Future	170
13	**A GIS Application for the Study of Beach Morphodynamics**	
	L.P. Humphries and C.N. Ligdas	
	Introduction	174
	Data Development and Analysis	175
	Results	179
	Discussion	184
	Conclusions	188
14	**Determination and Prediction of Sediment Yields from Recession of the Holderness Coast**	
	R. Newsham, P.S. Balson, D.G. Tragheim and A.M. Denniss	
	Introduction	191
	The Holderness GIS	193
	Calculation of Sediment Yield	195
	Calculation of Sediment Yields by Lithology	198
	Future	198
	Summary	198
	Acknowledgements	199
15	**Tracing the Recent Evolution of the Littoral Spit at El Rompido, Huelva (Spain) Using Remote Sensing and GIS**	
	J. Ojeda Zújar, E. Parrilla, J. Márquez Pérez, and J. Loder	
	Introduction	201
	Time Scales	202
16	**Littoral and Shoreline Processes in Large Man-Made Lakes**	
	A. Sh. Khabidov	
	Introduction	205
	Main Peculiar Features of Man-Made Lakes	206
	Surface Processes in the Coastal Zone	208

	Conclusions	211
	Acknowledgement	212
17	**Coastal Zone Management: The Case of Castellón** *A. Lloret Rodríguez, and J.M. de la Peña*	
	Introduction	213
	Location and Problem Aspects of the Castellón Coast	214
	Methodological Aspects	215

PART IV: HABITAT

18	**Evaluating the Coastal Environment for Marine Birds** *S. Wanless, P.J. Bacon, M.P. Harris, and A.D. Webb*	
	Introduction	221
	Study Area	222
	The Foraging Model	222
	Identification of Sea Areas of Importance to Marine Birds	225
	Indirect Effects of a Notional Oil-Spill	228
	Notional Effects of Reduced Food Supply	229
	Discussion	231
	Acknowledgements	231
19	**Initial Attempts to Assess the Importance of the Distribution of Saltmarsh Communities on the Sediment Budget of the North Norfolk Coast** *N.J. Brown, R.Cox, R.Pakeman, A.G.Thomson, R.A.Wadsworth and M.Yates*	
	Introduction	233
	Method	234
	Results	241
	Discussion	241
	Conclusions	244
	Acknowledgements	244
20	**Quantifying Landscape / Ecological Succession in a Coastal Dune System Using Sequential Aerial Photography and GIS** *S. Shanmugam and M. Barnsley*	
	Introduction	247
	Study Area	250
	Aerial Photography and Aerial Photo-Ecology	250
	Methodology	251
	Vector GIS and Dune-Landscape Succession Analysis	254

	Slack Hydrology as an Attribute	258
	Limitations of this Study	259
	Conclusions	259
	Acknowledgements	260
21	**Geomatics for the Management of Oyster Culture Leases and Production** *J. Populus, L. Loubersac, J. Prou, M. Kerdreux, and O. Lemoine*	
	Introduction	261
	Material and Methods	262
	The Lease Cadastral Maps	263
	The Depth Chart	266
	The Tidal Model	268
	Application to Dredging Plans	270
	Future Prospects	271
	Conclusion	273
22	**GIS and Aquaculture: Soft-Shell Clam Site Assessment** *A. Simms*	
	Introduction	275
	GIS and Aquaculture	276
	Spatial Data Structures	277
	Data Collection	278
	A GIS Approach to Aquaculture Site Assessment	280
	Site Characteristics	281
	Water Quality	282
	Mapping and Analysis of Soft-Shell Clams	284
	Aquaculture Management and GIS	290
	Conclusion	293
23	**Evaluation of Ecological Effects of the North Sea Industrial Fishing Industry on the Availability of Human Consumption Species Using Geographical Distribution Resource Data** *J. Robertson, J. McGlade and I. Leaver*	
	Introduction	297
	The North Sea Industrial Fishery and its Fish	300
	The Regulatory Framework	302
	Feeding Interactions and Stomach Content Data	303
	Method for Estimating Predator Consumption Rates	304
	Diet Content and Total Consumption 1983-95	305
	Biomass Consumption Comparison 1977-1986 and 1987-1994	318
	Ecopath Model	319
	North Sea Fleet Dynamics	325

	The Biological Effect of the North Sea Fisheries	333
	Discussion with Conclusions and Recommendations	334
	Acknowledgements	340
	Glossary of Terms	344
	Appendix 1	344

PART V: TECHNOLOGY

24 Digital Elevation Models by Laserscanning
U. Lohr

Introduction	349
Description of the System	350
DEMs for Various Applications	352
DEM Example of the Coastal Zone	352

25 Error Modeling and Management for Data in Geospatial Information Systems
M.A. Chapman, A. Alesheikh and H. Karimi

Introduction	355
A Decision Support Strategy	356
Error Source Recognition	356
Uncertainty Modeling of Geospatial Objects	359
Polygon Uncertainty Model	365
The Point-in-Polygon Problem	366
Conclusions	368

26 The Use of Dynamic Segmentation in the Coastal Information System: Adjacency Relationships from Southeastern Newfoundland, Canada
K.A. Jenner, A.G. Sherin and T. Horsman

Introduction	371
Mapping	372
Study Area	372
Dynamic Segmentation	375
Methods	375
Discussion	378
Conclusions	382
Acknowledgements	383

27 Consideration on Satellite Data Correction by Bidirectional Reflectance Measurement of Coastal Sand with a Remote Sensing Simulator
H. Okayama and J. Sun

Introduction	385
Experiments and Results	386
Correction of Reflected Intensity	393
Conclusions	395

28 Constructing a Geomorphological Database of Coastal Change Using GIS
J. Raper, D. Livingstone, C. Bristow, and T. McCarthy

Introduction	400
Data Sources for Spatio-Temporal Analysis of Changing Coastal Terrain	402
Integration of Coastal Spatio-Temporal Data Sources	407
Conclusions	412

PART VI: GEOGRAPHIC INFORMATION SYSTEMS AND DECISION SUPPORT SYSTEMS

29 Development of a DSS for the Integrated Development of Thassos Island
H. Coccossis and K. Dimitriou

DSS in ICAM	415
Description of the Case Study	417
Development of DSS for Thassos Island	421

30 Development of a Spatial Decision Support System for the Biological Influences on Inter-Tidal Areas (Biota) Project within the Land Ocean Interaction Study
N.J. Brown, R. Cox, A.G. Thomson, R.A. Wadsworth, and M. Yates

Introduction	425
LOIS	426
BIOTA	428
The Decision Support System - Design and Construction	428
The Decision Support System - Example Use	431
Discussion	433
Conclusions	434
Acknowledgments	435

31 User Assessment of Coastal Spatial Decision Support Systems
R. Canessa and C.P. Keller

Introduction	437
Questionnaire Design	438
Decision Making	439
Analysis	442
Data	443
Implementation	445
Conclusion	447

32 Internet-Based Information Systems: The Forth Estuary Forum (FEF) System
D.R. Green and S.D. King

Introduction	451
Access to Information	452
Networking, Communications and Information Technology	453
Decision Support Systems	454
The Internet	455
Internet and GIS	456
The FEF Pilot Information System	457
Some Basic Considerations	457
Potential Problem Areas	461
Summary and Conclusions	462

33 Mike Info *Coast* - A GIS-Based Tool for Coastal Zone Management
R. Andersen

Background	467
What is MIKE INFO *Coast*?	468
A Situation of Typical Use	469
Data Handled by MIKE INFO *Coast*	469
Functionalities in MIKE INFO *Coast*	470
Concluding Remarks	470

PART VII: REMOTE SENSING

34 Matching Issue to Utility: An Hierarchical Store of Remotely Sensed Imagery for Coastal Zone Management
S.D. King and D.R. Green

Introduction	474
The Importance of Geo-Spatial Information in Environmental Management	475
Using Remote Sensing to Collect Environmental Data	475

Scale and Spatial Resolution in Remote Sensing	475
Matching Issue and Utility: Using Coastal Zone Management as an Example	476
Moving Towards a Geo-Spatial Coastal Information System	480
Discussion and Conclusions	485

35 Predicting the Distribution of Marine Benthic Biotopes in Scottish Conservation Areas Using Acoustic Remote Sensing, Underwater Video and GIS
C. Johnston and A. Davison

Background to the Work	487
Description of pSAC Areas Surveyed	488
Surveys	490
Ground Validation	491
Interpolation of Acoustic Data	492
Matching of Acoustic and Ground Validation Data	492
Use of the Information	498
Acknowledgements	499

36 Submerged Kelp Biomass Assessment using CASI
É.L. Simms

Introduction	501
Data	502
Methods	503
Results and Discussion	504
Conclusion	507
Acknowledgments	508

37 Monitoring Coastal Morphological Changes Using Topographical Methods, Softcopy Photogrammetry and GIS, Huelva (Andalucia, Spain)
J.Ojeda. Zújar, L. Borgniet, A.M. Pérez Romero, and J. Loder

Introduction to the Study Area	511
Recent Foredune Evolution (1979-1996)	514
Objectives	515
Sources and Methodology	517
Results and Interpretation	518
Interpretation	519
Foredune Volume Calculation and Sediment Budget	520
Results	521
Interpretation	521
Volumetric Spatial Changes 1989-1994	522
Interpretation	522

	3D Graphic Presentation and Animation	522
	Conclusions	523
38	**Characterization of Coastal Waters for the Monitoring of Pollution by Means of Remote Sensing; the Use of Satellite Imagery to Establish the Appropriate Pattern for Timing and Location of Sampling in Coastal Waters**	
	J. Ojeda Zújar, L. Borgniet, A.M. Pérez Romero, and J. Loder	
	Introduction	525
	Study Area	526
	Plan for the Monitoring of Andalucian Coastal Waters	526
	Methods and Results	529
	Zoning the Estuary	531
	Conclusions : Proposals and Recommendations for Water Sample Collection	538

PART VIII: DEVELOPMENTS AND THE FUTURE

39	**European CZM and the Global Spatial Data Infrastructure Initiative (GSDI)**	
	R.A. Longhorn	
	Introduction	543
	History of the GSDI Initiative	544
	Coastal Zone Management and Marine Research in the GSDI Agenda	546
	International Initiatives/Programmes in Relation to the GSDI Discussion	548
	EU Initiatives in Relation to GSDI	551
	GSDI in the New Millennium?	552
40	**Access to Marine Data on the Internet for Coastal Zone Management: The New Millennium**	
	D.R. Green and S.D. King	
	Introduction	555
	Computer Technology	556
	Rapid Development of Technology	556
	Greater Public Awareness	556
	Access to Information	558
	Concerns	559
	GIS and Geospatial Data and Information Delivery	562
	Integrated Coastal Zone Management (ICZM)	564
	CZM Information System Portals or Gateways	564
	A National Coastal Data and Information Resource	564
	Progress to Date	569
	Summary and Conclusions	575

List of Contributors	579
Index	585

CHAPTER 1

The Coastal Zone Environment: A Place to Work, Rest, Play and to Manage

S.D. King

ABSTRACT: Over time growing environmental awareness has made us regard coastal landscapes as nationally and internationally important areas of scenic quality and natural habitat. Unfortunately the relatively recent desire to defend and to protect these particular aspects of the coastal environment does not necessarily fit in with the need to maintain the human aspect of the environment, e.g. our industrial, economic, and leisure activities. Maintaining a coastal environment to suit all the possible user requirements and interests and is difficult to achieve. It is vital to strike a balance between the natural and the man-made activities that operate within a coastal environment in order to sustain them for the benefit of everyone. Unfortunately, to date, an integrated approach to coastal management has not really been undertaken. The result of this has been that more often than not the continued use of a coastal environment has resulted in conflict, the consequences of which have been damaging to other users and aspects. The main problem with coastal management in the past is that it has been very reactionary. There has been little in the way of a planned response, and new ideas have only come about with hindsight. Now is the time to take stock of the situation and to clarify positions.

Introduction

Coasts by their very nature have historically always offered a place for humans to settle, to work, and to play. In order to maintain our coastal settlements, our investments and our pleasure (i.e. our way of life) has necessitated the implementation of various forms of coastal defence and protection from the sea and with it a recognition of the need to maintain an element of environmental constancy and stability for us to survive.

Over time growing environmental awareness, however, has made us regard coastal landscapes as nationally and internationally important areas of scenic quality and natural habitat, amongst many other things. Unfortunately the relatively recent desire to defend and to protect these particular aspects of the coastal environment does not necessarily fit in with the need to maintain the human aspect of the environment, e.g. our industrial, economic, and leisure activities. For example, local property owners may wish to see their land and buildings protected against flooding and erosion. At the same time, however, the conservation of natural features and habitats may depend upon the continued operation of natural coastal processes, which themselves can be affected by defence structures. This has currently made the issue of coastal defence and protection a very contentious one (Lee, 1993). Maintaining a coastal environment to suit all the possible user requirements and interests is difficult and in reality is achieved only by compromise. And yet it is vital to strike a balance between the natural and the man-made activities that operate within a coastal environment in order to sustain them for the benefit of everyone. They complement one another. However, to date, an integrated approach to coastal management has not really been undertaken. The result of this has been that more often than not the continued use of a coastal environment has resulted in conflict, the consequences of which have been damaging to other users and aspects.

The Need for Better Coastal Management

It has been argued by Carter (1988) for example that *"accelerated erosion, disappearing beaches, increased frequency of flooding, progressive siltation, degraded ecosystems and so on"* (p. 431) are all symptomatic of the inability of people to provide competent coastal land management, defence and protection. If this is so, the dilemma faced by coastal zone managers is how to provide areas which are, for example, susceptible to flooding with *"technically, environmentally and economically sound and sustainable defence [and protection] measures"* (MAFF, 1993, p. 3) in the future, while trying to maintain those of the past, and to deal with the problems they may have caused.

The problem currently faced is one of being able to maintain the coastal environment in such a way that it supports both the interests of the natural and the man-made environment, so that both can co-exist in such a way that our economic, social and industrial livelihood can be maintained but set within a natural environmental context. Sustainability is the key word here. In essence management of the coastal environment is about sustaining the natural environment.

It is now recognised that past practices at the coast, such as the construction of harbours, jetties and traditional defence systems may have contributed to the deterioration of the coast. English Nature (1992), for example, have argued that if practices and methods of coastal defence are allowed to continue, then coastlines would be faced with worsening consequences, including:

- **The loss of mudflats and the birds which live on them;**
- **Damage to geological Sites of Special Scientific Interest (SSSIs) and scenic heritage by erosion, due to the stabilisation of the coast elsewhere;**
- **Cutting of sediment supplies to beaches resulting in the loss of coastal wildlife;**
- **Cessation through isolation from coastal processes, of the natural operation of spits, with serious deterioration of rare plants, animals and geomorphological and scenic qualities. (English Nature, 1992)**

Managing the Coastal Landscape in the UK

Coastal landscapes are a result of the natural forces of wind, waves and tides, and many are nationally or internationally important for their habitats and natural features. Whilst there are many different forms of coastal management, which are well documented in the literature, and many of which have been implemented over the years, there is a growing realisation that we do not fully comprehend the coastal processes involved and the consequences of our actions on the coastal environment, either with regard to the natural or the man-made aspects, the results of which are only seen in the effects and changes that we now see taking place.

Coastal zone management has become, in recent years, a very prominent issue in the UK for example. This is demonstrated by the current literature available, including the Department of the Environment (DoE) review of coastal management and planning (1993), Ministry of Agriculture, Fisheries and Foods publications to guide Local Authorities in shoreline defence, conservation and management (1993, 1995, 1996), and a study by Healy and Doody (1995) on coastal management in Europe, many of whose examples come from the UK.

The need to deal with the actual and potential changes in the coast has been recognised and many different approaches have been tried. For example, a number of designations, provided by national and international legislation do exist to aid conservation. One of the most important in Britain is the SSSI "widely considered to be the cornerstone of conservation in Great Britain" (DoE, 1993, p. 71). However, coastal conservation is contentious and problematic and is a much broader issue common to nearly all coasts, particularly those that are mainly undeveloped. The problem the coastal manager faces here is how to balance the needs of the local community and the wishes of conservation groups, within the current legislation.

Defence, protection, and conservation are all-important issues in their own right. However, they are in themselves not necessarily management. An integrated approach to coastal zone management would bring these three issues together, for example, in an attempt to use the coast wisely, provide sustainable development and maintain biodiversity. Yet, in Britain at least, this is still only a political ideal to which there is, at the moment, no perfect solution.

It is generally accepted within the literature that coasts around the UK are physically unique and distinct, and have their own problems and conflicts. An important part of coastal management is the interpretation of national policy at local level.

Although coastal management needs to evolve further, there have, within recent years, been "remarkable changes in attitudes and practices of management of the shoreline" (Hooke and Bray, 1995, p. 336). Perhaps the most significant of these changes is the realisation that management can be based on the concept of the sediment cell. Indeed, MAFF sees this concept as the linchpin of modern integrated coastal zone management because it offers the opportunity for administrative structures to be much more closely related to natural forms and process units along the coast.

The major advances made with sediment cells are the setting up of voluntary coastal defence groups, and encouraging Coastal Protection Authorities (CPAs) to initiate Shoreline Management Plans (SMPs). The coastal defence groups have been able to provide a much needed regional contact for co-ordinating and exchanging information, so that the works of one CPA at the coast will not adversely affect the coast of an adjacent CPA. Likewise, SMPs have given Voluntary Management Groups the opportunity to develop sustainable coastal defence and conservation policies within a sediment cell, and to set objectives for the future management of the coast (MAFF, 1995).

The main focus of an SMP is on coastal defence and protection. It is recognised today that new defence and protection techniques must take account of natural processes, particularly sediment movement, which take place at the coast. The most successful of these techniques is arguably beach nourishment or recharge. Beach recharge appears to cause little disruption to natural processes, and evidence from the Netherlands, where nourishment schemes are actively used, suggest that the environment suffers few adverse effects.

As a nation, our awareness of conservation issues appears to have grown over the past fifty years. This view is justified by the numerous site designations that have been made to protect the environment, and the inclusion of the natural environment as an important consideration within SMPs. The growth of interest in these issues at the coast has caused controversy within coastal management, especially where conservation priorities have clashed with the needs and wishes of business and local residents at the coast. In many respects, such conflicts belie any advances in coastal zone management, as in essence, they are a consequence of poor planning, co-ordination, and hazard management, which have heightened the problems of flooding, erosion and threats to the built and natural environment. However, SMPs should provide a forum for all interests at the coast to arrive at an agreeable solution for all and it is important to recognise that it will be beneficial for conservation agencies, businesses and local residents, to discuss their concerns. At the same time, new defence and protection measures will be developed with respect to conservation priorities so that future conflicts may be avoided. In this way, coastal management is becoming properly integrated.

Some Problems with the SMP

There are, however, problems with basing coastal management solely on sediment cells. Hooke and Bray (1995) debate their usefulness, arguing that the cells are difficult to identify, and the concept hard to apply. This is especially true for finer sediment, as no one can be sure of its exact movements. Lee (1993) also argues that sediment cells will not be relevant to all aspects of coastal management, particularly where landward geomorphological systems are concerned. The ability of plans based entirely on sediment cells to integrate management strategies, may, therefore, be questioned; any framework not bringing all issues together is not totally integrating them.

Another dilemma faced by the SMP initiative is that it is non-statutory, although Jemmett (1995) argues that there are some advantages to this system. He suggests that it "places the responsibility for addressing individual issues on those affected by them and it fosters greater awareness and ownership" (Jemmett, 1995, p449). In other words, the plans, while based on national policy, can become very site specific, geared towards particular needs. However, he also acknowledges that some conflicts will only be resolved by statutory control (Jemmett, 1995). Also, being non-statutory, Local Authorities and consulting agencies do not have to initiate them, and even if they do, they remain little more than advisory measures that can only be enforced if made law within planning documents. Coastal management still relies heavily on the planning system that is less than adequate to deal with all its intricacies.

There is still a lot of room for improvement before it can be said that Britain has a framework for the wise use of the coast, its sustainable development, and maintenance of its biodiversity. The formation of voluntary coastal groups and non-statutory management plans are certainly a start, yet it has been seen how, based only on sediment cells, these may not account for all coastal issues. The other problem is that SMPs are not the only type of management plan, in force, at the coast. There are also estuary, heritage coast, and AONB management plans, which may all overlap. For example, an SMP for cell number 11 will cover the same area as the Solway Firth AONB management plan. Thus, two different methods of management and planning will be applied to the same area. Ultimately, this gives the impression that coastal management still lacks a co-ordinating body with responsibility for all issues.

A Different Approach

The magnitude of the problems associated with coastal management embrace a wide variety of impacts on different temporal and spatial scales (Carter, 1990). To develop studies further, issues such as pollution, recreation and education would also need to be included. Coastal zone management (CZM), could be more holistic, providing a deeper understanding and appreciation of all coastal issues.

Lee (1993) concludes that effective management in the future should be able to resolve conflicts between alternative demands on resources, and ensure that human activity does not significantly affect coastal systems or ecosystems. Carter, (1990) argues that the best way of achieving this will be through a framework which provides consistency and ease of communications. He suggests imposing a new structure, lying across existing areas of responsibility, hence, providing a lead group through which all existing interest groups can be co-ordinated. This is, perhaps, the best way of bridging the gap originally created by the division of coastal defence and protection.

On the other hand, it is hard to see how one body could possibly control the shear volume of work, and conflicts, arising from the many interests at the coast. The main problem with coastal management in the past is that it has been very reactionary. There has been little in the way of planned response, and new ideas have only come about with hindsight. This is partly why there are so many voluntary groups, management plans and conservation designations in operation. Perhaps now is the time to take stock of the situation and clarify positions. It may be that there is now a need for Coastal Authorities based on the boundaries of sediment cells, with statutory powers to govern the coast, but in close liaison with existing Local Authorities and their planners. In addition, new Acts of Parliament should bring coastal legislation up to date, especially that concerning protection and defence; and conservation legislation could be condensed to reduce confusion. Integration between fewer groups would seem to be a far easier prospect.

However, it remains to be seen whether the lack of money, bureaucratic inflexibility and a lack of co-ordination at all levels (Carter, 1990) will allow any such changes to take place. But co-existence can only be achieved through an improved knowledge and a greater understanding of the coastal environment.

Legislation

Traditionally, coastal management has centred on the problems of flooding and erosion, and this is recognised and reflected within the legislation governing the coast in England and Wales over the past century. Flooding has usually been covered within Sea Defence that may be defined as:

protection of human life and property in coastal settlements and industrial areas against sea and tidal flooding.

Similarly, erosion has been covered by Coast Protection, defined as:

protection of coastal settlements, industrial areas, bridges, roads and railway embankments from erosion and encroachment by the sea. (Solway Firth Partnership, 1996)

Thus, flooding and erosion have been treated as two different phenomena, although one may cause or lead to the other.

Sea Defence (Flooding)

In 1907, a Royal Commission was appointed to review land drainage. This resulted in the 1930 Land Drainage Act that included measures for defence against seawater. Forty-nine Catchment Boards were set up to tackle the more urgent problems arising from previous neglect of sea defences. However, these boards could only operate in connection with main rivers, and with the advent of World War II nine years later, defences again fell into neglect. In 1948, the River Boards Act reorganised the Catchment Boards, reducing their number to 34. Only five years later, one of the greatest storms to hit Britain in recent history occurred, with over 1200 breaches along 2000km of coast (Whittle, 1989). The 1953 storm raised important questions about the effectiveness of England's sea defences, and the Waverley Committee reporting on the storm recommended that future defences had to be built to try and withstand tides of similar magnitude to those of 1953. In 1961, the Land Drainage Act extended the River Boards powers to construct sea defence measures wherever the need arose. In 1973, the River Boards were integrated into ten regional Water Authorities with responsibilities for water supply, sewerage, river management and regulatory functions. Flood and sea defence became the responsibility of Land Drainage Committees funded by County Councils. In 1976, the Land Drainage Act gave permissive powers for sea defence against flooding to Local Water Authorities and Local Authorities which were overseen by the Ministry of Agriculture, Fisheries and Food (MAFF).

Coast Protection (Erosion)

Not unlike sea defence, legislation for coast protection was first started by a Royal Commission in 1906. It recommended that there should be controls on the removal of beach material, and that Local Authorities' works should be brought under central supervision. However, at that time they found no case for making coastal protection a Central Government responsibility. In 1939, the Coast Protection Act introduced provisions for the control and removal of beach material, and the Board of Trade was empowered to make orders restricting the removal of beach material wherever there was thought to be a danger of erosion. World War II saw the deterioration of protective structures, and by 1946, neglect and storm action had allowed erosive damage. It was apparent that private land owners lacked the funds to provide protection works, and the new Coast Protection Act of 1949 for the first time gave Local Authorities powers to carry out works under the general supervision of Central Government. This act was supervised by the Department of the Environment (DoE) until 1985, and is still used to administer funding today.

The Land Drainage Act 1976 and the Coast Protection Act 1949 appear to have separated flooding and erosion into two distinct issues. However, Carter (1988) argues that this historical dichotomy of interests may not have been the best way forward for coastal management. Far from being integrated, management in many ways has been fragmented and ill co-ordinated, with two similar problems being dealt with by two different government departments; MAFF and DoE. Park (1989) believes that this distinction may be particularly problematic where uncontrolled encroachment by the sea leads to flooding. In other words, a matter of erosion becomes one of flooding as well, and the difference between defence and protection becomes a grey area. In the wake of the 1953 storms, the Waverley Committee examined this point. Although they thought that there might be some advantage in bringing defence and protection together under one body, they believed that the River Boards and Coast Protection Authorities had different functions to perform, concluding that "there were no stretches of coast where interests were duplicated or left completely uncovered by either body" (Park, 1989, p. 14). By comparison, Carter (1988) maintains that the distinction served only to exacerbate conflicts at the coast, particularly where funding for coastal works was concerned. An excellent example of this, common in the literature, is known as the Whitstable Judgement.

In 1978, Canterbury City Council set out proposals for a coastal scheme at Whitstable, consisting of a 22-foot high wall, protecting against erosion and flooding. It was defined as a protection scheme and, therefore, applications for funding were sent to the DoE. However, the DoE determined that the main purpose of the wall was to prevent flooding, and was, therefore, more appropriately administered under the Land Drainage Act at a lower rate of funding. Canterbury City Council did not accept this judgement, and the disagreement reached the High Court, who in 1980 decided that erosion, encroachment and flooding are often inseparable. Therefore, works preventing both flooding and erosion could be considered under the Coast Protection Act.

The Whitstable Judgement led to a review of separate coast protection and sea defence schemes, and in 1985, a key institutional change brought defence and protection schemes under the sole jurisdiction of MAFF. Essentially this meant that defence and protection became combined under the term *coastal works*, to be administered by Water Authorities in collaboration with Maritime Councils, but funded at the lower sea defence rate.

This institutional change may be seen as an attempt at unified management strategies. However, the Coast protection Act 1949 and Land Drainage Act 1976 were still in force, and therefore sea defence and coast protection could still be judged as separate issues. Park (1989) argues that there is no evidence that this causes difficulties, the Whitstable incident being a rare situation. Despite this, others, such as Hooke and Bray (1995), conclude that the overall legislative procedure has resulted in a number of problems at the coast, and not just those of funding.

Hooke and Bray (1995) suggest that the two Acts have subdivided shoreline management according to the boundaries of district councils, which, they say, has resulted in a complex mix of authorities, each with responsibility for often short stretches of a shore. Such subdivisions have taken no account of coastal processes that may well have resulted in the coastal works of one authority affecting the coastline of their neighbouring authority.

As Sims and Ternan (1988) propose, there has been "no mechanism for overall concern or responsibility [at the coast]" (Sims and Ternan, 1988, p. 240) within the coastal legislation. In other words, the distinction between defence and protection, and the subdivision of shoreline management between local authorities has allowed coastal structures to be built without liaison between different groups, neglecting the importance of coastal processes.

These problems have been recognised to a certain extent. Now MAFF are advising all Coastal Protection Authorities (CPAs) to prepare Shoreline Management Plans (SMPs) which should aim to provide "a framework for the development of sustainable coastal defence policies within a sediment cell or sub-cell and to set objectives for the future management of the shoreline" (Solway Firth Partnership, 1996, p226). MAFF are, therefore, encouraging CPAs to liase with one another more closely and to take a greater account of coastal processes.

New Strategies for Defence

Since 1985, the responsibility for flood and coastal defence in England and Wales has been with MAFF, who administer the legislation that enables coastal works to be carried out. The following Acts are relevant for such work today:

- **Coast Protection Act 1949**
- **Land Drainage Act 1991**
- **(amended by Land Drainage Act 1994)**
- **Water Resources Act 1995**

MAFF contributes funding, national strategic guidance and specialist help, backed up by what they call a comprehensive research and development programme (MAFF, 1996). Individual flood and coastal defence works are designed, constructed and maintained by local operating authorities. These include:

The Coastal Zone Environment

- **THE ENVIRONMENT AGENCY** (Took on all the responsibilities of the National Rivers Authority NRA on 1st April 1996): They supervise all matters relating to flood and coastal defence in England and Wales.

- **INTERNAL DRAINAGE BOARDS (IDBs):** There are 235 IDBs in England. They have powers to carry out measures to alleviate flooding in districts with special drainage needs other than on main rivers.

- **LOCAL AUTHORITIES:** Can carry out works on water courses other than main rivers and those in IDB areas to alleviate flooding by rivers or by the sea.

- **MARITIME DISTRICT COUNCILS** (adjoin the sea): Have powers to protect the land against erosion or encroachment by the sea. (MAFF, 1996)

MAFF aims to reduce the risks to people and the developed and natural environment from flooding and erosion, their greatest priority being the protection of human life. Their proposals for succeeding in this include flood warning schemes, flood and coastal defence measures, and, perhaps most interestingly, discouraging inappropriate development in areas at risk from flooding and coastal erosion. As Lee (1993) argues "coastal planning has failed to take account of the dynamic nature of the coast" (Lee, 1993, p.172) and, therefore, development has occurred in hazardous areas necessitating defences which may not otherwise have been needed. Where defences are seen as necessary, MAFF hopes to promote new and improved defence measures, including natural defences such as beaches and sand dunes, although the term natural appears somewhat mis-used if the defences are still instigated by human activity.

Before approval, all defence schemes must meet three criteria:

i) **They must be sound in engineering terms.**
ii) **They must be environmentally acceptable.**
iii) **They must be economically worthwhile.**

All three criteria appear very subjective, yet they allow scope for adjustment to specific site needs and interests.

It has been seen that in the past, legislation and the lack of knowledge about coastal processes has allowed unsuitable defences to be built in many areas. "Nowadays, the idea that coastal processes do not operate conveniently within fixed administrative boundaries is central to modern shoreline management" (Lee, 1993, p.175). MAFF is, therefore, proposing a more strategic approach, so that authorities involved consider a wide range of possible approaches and the impacts defence measures may have on neighbouring authorities. They believe in two ways of achieving this; through voluntary coastal defence groups and shoreline management plans, both of which are based on the concept of sediment cells.

Sediment Cells

Sediment cells, as defined by Hooke and Bray (1995) are more or less self-contained circulations of coarse sediment along the coast, and are based on a systems approach of inputs, transfers, stores and outputs. Research undertaken on behalf of MAFF suggest that the coastline can be divided into 11 major cells, which can be further sub-divided. The boundaries of cells generally coincide with large estuaries and prominent headlands (MAFF, 1993). For example, cell no.11 covers the north west coast from Great Orme in North Wales to the River Eden in Cumbria. It is divided into five sub-cells. As sediment cells appear to reflect natural coastal process boundaries, many groups, such as MAFF, see them as the linch-pin of integrated coastal planning and protection. Sediment cells offer the opportunity for new administrative structures to be much more closely related to natural forms and process units along the coast. However, Hooke and Bray (1995) debate their usefulness, highlighting a number of problems:

i) *There are difficulties in identifying and applying the concept of sediment cells. MAFF themselves admit that "the boundaries of sub-cells are not definitive, they are based on the best available knowledge of large scale processes, and may need to be revised as further information becomes available" (MAFF, 1995, p1)*

ii) *The concept is not easily applied to finer sediments which are more easily transported offshore in suspension. Therefore, we cannot be sure of the exact movements of all coastal sediments.*

Hooke and Bray (1995) suggest other criteria for dividing the coast into units, such as economic, sociological and ecological, although they acknowledge that "a morphological basis is the least ambiguous and the most functional option for coastal defence" (Hooke and Bray, 1995, p. 363).

Voluntary Coastal Defence Groups

Over the past decade, MAFF has encouraged the formation of informal coastal defence groups to co-ordinate and exchange information between neighbouring authorities. Such groups provide a much-needed regional contact for coastal defence, trying to ensure that one authority's defence scheme does not affect adjacent authorities coastlines.

The House of Commons Select Committee on the Environment (1992) see the existence of such groups as the future for coastal management, recommending that they become statutory and formally integrated coastal authorities, thus facilitating management with nature rather than using artificial boundaries. Nevertheless, the Government has so far rejected this advice, preferring to stay within existing administrative units. This, perhaps, is because of two problems outlined by Lee (1993). First, the current defence groups are not entirely based on sediment cells, and secondly, they are not directly relevant to other aspects of coastal management, such as land use. However, as long as these defence groups remain non-statutory, they will continue to be little more than advisory quangos with no powers of their own.

The Coastal Zone Environment

Shoreline Management Plans (SMPs)

More closely related to the concept of sediment cells is the SMP, "a document which sets out defence strategy for coastal defence for a specified length of coast taking account of natural coastal processes and human and other environmental influences and needs" (MAFF, 1995, p. 1). SMPs are aimed at the local Coastal Protection Authorities (CPAs) and should provide the opportunity to develop sustainable coastal defence policies within a sediment cell, and to set objectives for the future management of the shoreline (MAFF, 1995). Coastal defence groups, in theory, provide a forum for CPAs and other interested groups to discuss the needs of a sediment cell and develop a forward-looking SMP.

In many ways the SMPs face the same problems as coastal defence groups, in that they are based on sediment cell boundaries and are non-statutory and not compulsory. They may, however, be useful in informing decisions on Structure, Local and Unitary Development Plans which do become law.

If a CPA does design an SMP, they are supposed to review:

i) **Coastal processes**
ii) **Coastal defences**
iii) **Land use and the human and built environment**
iv) **The natural environment**

Coastal Processes

Coastal processes are important because they can determine the past and future evolution of a coastline. Beaches are continually changing, and the processes of erosion and accretion are part of the natural tendency of the coast to find an equilibrium. Structures implemented at the coast by humans can have a significant affect on natural processes, particularly sediment transport; no matter the size of the structures, from groynes to harbours, they can still lock up supplies of sediments, affecting long-shore drift and rates of erosion and accretion.

Tide and wave-induced currents have a major bearing on the transport of seabed sediments.

Coastal Defence

In the DoE review of coastal planning and management (1993), it was suggested that the coastal zone of England has an inheritance of what were (in modern terms) poorly designed coastal defence structures.

Coastal structures such as harbours and jetties act as a dam to littoral drift. There is often extensive shoreline advance on the updrift side of the harbour, while erosion can occur on the downdrift side. Deposition and erosion can extend large distances from the structures (Komar, 1983).

Coastal Defence Schemes

There is a distinction between coastal defence and coast protection, yet most schemes do, in effect, serve to prevent both flooding and erosion. Nevertheless, there is one fundamental difference that should always be remembered. Whereas alleviating flooding can be permanent, "erosion is an on-going process that can normally only be delayed" (Penning-Rowsell, 1992, p. 10). Defence mechanisms may increase erosion rates, but they are not usually the sole cause.

Increased erosion rates can be a serious problem, and the poor design of defence structures is often to blame. "By tradition engineers have relied on interposing static structures between the sea and the shore" (Carter, 1988, p. 443) to prevent flooding and erosion . Concrete walls tend to be durable, immovable and capable of withstanding massive pressures. This type of defence is intended to be long-lived and should be carefully engineered to specific site factors. These are often termed hard defences, which armour the coast "and introduce materials alien [to it]" (Moutzouris, 1995, p.154). Although hard defences undoubtedly protect the land behind them from flooding and erosion, they can cause significant problems elsewhere. Sea walls are static and conflict with dynamic beach changes (Carter, 1988). Wakelin argues that "hard defences of all kinds carry within themselves the seeds of their own destruction" (Wakelin, 1989, p. 140). Their greater reflectivity of wave energy, even where stepped, causes greater turbulence under wave action, encouraging erosion of beach material on the foreshore. Not only does this damage the beach, but erosion can also expose the toe pilings and foundations of the wall. Solid sea walls also impede land-sea sediment exchanges, preventing the sea from winning material that would normally be transported further up the coast. Groynes "are wall like structures inserted perpendicular to the beach to arrest material drifting alongshore" (Carter, 1988, p. 452). Groynes are often seen as a panacea for all erosion problems, yet while they are useful under some circumstances, they can be "positively harmful" (Wakelin, 1989, p. 141) in others.

There are two functional types of groyne:

1) **The anchor groyne.**
2) **The terminal groyne.**
3) **Rock revetment**

The anchor groyne is designed to stabilise a length of shoreline by capturing a proportion of the material drifting along the shore, or by retaining a quantity of introduced fill. Once the material is caught, the design should allow for bypassing, either over or around the groyne, so as to maintain drift. Problems arise where bypassing does not occur, and sediment is prevented from reaching other parts of the coast. All groynes can act like small jetties, effectively stopping long-shore transport and enhancing erosion further down the coast. To date there appears to be little favourable evidence to support the use of groynes; "the inescapable truth that groynes are a static element within an essentially dynamic environment, mitigates against optimising their effectiveness" (Carter, 1988, p. 453).

Rock revetments are large boulders, placed at the shoreline as a barrier to the softer material behind. They have advantages in that they are not an alien material to the coast like concrete is (even though they are still placed there by humans) and their capacity to absorb allows the dissipation of wave energy. They also blend into the natural environment quite well. However, rock revetment schemes do have their disadvantages. While being cheaper to install than walls, they are more easily damaged by high energy tides and storms. They are also a form of coast armour, preventing the sea from removing material to transport to other coastlines, In a similar way to walls and groynes, they can serve to move the problem of erosion further up the coast. Erosion can still take place and, as "incident waves are quick to exploit the junction between protected and unprotected sections" (Carter, 1988, p. 448) it may be expected that soon there will be more demands for protective measures beyond where they already exist. Using guidance set out by MAFF, and with the help of SMPs, coastal managers have to find new techniques for defending the coast and as far as possible, they must operate in harmony with nature. New methods may still include hard defences, but increasingly, a soft alternative or natural solution is considered.

Offshore Breakwaters

While not being a truly soft or natural defence mechanism, breakwaters can be useful in protecting existing defences or vulnerable coasts from the full force of waves, and at the same time building out the shore, rather than maintaining its line like a solid wall. They may be part of the harbour or port developments, or integrated into urban shoreline protection schemes and are usually sited in such a way so as to suffer the full force of wave impacts (Carter, 1988). The most common design comprises a core of loose fill, overlain by a veneer of large interlocking riprap or armour units (Carter, 1988).

Breakwaters may be utilised in two ways:

1) **to dissipate and reflect wave energy.**
2) **to promote sedimentation.**

Breakwaters placed offshore in the front of a sea wall serve to reduce the force of the incoming waves, and may also prevent some beach scouring that reflection from walls can cause. Likewise, breakwaters in front of soft erodable beaches may lessen the force of the wave attack and hinder removal of material. More subtle use of breakwaters can encourage sedimentation at beaches. Strings of shore-parallel breakwaters interposed between the beach and incident waves create wave shadows, which may naturally or artificially infill to produce a stable beach (Carter, 1988). Breakwaters are not without their own disadvantages. Similarly, offshore breakwaters designed to trap sediment may contribute to erosion on adjacent shores. The problem of erosion merely shifts to another location. Carter (1988) also argues that breakwaters are subject to failure: *"They are exposed and prone to storm damage and may, therefore, be expensive to maintain".*

Beach Nourishment

A soft natural defence scheme is beach nourishment, the importing of foreign material to either directly fill a depleted beach zone or gradually feed a beach over time. This method is becoming extremely attractive because it does not involve the construction of costly structures, and can result in a more natural appearance. For example, in the Netherlands, sand nourishment has been chosen as the main method of alleviating erosion. Since 1991, about 7 million m^3 of sand has been added to the Dutch coast, and has proven to be a successful system (de Ruig, 1995).

Beach nourishment is not a straightforward process and "the techniques are more sensitive than is commonly envisaged" (Carter, 1988, p. 459). Sources of fill have to be selected with care so that they will merge with the indigenous material. It should also be of similar or slightly larger size so as not to initiate new and unstable conditions. The Dutch, for example, take sediment from the floor of the North Sea (de Ruig, 1995). There are considerable benefits to beach nourishment if it is carried out correctly. Adding sediment to the beach increases its natural volume, therefore counteracting foreshore erosion. Where used in front of a sea wall, it will protect the wall toe and foundations, preventing subsidence. In addition, control over longshore dispersal is possible. Beach nourishment provides sediment for longshore drift, which may benefit areas through natural accretion. In the Netherlands, where beach nourishment is now the main method of defence, it has been found to be "effective in coastline preservation and assists recreation, natural values and the flood protection system" (de Ruig, 1995, p. 254).

Beach nourishment may often be the best method of future defence as it fits many of the criteria of MAFF. However, although it seems to benefit natural processes such as sediment dispersal, it could have other environmental impacts and may be economically unacceptable. Both at the source site and the nourishment site, disturbance occurs which could be either temporary or permanent. For example, Loffler and Coosen (1995) suggest that extraction of sand increases suspended sediment that in turn increases turbidity. Prolonged turbidity affects the depth to which light can penetrate the water, inhibiting the growth of plankton and algae. Benthic organisms disappear during extractions, although mobile animals are less affected. At the nourishment site, replenishment material is generally deposited in a horizontal layer. If the layer is greater than 0.5 metres thick, the original benthic animals on the beach will die. This has an adverse affect on the availability of food for birds, yet eventual damage may not be too serious, and there are guidelines over the best time of year and the type of material for nourishment. Overall, de Ruig (1995) concludes that "the ecological effects of sand nourishments, both at the borrow and nourishment sites, seem to be minor" (de Ruig, 1995, p. 257). Economically, beach nourishment may be costly. The scheme allows for sediment dispersal, and, therefore, beaches require annual replenishment and maintenance, which means high labour costs.

In general "nourishment is an important competitor, it can be utilised virtually anywhere, is easy to adapt with regard to sand volumes applied and it allows spreading of costs of coastal protection" (de Ruig, 1995, p. 260). However, there needs to be careful consideration of its environmental consequences and possible costs before it is used.

Artificial Seaweed

This is a relatively new idea and involves implanting artificial seaweed fronds into the nearshore zone in order to reduce wave energy (Carter, 1988). Such a technique would, like offshore barriers, protect both sea walls and exposed coast from wave attack while also encouraging sedimentation. Costs of installing such a device like this have been calculated to be less than one quarter of breakwaters (Carter, 1988) and they are economically viable in the short term. Unfortunately, schemes have shown contrasting degrees of success. In Bridlington, northeastern England, seaweed mats were abandoned when they were dislodged (Carter, 1988). In reality artificial seaweed would only be another human structure cutting down on natural sediment movement.

Coastal structures and defences are problematic because they too often work against nature rather than with it. However, "since we have created assets along our coast and are reluctant to give them up, we are committed to defending the status quo over a large proportion of the total frontage [of our coastline]" (Wakelin, 1989, p. 150).

Yet, whatever the technique used for coastal defence against flooding and erosion, there will be no system which can overcome all eventualities (MAFF, 1993). As modern literature repeatedly points out, coasts are dynamic, they continually change, and no one can predict with absolute certainty what will happen. Flood defences are only designed to withstand a certain height and frequency of storm, and protection measures are unlikely to stop all erosion. In England and Wales, MAFF claims to ensure that adequate flood warning procedures exist (MAFF, 1996). They also fund the National Storm Tide Warning Service (STWS) operated by the Meteorological Office. The Environment Agency maintains local tide warning systems where required, and operate them in conjunction with STWS. Where floods are likely, people's lives can still be saved, even if they do eventually have to forfeit their property.

However, the real solution to protecting the human and built environment would be not to develop in marginal areas liable to flooding and erosion. As Moutzouris (1995) has argued "erosion [and flooding] only become(s) recognised as a problem when resources near the shoreline become threatened" (Moutzouris, 1995, p. 153). Humans have consistently built at the coast, usually for good reasons, but have just as often ignored the potential hazards of doing so. While the planning system has been very successful in arresting the spread of piecemeal development along the undeveloped coast, "it has had limited effectiveness in addressing issues related to coastal hazards such as erosion, deposition and flooding" (Lee, 1993, p. 170). Lee (1993) attributes this to:

1) **a lack of appreciation of the dynamic nature of the coastline.**
2) **a lack of co-ordination between land use planning and coastal defence strategy.**

In the future, MAFF will have to pursue their objective in discouraging inappropriate development, emphasised in the Government's Planning Policy Guidance Note on Coastal Planning (PPG20). It may also be useful for CPAs to include in SMPs a land use map, delimiting the areas they consider to be most at risk, indeed, a classification of land use is advised by MAFF (1995).

Summary

The emphasis now placed on the concept of sediment cells as boundaries for coastal defence groups, and the development of SMPs, should help CPAs realise the importance of natural processes at the coast when designing defence and protection schemes. However, this will only be the case where defence groups exist, and where CPAs take up the challenge of developing SMPs.

Coastal landscapes have been produced by the natural forces of wind, waves and tides, and many are nationally or internationally important for their habitats and natural features. Past practices at the coast, such as the construction of harbours, jetties and traditional defence systems may have contributed to the deterioration of the coast. English Nature (1992) have argued that if practices and methods of coastal defence are allowed to continue, then coastlines would be faced with worsening consequences, including:

- **The loss of mudflats and the birds which live on them**
- **Damage to geological Sites of Special Scientific Interest (SSSIs) and scenic heritage by erosion, due to the stabilisation of the coast elsewhere**
- **Cutting of sediment supplies to beaches resulting in the loss of coastal wildlife**
- **Cessation through isolation from coastal processes, of the natural operation of spits, with serious deterioration of rare plants, animals and geomorphological and scenic qualities (English Nature, 1992)**
- **A number of designations, provided by national and international legislation do exist to aid conservation. One of the most important in Britain is the SSSI "widely considered to be the cornerstone of conservation in Great Britain" (DoE, 1993, p. 71)**

Coastal conservation is a contentious and problematic issue. This is a broader issue common to nearly all coasts, particularly those that are mainly undeveloped. The problem the coastal manager faces here is how to balance the needs of the local community and the wishes of conservation groups, within the current legislation.

Defence, protection and conservation are all-important issues in their own right, however, they are in themselves not necessarily management. Integrated coastal zone management brings these three issues together in an attempt to use the coast wisely, provide sustainable development and maintain biodiversity. Yet, in Britain at least, this is still only a political ideal to which there is, at the moment, no perfect solution.

Coastal zone management has become, in recent years, a very prominent issue, in the UK. This is demonstrated by the literature available, including the Department of the Environment (DoE) review of coastal management and planning (1993), Ministry of Agriculture, Fisheries and Foods publications to guide Local Authorities in shoreline defence, conservation and management (1993, 1995, 1996), and a study by Healy and Doody (1995) on coastal management in Europe, many of whose examples come from the UK.

Whilst it is important to be aware of the national policies surrounding coastal management, it is no less essential to make a local study to give detail on how national policy is implemented at specific sites. It is generally accepted within the literature that coasts around the UK are physically unique and distinct, and have their own problems and conflicts. An important part of coastal management is the interpretation of national policy at local level.

Although coastal management needs to evolve further, there have, within recent years, been "remarkable changes in attitudes and practices of management of the shoreline" (Hooke and Bray, 1995, p. 336). Perhaps the most significant of these changes is the realisation that management can be based on the concept of the sediment cell. Indeed, MAFF sees this concept as the linchpin of modern integrated coastal zone management because it offers the opportunity for administrative structures to be much more closely related to natural forms and process units along the coast.

The major advances made with sediment cells are the setting up of voluntary coastal defence groups, and encouraging Coastal Protection Authorities (CPAs) to initiate Shoreline Management Plans (SMPs). The coastal defence groups have been able to provide a much needed regional contact for co-ordinating and exchanging information, so that the works of one CPA at the coast will not adversely affect the coast of an adjacent CPA. Likewise, SMPs have given CPAs the opportunity to develop sustainable coastal defence and conservation policies within a sediment cell, and to set objectives for the future management of the coast (MAFF, 1995).

The main focus of an SMP is on coastal defence and protection. It is recognised today that new defence and protection techniques must take account of natural processes, particularly sediment movement, which take place at the coast. The most successful of these techniques is arguably beach nourishment or recharge. Beach recharge appears to cause little disruption to natural processes, and evidence from the Netherlands, where nourishment schemes are actively used, suggest that the environment suffers few adverse effects.

As a nation, our awareness of conservation issues appears to have grown over the past fifty years. This view is justified by the numerous site designations that have been made to protect the environment, and the inclusion of the natural environment as an important consideration within SMPs. The growth of interest in these issues at the coast has caused controversy within coastal management, especially where conservation priorities have clashed with the needs and wishes of business and local residents at the coast. In many respects, such conflicts belie any advances in coastal zone management, as in essence, they are a consequence of poor planning, co-ordination, and hazard management, which have heightened the problems of flooding, erosion and threats to the built and natural environment. However, SMPs should provide a forum for all interests at the coast to arrive at an agreeable solution for all. While little information currently exists for any proposals of an SMP covering the study area, it is important to recognise that it will be beneficial for conservation agencies, businesses and local residents, to discuss their concerns. At the same time, new defence and protection measures will be developed with respect to conservation priorities so that future conflicts may be avoided. In this way, coastal management is becoming properly integrated.

There are, however, problems with basing coastal management solely on sediment cells. Hooke and Bray (1995) debate their usefulness, arguing that the cells are difficult to identify, and the concept hard to apply. This is especially true for finer sediment, as no one can be sure of its exact movements. Lee (1993) also argues that sediment cells will not be relevant to all aspects of coastal management, particularly where landward geomorphological systems are concerned. The ability of plans based entirely on sediment cells to integrate management strategies, may, therefore, be questioned; any framework not bringing all issues together is not totally integrating them.

Another dilemma faced by the SMP initiative is that it is non-statutory, although Jemmett (1995) argues that there are some advantages to this system. He suggests that it "places the responsibility for addressing individual issues on those affected by them and it fosters greater awareness and ownership" (Jemmett, 1995, p. 449). In other words, the plans, while based on national policy, can become very site specific, geared towards particular needs. However, he also acknowledges that some conflicts will only be resolved by statutory control (Jemmett, 1995). Also, being non-statutory, local authorities and consulting agencies do not have to initiate them, and even if they do, they remain little more than advisory measures that can only be enforced if made law within planning documents. Coastal management still relies heavily on the planning system that is less than adequate to deal with all its intricacies.

There is still a lot of room for improvement before it can be said that Britain has a framework for the wise use of the coast, its sustainable development, and maintenance of its biodiversity. The formation of voluntary coastal groups and non-statutory management plans are certainly a start, yet it has been seen how, based only on sediment cells, these may not account for all coastal issues. The other problem is that SMPs are not the only type of management plan, in force, at the coast. There are also estuary, heritage coast, and AONB management plans, which may all overlap. Thus, two different methods of management and planning could be applied to the same area. Ultimately, this gives the impression that coastal management still lacks a co-ordinating body with responsibility for all issues.

Lee (1993) concludes that effective management in the future should be able to resolve conflicts between alternative demands on resources, and ensure that human activity does not significantly affect coastal systems or ecosystems. Carter, (1990) argues that the best way of achieving this will be through a framework which provides consistency and ease of communications. He suggests imposing a new structure, lying across existing areas of responsibility, hence, providing a lead group through which all existing interest groups can be co-ordinated. This is, perhaps, the best way of bridging the gap originally created by the division of coastal defence and protection.

On the other hand, it is hard to see how one body could possibly control the shear volume of work, and conflicts, arising from the many interests at the coast. The main problem with coastal management in the past is that it has been very reactionary. There has been little in the way of planned response, and new ideas have only come about with hindsight. This is partly why there are so many voluntary groups, management plans and conservation designations in operation. Perhaps now is the time to take stock of the situation and clarify positions. It may be that there is now a need for Coastal Authorities based on the boundaries of sediment cells, with statutory powers to govern the coast, but in close liaison with existing Local Authorities and their planners. In addition, new Acts of Parliament should bring coastal legislation up to date, especially that concerning protection and defence; and conservation legislation could be condensed to reduce confusion. Integration between fewer groups would seem to be a far easier prospect.

However, it remains to be seen whether the lack of money, bureaucratic inflexibility and a lack of co-ordination at all levels (Carter, 1990) will allow any such changes to take place.

Bibliography

Carter, R.W.G. (1988) *Coastal Environments: An Introduction to the Physical, Ecological and Cultural Systems of Coastlines*, London, Academic Press.

Carter, R.W.G. (1990) 'Coastal zone management: comparisons and conflicts', In: *Planning - management of the coastal heritage*, Sefton Metropolitan Borough Council, 45-49.

Department of the Environment (1992) *Planning Policy Guidance: Coastal Planning, PPG20*, London, HMSO.

Department of the Environment (1993) *Coastal Planning and Management: A Review*, London, HMSO.

English Nature (1991) *Marine Conservation Handbook*, Peterborough, English Nature.

English Nature (1992) *Objectives for Coastal Conservation*, Peterborough, English Nature.

Hooke, J.M. and Bray, M.J. (1995) 'Coastal groups, littoral cells, policies and plans in the UK', *Area*, 27,4, 358-368.

Jemmett, A. (1995) 'Integrated coastal zone management: lessons from the Dee Estuary, UK', In: Healy, M.G. and Doody, J.P.(eds.) *Directions in European Coastal Management*, Cardigan, Samara Publishing Limited, 447-450.

Komar, P.D. (1983) 'Coastal erosion in response to the construction of jetties and breakwaters', In: *CRC handbook of coastal processes and erosion*, Boca Raton, CRC, pp, 191-204.

Lee, E.M. (1993) 'The political ecology of coastal planning and management in England and Wales: policy responses to the implications of sea-level rise', *The Geographical Journal*, 159,2, 169-178.

Loffler, M. and Coosen, J.(1995) 'Ecological impact of sand replenishment', In: Healy, M.G. and Doody, J.P. (eds.) *Directions in European Coastal Management*, Cardigan, Samara Publishing Limited, 291-299.

Ministry of Agriculture, Fisheries and Food (1993) *Strategy for flood and coastal defence in England and Wales*, London, MAFF.

Ministry of Agriculture, Fisheries and Food (1995) *Shoreline Management Plans: A Guide for Coastal Defence Authorities*, London, MAFF.

Ministry of Agriculture, Fisheries and Food (1996) *Flood and Coastal Defence*, London, MAFF.

Moutzouris, C.I. (1995) 'Shoreline structures for coastal defence', In: Healy, M.G. and Doody, J.P.(eds.) *Directions in European Coastal Management*, Cardigan, Samara Publishing Limited, 153-159.

Park, J.R. (1989) 'Legislation and policy', In: Maritime Engineering Board (eds.),

Coastal management, London, Telford, 11-20.

Penning-Rowsell, E.C. (1992) *The Economics of Coastal Management: A Manual of Benefit Assessment Techniques*, London, Belhaven.

de Ruig, J.H.M. (1995) 'The Dutch experience: four years of dynamic preservation of the coastline', In: Healy, M.G. and Doody, J.P.(eds.) *Directions in European Coastal Management*, Cardigan, Samara Publishing Limited, 253-266.

Sims, P. and Ternan, L. (1988) 'Coastal erosion: protection and planning in relation to public policies - a case study from Downderry, south-east Cornwall', In: Hooke, J.M. (ed.) *Geomorphology in Environmental Planning*, Chichester, John Wiley and Sons Ltd. 231-244.

Solway Firth Partnership (1996) *Solway Firth Review,* Solway Firth Partnership.

Wakelin, M.J. (1989) 'The deterioration of a coastline', In: Maritime Engineering Board (eds.), *Coastal Management*, London, Telford, 135-152.

Whittle, I.R. (1989) 'Technical overview', In: Maritime Engineering Board (eds.), *CoastalManagement*, London, Telford, 21-30.

CHAPTER 2

The North East Coastline of Scotland

J.S. Smith

ABSTRACT: The coastline of Northeast Scotland in plan is determined by faults running close inshore to both North Sea and Moray Firth coasts. Three broad sub-divisions can be identified where differences of lithology and orientation have created distinctive coastal scenery. The events of recurrent glaciation during the Pleistocene are much more pertinent to the soft coast geomorphology of Northeast Scotland. In terms of their current coastal dynamics, the beaches of Northeast Scotland can be subdivided into three groups. Within this developed region, the coastline has for the most part, remained relatively unaffected by intensive use. Shallow offshore gradients, potentially mobile surfaces, low quality soils and a tendency to poor drainage have preserved the dune and links areas from urban and industrial use. The broad planning strategy for future management might, therefore, appear to be that of preserving or maintaining the status quo whilst allowing local, well-managed developments to proceed, as long as they conform to a general regional plan for balanced coastal zone development.

The Geological Skeleton

The coastline of Northeast Scotland in plan is determined by faults running close inshore to both North Sea and Moray Firth coasts. Along the northern margin, an east-west trending system of faults can be identified by steep gravity gradients north of Fraserburgh. This line also coincides with the orientation of a prominent bathymetric deep. Off the east coast, recent geological work by the *Institute of Geological Science* has confirmed the pioneer Quaternary work of Jamieson which indicated that relatively recent geological strata of Permo-Triassic age lie close inshore. For both north and east coasts, the detail of the coastal plain, within which the soft coastal elements have accumulated, is governed by the varying lithologies of the solid geology along the land margin. Three broad sub-divisions can be identified where differences of lithology and orientation have created distinctive coastal scenery.

(a) **The Highland schists of the Banffshire coast**
(b) **The granites, gneisses and schists of the east coast of Aberdeenshire, to the north of the Highland Boundary Fault**
(c) **The younger sedimentaries downfaulted south of the Highland Boundary Fault, the outliers at Pennan, together with the downwarped sedimentary formations of the Moray Firth basin, west of Spey Bay**

The Banffshire coast consists of a generally high rocky coastline with small bays orientated in sympathy with the rapidly changing geological succession of the Highland schists. Beach units are small and generally separated from each other by rock cliff headlands and deep water. The bayhead orientation is continued inland by strike-stream valleys, often markedly overfit in relation to current fluvial activities, and the fluvial input, with the exception of the Deveron, is generally small. The coastal scenery is bedrock-dominated and rapidly changing in character along the geological succession.

On the east Aberdeenshire coast, the role of the underlying bedrock, although by no means uniform, is subordinate to Pleistocene and post-Pleistocene events. With the notable exception of the granite coastline north of Collieston (especially at Buchan Ness), and the intricate cliff coast between Girdleness and Stonehaven, the character of the coast is of extensive arcuate dune-backed embayments, hinged on low bedrock outcrops and lag glacial deposits. Fluvial inputs while locally significant, are not impressive in their supply of beach material.

West of the Spey and south of Stonehaven, the younger sedimentary rocks form substantial lengths of high cliff coast, with variation in character according to lithology. Particularly impressive are the Devonian conglomerate sea cliffs at Fowl's Heugh, just south of Dunnottar. On the north coast Permian sandstones form a very distinctive series of low sea cliffs between Covesea and Burghead, but west of this point, and between Lossiemouth and Portgordon, the coastline is dominated by major strandplains of late- and post-glacial age, again as in the case of the east Aberdeenshire coast, supported by bedrock outcrops.

In relation to the soft and therefore potentially dynamic coastal elements, the solid geology forms a varied platform on which beaches and landward blown sand deposits have accumulated. But over more than 80% of the sandy beaches, bedrock geology forms a minor element in their evolutionary history and present characteristics. Exceptions occur on the Banffshire, Kincardineshire and Moray coasts, where the beach units are set in small bayhead situations resting on rock platforms. Here the beach arc and its landward landform characteristics are closely governed by the geological framework within which it is lodged.

Glaciations, Deglaciation and Changing Land-Sea Relationships

The events of recurrent glaciation during the Pleistocene are much more pertinent to the soft coast geomorphology of Northeast Scotland. The glacial history of Northeast Scotland involved successive advances of ice both from the Moray Firth basin and from the Deeside-Donside area. In Banffshire, the cliffs are capped by till and by the enigmatic coastal gravels, both of which contain material of Cretaceous origin derived from offshore. On the east coast of Aberdeenshire and Kincardine a red till derived from Strathmore and from offlying submarine sandstone outcrops forms a prominent capping to the sea cliffs, as at Dunnottar and Hackley Head. At several points both on the east and north coasts, ancient elements of the preglacial coastline are plugged with glacial till even in relatively high energy situations, testifying to the relatively slow evolution of the inherited pre-Pleistocene bedrock coastal landforms. Examples of till-plugged sea stacks occur at Covesea and immediately south of the Bay of Nigg. Impressive sections of superimposed coastal tills were formerly exposed at the Bay of Nigg.

As the ice melted, large quantities of fluvioglacial materials were released into the coastal zone, the materials being transported down preglacial river valleys as in the case of the Spey and Findhorn, or more directly released from impressive meltwater channels as in the Pennan-Troup Head area. The high-cliff coast near New Aberdour is dissected by deeply incised coastal ravines which are partly of meltwater origin. In situations where ice-fronts rested near to the present coastline, as occurred between Aberdeen City and the Ythan, and at the Binn Hill just west of the Spey, sorted materials were released to be later incorporated into recent coastal strandplains. Large quantities of material travelled down the Spey and Findhorn which acted as major late-glacial spillways, as attested by the huge lateral fluvioglacial terraces of their lower and middle valley sections.

Land-sea relationships changed as a result of the interplay of isostatic rebound of the land and eustatic change in sea level. The result was to cause shoreline displacement, initially rapid and latterly slowing down. Where unconsolidated materials were subject to wave action, strandplains were formed representing the land-sea relationship at that time, and subsequent land rebound raised these to form the late-glacial shorelines which fringe the interior edges of the major coastal embayments. On occasion, the late-glacial sea overtopped the bedrock cliffs, and resorted the coastal gravels mentioned earlier to form beach gravels as at Portknockie. On the east coast of Aberdeenshire, late-glacial beaches at about 30 m. O.D. can be identified inland of the Loch of Strathbeg. Marine clays were deposited in sheltered coastal situations as at Cruden Bay and Tipperty, often associated with a cold water fauna. After this phase of relatively high sea level, it is evident that sea level dropped to a level below that of the present time, resulting in river and estuarine rejuvenation. As climate ameliorated, vegetation spread into the emerging coastal zone.

Shoreline Regularisation

Later in response to the warming world climate, sea levels rose rapidly, and the coastal zone was inundated by a major marine transgression (the Flandrian transgression) which caused a landwards migration of the shoreline, leading to the burial of peats and their capping with marine deposits. The highest point of the post-glacial transgression is marked by the well-marked sea cliff at about 6 m. O.D. which can be traced almost continuously from Strathbeg in the north to Aberdeen City in the south. It also formed the well-defined inland margin of the Moray strandplains as at the Binn Hill. Final land emergence caused a second seaward shoreline displacement, during which vast quantities of sands and gravels were made available for shoreline regularisation. Successive shingle bars hinged on till or isolated bedrock headlands mark the progress of this final period of shoreline displacement. In areas where substantial coastal embayments existed, as in the Moray coast or the east coast of Aberdeenshire, regularisation involved the construction of the wide strandplains on which blown sand accumulated. On occasion blown sand deposits extended landward of the marine limit forming a veneer on till and fluvioglacial landforms. In response to dominant wind and wave activity, shingle spits were built, deflecting the Ythan to the south on the east coast, and the Spey and Lossie to the west on the north coast. The smooth arcuate nature of the coastal plain between Aberdeen and Collieston, between Buchanhaven and Inzie Head, and between Spey Bay and Lossiemouth represent a final shoreline equilibrium form achieved through the recycling of sediment by wave and current. West of Burghead, active shoreline progradation continues through the building westwards of spits and bars. Much of the material is derived from re-cycling of existing forelands as in Burghead Bay, resulting in problems of coastal edge erosion, as waves and currents sweep the material westwards into the Inner Moray Firth.

In the case of the smaller bayhead beaches, the processes of change are less dramatic, as a balance appears to exist between material supplied from cliff and offshore sources, and that lost from the beach by downcombing during heavy swell. On the Banffshire coast, where each unit is separated from its neighbour by cliff coasts and stretches of deep water, each beach unit functions separately, but in the Inner Moray Firth, where large units are linked in arcuate coastal sweeps, coastal edge erosion is associated with deflation of the exposed dune faces. Backshore erosion at Burghhead Bay has in the past revealed exposures of the peat which formed during the shoreline regression phase preceding the Flandrian.

The Building of Dune Systems and Human Impacts

On top of the shingle forelands where the constituent material reflects the formerly abundant supply of fluvial and fluvial-glacial materials, sand deflated from the drying upper beaches has accumulated as dune systems. These vary locally in their form and complexity. The relatively simple successive foredunes of Strathbeg, St. Fergus and Lossie form massive ridges separated by slacks. At the Sands of Forvie, and to a lesser extent Culbin, the forms are much more complex, including parabolic dunes whose current activity and pattern of evolution makes them of great physiographic interest. Here the older late- and post-glacial landform elements are largely buried by a blown sand landscape carrying evidence of sequent occupance. In both Culbin and Forvie, the story of prehistoric and historic sand-blow cannot be separated from the activities of man. Historical and neo-historical sand-blow at Culbin resulted in 20th century afforestation of the dunes, but at Forvie, the dunes continue to evolve naturally.

The very large beach units and their backing forelands described above contrast in size with the small bayhead units set within a rock-girt coastline. Here the beaches rest on rock platforms and are generally backed by settlements nestling on the post-glacial ledges, at the base of a marked sea cliff, generally of bedrock, capped with till, and overlain by thin and often discontinuous blown sand deposits. The blown sand landforms are small by comparison with the North Sea coast beaches, and are generally inactive, partially masked and modified from their original characteristics. They can be regarded as degraded 'links'. These small beach units have frequently been modified in the 18th and 19th century through the construction of harbours, as at Hopeman.

In terms of their current coastal dynamics, the beaches of Northeast Scotland can be subdivided into three groups.

1. The massive arcuate plains consisting of long sand bays of gentle arc, backed by extensive foredune ridges, successively built outwards by shoreline regularisation. These are characteristic of the North Sea coast of the region north of Aberdeen. The blown sand deposits in this group have distinctive dune forms largely in their original state. The coastline is a relatively high energy one, with marked seasonal changes in the beach and coastal edge, but sand supplies both longshore and offshore appear to be available for coastal edge replenishment.

2. The beaches west of the Spey are equivalent in scale to those in the preceding group but differ in the relative proportion of sand and shingle involved in their construction. Fluvial inputs have been more significant in their evolution, together with a uni-directional pattern of sediment movement westwards. The thickness of blown sand resting on the shingle forelands is variable, with on occasion only vestiges remaining as in the case of Spey Bay links. Recurrent sand-blow in the immediate past has been countered through afforestation, with the result that much of the original mobility of the dune landforms has ended. As the offshore gradient shallows westwards, mud- and sand-flats appear, and spits and offshore bars have formed. The ample supplies of sediment available during their construction have apparently declined in recent times, and the distal portions of the arcs are tending to build at the expense of the proximal (updrift) portions with the result that beach material starvation is leading to considerable coastal edge erosion as at Burghead Bay and the proximal portions of Whiteness Head.

3. The bayhead beaches of Banffshire and Kincardineshire form a distinctive third group of beaches. These are relatively small units, often partially cliff-girt, and their stability is generally assured as a result of their enclosed location and superimposition on pre-glacial bedrock platforms and ledges.

Offshore Topography and Deposits - The Coastal Sediment Bank

A supply of materials suitable for transport and capable of being moved by waves and to a lesser extent coastal currents is necessary for any beach form to be created or maintained. This source may be material removed from another coastal location and re-deposited or may be derived from an offshore source. Increasing information on the offshore topography and sub-sea deposits are becoming available through analysis of Admiralty hydrographic surveys and charts. The distribution of offshore sediments as far as they can be constructed show clearly that reservoirs of all sizes of material lie offshore derived from deposits released during deglaciation and glaciation. These deposits rest on a complex basement with a marked break of slope about 6 km. offshore marking the junction between the crystalline and sedimentary rocks. The coastal platform thus identified presumably represents, in part, a lithological boundary and in part a process-divide, related to marine activities during low glacial sea levels. Notable bulges of the inshore submarine contours opposite major river mouths represent the drowned delta fans of the fluvial inputs during low sea levels. In relation to current fluvial inputs of materials of sand and coarser sediments, it is almost certain that only the rivers of the Moray Firth coast, notably the Spey, contribute much to the offshore sediment bank, the lower courses of the others being characterised by highly efficient sediment traps in the form of estuaries.

The period of shoreline regularisation described earlier in this chapter was characterised at first by a rising sea level, with subsequent transportation of sea bottom materials over a wide zone available within wave base. As the forelands and strandplains were built, and as the shoreline was displaced seawards through land uplift, so the quantity of material within the reach of waves decreased. Thus the scale of the coastal forelands reflects large quantities of Pleistocene sediments re-worked by wave activity over a period of several thousand years. As land-sea relationships became stabilised at their present position, materials within wave base, even allowing for storm and surge activity, decreased, as the fresh supplies from cliff erosion and fluvial inputs are confined to a few specific locations.

The morphological evidence along the coast can be taken as indirect evidence of the current state of the offshore sediment bank. In interpreting this evidence, care must be taken to allow for the exceptional storm tide activities likely to result from temporal correspondence of low barometric pressure and spring tides (the so-called barometric tide or storm surge), as well as the characteristic seasonal beach profile cycle, where winter downcombing of the dune edge at the beach back is healed by summer deflation from the drying upper beach, forming a generally neutral coastal edge.

In addition, there are considerable variations in the exposure of beaches, i.e. the orientation of the beach arc in relation to dominant wave trains. The North Sea beaches, for example, are normally set in arcs reflecting the dominant fetch of southeasterly wind-generated waves, and thus contrast with the more sheltered arc of Burghead Bay. Special attention should be paid to the area between Fraserburgh and Peterhead where open fetches to the north and northeast create particularly exposed and potentially high energy conditions. Certain beaches like Stotfield Links are partially protected by off-lying skerries, others have very narrow exposure arcs governed by the relatively rigid geological framework as in the case of much of the Banffshire coast. As if to compensate for the high energy status of the North Sea coast, the beaches of the Inner Moray Firth are considerably affected by the strong tidal currents settling westwards into the Moray Firth, and accentuated by its decreasing width in the same direction.

Setting these constraints aside, it is possible to characterise the North Sea and Moray Firth coasts as respectively actively building or balanced beaches, and retrograding or redistributing beaches. On the North Sea coast, beaches change their characteristics frequently in response to weather and tidal circumstances, but overall tend to be building beaches. Their landward dunes confirm this continuing supply of material. Because these beaches face the North Sea, the pattern of wave activity is multi-directional, and material can move in quite contrary directions on different occasions. Stream deflection and spit development seems to indicate a north-going trend, and this has been substantiated by wave climate calculations for Aberdeen Bay. In the Moray Firth, only parts of the beach complexes appear to be building systems, and usually the source of the material is derived from immediately updrift, often on the same physiographic unit. Whiteness Head, for example, is building at its western end, and suffering progressive erosion at its eastern 'up-drift' end. One part of the coast is in a sense evolving at the expense of the stability of another (basically beach cannibalism), with the net result being the infilling of the Inner Moray Firth. Sediment transport is notably uni-directional.

Thus, the variety of materials identified offshore below wave base has only a hypothetical interest, at least in the Moray Firth coastal situation, as the evidence seems to suggest that very little of the remaining offshore sediment is finding its way into the wave-break zone. In this sense, periods of substantial beach replenishment and construction may follow glacial episodes and these in their turn may be followed by periods of relative beach starvation. Nevertheless, these are superimposed onto the short-term cycles of beach and shoreline change related to the weather variables which affect coastline evolution.

TABLE 1.
Tidal data for specific ports
(*Differences in metres from standard port, Aberdeen*)

Port	MHWS (metres)	MLWS (metres)
Aberdeen	4.3	0.6
Stonehaven	+0.2	0.0
Peterhead	-0.5	-0.1
Fraserburgh	-0.4	0.0
Banff	-0.8	-0.2
Buckie	-0.2	+0.1
Lossiemouth	-0.2	0.0
Burghead	-0.2	0.0
Nairn	0.0	+0.1
Inverness	+0.5	+0.1

Note: Actual water levels may exceed the predicted tide levels by several metres if onshore storms and North Sea surge conditions coincide. Source: Admiralty Tide Tables

Almost half the total length of coastline in the area of study consists of some form of sand accumulation. Most of these sandy coastlines take the form of extensive systems several kilometres long. The bayhead beach with its sheltering headlands is rare. Major rivers also cross these lowland coasts and invariably produce dynamic spit, bar and beach features. With high tidal ranges (3 to 5 m. range) and periods of strong wave action, the open beach environments are exposed to relatively high energy conditions. Beach and nearshore processes are characteristically variable in direction and strength. Nevertheless, it is possible to recognise a general north-going direction of coastal processes on the east coast and a west-going direction along the north (Moray Firth) coast. The east coast (North Sea coast) tends to have narrower beach, dune and links forms. The characteristic features are linear and essentially parallel to the general trend of the coastline. Against this trend, dune ridges and erosion hollows are normally oblique and it is the intersection of these oblique trend lines with the general parallelism of the main dune and beach features that produces variety and complexity into the beach/dune/links topography. On the north coast the characteristic form is the strandplain which has been formed in extensive low-lying basins by a falling sea level. Dunes and links tend to stretch further inland over raised shoreline basements which are usually composed of shingle. Another contrast between the north and east coasts is in the presence of shingle. Shingle normally forms the backshore material of most Moray Firth beaches whereas shingle storm beaches are almost unknown except as cliff-foot forms on the North Sea coast. There are, however, local exceptions, e.g. Stonehaven has mainly a gravel and shingle beach.

Sustainability of the Beach/Dune Resources

Throughout the region it would appear that extensive glacial and fluvioglacial deposits on the continental shelves are the ultimate sources of beach and sand dunes. It is also of interest that there are higher shell sand contents on beach sands between Rosehearty and Peterhead. It has not been possible so far to estimate the contribution being received from current processes such as cliff erosion or river discharge. Similarly it has proved impossible to determine whether or not the coastal dunes are at a stage of net retreat or net advance. Examination of the detailed morphological evidence suggests that the steep, undercut dune face is most common especially on the North Sea coast of North East Scotland so that the balance of opinion favours coastline retreat. Nevertheless, there are significant areas of accretion especially near the river outlets, at places between Fraserburgh and Peterhead and in the Inner Moray Firth west of Spey Bay. In the last named area, however, it is thought that erosion along one section of the coast is balanced by redeposition further along the coast.

On both coasts there are good examples of the problem of differentiating between long-term changes and short, violent events. In January 1978, for example, a particularly severe storm coincided with elevated sea levels produced by a small North Sea surge and several hundred thousand pounds worth of erosion damage occurred between Spey Bay and Peterhead. Significant local changes in beach and dune morphology were also produced by this storm. It is important to stress that these 'exceptional' storms and high water levels occur with a high statistical frequency. Research programmes, which have been undertaken as design studies for offshore North Sea oil and gas installations, have calculated that the 'ten-year' storm or the 'hundred-year' high water level may produce wave heights several metres above the average along any part of the northeast coastline.

Weather changes have the same degree of variability. Precipitation, for example, has varied from near drought in summer 1976 to excessively high winter values in 1977. River discharges vary accordingly and cause significant changes to occur on adjacent spit and beach features as for example at Speymouth, Banff Bay or Donmouth. Similarly the characteristic winter flooding which converts most of the low-lying links and dune areas to temporary lakes varies both in timing and depth of flooding. This seasonal flooding has important physiographic and ecological effects.

As explained earlier in the chapter, the dunes and links are generally more stable on the north coast. There are presently no expansive bare sand waves comparable to the great spreads of bare sand that are found between Strathbeg and Aberdeen. To some extent this contrast is explained by different land use. Culbin and Burghead Bay, for example, were described as consisting of extensive migrating bare sand dunes well into the 20th century, but with afforestation they are now fixed landforms.

Morphological evidence suggests that the dunes between Fraserburgh and Aberdeen are more active than those elsewhere in the region. Erosion forms are more frequent and the landforms are higher and steeper than those found on the Moray Firth coast. Free-standing dune ridges and sandhills attaining more than 16 m. O.D. are common on the North Sea coast but rare along the Moray Firth coastline. Detached sandhills and dunes that run relatively far inland from the coast are also characteristic of the east coast whereas the Moray Firth systems tend to have lower, ridged or undulating plains behind a single foredune ridge.

Both areas show examples of old links on the face and on top of pre-existing abandoned clifflines and plateaux. The morphological (and subsequent land use) control exerted by the position and elevation of such raised shoreline features is extremely important throughout the region; with the main contrast occurring where the old cliffline is close to the present coast, or alternatively, where landforms associated with a former sea limit extends far inland and take the form of a degraded low angle slope or, occasionally, an inland basin derived from a former marine embayment.

Coastal Zone Land Uses

Related to the physical nature of the resource are the various forms of land use. As illustrated by the table, the main land use is broadly defined as agriculture. Grazing of sheep and cattle is common in most links areas. The dune and sandhill areas are only rarely used and it is on the various types of links and transitional surfaces that agricultural use is concentrated. Another significant land use is golf links. The concentration of golf courses along the entire coastline is remarkable. There is, for example, about 50 km. of dunes and links between Fraserburgh and Aberdeen and about 20% of this length is occupied by eight main golfing areas. Almost without exception every major coastal settlement between Nairn and Aberdeen has one or more golf courses at the coast. Another land use which has a distinctive distribution is forestry; almost half the area between Portgordon and Inverness has this as the main land use. Industrial use is minimal. Only Whiteness Head (oil rig fabrication yard) and St Fergus (North Sea Gas Terminal) have significant areas of industrial use. Similarly military use also occupies very small but locally significant areas.

Recreational use takes several forms. Caravans are almost exclusively a Moray Firth phenomena, especially west of Burghead. There are also town beaches with built facilities since most of the coastal settlements have a degree of dependence on the tourist industry. It is of some interest to note that the greatest concentrations of recreational activity are on the smaller beach units which are located close to or within towns. Beaches within harbours, e.g. Collieston, Banff, Peterhead, Aberdeen are often used for recreational purposes. Many of these beaches are more attractive because of the variety of scenery nearby, and houses, cliffs, caves, skerries, etc. give a visual richness that is lacking from the more extensive beach and dune areas which are often monotonous and drab. Because of

TABLE 2.
Dominant coastal zone land use - beach units
(expressed as percentages)

	Total Length
Industrial/Settlement	5.8
Forestry	17.4
Recreation	22.6
Nature Conservation	15.5
Agriculture	37.3
Scrub/Other	1.6

Note: Northeast Scotland defined here as from Inverness to the River North Esk (Angus).

the extensive scale of the large units, i.e. Ardersier to Burghead, Lossiemouth to Portgordon, Fraserburgh to Peterhead and Collieston to Aberdeen, recreational use is concentrated in relatively few localities. These locations are beside towns and villages, or at specific access points which have often been provided for the purpose of permitting informal day recreation at and near the coast.

Areas where recreational use appears to be having some adverse effect on the resource occur but are relatively restricted. This may take the form of a physical impact, e.g. erosion, as at Bridge of Don, Balmedie, Findhorn or at the forest access at Roseisle near Burghead, or an aesthetic effect, as in the beach huts and caravans at Findhorn, Nairn (East), Hopeman, Covesea or Inverbervie.

Attention should be drawn to the relatively high proportion of the coastline, especially dune and links areas, that are conserved as some form of nature reserve. In addition, although significant exceptions occur at Blackdog north of Aberdeen and between St Fergus and Peterhead further north, the relative absence of dune sand extraction pits is a welcome feature of the region.

Perhaps the most significant observations in any objective appraisal of the resource base of beaches, dunes and links in North East Scotland is the high level of availability of the resource. None of the large towns in North East Scotland are more than about a 30-minute drive from a sand beach. Essentially the beach sand resource is abundant. The beaches and dunes may lack the scenic quality of many Scottish Highland and Island locations but they are located close to a dense settlement pattern with a developed infrastructure of roads and services. Numerous settlements are actually located on or beside the beach zones. No beach is more than a few kilometres from a reasonable road. Accommodation in hotels, boarding houses, chalets and caravans is readily available. It is not surprising that some beach areas are heavily used nor that a proportion of local government resources are currently used in managing specific areas and in providing a range of different recreational facilities. Several types of reports and plans exist for the conservation and utilisation of parts of the coastline. At the other extreme, there are several stretches of beaches and dunes that are as empty and unaltered as any to be found elsewhere in Scotland, including the Outer Islands. This is particularly true of parts of the more extensive beach and bay systems. This can be explained by the fundamental control that is exerted by access. Normally, where the beach and dune zones are 'insulated' from public roads by a belt of fenced agricultural land or, along the Moray Firth, by forest, public access is effectively denied. In contrast, where access is easy there are nodes of pressure where tracks and roads lead directly to the dune zones and the pressures that are generated may be related directly to the area of car parking space available. Town beaches are obviously a special form of this general rule relting utilisation to access conditions. In contrast, coastal sections between these nodes of activity are zones which are visited only by a very few people - specifically field scientists, walkers, naturalists, salmon fishermen and local inhabitants.

Another aspect of the developed nature of the coastal hinterland is the response time in the utilisation of the more accessible beach and dune areas, especially near the large towns. Within a few hours, in all seasons, these areas can become relatively busy with day visitors, particularly at weekends. Obviously peak usage is in July and August but it is important to stress that in mid-winter many of these beaches are used and provide valuable recreational and amenity areas for a considerable number of people.

Conclusion

Within this developed region, the coastline has for the most part, remained relatively unaffected by intensive use. Shallow offshore gradients, potentially mobile surfaces, low quality soils and a tendency to poor drainage has preserved the dune and links areas from urban and industrial use. Essentially conservational use or, effectively, minimal use in the form of light grazing or low density recreational activity, has protected the majority of dune and links areas. Outwith the coastal towns there are few signs of direct alteration to the coastline in the form of groynes, sea-walls, piers and harbours. There is little pollution or effluent discharge. Moreover, there is little indication that this condition will alter; if anything the forces and mechanisms for coastal management and conservation embodied in planning strategies is becoming stronger. Thus, bearing in mind that this is part of Lowland Britain with a developed economy and a moderate population density (much of which is located along the coast), it is fortunate that such a high proportion of the coastal resource base remains remarkably intact and conserved. The broad planning strategy for future management might, therefore, appear to be that of preserving or maintaining the status quo whilst allowing local, well-managed developments to proceed, as long as they conform to a general regional plan for balanced coastal zone development.

Selected Bibliography

Firth, C.R. (1988) Devensian raised shorelines and ice limits in the inner Moray Firth area, Northern Scotland, *Boreas* 18, 15-21.
Haggart, B.A. (1987) Relative Sea Level Changes in the Moray Firth area, Scotland, in Tooley, M.J. and Shennan, I. (eds.) *Sea Level Changes*, Institute of British Geographers Special Publication 20, 67-108.
Jamieson, T.F. (1882) On the Crag Shells of Aberdeenshire and the Gravel Beds containing them, *Quart. J. Geol. Soc. London* 38, 100-177.
Kirk, W. (1955) Prehistoric sites at the Sands of Forvie, Aberdeenshire, *Aberdeen Univ. Review*, 35, 150-171.
Long, D., Smith, D.E. and Dawson, A.G. (1988) A Holocene Tsunami deposit in eastern Scotland, *Journal of Quaternary Science* 4, 61-66.
Owens, R. (1977) Quaternary Deposits of the Central North Sea, *I.G.S. Report* 77/13.
Ritchie, W., Smith, J.S. and Rose, N. (1977) *The Beaches of Northeast Scotland*. Perth: Countryside Commission for Scotland.
Ritchie, W. (1992) Scottish Landform Examples 4 - Coastal Parabolic Dunes of the Sands of Forvie, *Scott. Geog. Mag.* 108, 39-44.
Ross, S. (1994) *The Culbin Sands - Fact and Fiction*. Aberdeen: Centre for Scottish Studies.
Synge, F.M. (1956) The Glaciation of Northeast Scotland, *Scott. Geog. Mag.* 72, 129-143.
Walton, K. (1956) Rattray: a study in Coastal Evolution, *Scott. Geog. Mag.* 72, 85-96.
Walton, K. (1989) Ancient Elements in the Coastline of Northeast Scotland, in Miller, R. and Watson, J.W. (eds.), *Geographical Essays in Memory of A.G. Ogilvie*, 93-105. Edinburgh: Thomas Nelson.

CHAPTER 3

Coastal/Marine GI/GIS - A Pan-European Perspective

R. A. Longhorn

ABSTRACT: This chapter reviews the status of coastal and marine GI (geographic information) and GIS (geographic information systems) with regard to trans-national projects or programmes wholly or partly funded by the EU Institutions and other international bodies. The chapter examines the degree to which GI and GIS are taken into consideration at planning stages in new pan-EU initiatives for coastal and marine use, such as the current focus on the Baltic and regional planning in the Maghreb (Mediterranean basin) and European participation in various global climate change actions. The chapter concludes with recommendations for national and international GI/GIS organisations, public and private, to ensure that their voices are heard at the highest political level in the European Institutions in the current Information Society debate. Important initiatives are now underway, such as the EC Communication to the Council of Ministers and European Parliament "GI2000: Towards a European Policy Framework for Geographic Information", the development of the "EGII - European Geographic Information Infrastructure" working document by EUROGI, GISDATA and other pan-European organisations, and the degree to which GI will be covered in the EU's 13 billion ECU Fifth Framework Programme for Research and Technological Development, especially for coastal and marine research.

Introduction

Geographic information is usually described as any information which can be related to a location on the Earth, particularly information on natural phenomena, cultural and human resources, or pertaining to the natural or man-made environment. Historically, geographic information was represented by two-dimensional maps which focused on the ownership or control and management of territory (land and sea). The more modern definition of GI covers a very wide spectrum of information which is gathered, processed, disseminated and used in daily life in countless practical applications for governing society, e.g. land registration to record ownership details and for taxation purposes, environmental monitoring, urban and rural development planning, for planning and maintaining transport corridors (land, sea and air) and in many other ways.

Carrying out practical projects in coastal zone management (CZM) or other ocean-related research would be almost impossible without geographic information, i.e. recording the spatial or location attribute of each piece of data collected - often in 3-dimensions! Coastal and deep-ocean marine geographic information offers considerable opportunity for expanded use of GIS (geographic information systems) tools and modelling techniques to make better and more rapid use of the data collected. Marine GI often requires the latest technology to provide the position attributes for the data being collected, including remote sensing, GPS for data collected in the field, and sonar for determining depths or locations of objects. Data integration is difficult and expensive, especially across scientific disciplines, across projects, across data collection and management technologies and across country borders.

This chapter attempts to briefly introduce major initiatives relating to coastal zone research and management of the coastal regions which are pan-European in nature, i.e. inherently trans-national or cross-border, rather than looking at specific programmes at national level. Table 1 lists the many countries in the EU, the EEA - European Economic Area, East and Central Europe, plus the Russian Federation and Maghreb and Mediterranean regions, all of which participate in coastal projects which are often partly or wholly funded by the EU Institutions or by EU Member States via international aid and development programmes.

Note that several countries span "regions" and thus may find themselves participating in several different cross-border programmes at one time, often executed under different rules and/or with different objectives, while other countries may share a common border on a single ocean or sea, yet be unable to participate in a single programme because of funding restrictions, which are often political and geographical in nature. A typical example is funding to the Russian Federation under TACIS but not under Phare, even though a major Phare project might be enacted in the Baltic.

Diverse Coastal Zone Activities Require Diverse Programmes of Research

Table 2 attempts to categorise the diversity in types of activities relating to coastal zone management, to remind the reader of the wide range and diverse nature of projects and programmes which are conducted at national or international (cross-border) level. These go beyond pure scientific research, and require quite different types of funding and forms of agreement or rules of participation, depending upon the type of activity and scope of the project or programme. Knowing that many of the policy makers who set up these rules or project/programme skeletons have little or no experience in CZM or of GI/GIS, you begin to realise the importance of education,

TABLE 1
Countries sharing coastlines in "Europe"

Grouping	Country	EU Progs	Coastal "Regions"
EU	Belgium	All	North Atlantic, North Sea
	Denmark	All	Baltic Sea, North Sea
	Finland	All	Baltic Sea, Nordkap
	France	All	Mediterranean, North Atlantic, English Channel
	Germany	All	Baltic Sea, North Sea
	Greece	All	Mediterranean
	Ireland	All	North Atlantic
	Italy	All	Mediterranean
	Netherlands	All	North Atlantic, North Sea
	Portugal	All	North Atlantic
	Spain	All	Mediterranean, North Atlantic
	Sweden	All	Baltic Sea, Nordkap
	United Kingdom	All	Atlantic, North Sea, English Channel
EEA	Iceland	All	Atlantic, Norwegian Sea
	Norway	All	Baltic Sea, Norwegian Sea, North Sea, Nordkap
East & Central Russia	Albania	Phare	Mediterranean
	Bulgaria	Phare	Black Sea
	Croatia	Phare	Mediterranean
	Estonia	Phare	Baltic Sea
	Georgia	Phare	Black Sea
	Letland (Latvia)	Phare	Baltic Sea
	Lithuania	Phare	Baltic Sea
	Poland	Phare	Baltic Sea
	Romania	Phare	Black Sea
	Russia	Tacis	Baltic Sea, Nordkap, Black Sea
	Slovania	Phare	Mediterranean
	Ukraine	Tacis	Black Sea
Maghreb (M) Mediterranean	Algeria (M)		Mediterranean
	Cyprus		Mediterranean
	Egypt (M)		Mediterranean
	Israel	Frame-work	Mediterranean
	Lebanon		Mediterranean
	Libya (M)		Mediterranean
	Malta		Mediterranean
	Morocco (M)		Mediterranean
	Syria		Mediterranean
	Tunisia (M)		Mediterranean
	Turkey		Mediterranean, Black Sea

TABLE 2
Different aspects of activities related to Coastal Zone Management

Work type	Examples
Pure research	• marine sciences (investigating flora and fauna) • geophysics and oceanography topics
Practical research and applications development	• managing and protecting marine flora and fauna • bottom topology, causes of erosion • hydrology, land/sea interface, river/estuary/ocean interface • pollution control
Activities related to the economy	• fisheries and aquaculture • harvesting marine products from the coastal zone • pollution monitoring and control • coastal/harbour/sea lane navigation • beach erosion control • leisure use of the coastal zone
Work of importance for social reasons	• environmental monitoring and global climate change • impact on human and animal health of the misuse of the coastal zone • offering safe amenities and leisure activities affecting both commerce and quality of life
Legal issues	• regulations for pollution control • regulations controlling use of coastal regions, i.e. planning restrictions for developers, etc.
Political issues	• establishing instruments for cross-border cooperation on legal or regulatory issues • settling cross-border disputes • participating in regional or global initiatives, including education, research and raising awareness of problems facing use and users of the coastal zone
Education and Training	• higher education in disciplines pertinent to CZM research • training in use of CZM "tools" to provide a better environment

awareness and wider dissemination of information on key issues, at all levels, from secondary school children to national Ministers and EU Commissioners.

Pure and applied research into coastal zone issues is supported by a range of EU funding instruments and programmes available from EU Institutions, including the European Commission, European Environmental Agency (EEA) and EUREKA. Scores of projects with CZM objectives, or which create or further develop new technologies or techniques useful to CZM, or which effect new research into CZM/marine issues, are enacted via pan-European partnerships under many different RTD (Research and Technological Development) programmes. Unfortunately, there is seldom (if ever) a specific GI/GIS focus in any of these programmes, except in the sense that it is almost impossible to do CZM research without using GIS tools in one form or another. For that reason, it is important that the marine and CZM research community make its special needs heard in the current EGII and GI2000 debates.

Initiatives from EU Institutions

The major programmes for effecting research or technological development with EU financial support include:

- MAST (I, II & III) - Marine Science and Technology Programme, under DG XII (Science, Research and Development), has three main areas of scientific content: Marine Sciences, Strategic Marine Research and Marine Technology. Because MAST is probably the most focused "marine" scientific programme of the EC, it is described more fully in a later section.

- The COST research support programme (European Cooperation in the field of Scientific and Technical Research) of DG XII includes among its 15 main domains the following areas of possible interest to CZM researchers: oceanography, environment, meteorology. Typical projects from these domains are described in a later section.

- Subsidiary actions or programmes which are also effected under DG XII include seven special Task Forces created in the final years of the Fourth Framework Programme, of which "Maritime Systems of the Future" and "Environment - Water" have some relevance to both marine and CZM research. There is also the RTD International Cooperation initiative (INCO) which involves non-EU members in technology development and transfer projects.

- The ESPRIT RTD programme, overseen by DG III (Industry) has nine different research domains, four cross-domain research "themes", nine "preparatory, support and transfer activities" and several "Joint Calls" with other programmes, such as Telematics, Socrates, Leonardo da Vinci, TEN Telecom, etc. There are no specific domains or themes for GI/GIS or CZM in the current ESPRIT programme, which ends with a final Call for Proposals in September 1997.

- Education and exchange of academic research staff are/were effected under the COMETT 1 & 2 programme(s), and its newer replacements, SOCRATES (European Community action programme for co-operation in the field of education) and LEONARDO DA VINCI (for part-funding of trans-national partnerships for projects in training). The education and training initiatives are often the focus of joint calls from EC Directorates, especially DG III, DG XII and DG XIII, in the technology or research areas.

- The Telematics Applications Programme (TAP), coordinated by DG XIII, was divided into 17 different streams of activities, including Telematics for Research, Telematics for the Environment, Telematics for Education and Training and Research Networks, Telematics Engineering, Information Engineering and more, for which individual calls for proposals were issued and certain projects relating to CZM or other marine research were considered.

- INNOVATION, the follow-on to the DG XIII SPRINT programme (Strategic Programme for Innovation and Technology Transfer), focuses mainly on creating and exploiting networks of research and industrial partners to transfer and develop existing technology or emerging technologies resulting from advanced research effected under other EU or nationally funded R&D programmes.

- Specific information and communications technologies related programmes such as IMPACT (Information Market ACTions) and INFO2000 were enacted by DG XIII/E and both included a GI/GIS component for pilot projects, several of which were in the area of metadata and/or GI data provision, including in the marine area (e.g. funded projects such as ERGIS - European Marine Resource Geographical Information Service and ENGINE, and proposals such as COASTWEB, Arctic2, EMMA - European Maritime Multimedia Data Agency for the North Sea region, etc.). Programmes such as IMPACT and INFO2000 will not fund pure research or "science", but they can fund supporting initiatives in other action areas necessary to advancing the objectives of marine and coastal zone researchers, i.e. awareness, information networks, etc.

- At the EC's DG Joint Research Centre (JRC) in Ispra, Italy, two initiatives exist which have a high GI/GIS and/or marine component. These are the Centre for Earth Observation (CEO) and the Marine Environment Unit of the Space Applications Institute.

- The EU Directorates already mentioned also sometimes "mix and match" their budgets to achieve a special result, perhaps in an area not foreseen when the original programme budgets were set. A typical example is the "Blue Seminars" on "aquatic processes and water technology" of the JRC which were allied to the DG XII Task Force on "Environment - Water".

The programmes mentioned above are the prime funding sources for marine and coastal research from the EU Institutions. However, there are other GI/GIS and marine/coastal related projects, which are effected under various programmes of other EU Directorates or as part of their operational remits. Examples are certain projects executed under the PHARE and TACIS programmes of DG I (International Affairs) for East and Central Europe and the Russian Federation, in which the EEA (European Environment Agency) sometimes also plays a part. DG XI (Environment) has already been mentioned and effects joint projects with the EEA. GI/GIS are coming into wide use in the work of DG VI (Agriculture) and DG XIV (Fisheries), especially in planning and remote monitoring, and marine/coastal zone issues are of direct interest to both these Directorates General.

DG XII Programmes with Marine and Coastal Zone Related Actions

DG XII oversees several different programmes, either totally, or in cooperation with other parts of the Commission, such as DG III (Industry), DG XIII (Telecommunications), DG XI (Environment).

MAST

The MAST - Marine Science and Technology - Programme began in 1989 with MAST I (1989 - 1992), followed by MAST II (1991-1994) and now MAST III (1994-1998), that is funded by the EU at 243 million ECU, under supervision of DG XII. There are four main activities:

- **marine science, which covers all seas in the European Economic Area, involving much multi-disciplinary research which is often specific to regional seas and/or to extreme marine environments;**
- **strategic marine research, emphasising the coastal zone and socio-economic impacts;**
- **marine technology; and**
- **a range of supporting initiatives.**

Activities conducted under the MAST banner liaise with many other international programmes, including the International Geosphere-Biosphere Programme (IGBP), the World Climate Research Programme (WCRP), the Human Dimension of Global Environmental Change Programme (HDP), the Global Ocean Observing System (GOOS) in cooperation with international bodies such as the Intergovernmental Oceanographic Commission (IOC) of UNESCO, the International Council for the Exploration of the Seas (ICES) and the International Commission for the Scientific Exploration of the Mediterranean (ICSEM). Most projects are conducted on a shared cost basis (up to 50% funding) and include consortia of partners from more than one EU of EFTA/EEA Member State.

To date, MAST has supported 23 projects in the Marine Science area, including four regional sea projects (Ocean Margin Exchange, Baltic Sea Study, Mediterranean Targeted Project, Canary Islands Azores Gibraltar Observatory); 17 in Strategic Marine Research (all in coastal and shelf sea research and coastal engineering); and 18 projects in Marine Technology. Full details of the MAST programme and all projects currently underway or completed is available from the MAST Programme office at DG XII or from mast-info@dg12.cec.be or from the DG XII/MAST Website on the Europa server at URL http://europa.en.int/en/comm/dg12/. MAST remains the premier programme under which the marine and coastal zone research community can access EU support.

ENVIRONMENT AND CLIMATE

A special programme in Environment and Climate was created by Council of Ministers decision at the end of 1994, to provide 566.5 million ECU for shared cost (50% part-funding) and concertation actions through the end of 1998. The main areas of research include: research into the natural environment, environmental quality and global change (247.75 million ECU); environmental technologies (131.25 million ECU); space techniques applied to environmental monitoring and research (107.6 million ECU) including activities with the CEO - Centre for Earth Observation of DG JRC; and human dimension of environmental change (39.4 million ECU).

Since much marine and coastal zone work is directly related to environmental research or climate/earth observation, there are numerous opportunities for researchers to participate in this programme. Specific to the CZM community are the projects ELOISE (European Land-Ocean Interaction StudiEs) and the Aquatic and Wetland Ecosystems study, carried out under the Biospheric processes action line. ELOISE is implemented in cooperation with MAST III.

A second joint action between E&C and MAST is ENRICH - the European Network for Research into Global Change - which runs through 1998. Full details on the programmes, past and current projects, and application procedures are available from the E&C help desk at DG XII or from environ-infodesk@dg12.cec.be or from the DG XII Web site at URL http://europa.en.int/en/comm/dg12/.

COST

DG XII is also heavily involved in COST - European Cooperation in the field of Scientific and Technical Research - a framework that permits coordination of national level research on a European scale. COST Actions support both basic and pre-competitive research. COST began in 1971 via a Ministerial Conference supported by the original 19 members of the planned consortium. Today's 25 member countries include all 15 EU Member States plus Iceland, Norway, Switzerland, Czech Republic, Slovakia, Hungary, Poland, Turkey, Slovenia and Croatia. COST began with 7 individual actions in 1971 and has grown to more than 115 today. Any COST country can join any Action by signing a Memorandum of Understanding (MoU) which is the legal basis for each Action. An Action comes into force when at least five signatory countries have signed the MoU relating to that Action.

The COST administrative structure includes a Committee of Senior Officials (CSO), the highest decision making body of the framework, with representatives from all 25 member countries and the European Commission. Each member also has a COST National Coordinator. The Secretariat consists of a Council COST Secretariat, representing the Council of Ministers of the European Union, and a Commission COST Secretariat, representing the European Commission. The EC provides the scientific secretariat for the Technical Committees, primarily via DG XII, but also from DG VII (COST Transport) and DG XIII (COST Telecommunications).

As of June 1997 there were COST Actions operating in 18 research areas, including Oceanography, Environment and Meteorology. In Oceanography, two main activities are the European Sea Level Observatory System (EOSS) and Use of Marine Primary Biomass. In Meteorology, one activity is Measurement and Use of Directional Spectra of Ocean Waves.

Because COST is spread across multiple Directorates and Commission and national secretariats, the best place to find information is from the CORDIS Web information service of the European Commission at URL http://www.cordis.lu/cost/src/int_an1.htm.

INCO

A separate programme for International Cooperation in RTD (INCO) is also managed by DG XII, with support from other Directorates as necessary. INCO strives to strengthen links with international organisations and non-EU Member States, especially in East and Central Europe and Russia, to promote scientific and technological cooperation across Europe. As such, INCO supports projects across a wide range of RTD activities, spanning the Fourth Framework Programme. Of perhaps most relevance to the marine/coastal zone research community are projects in the area 3.1.1 of the INCO-DC programme - Sustainable Management of Renewable Natural Resources or certain themes in area 3.1.2 - Sustainable Improvement of Agricultural and Agro-industrial Production. The September 1995 Call for Proposals resulted in 65 projects being funded in these two scientific content areas, including many dealing with wetlands, aquatic ecosystems, coastal zone (nearshore) ecosystems, aquaculture, CZ risk management, etc.

Full details of the programme are available from the INCO Programme help desk at DG XII in Brussels, from inco-desk@dg12.cec.be or from the DG XII Web site http://europa.en.int/en/comm/dg12/.

DG Joint Research Centre (JRC), Ispra, Italy

In 1996, the Joint Research Centre, formerly part of the EC's Directorate General (DG) XII (Science, Research and Development), became the Commission's newest Directorate General in its own right, now known as "DG JRC".

CEO (CENTRE FOR EARTH OBSERVATION)

The Centre for Earth Observation Unit of the Space Applications Institute of the EC's Directorate General JRC (Joint Research Centre) is located at Ispra, Italy. It is one of the largest EU funded GI-related metadata projects ever launched by the Commission, with full Member State support, on a total budget of 100 million ECU (50% from Member States and 50% from the EC budget). In the CEO programme, the main on-line information resource is the EWES server, containing hundreds of references to Coastal Zone research and management projects, as well as access to expertise, to other researchers, to institutes and other lists of CZM resources. There are case studies of interest to marine/coastal researchers, as well as a coastal zone demonstrator. EWES can be reached at URL http://ewse.ceo.org/ and the CEO home page is at URL http://www.ceo.org/.

The CEO places heavy reliance and emphasis on GI/GIS as a key element to all future work in earth observation, including ocean and coastal observation, both from space and land/sea based. Representatives from CEO have participated in the consultation process for both the EGII and GI2000 Communication documents.

MARINE ENVIRONMENT UNIT, SPACE APPLICATIONS INSTITUTE, JRC

The JRC SAI is also the home of the Commission's Marine Environment Unit, whose aim is to "develop, demonstrate and validate methodologies for the use of data from space and airborne platforms in both operational applications and scientific investigations related to the marine environment." (EC EUR 16384, 1996) Because of the number of different programmes in which the Unit operates, at EU, regional and global level, it is difficult to review them all in this chapter. Rather the reader is directed to the excellent Web site maintained by the Unit at the URL given at the end of this section. Two main projects now underway include:

- *Project OCEAN* - **Ocean Colour European Archive Network, established in 1990 as a joint effort between JRC and other EC directorates, supported by DG XI (Environment) and the European Space Agency (ESA). The main objective is to create Coastal Zone Colour Scanner (CZCS) data for the European seas, resulting in final data products at four levels, from raw data (level 0) to fully processed, re-mapped data covering fixed geographical areas (basins) using orbital and geo-referencing data (Level 3). Since its start in 1990, more than 15000 CZCS Level 1 images have been captured, and over 3500 Level 3 images created. Use of GI and GIS tools is of crucial importance to this effort.**

- *Project CORSA* - **Cloud and Ocean Remote Sensing around Africa, "aims to provide a quality controlled data set of surface, atmospheric and cloud parameters over a time period and resolution not available from any other data source". (CORSA Web homepage) The project has derived sea surface temperatures at weekly and monthly intervals from August 1981 to December 1991, now being extended in time, while cloud classification products are also being derived, all based on analysis of NASA AVHRR GAC level 1b data products, more than 13000 of which have been processed.**

The Marine Environment Unit at JRC SAI can be reached via URL http://me-www.jrc.it/. Information on OCEAN is available from vittorio.barale@jrc.it and on CORSA from leo.nykjaer@jrc.it.

EUREKA

EUREKA - Technological Cooperation in Europe - is not a programme of the European Commission, but rather is an initiative created at Ministerial level by 22 European member states, including all EU Member States, for the purpose of encouraging cross-border technology transfer, research and development in a range of technologies. The EU Institutions may participate in EUREKA projects via their own research capacity (such as at JRC) or via R&D programmes (such as Esprit and Telematics Applications) and by other financial facilities (such as support to third countries for technology transfer projects). The technologies and projects are organised into various multi-project "umbrella" actions, such as EUROENVIRON and EUROMAR. The latter umbrella has 22 associated marine research RTD projects either already completed or underway.

The EUROMAR Board, responsible for monitoring the activities of the umbrella projects, cooperates with the EU's MAST programme in developing the basic research supported by MAST. The EUROMAR projects, some of which are already finished, are listed in Table 3.

A typical success story from the EUROMAR stable of projects is SEAWATCH, a six-metre high buoy-based marine monitoring system able to measure 22 environmental parameters every three hours, transmitting the data to land by satellite. A 10-buoy network is already in operation from the coast of the Netherlands to the Barents Sea off Russia, an entire network has been sold to Thailand, and the Global Ocean Observing System (GOOS) may use SEAWATCH as a demonstration project. In any event, the EUROMAR projects demonstrate that industry can work effectively with research institutions in the area of developing marine and coastal zone management and research techniques, tools and systems, for which there is a ready market.

European Environment Agency (EEA)

The EEA, whose headquarters is located in Copenhagen, Denmark, is "the environmental information reference centre of the European Union" (EEA, June 1997). EEA's mission is to monitor the state and trends of the environment in Europe and to support the EU, EU Institutions and Member States in improving their environmental policy and its implementation. A main initiative of EEA is to provide wide access to various levels of environmental information, including marine and coastal data, to as wide an audience as possible, using various means, including the Web, CD-ROM products, and printed material.

To that end, the EEA has created eight European Topic Centres (ETCs), each with from three to 14 partner organisations, helping to collect information that is then made available via the EEA's Web server and via EIONET - the European Environment Information and Observation Network, created in response to Article 1 of Regulation 1210/90 which established the EEA. Full details of EIONET are available elsewhere. (see EEA Newsletter 9). The eight ETCs and hundreds of EIONET partners address 23 projects from the current work programme, including ETC Marine and Coastal Environment MW6-7, which is conducted under the leadership of ENEA, Italy.

This ETC will convene "an inter-regional Forum to facilitate exchange and possible integration of existing data and information produced by regional and international organisations/conventions ... to improve working complementarities and task-sharing." (EEA Newsletter 9) It will also develop a range of data and data products provided by national sources that can be used as indicators of the state of the coastal environment and pressures on the environment.

A major output of the EEA's activities is the forthcoming "Dobris+3 Report", the EEA's second Pan-European State of the Environment report, first available in September 1997, as input to the next Conference of European Environment Ministers to be held in Denmark, June 1998. The Conference is being prepared by an Ad Hoc Working Group of the UN's Economic Council for Europe (ECE) Committee on Environmental Policy (CEP). There is a main chapter on "Marine and Coastal Environment" in the report now in preparation. Those members of the marine and CZM research community engaged primarily in environmental monitoring and/or climate change should be aware of this important document and be prepared to comment on the early drafts.

In May 1997, the EEA also released its official position statement on Public Access to Environmental Information that should be of interest to most of the research community. (Hallo 1997) This document is important input to the larger debate now taking place in regard to public access to GI collected by public agencies, and the Commission's own Green Paper on Access to Public Information. These issues are of importance to all members of the research community.

Databases for the seven Topic ETCs (excluding the ETC "Catalogue of Data Sources") are available from the EEA's Web server at URL http://www.eea.eu.int/. Full annual work programme details, general information and newsletters can also be found at this site, and information is also available on request from eea@eea.eu.int.

Initiatives from Non-EU Institutions

There are many initiatives sponsored by international organisations such as the UN or international associations of research institutions. In this chapter, only certain initiatives that focus primarily on pan-European marine or coastal areas will be presented. A second chapter by the same author, also presented at this symposium, will look at the more global programmes which include pan-European marine GI coverage, but which do not focus exclusively on a European area.

TABLE 3 EUROMAR Projects in EUREKA

(F=Finished; O=Ongoing)

Acronym	Proj. No.	Name	F/O
ATOMAR	EU450	Development and application of hardware and software relevant to atmospheric input into European seas.	F
BIMS	EU408	Benthic instrumentation and monitoring system	F
BISCUIT	EU1281	In-situ dynamic benthic chamber	O
CARIOCA	EU819	Carbon interface ocean atmosphere	O
CHARISMA	EU344	Characterisation by remote and in-situ sensing of marine sediments	O
CUPIDO	EU1129	Multifunction acoustic current profiler	O
DISC	EU372	Directed sensor carrier system for seabed surveys	F
ECHOSEA	EU628	High resolution, high acoustic power echographic system for exploration of the sea bottom and sediment structure	O
ELANI	EU493	Electroanalytical instrumentation development for physico-chemical characterisation of trace metals/marine environment	O
FIESTA	EU449	Rules for field data quality standardisation	F
MAROPT	EU413	Marine optical recording system	F
MERMAID	EU417	Remote-controlled, modular measuring system for sampling/in-situ toxic contaminants analysis in estuaries/coastal waters	O
MICSOS	EU1246	Micro structure ocean sonde	O
MOSES	EU410	Mobile station for environmental service	F
MURPOS	EU1657	Development of a multi-purpose buoy	O
OPMOD	EU429	Operational modelling of regional seas and coastal waters	O
PROBIO	EU729	A programmable/event-controlled large volume filtration system for long term biological sampling	F
SEASTAR	EU494	System for airborne remote sensing of the sea	F
SEAWATCH	EU453	Experimental operational marine environmental surveillance and information system for European seas	F
SMURV	EU409	Swath multipurpose research vessel	F
STIRLING-AUV	EU1249	Stirling-electric powered autonomous underwater vehicle	O
VISIMAR	EU495	Visualisation and simulation of marine environmental processes	F

The main pan-European actions focus on the Baltic Sea, the Mediterranean and Black Sea, and the Arctic Ocean. The sponsored activities are of various forms, but normally cover:

- **provision of a wide range of information via networks of research institutes or libraries, or by on-line means, including Web sites and e-mail discussion fora;**
- **major conferences focusing on specific research or environmental themes relating to specific geographic areas, such as Baltic or Mediterranean;**
- **conferences focusing on technologies applicable to marine and coastal research, e.g. remote sensing as applied to marine research, modelling techniques for coastal research, etc.;**

- purely educational actions, such as summer schools or extended workshops (a week or more), which provide special training relating to GI/GIS and marine research;
- awareness raising actions targeted at national or regional governments or at international organisations such as various divisions of the UN

Unfortunately, space restrictions limit this chapter to providing only a descriptive paragraph or two for each institution, action or initiative.

Initiatives and Actions for the Mediterranean and/or Black Sea

MEDITERRANEAN ACTION PLAN (MAP) OF UNEP (UN ENVIRONMENTAL PROGRAMME)

The MAP - Mediterranean Action Plan - has carried out pilot programmes on coastal management based on co-operation between UNEP, local and national organisations. Direction is provided by a Co-ordinating Unit for the MAP at UNEP.

INTERNATIONAL CENTRE FOR COASTAL AND OCEAN POLICY STUDIES - ICCOPS

ICCOPS is located at the University of Genoa, Italy. It focuses on training and education for integrated coastal management (ICM) especially in relation to the Mediterranean area. ICCOPS is also the home of the Ocean and Coastal Management Archives (OCMA) multimedia project with support from the Department Polis and Library Service Centre "N. Carboneri" of the Univ. of Genoa, plus the Chamber of Commerce of Genoa. Typical of ICCOPS activities are the forthcoming international conference on Education and Training in Integrated Coastal Area Management - the Mediterranean Project, held in Genoa, Italy in May 1998.

EURO-MEDITERRANEAN CENTRE ON INSULAR COASTAL DYNAMICS - ICOD

ICoD is situated at the Foundation for International Studies at Valletta, Malta, where it has been operational since May, 1987, although it took its present name in 1993. ICoD is involved in both research and workshops on risk modelling, assessment and management, application of space technology for prevention and management of insular coastal hazards, training courses in GIS, the proposed Seaweb network for information provision and the Mediterranean Marine Information network - MEDINFO. MEDINFO is a "cooperative regional network of marine libraries, information and documentation centres located in the Mediterranean and Black Sea areas, which is available on-line via the Web. The database of contacts contains a 140 page directory listing over 85 institutions involved in marine or environmental research throughout the region.

MEDISLE - Mediterranean Island Coastal Network - is another ICoD initiative, proposed in 1996, with support from the Council of Europe, to promote the sustainable development of Mediterranean islands. MEDISLE will focus squarely on the often unique problems relating to the land-sea interface in the planning and management of small islands, which are especially vulnerable to natural disturbances. They have extensive and complex coastal zones, the health of which is often of critical importance to the inhabitants, whether human, animal or flora. The MEDISLE concept follows the recommendations of the Tunis Declaration on Sustainable Development in the Mediterranean Basin and with Agenda 21 for the Mediterranean and with the UN Convention on the Law of the Sea.

MEDCOAST INSTITUTE

The MEDCOAST Initiative started with its first International Conference on Mediterranean Coastal Environment in Antalya, Turkey in November 1993 with the objective of contributing to environmental management in the Mediterranean and Black Sea via research, human resources development, information exchange and technology transfer. MEDCOAST Institute 1994 focused on Coastal Zone Management in the Mediterranean and Black Sea. MEDCOAST Institute 95 was held in Ankara and Marmaris and concentrated more on real case studies of coastal environment problems in the Mediterranean and Black Sea rather than on theoretical lectures. MEDCOAST 97 was held in Qawra, Malta in November 1997. The MEDCOAST initiative (the study institutes and shorter conferences) was partly funded by the EC's Med-Campus programme.

EUROPEAN DIRECTORY OF MARINE ENVIRONMENTAL DATA (EDMED)

EDMED is a computer searchable directory of data sets relating to the marine environment created under the EC's DG XII MAST programme (see reference earlier in this chapter), and is a collaborative venture 11 EU Member States (Belgium, Denmark, France, Germany, Greece, Ireland, Italy, the Netherlands, Portugal, Spain and the UK). The British Oceanographic Data Centre (BODC) in the UK maintains EDMED.

Initiatives and Actions for the Baltic Sea

The two major actions relating to GI/GIS and marine/coastal zone research in the Baltic Sea are the Baltic Sea Drainage Basin GIS Project of UNEP/GRID-Arendal and the Baltic Marine Environment Protection Commission (HELCOM), which are described more fully below. There are other initiatives that include CZ research projects in the region, but they are part of global programmes that are covered elsewhere in this chapter (see "International Initiatives/Programmes Active in Europe").

BALTIC SEA DRAINAGE BASIN GIS PROJECT (BGIS) - UNEP/GRID-ARENDAL

Baltic Sea projects often include some countries which have no access to the coastline, simply because they are part of the larger drainage basin which feeds the Baltic and which certainly affects the coastal zone. For example, the countries participating in the Baltic Drainage Basin - Baltic Sea Region database include the coastal states Finland, Russian Federation, Estonia, Latvia (Letland), Lithuania, Poland, Germany, Denmark, Sweden and Norway, plus landlocked Belarus, Ukraine, Slovakia and the Czech Republic.

Within the BGIS Project is the project *MapBSR - Map Data Sets in the Baltic Sea Region* - coordinated by the National Land Survey of Finland, to oversee creation of the basic map data sets for the Baltic Sea drainage area, using input from the National Mapping Agencies of the project's 14 member states. Elements included in the map data sets include boundaries, hydrography, transport, settlements, geographical place names, elevation, nature and land use, forming the first uniform, reliable map data set for the Baltic Sea drainage area and the countries around. Other thematic data can be added to the database at a later time, such as water quality. Three of the many themes directly of interest to marine and CZM researchers are bathymetry (Theme 301), sea area division (302) and coastline (303), joined by rivers (306), lakes (307) and drainage basins (308). (Langaas, 1993) Details of the MapBSR project, plus the Nordic map database and Barents Geographic Information Technology database, are available from the Webmaster at the National Land Survey of Finland, e-mail webmaster@nls.fi.

BALLERINA - UNEP/GRID-ARENDAL.

The umbrella activity for providing comprehensive environmental information about the Baltic region began in 1996 as BALLERINA - BALtic Sea Region On-Line Environmental Information Resources for Internet Access. The main goal of this ambitious initiative was "to bring more substantive and relevant environmental information from and about the Baltic Sea region to the Internet ... and to develop a personal and institutional network of environmental information providers in the Baltic Sea region..." (UNEP/GRID 1996).

Full information is available on the UNEP/GRID-Arendal activities, including the Baltic Sea Drainage Basin project and BGIS from URL http://www.grida.no/baltic/welcome.htm.

HELSINKI COMMISSION - BALTIC MARINE ENVIRONMENT PROTECTION COMMISSION (HELCOM)

HELCOM is the acronym for the Helsinki Commission - Baltic Marine Environment Protection Commission, the governing body for the first Convention on protecting the Baltic Sea area, signed in 1974 by the Baltic coastal states. The first Convention was joined by the EU, represented in 1992. Present contracting parties to

HELCOM include Denmark, Estonia, European Community, Finland, Germany, Latvia, Lithuania, Poland, Russia and Sweden. The main aim of the Convention is to protect the marine environment of the Baltic Sea from all sources of pollution, whether from land, ships at sea or airborne.

The Commission meets annually with occasional meetings held at Ministerial level, reaching unanimous agreement on actions to be taken to achieve the aims of pollution prevention, which are then regarded as recommendations to the governments concerned. There are four Committees (Environment, Technological, Maritime, Combating), the Programme Implementation Task Force and an Administration Unit. As of March 1996, there were 116 separate Recommendations accepted by the signatories to the Convention - far too many to even begin to review in this chapter. However, as one might expect from a major convention regarding marine pollution, very many of these Recommendations bear either directly or indirectly on coastal zone management and research.

Full details of all activities of HELCOM and texts of all Recommendations can be accessed from their Web site at URL http://www.helcom.fi/.

International Initiatives/Programmes Active in Europe

As to international programmes, UNEP has already been mentioned in relation to the Baltic Sea programme. Other UN organisations involved in marine or coastal research, training or education include UNESCO, the WMO - World Meteorological Office, FAO - Food and Agriculture Organisation, UNIDO - UN Industrial Development Organisation, IMO - International Maritime Organisation and more. The three institutions briefly described below are included separately in this chapter, as they strongly support various initiatives in the European region and/or work jointly with other EU Institutions in the various European coastal areas.

INTERGOVERNMENTAL OCEANOGRAPHIC COMMISSION (IOC) - UNESCO

IOC, founded in 1960, comprises an Assembly, Executive Council and Secretariat (based in Paris) representing 125 member states and has established several subsidiary bodies. Because IOC's activities are mainly global, with regional subsidiaries in the major ocean areas of the world, its activities and objectives will be reported on more fully in an allied paper from the same author. The IOC lends support to numerous national and regional programmes and initiatives, such as ICoD, ICCOPS, MEDCOAST and others in the Mediterranean region, and has a separate Regional Committee for the Black Sea. It also supports TEMA - Training, Education and Mutual Assistance in the marine sciences. IOC is also responsible for the Global Ocean Observing System (GOOS), including coastal zone activities.

INTERNATIONAL COUNCIL FOR THE EXPLORATION OF THE SEA (ICES)

Founded in 1902 and with current membership representing 19 countries from both sides of the Atlantic, including all European Coastal states (except the Mediterranean countries east of, and including, Italy), ICES is the oldest intergovernmental organisation in the world focusing on marine and fisheries science. Its multi-disciplinary work programme concentrates on hydrography, physical oceanography, population dynamics of fish stocks, standards of quality and comparability of ocean-related data. The ICES secretariat is located in Denmark, from which site three databanks are maintained for oceanographic, fisheries and environmental (pollution) data. More than 100 meetings are held each year by ICES working groups, study groups, workshops and committees, the latter of which advise national Member Country governments, international regulatory commissions and the European Commission. Many of the workshops deal with coastal or estuary problems, not only deep ocean research, and focus on regions such as the Baltic Sea and Mediterranean.

WORLD WIDE FUND FOR NATURE (WWF) INTERNATIONAL

The WWF's Europe/Middle East Programme has focal activities in the Mediterranean, the Baltic Sea, the North Sea, the Arctic and a unit for European Policy development which often works with relevant directorates at the European Commission on agriculture, environmental and development policy. The Europe/Middle East Programme stretches from Portugal to the Bering Strait, from the Arctic Ocean to the Persian Gulf, fielding National Organisations in 14 countries, supporting conservation measures in 18 more, for a total of 120 projects. WWF considers oceans and coasts to be one of the three major "priority biomes" to be protected and properly managed for future generations.

In the Mediterranean, wetlands are a key issue and concern, and WWF participates in the EU-funded MedWet initiative. WWF's marine strategy in the area is concentrated on integrated coastal management and conserving wildlife (sea turtles and monk seals).

In the Baltic area, WWF's Baltic Programme succeeded in persuading the countries of the Baltic Sea region to include a special article on nature conservation and biodiversity protection in the revised 1992 Helsinki Convention (HELCOM) referred to earlier in this chapter. HELCOM adopted WWF's proposal to establish more than 60 Baltic Coastal and Marine Protected Areas (BSPAs). In 1993, as part of the Baltic Sea Joint Comprehensive Environmental Action Programme, the member states asked WWF to coordinate the development of integrated coastal management plans for five large coastal lagoons and wetland areas on the south-southeastern Baltic coasts.

In the Arctic region, WWF projects support AEPS (Arctic Environmental Protection Strategy) priorities, including conserving marine resources and developing coastal zone and management schemes.

Full information on the WWF and its extensive global programmes is available from their excellent Web site at URL http://www.panda.org/wwf/.

Potential Impact of the EU's Fifth Framework Programme for RTD (1998-2002)

All members of the European marine and coastal zone research community should be very concerned about the development of the EU's Fifth Framework Programme (5RWP) for research and technology development, whether they are engaged primarily in local or national research or participate in pan-European initiatives. Nearly all the EU funded programmes mentioned in this chapter have been funded in the past by the many actions that constitute the EU's Framework programmes. The current 5FWP working documents have attempted to adopt a slightly more radical approach to structuring the future of sponsored R&D in the EU.

The Commission proposes three broad "Thematic" programmes and three "Horizontal" programmes:

- **Unlocking the resources of the living world and the ecosystem,**
- **Creating a user-friendly Information Society, and**
- **Promoting competitive and sustainable growth.**

- **Confirming the international role of European research,**
- **Innovation and participation of SMEs (Small to Medium Enterprises)**
- **Improving human potential.**

The Thematic programmes will be executed via 16 "Key Actions". Meanwhile, coordination will continue to be maintained with allied programmes such as PHARE, TACIS, COST, EUREKA, activities of the JRC, etc. The "new approach" is more than just a little confusing, to say the least". The explanatory documentation runs to more than 100 dense pages of text across several documents, to which over 18 official replies from Member States and organisations such as EUREKA and the EEA have been made, totally yet more hundreds of pages of print. All of this is available on the Web from the main CORDIS documentation delivery service at URL http://www.cordis.lu/.

However, readers should know that the process of agreeing a major Framework Programme for RTD is not quick or easy. The official work programme as presented by the Commission must be sent to both the Council of Ministers and the European Parliament for approval, both as to content and budget. Many comments are now being fielded by the European Parliament's Committee on Research, Technological Development and Energy as well as by national Ministers for Research at Member State level (who represent their countries on the Council of European Research Ministers).

There is still quite a long way to go in agreeing the final content of the 5FWP. One fact is clear at the moment though. GI and GIS do not figure predominantly in any of the programme actions. Rather, the role of GI as a whole, without regard to the discipline in which it may be necessary or used, is being treated as "geo-publishing" under the information market banner. This needs to be redressed very soon by the whole scientific research community.

GI2000: Towards a European Policy Framework for Geographic Information

Following more than two years of wide consultation with the European geographic information community, the EC recognised the need to formulate a European policy framework for geographic information which would permit European GI to be created, combined, marketed, used, reused and shared in a cost effective manner for the benefit of the European information society. Studies sponsored by DG XIII/E indicate that the major impediments to wider and more efficient use of geographic information in Europe are political and organisational, not technical. The lack of a formal mandate at European level, to address the needs of the GI community, is retarding development of harmonised GI strategies, across national boundaries and across disciplines.

To redress this situation, in October 1997 the Commission sent a Communication to the Council, to the European Parliament, to the Economic and Social Committee and to the Committee of Regions titled "GI2000: Towards a European Policy Framework for Geographic Information". Quoting from that document, the Commission states:

"What is required is a policy framework to set up and maintain a stable, European-wide set of agreed rules, standards, procedures, guidelines and incentives for creating, collecting, updating, exchanging, accessing and using geographic information. This policy framework must create a favourable business environment for a competitive, plentiful, rich and differentiated supply of European geographic information that is easily identifiable and easily accessible."(DG XIII/E, 1997)

The Communication predicts benefits which include greater efficiencies of scale in a more unified market, reduced problems for cross border and pan-European projects, efficient technical solutions for future growth, increasing use of European skills, improved market position in geographic information and better results of European-wide planning and decision making. The policy framework must include the legal aspects of geographic information to ensure the creation and use of EU-wide datasets and standards. It must also stimulate and challenge private companies and public bodies to invest in the creation of such datasets and to cooperate where appropriate.

The EU Member States should agree to set up a common approach to create European base data, and to make this generally available at affordable rates; to set up and adopt general data creation and exchange standards and to encourage their use; to improve the ways and means for both public and private agencies and organisations to conduct European-level actions (including cross-border and cross discipline initiatives); and to ensure that European solutions are globally compatible.

The Commission has created a GI2000 High Level Working Party, comprising representatives from the leading players in the public and private sectors of the wider GI community, including strong user representation, to be chaired by the Commission. This Working Party would elaborate a detailed action plan to implement the policy and provide political leadership and vision to guide implementation of the action plan. "The High Level Working Party will also provide the focal point for promoting a sense of unity across disciplines and national borders."

The Commission's role would be to provide the European dimension to actions at national level, acting as a catalyst, to coordinate Member States' policies, always building on existing national information holdings and structures. No new European organisational structures nor any form of central GI data storage is proposed. Collection and storage of GI, creating and disseminating GI metadata, remain national tasks. The Commission will attempt coordination in regard to global geographic information policy and projects, such as those proposed via the G7 and discussions already initiated at global level for the Global Spatial Data Infrastructure. (Chenez 1996)

Where does this leave the marine and coastal zone research and applications development community in regard to GI/GIS and marine/CZM data? The short answer is without any specific focal point at political level, where the major decisions will be made regarding "infrastructure". The Commission's Communication, in its current form, focuses very heavily on "information market" issues, which is not surprising considering that it comes from the Directorate (DG XIII/E) which is responsible for this aspect of the Information Industry. Research, education, training and related issues are not in the remit of DG XIII/E and, although mentioned in the Communication, do not receive the degree of attention deserved for so important a document. The marine/CZM research community should be making itself aware of the content of the Communication and expressing their concerns in the strongest manner possible to their national associations, relevant national Ministries, to EUROGI - the European Umbrella Organisation for Geographic Information and perhaps to GISDATA, the Geographic Information Systems Data Integration and Data Base Design initiative within the European Science Foundation Scientific Programme.

Both EUROGI and GISDATA have recently taken their concerns to the European Parliament to try to ensure that the needs of the whole European GI community are considered when the Commission Communication reaches Parliament in October 1997. EUROGI is also playing a leading role in the global GSDI discussions, the next high level meeting of which will be held in North Carolina, USA, in September 1997. As regards the needs of the research community, as the GISDATA initiative comes to a close this year, a new association is being formed, as described in the next section of the chapter. Members of the European-wide marine and coastal zone research community are strongly urged to consider the benefits of joining such an association, which could act in future as their spokesman to the EU Institutions for concerns in their special field(s) of endeavour.

GISDATA and AGILE

The GISDATA programme has operated a series of specialist meetings, focusing on pre-selected problems in the social sciences, from the viewpoint of use of GI/GIS to address a wide range of social issues. Environmental issues were included in those meetings. As the GISDATA has come to an end (a five-year funding effort from the ESF which began in 1992), the members of GISDATA have proposed creating a new association with a wider brief called AGILE - Association of Geographic Information Laboratories in Europe (Burrough et al., 1997). This new association would move beyond examining only social science issues and would represent the views and needs of the whole European GI research community in its widest definition, which certainly includes the practical needs of the marine and coastal zone research community.

A survey carried out in February 1997 by questionnaire sent to more than 50 GI research establishments in 13 EU Member States, from which 25 positive responses were returned, indicated that:

- **there is unanimous agreement that an organisation such as AGILE is needed in order to continue the previously important and valuable work of GISDATA and the EGIS/JEC conferences;**
- **the most important task for such an organisation would be to organise focused meetings on key research issues;**
- **the organisation should also organise an annual or bi-annual geographic information research conference and set up working groups intended to influence the future European geographic information research agenda.**

If these views are regarded as reasonably representative of other GI research groups in Europe, or of multi-disciplinary research groups which rely heavily on GI to effect their research, then there is a strong case for establishing an organisation such as AGILE. To that end, a small group of representatives from research centres in different parts of Europe has been created to prepare proposals with respect to the title of the organisation, its mission statement, its organisational structure, the procedures for the election of officers, the criteria for membership and subscription fees and to define a position with respect to other European level bodies such as EUROGI.

AGILE can provide an important role in furthering GI/GIS research and education in Europe, at a pan-European level. It will most probably be able to launch itself, with wide European participation of major actors in GI/GIS R&D and academia, based on resources generated from low annual subscription fees. It is up to members of allied research communities which rely heavily on GI and GIS, especially those involved in cross-border projects or programmes, to make their concerns known to the policy makers at national and EU level. AGILE could provide a useful vehicle for achieving this goal.

Conclusion and Recommendations

What the reader will no doubt by now have realised is that there is no single focal point nor source of EU funding for research or technology projects relating specifically to "coastal zone management". There never has been nor (probably) ever will be a single focal point for all the diverse actions that need to be undertaken in regard to CZM at transnational (cross-border) level, simply because these activities are so diverse in nature. The MAST programme of DG XII and environmental actions of the DG Joint Research Centre (JRC) are the most relevant actions for CZM researchers.

The CZM research and technology development and applications community is required to make themselves fully conversant with dozens of EU sponsored initiatives in order to be in the best position to apply for research support. Also, there is no central body that makes any attempt at "coordinating" such research across national boundaries or across disciplines or application areas. No one speaks for "coastal zone management" at the decision making level of the EU institutions - although many Directorates will claim "to represent the needs of the CZM community".

Many of the EC directed programmes listed above are also in their final stages of funding, due to the expiration of the Fourth Framework Programme in 1998, from which budget line they were primarily resourced. Further delays can be expected in adoption of the Commission's proposals for the Fifth Framework Programme due to objections from the European Parliament and the Council of Research Ministers that are still being debated. The implication for CZM researchers is that, with no or few new calls for proposals expected in the dying stages of the Fourth Framework Programme, no new funding was available until late 1998, for programmes or projects to start in 1999.

There has been a steady shift away from "research" towards "development" and now "near market actions" across the Third, Fourth and Fifth Framework Programmes, respectively. Many of the "Key Actions" that are the focus of the Fifth Framework Programme will focus on "Information Society" issues, rather than pure research. The GI/GIS community is already concerned about this. An approach to the European Parliament's Committee on Research, Technological Development and Energy was made to ensure that GI is not buried in the programme of work as a "geo-publishing" issue (which was the thinking). The CZM research community needs to consider a similar approach, before it is too late.

Therefore, it is up to the CZM research community itself to:

- **know whom to approach (senior Commission officials and European Parliamentarians),**

- **when (during major programme development exercises, such as those underway for the Fifth Framework Programme),**

- **with what objective(s) (seeing that any special needs in relation to CZM research are not forgotten in the clamouring of hundreds of other special interest groups), and**

- **how (by understanding the programme development and full project implementation cycles, the key players who determine budgets, how the budgetary system of the EU Institutions works, and how to join forces with**

larger organisations such as industrial research councils, the European Science Foundation, etc.).

This is an area in which the AGI's Marine and Coastal Zone Management SIG (Special Interest Group) could become usefully active, in cooperation with like-minded groups or associations at national or pan-European level. Without concerted action now and in the formative stages of the detailed work plans for the Fifth Framework Programme, marine and CZM research may not be progressed.

To complicate matters, there are many other pan-European initiatives relating to CZM, IT research, training and applications development requirements, which are not directly sponsored by the main EU institutions. (Remember that we are not considering here purely national initiatives, or those of any single educational or research institution, unless it has a primary trans-national element).

The whole community of actors involved in coastal zone research and management must strive to ensure that the developers of such international programmes or initiatives receive learned, useful, timely input from the national academic and political representations who work with the various implementing organisations, e.g. divisions of the UN (UNEP, UNDP, UNESCO, FAO), the World Bank, and the EU Institutions (European Commission, European Parliament, Councils of EU Ministers, European Investment Bank, European Bank for Reconstruction and Development, European Environment Agency, etc.). Is there a mechanism for such technical and political input today? I believe not, and this should be considered for future implementation.

As regards participation in the preparation for the Fifth Framework Programme for RTD of the EU, it may already be too late to exert any major influence on development of the initial work plans, as these will have been decided at Commission, Council and Parliament level by September 1997. However, as has happened in both the preceding Framework Programmes (3rd and 4th), changes or additions are not only possible, but highly likely as the work programme proceeds. To effect such changes, the marine and coastal zone research community, which covers a wide range of disciplines, must find a means to examine the individual work programmes in detail, then pool their concerns and appoint a spokesman/group to act on their behalf at high government level. Organisations such as EUROGI and AGILE stand ready to assist in this non-trivial task.

References

Burrough, P, Masser, I., and Craglia, M. (GISDATA Steering Committee) and Longhorn, R.A. (May 1997). *European Research and Education in GI and GIS: AGILE - Association of Geographic Information Laboratories in Europe: A Proposal*

Chenez, C.C. (1996). *Proceedings of a Conference on the Emerging Global Spatial Data Infrastructure,* held under the Patronage of Dr Martin Bangemann, European Commissioner for Industrial Affairs, Information and Telecommunications Technologies, Bonn, 4-6 September 1996.

European Commission, DG XIII/E (1997). *GI-2000: Towards a European Policy Framework for Geographic Information.* DG XIII/E, Luxembourg.

European Commission, EUR 16384, A. Belward (1996). *Institute for Remote Sensing Applications / Annual Report 1995,* Chapter 4, Office for Official Publications of the EC, Luxembourg.

European Environment Agency (EEA) (June 1997). *Newsletter(s) 9, 12 and 13.* p. 4.

European Environment Agency (EEA) (1997). *Annual Work Programme - 1997.*

Hallo, R. (1997). *Public Access to Environmental Information, Expert's Corner publication no. 1997/1,* EEA and Stichting Natuur en Millieu (Netherlands Society for Nature and Environment).

Langaas, S. (1993). *Global GIS Data Made Regional.* UNEP/GRID-Arendal, Stockholm, Sweden.

UNEP/GRID-Arendal (1996). *Annual Report 1996,* GRID-Arendal, Stockholm Office, Sweden.

CHAPTER 4

Plans for the Coastal Zone

P.A.G. Watts

ABSTRACT: Coastal Zone and Shoreline Management Planning will benefit from continued development of "best practice". High quality, consistent topographic information is key to the effective integration and interpretation of the varied data that are essential components in Shoreline Management Plans. These data will deliver greater benefit if they are used within a GIS that can offer flexibility and the capability to deliver dynamic planning processes rather than static documentary plans. Ordnance Survey® (OS), provides key "framework" topographic information for many applications, through a portfolio of complementary, digitally-based products and supporting services. Recently OS has been refining a selection from this portfolio to meet the special needs of Coastal Zone and Shoreline Managers. The chapter describes these products and discusses their contribution to Coastal Management "best practice".

Introduction

It is perhaps purely coincidence that within a couple of miles of the OS HQ building where this chapter was written, is Canute Road, Southampton. This is claimed as the spot where, according to tradition, King Canute attempted to drive back the sea by his command. Depending on your perspective and on which version of this story you read, he was either convincing his courtiers of his impotence, even as a king, to command the great forces of nature, or alternatively, he was actively involved in an early example of Coastal Zone or Shoreline Management Planning! That he was not as successful as other famous (Biblical) predecessors, Moses and Elijah, is perhaps not so surprising given the "power" they had on their side.

      Nevertheless Canute would probably be surprised if he were to return to the same spot today. He would now find a significant area of well established land reclamation, the coast having retreated, under human management, by several hundred metres from the shoreline he knew - such is the "power" that man can now harness.

On reflection this is perhaps a trite example, but it does illustrate the dynamic nature of our landscape, and the extent to which man has influenced it, particularly in the past couple of centuries. Our attempts to change and control the natural processes of coastal erosion, inundation and deposition are now being re-evaluated There is a growing need to manage "hard cash" investments in "hard landscapes" with much greater sensitivity and with increased regard for the environmental, economic and socio-cultural impacts of these activities. Nevertheless, whatever the regime employed, whether managed retreat, natural accretion or "hard engineering", the need for reliable up-to-date topographical information is a vital component of the decision informing, decision-making and decision-taking processes that are at the heart of properly managed Coastal Zone activity. This is a sphere where Ordnance Survey has over 200 years of experience, and where, in today's Information Technology-led environment there is scope for OS to make a significant contribution to best practice.

A Map-Oriented Appraisal for the Coastal Zone

From a surveying and mapping perspective the Coastal Zone represents a critical interface between the technological (and user) facets of the topographic map and those of the nautical chart. Inevitably then, one of the problems facing those who operate in the Coastal Zone is that of managing the disparity between these two, often overlapping, information sources. The map is principally land based and topography-oriented. It is underpinned by surveys such as those undertaken by OS, that extend typically down to mean low water, and tend to reflect the sea as a limit to the land.

By contrast, the charting of the seaward dimension, ostensibly from above high water out to whatever offshore limits are defined for a particular Coastal Zone, treats the land as "bounding" the sea. In Great Britain the recording of this marine "landscape" is primarily the domain of the Hydrographic Office, supported according to local circumstance by port, harbour and other inshore management agencies.

The Coastal Zone is, of course, also the interface where land-based activities deliver their impact to the sea, and where in-shore activities and coastal hydro-dynamics contribute an influence on the coastal landscape and what happens on and near to it. This zone is therefore, by definition, one of intense activity and sustained pressure for change. In the contemporary environment the growing threat of global warming is pointing to a considerable likelihood of rising sea levels. Allied changes to climatic conditions may also be leading to alterations in marine currents and flows, and to the increased incidence of storm and severe storm conditions. All of these circumstances pose threats to the stability or predictability of coastal geography and its evolution.

Simultaneously there are continued and growing pressures for man-made development and more intensive use of near-coast land. Demands for reclamation, development exploitation and new uses for land at the coastal margins and for the construction of "land protection" engineering to safeguard these developments are constant issues and concerns for those involved in the management of the coastal zone.

Conversely there is also a growing awareness of the sensitivity of the coastal environment. This viewpoint is backed by lobbying that now influences the highest levels of organised society. It demands a more sympathetic, "natural-process-tolerant" approach to managing what are frequently fragile coastlines. Often these "green"

Plans for the Coastal Zone 63

arguments advocate "positive discrimination" in future approaches to environmental and ecological matters. These must be accommodated within the overall framework of Coastal Zone management strategies.

Taken together these many conflicting demands raise growing challenges to the management of the Coastal Zone. Coastal Zone Managers need increasingly wide-ranging and sophisticated knowledge about the landscape, environment and socio-cultural circumstances of their area of interest if they are to meet these complex challenges and respond to them effectively. It follows that, at the heart of any effective "best practice" management and executive decision-making system must be sound, reliable and comprehensive geospatial and topographic information.

Mapping (Geographic Information) and the Coastal Zone

In the context of the Coastal Zone, and seen from the mapping and GIS perspective, it is possible to crystallise three key high level groups of processes or activities:

- **Understanding what has happened in the past and up to the present**
  i.e. Modelling and understanding the current coastal zone situation, and particularly the interaction of the maritime and terrestrial domains
- **Deciding what should happen in the future**
  i.e. Establishing the objectives, and planning the strategies that are most appropriate for the future management of the particular coastal zone
- **Making it happen**
- **i.e. Managing and monitoring the implementation of these plans and strategies over time**

In these circumstances it may be perhaps fortuitous that, within any Coastal Management sector, (be it a Management Unit, Sub-cell or Cell), many of the management processes and initiatives that will be planned must, perforce, be implemented from a landward perspective. Further, a large proportion of the agencies involved in devising, implementing and managing these schemes are land-based. Hence there is significant potential value to be derived from a "best practice" Coastal Zone Planning and Management regime that exploits the ready availability of high quality, consistent land-based topographic information, such as that offered by agencies like OS.

The advent of digital technology and more importantly the recently promoted concept of a (UK) National Geospatial Data Framework (NGDF) further enhance the potential for helpful solutions for the Coastal Zone Manager, in the UK at least. NGDF particularly, offers the potential to facilitate the exchange, integration and value-added analysis of disparate geospatial datasets such as those likely to be relevant to Coastal Zone Planning and Management functions. The potential of "tools" such as GIS and "protocols" like NGDF as aids to developing "best practice" for shoreline management at the "map/chart interface" should not be underestimated. The potential for NGDF in this arena in particular is discussed later in this chapter.

Mapping (Geographic Information) for the Coastal Zone

There is a well worn adage in OS, taken from the report of the Davidson Committee of 1938 (Davidson, 1938): *"There is no question that what is wanted by Government Authorities, and by the public, is a map which represents the relevant facts as they are on the day when the map is required for use"*. Though this statement was initially made with regard to map currency, the philosophy may be applied with equal pertinence to the issue of what are *"the relevant facts"*. They will, of course, be determined principally by the user, and by the user's interest or application for the map in question. Thus, in devising the relevant mapping and geographic information for the Coastal Zone, it is necessary to define the users and their needs/applications.

Potential Coastal Zone users of geographic information may include:

- **Specialist Planners**
- **Coastal Engineers**
- **Those concerned with ownership, occupation, development or use of coastal real estate, (and their management agents)**
- **Industrial, Commercial and Financial interests**
- **Environmentalists and Ecologists**
- **Marine users including shipping, port and fishing applications**
- **Leisure and Tourism**

Superimposed upon these applications sectors are the "public service" needs of Local and Central Government Agencies and some major Utilities. The list is large and wide ranging with many applications and disparate requirements.

There are of course many sources of geo-topographic information potentially available to the Coastal Zone community. These include a variety of specialist datasets captured over time by local, regional, national or specialist agencies. Many will have a spatial index or reference, and frequently the user information will be presented as an overlay on a general-purpose map, or in the form of a specialist thematic map. General topographic information is also available from remotely sensed imagery, either aircraft borne (air-photography) or satellite sensed. Increasingly this information is available as digital imagery that lends itself to specialist analysis using GIS. In the coastal arena infrared aerial photography has long been a key tool in the mapping of low water and the foreshore.

Examples of these specialist datasets include:

- **Shoreline Physical Process data:- e.g. foreshore characteristics, rates of coastal change through time, beach erosion and accretion data and flood monitoring information**
- **Coastal Engineering information:- e.g. sea defence surveys (Environment Agency, Halcrow), coast protection Surveys (MAFF)**
- **Natural Habitat and Environment data:- e.g. land cover, designated sites and areas (SSSI, NNR, LNR, AONB, county reserves, protected sites) etc.**
- **Man-made Environment data:- e.g. county and regional structure plans, L.A. local plans, urban and settlement geography, communications infrastructure,**

Plans for the Coastal Zone 65

land use etc.
- **Heritage:-** e.g. landscape, historical features, places of interest, landscape and marine archaeology
- **Marine Process data:-** e.g. tidal flows and current movements, sediment transportation, navigational engineering information etc.

More traditional topographic mapping, together with complementary digital map data is available from a number of local and national sources. In Great Britain this includes specially commissioned material acquired by Local Government and other public agencies, and commercially supplied information from organisations such as Bartholomew, The Automobile Association and from OS itself. Typically these products will include:

- **Strategic and national mapping:-** scales of 1:250,000 to 1:1,000,000 for overview, indexing and national site definition and evaluation
- **Regional and coastal cell/sub-cell cover:-** scales of 1:25,000 to 1:250,000 for process studies, land cover analysis, Cell and Sub-cell high level management
- **Local, estuarine, and coastal management unit mapping:-** scales of 1:1250 to 1:25,000 for detailed scheme strategy design, implementation and management monitoring

However, as the widespread Coastal Zone user-community has such a wide range of specialist needs, satisfying them requires access to a wide portfolio of geo-topographic products. In a "best practice" scenario, the ability to focus and customise this portfolio to meet special interest needs is critical. However, there is also considerable merit in basing information systems upon topographic material that also has parallel, often complementary applications in other land-management processes. This is particularly beneficial where these are operated by the same agencies and/or they cover related or "close proximity" areas of interest. Here there is significant benefit to be gained from common technologies, common standards and a common use and understanding of outputs.

This is where major mapping organisations such as OS can offer a significant contribution. In UK we have the added advantage that the national mapping agencies, (Ordnance Survey [of Great Britain] and its counterpart Ordnance Survey Northern Ireland), offer a full range of integrated geo-topographic products and services from one agency source. These products are wide ranging. They include detailed "cadastral" large scales plans based upon surveys at nominal scales of 1:1250 (urban) and 1:2500 (rural), medium scales landscape mapping (typically at scales of 1:10,000 and 1:25,000 to small scale leisure, and general purpose topographic maps in the scale range 1:50,000 to 1,625,000. There are significant benefits from this almost unique situation of a single source for all mapping needs. The public service status of both agencies' (OS GB is a Government Executive Agency) also offers the benefits of full national coverage of all products and services. OS offers consistency of specification and of underlying currency, ready availability, and a strong commercial culture that ensures a customer-focused and cost-effective response to special local requirements or customisations.

The GIS Dimension

In the last quarter of this century, Ordnance Survey has been in the vanguard of digital conversion. The Agency can now offer a full and comprehensive range of nationally available products in computer readable form. These are ideally suited, inter alia, to the "best practice" needs of Coastal and Shoreline applications, particularly within a GIS environment. Taken together, the surveys and digital topographic data that underpin these products and services are widely recognised as forming the National Topographic Database of Great Britain (NTD). Today NTD provides the most comprehensive and nationally consistent record available of the landscape of Great Britain. Data extends down to Mean Low Water for England and Wales, and to Mean Low Water (Springs) in Scotland, and includes almost all of the "detached" terrain that remains exposed at Mean High Water (Springs).

While each of the many available representations of NTD will be critical for one or more specific Coastal Zone applications, there is also value in having the ability to model change over time. This provides a means to deduce historical shoreline evolution, and can aid forward modelling scenarios. In Britain the historical dimension is well catered for through increasing use of historical mapping such as that available from OS' extensive archives.

OS has been mapping the landscape of Great Britain continuously since AD 1791. Since the 1850s a range of large scales map series and editions (i.e. at 1:10,560 scale and greater) have been available across the whole of Britain. They depict the then contemporary topography in considerable detail, right down to low water mark, and over time have been supplemented by a full portfolio of smaller scale general topographic mapping offering a regional and national perspective. These historical mapping epochs are rapidly becoming available from OS in raster form. Together with other historical mapping at 1" to 1 mile scale and smaller (particularly at 4" to 1 mile and 10 miles to 1"), they offer a particularly valuable audit of historical changes to the shoreline and its hinterland. This may in turn give useful indications of possible continuing trends of shoreline evolution, especially for areas where a "natural" management regime is adopted.

Mapping the Changes

In the last fifty years OS' complete portfolio of maps and plans has been overhauled and regularly revised. It has also been supplemented by wholly new post-war surveys of urban areas at 1:1250 scale, and of mountain and moorland areas at 1:10,000 scale. In addition, since 1945 new maps have been produced and maintained at 1:25,000 scale. More recently, during the 1970s, the famous 1" to 1 mile, general purpose topographic mapping of Britain (OS' "first published" series tracing its origins back to the early 1800s), was superseded by a new mapping series at 1:50,000 scale.

At the heart of this post-war activity, and at the core of OS' current and future business, is the continuous and consistent maintenance of the currency and content of NTD and consequently the many mapping and latterly digital data products derived from it. This work has been driven primarily by the philosophy so eloquently outlined by the Davidson Committee back in 1938. Continuing the quotation mentioned above .

Plans for the Coastal Zone   67

..*"The governing factor is not therefore, the age of the map, but the extent to which the facts shown on it have changed since the date of its publication."* Today OS maps nationally are as up-to-date as they have ever been. This reflects not only more efficient and effective operational regimes, but also a recognition by the OS that today's world is one of "instant recall" and that the GIS environment is often data rich. Consequently there is key process necessary to extract the most meaningful information from the available data. This process, and the quality of the decisions that will ensue, demands that the topographic "template" provided by OS as vehicle for correlating other user data is, as far as possible, wholly fit for purpose (within the limits of the operational constraints applied to OS).

OS currently operates a system of continuous revision of the core NTD large scales database. This ensures that virtually all detected major and significant landscape change is captured within six months of construction. Around 500 field staff undertake this work. They are based in approximately 100 OS field survey offices around Great Britain. In addition to the gathering of "intelligence" about change from many disparate sources, and the survey and recording of that change into the NTD, these field staff also operate a programme of cyclic revision. This is designed to ensure that less significant local change, particularly in more open areas, is mapped on a regular cycle; five years for rural mapping and ten years in mountain and moorland areas. Coastal zones will be visited at a frequency dictated by nature of the change, or by default, according to the cycle that applies to their hinterland landscape.

Supplementing these programmes of topographic revision, OS is making significant investments in the development of new technologies and new applications. These investments include growing use of pen computers for real-time updating of changes to the topography, on site by the surveyor; increased and more sophisticated use of GPS and other positioning and ranging devices for location and data capture. Other work involves adding new information and more sophisticated data structures within NTD, to ensure that the information held by OS becomes more accessible, more flexible and more appropriate to the changing needs of the GIS community. Such developments are also taking full account of the specialist needs of users such as those operating in the coastal zone. As with any other well focused, customer-oriented supplier, OS is always ready to respond to changing needs for information, and to map the new phenomena that appear in the topography from time to time, particularly those that result from other technologies impacting on the landscape.

These ongoing developments manifest themselves from time to time in new or revised products and services. Meridian™ for example is an OS digital data product launched in the autumn of 1995. It represents an innovative use of digital mapping technology to integrate a variety of topographic data themes from existing OS databases. This new, flexible "mid-scales" dataset is particularly suitable for a wide range of GIS-based coastal applications at the sub-cell level of shoreline management.

Overall, OS intends to continue to evolve NTD with the aim of delivering even more relevant information about the real world. To ensure that our users (and our stakeholders - the taxpayers and customers who together fund OS) benefit to the fullest extent from these investments, OS is increasingly moving towards a "Solutions" culture. This promotes the concept of standard portfolio products being supplemented or complemented as necessary by a growing range of "bespoke" services. These may range from database management, through GIS and geospatial data Consultancy to specialist data acquisition and project management. In this way we hope to ensure that future generations of customers, including those in the coastal zone can acquire not just raw data, but more importantly, refined information to support their core business processes and decision-making/taking activities.

Matching Mapping (Geospatial Information) to Applications

One of the problems facing new or developing users of GIS and related "tools" is how to gain an appreciation of what is most relevant and suitable for a specific application, from the plethora of products and services available. In the coastal zone, where planning and management processes relate to a wide range of activities and can potentially vary from a short length of leisure beach, or a marina and its environs, to many tens of miles of dynamic coastline, the problem becomes more significant.

Adding to this situation is the rapidly changing environment in which coastal zone users are operating. For example many Coastal Zone organisations are facing new objectives, particularly the need to respond to national and international initiatives such as devising strategies for Sustainable Development and action plans on Bio-diversity. These changes are leading to new functions such as Shoreline Management Planning and Coastal Zone Planning. There are also new users who have acquired an interest in the coastal zone directly because of such changes in the circumstances of the zone. They include, for example the Insurance Industry, Environmental agencies, some Utilities and those involved in Contingency and Emergency Response issues.

The needs of this new and evolving user base in the Coastal Zone are also leading to the requirement for, and adoption of, new products and systems. New data requirements include, for example, Digital Elevation Models of increasingly higher resolution, Imagery of various kinds and data offering a strategic dimension. Demands for new delivery systems are also emerging. In addition to traditional paper-based products and latterly magnetic tape and disc media, there is now a growing requirement for On-line access to regional and national databases, and for access to data, and to refined information via the World Wide Web. The transience of processes for some users is also leading to a growing need for new services such as customised maps, often with ephemeral content. Many of these demands are emerging as a direct result of the growing use of CAD and GIS in the routine business operation of organisations that are involved in the Coastal Zone.

Looking at a selection of Coastal and Shoreline applications may help to crystallise some of the principles that need to be encapsulated in a "best practice" approach.

Shoreline Management Planning

In Great Britain Shoreline Management Planning is primarily the responsibility of the coastal Local Authority that is responsible for sea defence and shoreline protection. Lengths of shoreline (Management Units) are defined, where a consistent defence and protection management policy is envisaged. The role of the coastal local authority in co-ordinating the preparation of the SMP ensures that it is complementary to adjacent onshore strategic development and control plans such as the "County" Structure Plan and "District" Local Plan.

OS maps (particularly at scales of 1:50,000, 1:250,000 and 1:625,000) together with their digital counterparts - 1:50,000 Raster Data, Strategi™ and BaseData.GB™, are important in underpinning the definition of the Management Units. They also provide a link to the mapping products used in the local authority landward planning systems. These products depict the human and natural environments, and (through successive epochs and editions) coastal change, in sufficient detail to identify constraints to, and limitations on, shore defence options.

The next management phase is the development of Scheme Strategy and Beach Management Plans for implementing coastal defence and shoreline use options. These are well supported by the comprehensive detail of OS mapping products at 1:10,000 and larger scales. Here the Land-Line.Plus® and Superplan™ vector data products (nominally at 1:1250 or 1:2500 scales) and the 1:10,000 scale raster data together with Superplan™ hard copy graphics, offer the ideal integrated solution as backdrop products for design and monitoring of these plans.

Coastal Communications

In the coastal zone the importance of communicating with other bodies and with the public on proposals and management schemes should not be understated. Whether providing information for statutory or voluntary consultation on proposals for managing fragile coastal environments, or in promoting the benefits of specific coastal uses such as leisure and tourism, the benefits of using well presented map-based information are significant. In Great Britain the general public are most familiar with OS maps at 1:10,000 and 1:50,000 scales. OS "Solution Centre" is well able to support specialist customising of local information, particularly for these types of applications using our famous 1:50,000 Landranger™ map series. The raster data variants of these products offer the potential for complementary local design, integration and presentation of high quality map-based material.

Contingency Planning

As we have seen all too frequently around the shores of Great Britain, being at a major crossroads of the shipping arteries of the world brings the ever-present risk of accident and potential disaster. Responses to incidents such as oil spills and in-shore maritime emergencies call not only for effective and targeted contingency plans, but the mechanisms to communicate unambiguously scoping, disaster management and recovery strategies. This is where comprehensive nationally recognised and consistent topographic information really comes into its own.

The familiarity of OS maps to those involved in coastal zone contingencies, derived from their use of these products in their day to day activities, makes these products the obvious choice for presenting and communicating response plans and actions. The comprehensive nature of the OS portfolio enables the full range of tracking, monitoring, response and reporting processes to be undertaken in a fully integrated way, both via traditional hard-copy media and using the best of modern information technology. Again the nature and spread of impact of the emergency will dictate the detailed choice of products, but for the majority of contingency planning purposes the 1:50,000 colour raster, 1:10,000 raster and large scales Land-Line.Plus® product suite will form the most appropriate core to a "best practice" system.

Flood Risk and Hazard Analysis

As concern over global warming and changing sea levels increases, so the need to consider the consequences for real estate insurance has been brought into sharper focus. This is an issue that impacts on both landowners seeking to insure, and those providing the cover. Assessment of the risk for insurers and therefore the likely premiums and caveats imposed upon those seeking cover require increasingly sophisticated spatial analyses to provide predictions of (future changes to) the likely frequency and magnitude of any coastal flooding threat. These analyses rely on the combination of socio-economic and demographic data about the residential and commercial environment, with the geospatial and topographic location and aspect of individual or groups of properties.

OS has a number of relevant products here, including ADDRESS-POINT™ and DATA-POINT™. These are national postal address and postcode unit point reference gazetteer products respectively. These products offer property geo-location to 0.1m and to circa 100m respectively. They can be effectively integrated with Land-Form PROFILE™, (the data product derived from the OS' National Height Database) to provide a three-dimensional geo-reference for individual or grouped properties. This can be used as a geospatial template against which socio-economic data, and hence risk, can be modelled. OS is also working with a number of specialist Value Added Resellers to develop these product integrations so that refined risk models rather than raw data can be delivered to users.

For those who wish to represent risk or hazard zones geographically, BaseData.GB (1:625,000) and Strategi provide excellent national and regional base templates, while 1:50,000 Colour Raster (Landranger) and 1:10,000 Raster provide a regional and local perspective. If detailed flood management planning is required, OS' Solution Centre can supplement the general purpose Digital Terrain Models supplied under the Land-Form PROFILE product brand, by the addition of extra height data. This enables much more sensitive high-resolution terrain models to be generated. These, linked to Land-Line.Plus data, offer a very powerful tool for scenario modelling and visualisation.

Why GIS for Best Practice?

As discussed above, the Coastal Zone Community is multi-user and multi-agency. There is a requirement to manage ongoing rather than one-off processes. These functions demand storage and presentation of information at a variety of scales. Such activities will be best served if the management systems underpinning them offer dynamic data management, the ability to facilitate customised presentation and display by the user, and an ability to combine, aggregate, integrate and analyse disparate data in a variety of different "value added" ways. These facilities will enable decisions to be taken in the light of the best possible interpretation of the available information.

In a "best practice" environment shoreline management activity should be focused on Shoreline **Planning** rather than Shoreline **Plans**. Plans are static, authoritarian, imposing documents that imply discrete ownership of the content and message. They tend to lend themselves to becoming "shelf-ware". By contrast *planning* is a process! It implies a hands-on, dynamic and responsive activity with shared responsibility.

The modern "best practice" tool that gives life to this change of emphasis is the (computer-based) Geographic Information System. In a GIS, data can be integrated, analysed and used for effective modelling. GIS, by virtue of its flexibility and dynamic functionality enables information and data sources to be kept up to date very efficiently. New information such as datasets on environmental, socio-economic or topographic aspects can be incorporated and integrated effectively. New users and new applications can be integrated into existing systems with relative ease, allowing expansion and inclusiveness rather than status quo and exclusivity. These must be the characteristics of a successful and effective operational environment for coastal zone agencies.

The National Geospatial Data Framework and the Coastal Zone

This concept was initially proposed by Ordnance Survey in the Autumn of 1995 under the title of the National Geospatial Database. At that time there was a "mind's eye visualisation" of a "virtual database" of (principally) publicly gathered, owned or administered geospatial datasets. It was envisaged that if these data could be made more easily available and accessible through a range of protocols and data exchange standards and mechanisms, this would encourage wider and more effective use of the data. The benefits would be derived from offering a better return on the original investment in data acquisition, through more extensive use. NGD was also seen as offering the potential for improved correlation, integration and analysis of (often disparate) data. This would promote better informed decision processes and possibly "value added" applications, and income for the data "contributor! Obviously the Internet and World Wide Web were seen as important mechanisms for the identification, procurement and supply of these data for potential "value adding" and end users.

In the ensuing eighteen months, and following a number of fruitful and constructive meetings, seminars and dialogues among the UK geospatial community, the concept of a "virtual NGD" has been refocused. The short-medium term objectives have evolved toward the establishment of a National Geospatial Data Framework (NGDF). This will embrace, inter alia, the standards, data indices, metadata catalogues and other relevant protocols and infrastructures necessary to facilitate and encourage data exchange and use. This focus is intended to encourage those bodies who have data to offer it to a wider user community (if appropriate on a commercial basis). Those who are "Value Adding" organisations wishing to combine and integrate data for onward distribution, and those organisations who are end users seeking to exploit such data combinations for practical applications, will have a clear and unambiguous technical environment in which to operate and "trade" their data and their skills.

The NGDF concept offers great potential for the coastal zone community where there are many, frequently consortium, players who need to contribute and exchange both data and information about coastal zone phenomena in the design and execution of their business. In this regard, there is much to be learnt from the pioneering work already done by David R. Green and Stephen D.King at the University of Aberdeen on using the Internet, World Wide Web and Browsers to create a prototype "network GIS" for coastal zone applications (Green, 1995). This "Coastal Zone NGDF" model provides a particularly pertinent example of the sort of lateral thinking that will be needed for "best practice" to evolve toward the next millennium in the collation and integration of disparate but related data about the coastal zone.

Coastal Map Data Availability and Accessibility

None of the foregoing discussion is of practical relevance to practitioners in the Coastal Zone arena unless the data described are available and accessible.

Fortunately in Great Britain, these issues are already well addressed. Widespread information about the design, content and technical specification of all Ordnance Survey products can easily be obtained. It is available either from OS stands at many trade and sector specific conferences and exhibitions, or else direct from the Customer Services Sales Desk at OSHQ in Southampton. Comprehensive catalogues and price lists, and access to technical and sales support are readily available on request.

For many coastal zone users, OS data is already easily accessible within their own organisations. For example, the Local Authority community already has a long standing centrally negotiated Service Level Agreement (SLA). This was established with OS to provide for the comprehensive supply of a "basket" of maintained digital and graphical products, in return for a standardised annual payment covering data supply, use and copyright royalties. These payments are collected from all Local Authorities. OS is always willing to consider proposals from LA users to extend this "basket" of products to meet the specific needs of specialist users within Local Government.

Similar SLA arrangements exist with Central Government in Scotland under the auspices of the "Scottish Executive" SLA, and (covering copyright royalty payments only) with the major Utilities. Negotiations are in hand, or under active consideration, across a wide range of other customer sectors. In return for stable revenue streams, OS is able to offer favourable "consortium terms" to large users and user communities. It follows that if there are specific consortium interests or customer groupings who can come together for mutual benefit, such as those collectively managing the coastal environment, OS will be willing to consider how best to respond favourably to their common needs.

Summary

Coastal Zone management processes will benefit from the evolution and adoption of "best practice" approaches and methodologies. In these GIS has a key role to play, and reliable, consistent, high quality topographic information is a key decision-informing component. Digital Mapping and GIS products will offer the flexibility and functionality to enable Coastal Management Planning to become dynamic, responsive and inclusive.

In Great Britain OS, has a wide and appropriate range of GIS topographic products that are particularly pertinent and relevant to coastal zone applications and readily available and accessible. Current and future investments and developments in data, data technology and in infrastructures such as NGDF will ensure that there is considerable potential for future involvement of GIS in the Coastal Zone.

References

Davidson, Visc. J. (Chairman) (1938). *Final Report of the Departmental Committee on the Ordnance Survey.* HMSO London. 44 pp.
Green, D.R. (1995). Internet, the WWW and Browsers: the basis for a network-based Geographic Information System (GIS) for coastal zone management. *AGI '95 Conference Proceedings.* Association for Geographic Information, London.
Nanson, B., Smith, N., and Davey. A. (1995). What is the British National Geospatial Database? *AGI '95 Conference Proceedings.* Association for Geographic Information, London.
Clark, M (1996). Terrain Data Requirements for Coastal Zone Management Applications. *[National] Height Resolution Height Data Seminar Proceedings.* Ordnance Survey, Southampton.

CHAPTER 5

Hydrographic Data and Geographical Information Systems

P. Wright

ABSTRACT: The take-up of Geographical Information Systems (GIS) for coastal zone and marine applications has been comparatively slow in the United Kingdom (UK). Several factors have contributed to this, one of which has been the lack of suitable data for use in these systems. Recognising this, and spurred on by a rising tide of concern for the coastal zone and the recommendations of past Government Select Committees (House of Lords Select Committee on Science and Technology, Remote Sensing and Digital Mapping, 1983 and Ordnance Survey Review Committee, 1979), the UK Hydrographic Office (HO) and the Ordnance Survey of Great Britain (OS), joined forces in 1992 to produce a prototype coastal zone map for the UK. This chapter describes the outcome of this project. In particular it identifies some of the many applications for GIS in the coastal zone and offshore, some of the technical issues which need to be considered when creating data for these systems, and a personal view on the way ahead for coastal zone mapping in the UK and the prospects for supplying hydrographic data for use in GIS.

Introduction

The take-up of Geographical Information Systems (GIS) for coastal zone and marine applications has been comparatively slow in the United Kingdom (UK). Several factors have contributed to this, one of which has been the lack of suitable data for use in these systems. Recognising this, and spurred on by a rising tide of concern for the coastal zone and the recommendations of past Government Select Committees (House of Lords Select Committee on Science and Technology, Remote Sensing and Digital Mapping, 1983 and Ordnance Survey Review Committee, 1979), the UK Hydrographic Office (HO) and the Ordnance Survey of Great Britain (OS), joined forces in 1992 to

D.R. Green and S.D. King (eds.), Coastal and Marine Geo-Information Systems, 75–84.

produce a prototype coastal zone map for the UK. This chapter describes the outcome of this project. In particular it identifies some of the many applications for GIS in the coastal zone and offshore, some of the technical issues which need to be considered when creating data for these systems, and a personal view on the way ahead for coastal zone mapping in the UK and the prospects for supplying hydrographic data for use in GIS.

Stage 1 of the HO/OS Coastal Zone Mapping Project

During the first stage of the project a desk-top mapping system was used to produce a prototype paper map of Langstone and Chichester Harbours on the south coast of England. The map was widely distributed to potential customers and was well received.
Two distinct markets were identified for the product, one with leisure users and the other with planners. The prototype paper map was particularly popular with yachtsmen, boat owners and watersports enthusiasts. However, market research indicated that demand for the product was unlikely to be sufficient to recover the full costs of production. This phase of the project has been reported on in detail by Harper (1993a and 1993b) and Curtis et al. (1993).

Stage 2 of the HO/OS Coastal Zone Mapping Project

Although many planners with responsibilities for the coast said they would welcome a paper map that spanned the coastline and combined the content of an OS map with that of an Admiralty chart, it was clear from their replies that such a map would not fulfil all of their requirements. Planners need to be able to overlay their own specific information on some form of base map and to be able to regularly update this information. A highly detailed and full colour paper map is not an ideal document on which to do this. Planners require a simple and unobtrusive base map upon which other information can be easily displayed, amended and erased. The intention of the second stage of the coastal zone mapping project was to create just such a map in digital form for use in GIS and from this to derive one or more paper map products.

Applications

Applications for GIS in the coastal zone and offshore extend well beyond planning. Listed in Table 1 below are the main areas of activity which can benefit from using GIS.

TABLE 1
Marine and coastal zone activities that can benefit from the use of GIS

Transport

- Navigation
- Vessel Traffic Services (VTS, VTMS and VTMIS)
- Search and Rescue
- Vessel Tracking

Environment

- Planning
- Estate Management
- Conservation
- Coastal Protection
- Flood Prevention
- Insurance Services
- Waste Disposal
- Pollution Control

Trade and Industry

- Fishing
- Aggregate Extraction
- Oil and Gas Extraction
- Construction

Education and Science

- Teaching
- Research

Defence

- Command and Control
- Amphibious Operations
- Mine Countermeasures
- Anti-Submarine Warfare

TRANSPORT

Hydrographers are generally most familiar with the transport related applications. All four of the applications listed in Table 1 have broadly similar requirements for electronic chart data. Systems that hold this data in a database and allow users to interrogate the data are GIS, although they are more commonly known as ECS (Electronic Chart Systems) or ECDIS (Electronic Chart Display and Information Systems). These systems are specifically designed to be used for navigation, but they can also form the basis of systems required for other transport related applications.
With the aid of radar and radio communications, operators of vessel traffic services and port and harbour authorities can use these systems to monitor the behaviour of vessels in traffic separation schemes and manage vessel movements in port approaches. Coastguards can use electronic chart display systems to plan and execute search and rescue operations. Shipping companies may use shore-based GIS to track and manage vessels, if they are fitted with transponders or use some other means of regularly reporting their positions to head office.

ENVIRONMENT

Most uses of GIS in the coastal zone are for environmental purposes. In addition to the specific activities listed in the second section of Table 1, several of the applications listed elsewhere in the table also have implications for the environment or involve environmental studies. Within this sector, planners represent the single biggest group of users of GIS. Planners at local, regional and national levels can use GIS to develop, assess, record and maintain plans relating to the use not only of the land but also the sea in the coastal zone. For example, in inshore waters GIS can be used to develop management plans which attempt to resolve the conflicts which can exist between the many users of these waters.

GIS also has important applications in resource management and environmental monitoring. GIS can be used to assist with the management of large estates, both on land and at sea. For example, they could benefit organisations like the National Trust, who own and manage large tracts of the coastline of England, Wales and Northern Ireland; and the Ministry of Defence, which has a large number of ranges and bases on land and extensive practice and exercise areas at sea. GIS can be used by conservationists, water quality inspectors and others to plan and record the results of surveys, and to monitor changes in the environment.

Engineers, planners and insurance companies can use GIS to identify parts of the coast that are liable to flood or are at risk from coastal erosion. Engineers can take this work a stage further and use these systems to design coastal protection schemes, based on studies of tides, waves and sediment transport. GIS can also be used extensively to help prevent and combat pollution. By modelling patterns of water circulation and sediment movement, GIS can be used to predict the effects of dumping waste at sea. Using similar techniques, coastguards and others can improve the efficiency and effectiveness of clean-up operations, by using GIS to predict the patterns and rates of dispersal of pollutants that are accidentally spilled at sea.

TRADE AND INDUSTRY

For each of the four activities listed in the trade and industry section of Table 1 there may be several different but inter-related requirements for GIS. For example, marine biologists may require GIS to study changes in fish stocks. Fishermen may use GIS, loaded with electronic charts and extra information about fishing grounds, for navigation and to improve the quality and efficiency of their catches. Fisheries protection officers may use similar systems aboard vessels at sea to enforce fishing regulations. In future fishing vessels may be fitted with transponders to allow shore-based fisheries inspectors to monitor and record the time these vessels spend at sea. Marine sedimentologists may use GIS to model and monitor the effects dredging the seabed for aggregates may have on adjacent coastlines. Regulators may use GIS to record details of licences issued for dredging and to monitor whether companies adhere to the conditions of these licences. Dredging companies may use GIS aboard their vessels and ashore to control and monitor dredging operations. In the oil and gas industry GIS may be used by oil companies to determine the extent of oil and gas fields and to control drilling operations, whilst government officials may use GIS to record details of licences issued for oil and gas exploration and extraction. In the construction industry GIS can be used to assist engineers to find optimum locations for new outfalls, jetties, submarine cables and pipelines, shipping channels, ports, marinas, power stations, oil refineries, bridges, tunnels and barrages, etc. These systems can also be used by engineers, planners and environmental consultants to determine the effects these structures will have on the environment.

EDUCATION AND SCIENCE

GIS are widely used in universities and colleges. Environmental sciences departments use GIS to teach students about the environment and as tools for environmental research. GIS are similarly used in many computer science, oceanography and civil engineering departments for teaching and research purposes. These systems are also extensively used in many research institutes.

DEFENCE

Defence applications of GIS extend beyond the coast, across the continental shelf and into the deep oceans. Standard ECDIS equipment can be used as the basis of systems for navigation, as well as command and control. Other applications require specialist geospatial databases. Systems required to support amphibious operations can be based on ECDIS technology, but they also require a specialist database that combines standard large scale map and chart data with additional information on ground conditions throughout the coastal zone. Minehunters require shipboard systems that access detailed information on the composition and texture of the seabed, and especially details of mine-like objects in the vicinity of shipping lanes. Large databases of oceanographic information are needed for anti-submarine warfare to predict acoustic conditions in the deep oceans and surrounding shelf seas.

Each application listed in the preceding sections will have its own unique requirements for data, but most will require some form of base data, either in the form of a digital map or chart, to which other data can be referenced, or a digital elevation model of the seabed.

TECHNICAL CONSIDERATIONS

The aim of the second stage of the HO/OS Coastal Zone Mapping Project was to create a simple base map of the type referred to in the last paragraph of the previous section. An area of Swansea Bay in South Wales was chosen for the experiment. The map straddled the coastline and was created from existing map and chart data. Many of the technical issues that had to be considered during this stage of the project apply to the creation of any digital map or data set for use in GIS. These issues are considered further in this section.

DATA AVAILABILITY

Very little hydrographic data is currently available for use in GIS, and still less is marketed specifically for this purpose. This is not because there is no demand for the data, but because at present few hydrographic offices have data readily available in a form that is suitable for use in these systems. At present hydrographic offices are too busy grappling with the difficulties of developing digital versions of their core navigational products to be able to devote much time to creating datasets for use in GIS. Priority is having to be given to producing digital data for use in electronic chart display systems. However, in much the same way that raster electronic charts can be produced as by-products of paper chart production, in future it should also be possible to produce intelligent vector data for GIS as a by-product of vector electronic chart production. In this way it should be possible to produce the data more cheaply than would be the case if it was necessary to digitise the data specifically for this purpose. Swansea Bay was chosen for the trial with the OS because at the time it was one of a few areas for which the HO had digital data.

DATA QUALITY

Data quality is an important issue for most users of GIS. Reliability, accuracy, currency, completeness and consistency are all aspects of quality that are important in this context. GIS users require data that they can trust to be correct. All data must therefore be authenticated before it is supplied for use in GIS. Whether data is reliable or not will depend to a large extent on its source.

Hydrographic Data and GIS

Hydrographic offices are used to receiving data from many different sources and have well established procedures for assessing its reliability. All data used in the trial with the OS was assessed as part of the standard screening process used to ensure that only valid data is included in the HO's navigational products.

Most users will wish to know something about the accuracy of the data they buy. Many users will want the data to meet certain minimum accuracy standards. For some applications meaningful results can only be achieved with very accurate data. For example, if hydrographic surveys are used to monitor the volumes of spoil removed from a navigational channel, or the quantity of aggregates dredged from an offshore bank, depths have to be measured to within +/- 0.1 metre and fixed horizontally to within +/- 1 to 3 metres. Most applications, however, have less stringent accuracy requirements. Less accurate data may be used for most modelling and mapping applications.

Most users require data that is up-to-date. If the data is not up-to-date, users generally require an assurance that it is still extant or, failing this, an indication of how much change may have taken place since the data was collected. Some applications require historic as well as current data, for example, to monitor seasonal or annual changes in sea water quality, or long term changes to the position of a coastline or offshore features such as sandwaves. Users also require data to be complete. As far as possible there should be no gaps in the area covered by the data. Finally, the data should be consistent. Ideally all of the data should be equally reliable and of the same scale, accuracy and age.

Data Compatibility

Data has to be compatible with the GIS in which it is to be used. This is not a major issue for serious producers of data for GIS. Having created the basic data it is a relatively easy task to develop or acquire the software needed to output the data in a range of different formats, such as NTF (National Transfer Format - British Standard 7567), DXF (Data eXchange Format) and a variety of other proprietary formats. Creating the basic data can be far more problematic, especially if the dataset has to be created by combining data from a number of different sources. This was the case during the second stage of the HO/OS coastal zone mapping project. During this stage of the project, hydrographic data from two different sources was combined with topographic data from another six sources.

The aim of the project was to create a digital coastal zone map made of intelligent vector data. None of the data used in the trial was already in this form. Intelligent vector data had to be created either by reprocessing existing digital data or by digitising the data directly from repromat. With the exception of some additional drying heights, all of the hydrographic data had been digitised as part of the normal production process for new paper charts. Although feature codes were included in the data, these referred to the linestyles that were to be used for reproducing the chart. In this form the data could only be used within a GIS to create a digital image of the original paper chart. To enable the data to be displayed selectively, queried and used for analysis it had to be recoded as a series of real world objects. Linework broken for text and other features had to be made up and polygons constructed. Tables of attributes had to be assembled from the captions, legends and notes shown on the chart.

The topographic data used in the trial was derived from a more diverse range of sources. The map outline and woodlands were digitised from repromat. Contours, rivers, roads and urban areas were derived from existing digital data. The topographic data required less reprocessing than the hydrographic data, because in effect it was already separated into layers. But it was deficient in various other respects. The rivers, woodlands and outline datasets contained no feature codes or attributes. Linework in the rivers dataset was discontinuous, and contours were broken where form lines and other features were present on the paper map from which the data was derived. The geometry of some of the road junctions and roundabouts was too complex to be displayed at a scale of 1:25,000, the scale of the prototype digital map.

Having created or reprocessed the data in each of the eight individual sets of data, these then had to be merged to form a single multi-layered dataset. This involved clipping and erasing surplus data from each of the datasets. Data landward of the Mean High Water Springs mark was taken from the topographic datasets of the OS and data seaward of the Mean High Water Springs mark from the hydrographic datasets of the HO. All positions were transformed to a common reference system, in this case the UK National Grid. The datasets were registered one with another and any apparent mismatches in the data were investigated. Although it was recognised that there would be benefits in referencing all heights and depths to a single datum, this was not done during the trial. In accordance with the standard practices of the OS and HO, all heights above Mean High Water Springs were referenced to Mean Sea Level (Ordnance Datum Newlyn) and all depths and drying heights below Mean High Water Springs to Chart Datum, the level of the Lowest Astronomical Tide. One of the biggest difficulties arose because the datasets were not based on a common data model.

Some of the feature codes were not unique and the datasets could only be merged if the tables of attributes were identical.

Although raster data was not used in the trial, some consideration was given to the processes that would be involved in combining this type of data for use in GIS. These processes included clipping and erasing any surplus data, transforming the data to a common projection, spheroid, datum and scale, edge matching adjoining datasets and converting all the data to a common format. Apart from the limited use that can be made of raster data in GIS, one of the main drawbacks of this type of data is that if it is necessary to enlarge or reduce the scale of adjoining datasets in order to combine these into a single map or chart, all text, numerals and symbols are enlarged or reduced in the same proportions. A user panning about the resulting image of the map or chart may easily gain the false impression that the scale of the data varies from one part of the map to another.

Data Suitability

Data may be readily available, of excellent quality and compatible with the leading makes of GIS, but it may still not be "fit for purpose". Data not only has to be compatible with the target GIS, it also has to be compatible with the application with which it is to be used.

Two main aspects are important in this context, content and structure.

For data to be fit for purpose it must contain all the elements needed for the application. For example, a dataset might contain detailed information on the composition of sediments on the surface of the seabed, but it could not be used to calculate the volume of sediment available for extraction as aggregates if the dataset contained no information on the depths of the sediments. Similarly, most surveys collected by national hydrographic offices are at scales that are too small for many engineering applications. Large scale surveys are needed, for example, to plan coastal defence schemes and to determine the most suitable location for submarine pipelines, cables and outfalls.

Different applications also require data to be structured in different ways. Most applications require geospatial data for one of three purposes: to create a map or chart as a back-drop to other information, to monitor change or to model processes. Maps and charts can be constructed from dumb raster or unstructured vector data if users do not need to be able to de-select data or interact with any of the elements shown on the map. Intelligent vector data will be required if users do need to do either of these things. Similarly, for monitoring change, both raster and vector data can be used. Changes to spatially continuous data, such as sea surface or ocean temperatures, are most easily compared using raster or tessellated data. This applies especially if remotely-sensed data is used. Vector data is more suitable for monitoring changes to discrete features, such as the positions of coastlines, sandwaves, longshore bars, offshore banks, etc. In addition to raster and vector data, gridded data is often needed to model processes. For example, most computer models of tides, waves and sediment transport require gridded bathymetric data as one of their main inputs. These data sets can contain actual or interpolated depths and can be based on regular or irregular grids that are rectangular, triangular or curvilinear in shape.

Way Ahead

Most users of GIS either cannot afford or are unwilling to pay a high price for data. The UK Government has made it clear that any initiative aimed at providing coastal zone mapping for use in GIS must be self-financing. This also applies to the supply of any other data for which there is no endorsed defence or other national requirement. If the OS and the HO are to satisfy the requirements of the UK Government and most users, data must be cheap to buy and inexpensive to produce. At present it is not possible to satisfy these requirements. The Coastal Zone Mapping Project showed that although it is feasible to combine topographic and hydrographic data from the OS and HO, this cannot be done easily or cheaply. Less complex datasets, such as digital elevation models of the seabed, can be produced more easily but not necessarily at less cost, because in most cases not all of the data is readily available in digital form.

Four developments should, however, provide the HO with the opportunity to supply data for use in GIS more cost effectively in the relatively near future. Firstly, the HO is in the process of digitising its charts in the IHO S57 data format for use in ECDIS. In this format the data will be held as real world objects not as cartographic features. This will overcome many of the difficulties referred to in the earlier section of this paper on data compatibility. Secondly, a number of emerging defence requirements should result in the creation of a series of rich datasets covering the coastal and offshore zones. It may be possible to make some of this data available to GIS users. Both of these developments should enable data for GIS to be produced more cheaply than it can be produced at present. Two other developments could increase the demand for data, which would also assist the HO to sell data at a lower price than it could be sold now. Firstly, the Department of the Environment has established the Coastal Forum (Department of the Environment, 1995) to encourage greater co-operation and best practice in coastal management. Secondly, the OS has recently proposed the concept of a National Geospatial Database (Nanson et al., 1995). Both initiatives should raise people's awareness of the potential value of GIS and the potential for the HO to provide data for use in these systems. In addition, the former provides the HO with the opportunity to identify customers and customer requirements and the latter the opportunity for the HO to add value to any service it provides to the GIS community by linking its datasets with those produced by other organisations.

References

Allan, P. (1994). "Coastal Zone Mapping - A Digital Data Trial", *Ordnance Survey Internal Report,* E481, 19pp.
Curtis, M., Harper, B., and Fellingham, W. (1993). "Is there a Mart for these Chaps?", *GIS Europe,* Vol. 2, No. 3, pp. 26-28.
Department of the Environment (DOE) (1995). "The Coastal Forum", *Wavelength,* Issue 1, 4pp, DOE, London.
Harper, B. (1993a). "The Report on the HO role in the Joint HO/OS Coastal Zone Mapping Project 1992 - 93", *Hydrographic Office Internal Report,* pp 30.
Harper, B. (1993b). "A Coastal Zone Mapping Project in Great Britain", *Proceedings* of the Canadian Hydrographic Service 1993 Surveying and Mapping Conference, pp. 220-223.
House of Lords Select Committee on Science and Technology, Remote Sensing and Digital Mapping (1983). "Report of the House of Lords Select Committee on Science and Technology, Remote Sensing and Digital Mapping", Vol. 1, Her Majesty's Stationery Office (HMSO), London.
Nanson, B., Smith, N. and Davey, A. (1995). "What is the British National Geospatial Database?", *Proceedings* of AGI'95, pp. 1.4.1-1.4.5.
Ordnance Survey Review Committee (1979). "Report of the Ordnance Survey Review Committee", Her Majesty's Stationery Office (HMSO), London.

CHAPTER 6

GIS for Sustainable Coastal Zone Management in the Pacific - A Strategy

B. Crawley and J. Aston

ABSTRACT: The Pacific islands region has more than 22 island countries in an area of approximately 30 million square kilometres. It is home to nearly 6 million people and has a complex of islands and habitats. Pacific Island countries are largely coastal in nature. These areas play a major role in the daily existence of Pacific Islanders. SPREP, in collaboration with the United Nations Environment Programme - East Asia Program for Asia Pacific (UNEP EAP-AP), have initiated a program to provide the Pacific community with improved access to meaningful environmental data and information through establishment of GIS in several of the countries. GIS is still new in the Pacific and is only rarely used as a tool for coastal zone management. This chapter outlines a strategy to assist SPREP member governments use GIS and environmental information for the planning and management of coastal areas.

Introduction

The South Pacific Regional Environment Programme (SPREP) was set up as an intergovernmental organisation in 1986 to co-ordinate and facilitate environmental sustainable development in the Pacific region. SPREPs mandate covers 30 million square kilometres. Within that area there are some 22 Pacific Island nations and territories (all members of SPREP) with a combined population of about 6 million people.

SPREPs programmes are guided by the *Action Plan for Managing the Environment of the South Pacific* (SPREP, 1996). The overall aim of the Action Plan is to build national capacity to protect and improve the environment of the region for the benefit of Pacific Island people, now and into the future.

One of the goals of the *Action Plan* is to develop Integrated Coastal Management (ICM) approaches for Pacific Island situations (SPREP, 1996). ICM is accepted world-wide as a comprehensive, multi sectoral integrated approach to the planning and management of coastal areas. ICM is also recognised as an appropriate process for dealing with coastal problems in the Pacific (Fauvao, 1995). It is particularly suited to the small Pacific Island countries because of the interconnectedness of the coast and terrestrial areas.

The islands are a scattered community, isolated by hundreds if not thousands of kilometres (SPC, 1992). The *smallness* of the islands has a profound influence on their social goals and development options, as well as magnifying the impacts of poor decisions (Fauvao, 1995). Therefore, decisions regarding management of coastal resources must be based on the best available information.

The *Action Plan* recognises the importance of improved availability of information on all aspects of environment and development for decision making towards sustainable development (SPREP, 1997). This is supported by Agenda 21, Chapter 40 that emphasises the need for improved collection as well as presentation of data and information.

International and Regional Initiatives

THE ACTION PROGRAMME FOR THE SUSTAINABLE DEVELOPMENT OF SMALL ISLAND DEVELOPING STATES

At an international level, the particular dimensions of the "coastal management problem", recognised in the *Action Programme for the Sustainable Development of Small Island Developing States*, are:

1) the paucity of "data" necessary to manage the coastal zone"
2) the lack of institutions at the national level and;
3) the need for an approach to Integrated Coastal Zone Management (ICZM) that is relevant to islands.

In recognising these constraints, the *Action Programme for the Sustainable Development of Small Island Developing States*, considered and endorsed by the Barbados Conference, calls for better information to assist decision-making, the strengthening of appropriate institutions and legislation, and the development of island-appropriate methodologies for ICZM.

Regional Priority Setting Exercises
The issues and problems facing the small island developing states in the management of their coastal areas have been prioritised over the last ten years through ad hoc needs analyses and priority setting exercises. Such priority setting exercises have identified

the need for cross linkages between sub projects including regional fora for communication and exchange of data and information. Several programmes aim to address those issues.

National Environmental Management Strategies
In 1991, National Environmental Management Strategies (NEMS) were developed in 15 of SPREPs' member countries through a process of extensive in-country consultation and gathering of relevant background information. The NEMS lay out a blueprint for environmental priorities to the end of the decade and outlines the major environmental issues faced by each country including the steps required to address them. Many NEMS have an information management strengthening component.

Coastal Protection Meetings
In 1994, a series of *Coastal Protection Meetings* assessed the key needs and actions of the region required to provide effective coastal protection (SPREP/SOPAC, 1994). During two of these meetings, eight general areas of needs were identified. Two of the eight needs related to mapping and data collection to better understand physical processes in coastal environments and Integrated Coastal Zone Management.

International Coral Reef Initiative (ICRI)
In 1995, the International Coral Reef Initiative (ICRI) Pacific Regional Strategy identified regional priorities to address pressing coastal management issues including capacity building; research and monitoring; and management. One of the capacity building elements of the *ICRI Pacific Regional Strategy Framework for Action* is to improve co-ordination and co-operation of activities and information exchange.

The Pacific Environmental Natural Resource Information Centre
The Pacific Environmental Natural Resource Information Centre (PENRIC), is part of a network for environment assessment in the Asian and Pacific Region. It was initiated through collaboration of SPREP/PENRIC and other sub-regional institutions viz. a viz. the Association of Southeast Asian Nations (ASEAN); the International Center for Integrated Mountain Development (ICIMOD); the Interim Committee for Co-ordination of Investigations of the Lower Mekong Basin (MEKONG); and the South Asia Co-operative Environment Program (SACEP).

PENRIC's overall objective is to generate, analyse and disseminate information to member countries for environmental assessment. The specific aims are to:

(1) **Promote timely and scientific environmental assessment within the region and support effective management of the region's natural resources**
(2) **Produce information needed for the region in a standard format; and provide timely and easy access to the information required by users**

and

(3) Address, on a priority basis, national and international environmental concerns with a knowledge base for project implementation and follow up (e.g. NEMS)

PENRIC's implementation requires:

- **Institutional strengthening through the provision of training, expertise and development of facilities.**
- **Integration of scattered institutions, experts and data to maximise economies of scale and avoid duplication of effort.**
- **Building data and databases which are compatible to facilitate exchange of data.**
- **Strengthening of human resource development and institutional capacity building in country.**
- **Set up of national/regional networked and decentralised environmental resource information systems.**

Within the past 3 years, the PENRIC program has devoted much to making its member countries aware of the importance of information and how tools like GIS and Remote Sensing can be used to catalyse and formulate decisions on environment and development issues. Various levels of applications have been developed in the countries of Fiji, Tonga, Cook Islands, Western Samoa, Vanuatu, Solomon Islands, PNG, Kiribati and Niue. For example, in 1994, GIS was utilised to develop a Nation-wide Vulnerability Assessment for Fiji and Western Samoa as part of a joint venture between SPREP, Environment Agency of Japan(EAJ) and the Overseas Environmental Co-operation Center of Japan (OECC).

Regional Institutional Collaboration
PENRIC collaborates with other regional institutions to maintain personal networks and ensure a level of compatibility amongst the various programmes. Collaboration between United Nations Environment Program - East Asia Program - Asia Pacific (UNEP EAP-AP) and SPREP/PENRIC led to the formation of the Regional GIS Training Centre within the University of the South Pacific in Fiji. Currently, GIS courses have been offered as pre-requisite degree courses in Geography and Land Management. This year, the course has been further upgraded to the Diploma level. The University has plans to develop an exchange scheme with the Asian Institute of Technology(AIT) for integrating GIS into other disciplines. The University of Papua New Guinea, University of Hawaii, University of Tonga, University of Hawaii, East West Center and Landcare New Zealand will be consulted to strengthen this initiative.

Other institutional initiatives include: PNGRIS - the Papua New Guinea Resource Information System developed by CSIRO; INFORMAP with geographic mapping capabilities at the native Land Trust Board (NTLB) in Fiji; a Land Use Inventory Asian Development Bank (ADB) Project in Western Samoa; an ARCINFO system for the Cook Islands; the Solomon Islands Resource Information System (SOLRIS) and; the Vanuatu Resource Information System (VANRIS).

Capacity Building
Already, there has been a lot of work done in the Pacific ranging from research and pilot studies through to detailed surveys. The work required to produce these data are expensive and time consuming. To overcome these problems, a regional directory of environmental institutions, experts and data has been compiled. The directory can help identify information gaps and contains knowledge of what has been done and what data is available in the region. The Directory can: reduce duplication of effort; identify the level of expertise and how it can best be utilised and; provide a basis for more detailed assessment of the needs of the governments in the field of modern GIS technology. Used in conjunction with the NEMS, this document will help national governments in the development of capacity building proposals for the donor community.

State of the Environment Report
The SPREP Secretariat, in collaboration with other regional bodies and donors, are continuing to develop and modify a framework to assess the 'State of the Environment' in the face of threats by man or indirectly by natural disasters. As part of this process, the UNEP Environment Assessment Program for Asia and Pacific (UNEP- EAP AP), together with SPREP and its member governments, are developing procedures for environmental assessment. This framework will: support the generation of environmental indicators and indices; help integrate multi-sectoral data; identify areas where data are inadequate and; indicate the weak links in institutional networks that need strengthening. Most importantly, the framework is designed to enable governments to meet their environment reporting obligations and formulate action strategies efficiently and effectively.

Four categories of databases have been developed under the State of the Environment Reporting System. These are:

1) **Biophysical Environment including land use, topography, land tenure, soil types, village boundaries, conservation areas and key flora and fauna**
2) **Socio-economic Environment including human activities; agriculture, forestry, transportation, energy and tourism, population parameters (e.g. size, growth and distribution) and health issues**
3) **Natural Disasters including floods, droughts, susceptibility to cyclones and earthquakes**
4) **Policies and Institutions including responses by governments and government agencies**

State of the Environment databases are now operational in 9 countries; Cook Islands, Fiji, Federated States of Micronesia, Kiribati, Niue, Solomon Islands, Tonga, Tuvalu, Vanuatu and Western Samoa. These databases will also serve as a tool in implementing and following up NEMS that were produced for the member countries. To ensure the continued participation of all sectors, the framework has been broken down in to 7 components.

1) Data processing and information flow
2) Integration and linkages of bio-physical and socio-economic data
3) Technological support
4) Development of indices and identification of issues to

- reflect the quality of the environment
- indicate the impact or stress on the environment resulting from human actions;
- evaluate the cost-benefit of environment measure; and
- indicate sustainable development trends

5) Outputs-inputs for State of Environment reporting, legislation and Action Plans
6) A decentralised network of institutions
7) National perspectives

At the national level the framework aims to:

a) ensure an integrated national systems for measurements of environment quality
b) maintain a dataset to assess the State of Environment
c) develop national baseline data to evaluate the effective integration of environment and development information

Data Requirements for Coastal Management

At a conceptual level the data needs for coastal management have been identified by Scura et al. (1992) in Table 1. At the country level, the information needs for coastal management will need to be determined by those that have vested interests in the coastal areas. Although there are needs common to all Pacific island countries, the data requirements of particular countries will vary according to the scope and extent of the problems to be addressed.

TABLE 1
The information needs of importance to coastal management

Biophysical and Environmental Aspects

- Resources inventories
- Determination of environmental linkages and processes
- Identification, monitoring and evaluation of environmental change
- Physical quantification of environmental impacts

Social and Economical Aspects

- Social, cultural and economic characterisation of coastal communities
- Estimation of demand and supply of coastal resources, and projection of future demand and supply
- Identification of current and potential future resource conflicts
- Identification of market and policy failures
- Economic valuation of coastal resources including non-market valuation
- Evaluation of alternative policy options and management strategies

Institutional and Organisational Aspects

- Rights and obligations with regard to coastal resources use
- Organisational jurisdiction, responsibilities, structure and co-ordination

Opportunities for Management Interventions

- Evaluation of opportunities for and efficacy of interventions to influence behaviour
- Evaluation of opportunities for and efficacy of direct public involvement or investment

Adapted from Scura, L.F et al. (1992) Lessons for integrated coastal zone management: the ASEAN experience. In *Integrative Framework and Methods for Coastal Area Management* eds. T.E Chua and L.F. Scura. ICLARM Conference Proceedings 37, pp. 1-70.

Constraints to Coastal Management in the Pacific

The full functionality of GIS for coastal zone management has not been realised in the South Pacific. In general, there is a non strategic and uncoordinated approach to information acquisition, management, analysis, interpretation, dissemination and application at the national and regional level. Many government decisions impinging on the coast are not grounded on established policy or the best available information. All too often management priorities are set to address problems for which there is an immediate fix but which may be unsustainable in the longer term. Strategic approaches are required to address more complex or resource demanding management concerns. Implementation of such actions to address regional priorities will be at the national level although the specific priorities will vary among countries.

Experience thus far has shown that the transfer of GIS technology for coastal management in the Pacific is dependent on a plethora of social, cultural, political and institutional factors. The most influential factors affecting the viability of coastal management in the Pacific region are related to the population, economy, culture, institutional setting, funding and support, training and information.

POPULATION

A common feature of many Pacific islands is the concentration of people in small, predominantly coastal, areas as a result of migration from rural to urban areas, outer islands to main islands, or small to large islands, compounded by often high rates of natural population increase. Large families are the norm and, as a result, population growth is commonly between two and five percent.

ECONOMY

Many Pacific islands have economies that are still largely subsistence based, but with a slowly growing cash economy. Few have economies that could continue to expand without on-going substantial aid inputs or remittances. Nevertheless, there is a trend to develop in-country capacity for self-sufficiency, stimulation and nurturing of small private sectors, and a growing emphasis on human resource development.

CULTURE

The Pacific Islands have separate identities and cultures, although there are regional similarities. Most cultural traditions are centred on the extended family. In many cases, the resources are concentrated on immediate, up-coming key events of local and cultural significance (Fauvao, 1995). Up until recently, these societies have not had to 'plan for the future' because traditionally there has been an abundance of natural resources requiring minimal strategies to ensure continuity of supply. Pacific communities are also used to managing their affairs day by day and week by week through meetings and verbal consensus (Fauvao, 1995).

INSTITUTIONAL SETTING

Some general characteristics of Pacific islands have particular implications for the development of appropriate institutional arrangements for sustainable ICZM. These include:

- the generally small size and limited ability of government to administer and manage complex new programmes
- the often ex-colonial nature of many bureaucracies
- heavy dependence on the support of aid programmes, both financially and in terms of expatriate staff and advisers
- often disproportionately large bureaucracies in relation to the size of the population and the economy; and
- disjunction between government and its bureaucracy. In some cases, governments exist to carry out government and ministerial decisions rather than to participate in the decision making process

FUNDING AND SUPPORT

The nature of GIS may preclude it as an appropriate tool for all of the Pacific island countries. One way of integrating GIS into organisations is to build it into the work program (Davis, 1996). Part of this process is to conduct a needs assessment, prior to the installation of particular hardware and software and evaluate the success of that program at a later date. The output of this exercise should clearly identify what resources are available in terms of personnel and expertise, data and information and most importantly the level of commitment from the country.

TRAINING

In the Pacific there is still a reliance on recycling the first generation of professionals to keep GIS moving (Davis, 1996). Some training has been conducted, either within the recipient country or abroad. The results have been highly variable depending on the personal characteristics of the expatriate trainers involved including their knowledge of the country and its needs.

INFORMATION

The major obstacle to the inventory, development, management and conservation of living and non living resources in the coral reef environment are:

1) the lack of base maps of the coastline
2) the water depth
3) the nature of the sea bottom in the shallow lagoon (0-6m)
4) the nature of the topography
5) the density of vegetation along the coastline

The information that has been collected is not always available in a form that can be used during the decision making process. Nevertheless, GIS has shown to be an ideal tool to conceptualise and integrate data at a range of scales. Sometimes the scale, connectivity of the marine environment and the nature of the life cycles of marine organisms is so large that one department or one country cannot manage the area (see Kenchington, 1990).

The free exchange of information can facilitate the development of GIS systems. Mechanisms to enhance the exchange of data should be placed as a priority. In these cases international and regional agencies such as Forum Fisheries Agency and the South Pacific Commission need to co-operatively deal with these problems.

Information Constraints and Sources

Potential sources of spatial data are traditional aerial photography, small-format aerial photography, airborne radar and satellite imagery. Aerial photography in particular can be an appropriate data source given the small size of the islands. Nevertheless, selection of an appropriate image source should be evaluated with other mapping techniques and sources of information.

Remote Sensing via satellite has become increasing useful for mapping. These sources of data are considerably more effective compared to traditional techniques such as plane table survey for mapping and other ground survey techniques which tend to be time consuming, costly and cumbersome.

In the Pacific, the most appropriate image sources are LANDSAT and French SPOT. Their usefulness depends on the desired resolution of the image, size of the study area and purpose of the study. The main limitation of satellite images is due to the high convectivity from the vast ocean and geographic location that separates the island countries. This has created difficulty in receiving cloud free images for delineation of features. On the other hand, users will have access to more frequent data for monitoring changes over time where there are frequent passes of the satellite over a specific area.

Conclusion

GIS IS AN ENABLING TECHNOLOGY

GIS is an enabling technology, the major functions are to empower people to achieve goals that incorporate geographic spatial data (Davis, 1996). Therefore, primary consideration should be given to the social, cultural, political and other circumstances during the planning, implementing and operating GIS rather than the data or software. It will be important to build on the existing foundation and local innovations.

The ability of the Pacific islands to adopt GIS technologies will depend, not only on the appropriateness and type of GIS technology, but on how the technology is transferred. The successfulness of the transfer of GIS technology will depend on raising awareness of the individuals or groups expected to adapt or adopt the new technology; having an understanding of the physical and cultural environment where the new technology is to be adopted; demonstrations of GIS technology applications and the benefit it provides; whether the technology is made to look attractive from the perspective of those individuals or groups; building necessary human resources to support the integration of technology into local culture; making a commitment to transfer of technology and; whether the technology is readily available under conditions acceptable to the target individuals or groups (Commonwealth of Australia, 1991).

Technological tools such as GIS can help identify indicators, linkages and indices for the managing coastal environments of the Pacific. This information will form the backbone of the decision making process in implementing the *Action Plan*. Given the reliance of the Pacific peoples on the coastal resources and considering the size and isolation of the Pacific island countries, GIS plays a major role in ensuring its sustainability.

References

Commonwealth of Australia (1991). *Final Report* on Technology Transfer Opportunities for Reducing Greenhouse Gas Emissions. Prepared by Coffey, MPW Pty Ltd for Department of the Arts, Sport, The Environment, Tourism and Territories.

Davis, B. (1996). GIS Human Resource Development in the Pacific. *Geographical Information Systems and Remote Sensing Networks*, No. 14, 1996.

Fauvao, V. (1995). *Coastal Management in Small Island Developing States*. Paper presented at the Global Conference on the Sustainable Development of Small Island Developing States, Barbados, May 1994.

Kenchington, R. (1990). *Managing Marine Environments*. Taylor and Francis, New York.

Scura, L.F. et al. (1992). Lessons for integrated coastal zone management: the ASEAN experience. In *Integrative framework and methods for coastal area management* eds. T.E Chua & L.F. Scura. ICLARM Conference Proceedings 37, pp. 1-70.

Smith, A. (1996). Pacific Regional Report on the Issues and activities associated with coral reefs and related ecosystems. - Apia, Western Samoa: SPREP.

SOPAC/SPREP (1993). Coastal Protection in the South Pacific. Prepared for the South Pacific Forum in 1993. Unpublished Paper.

SPC (1992). The Pacific Way. Pacific Island Developing Countries' report to the United Nations Conference on Environment and Development. PIDC report to UNCED. Prepared by the South Pacific Regional Environment Programme.

SPREP/SOPAC (1994). Coastal Protection in the Pacific Islands: Current Trends and Future Prospects. *Proceedings* of the First and Second Regional Coastal Protection Meetings held on 21 to 23 February 1994 in Apia, Western Samoa. SOPAC Miscellaneous Report 177.

SPREP (1993). Assessment of Coastal Vulnerability and Resilience to Sea Level Rise and Climate Change. Case Study: Upolu Island, Western Samoa. Phase 1: Concepts and Approach, Apia.

SPREP (1996). Action Plan for Managing the Environment of the South Pacific Region 1997 - 2000. South Pacific Regional Environment Programme.

SPREP (1996). Coastal Vulnerability and Resilience in Fiji - Assessment of Climate Change Impacts and Adaptation, Phase IV of the Integrated Coastal Zone Management Programme for Fiji and Tuvalu.

CHAPTER 7

Managing Marine Resources: The Role of GIS in EEZ Management

S. Fletcher

ABSTRACT: Following the agreement of UNCLOS III, coastal states have new powers and responsibilities relating to the use and management of their Exclusive Economic Zones (EEZs). The exploitation of these zones is diverse and of increasing economic importance, particularly to developing countries. In order to encourage long term economic development, integrated management policies to ensure the sustainable use of EEZ resources are required. With reference to selected EEZs, this chapter examines the actual and potential resource conflicts, and proposes Geographical Information Systems (GIS) as a key tool for the integrated and sustainable management of these zones.

Introduction

The oceans and coastal zones surrounding coastal states contain a range of living an non-living resources which are increasingly exploited for economic gain. This is particularly the case for developing countries and small island states that may be reliant upon marine resources due to a poor terrestrial resource base. Ocean uses typically include the exploitation of fishery, recreational, biotechnological and industrial resources, renewable energy production, hydrocarbon exploration and waste disposal (Champ and Ostenso, 1985).

Many of the uses made of marine resources are prone to conflict as increasing pressure is exerted on a limited resource base. Such conflicts result in the misuse and over exploitation of resources which undermines the sustainability of the resource base itself. The ocean area available for exploitation reflects the maritime boundaries of coastal states, derived from international agreements and conventions. The vast majority of the oceans are traditionally 'high seas' under control of no individual state, but in recent history national jurisdiction has increasingly encroached on this (Gubbay, 1993). This is primarily due to the continued efforts of the United Nations to develop a Law of the Sea and more recently UNCLOS III.

United Nations Convention on the Law of the Sea

The third United Nations Convention on the Law of the Sea (UNCLOS III) was the most recent in a series of international conferences seeking to develop a regulatory framework for the use and exploitation of the oceans. The first conference in this process was the League of Nations Codification Conference in 1930 which sought to formalise existing customary law relating to the use of the sea. Following the Second World War, negotiations concentrated on the contentious issue of control over continental shelf resources. However, agreement was not achieved and the view developed that only a complete package of new measures would suffice to resolve the situation. This resulted in the convening of UNCLOS III in 1983 and the subsequent Law of the Sea Convention which represented such a package (Haines, 1995).

UNCLOS III entered into force on 16 November 1994, when the 60th nation ratified the convention. The preamble to the Convention describes the document as "a legal order for the oceans which will facilitate international communication, and will promote the peaceful uses of the seas and oceans, the equitable and efficient utilization of their resources, the conservation of their living resources, and the study, protection and preservation of the marine environment" (UN, 1983). UNCLOS III recognises 5 maritime zones over which coastal states have jurisdiction (Herriman, 1992). The most significant are Interior Waters, Territorial Waters and the Exclusive Economic Zone:

Internal Waters: These are defined as waters landward of the baseline. They may include estuaries, bays, or areas landward of nearshore islands.

Territorial Waters: (baseline - 12 nautical miles) these are considered to be the territory of the coastal state, however, foreign vessels are given the right of innocent passage;

Exclusive Economic Zone: (edge of territorial waters - 200 nautical miles) This is an area of claimed ocean "lying seawards of and adjacent to territorial waters, within which the claimant state has exclusive authority over the resources of the sea and the seabed" (Prescott, 1985). The powers conferred in the EEZ are a sub-set of those in territorial waters, therefore it is commonly stated that the EEZ extends 200nm from the baseline.

The declaration of an EEZ represents a significant increase in the rights and responsibilities of coastal states and a major expansion in the area of ocean falling under the jurisdiction of individual states. The EEZ is therefore considered to be the most significant provision within the Law of the Sea relating to marine resource exploitation and as such requires further scrutiny.

Rights and Responsibilities in the EEZ

Article 56 of UNCLOS III states that within their declared EEZ, coastal states have "sovereign rights for the purpose of exploring and exploiting, conserving and managing the natural resources, whether living or non-living, of the waters superjacent to the seabed and of the seabed and its subsoil, and with regard to other activities for the economic exploitation and exploration of the zone, such as the production of energy from the water, currents and winds" (United Nations, 1983).

The declaration of an EEZ therefore provides an entitlement to explore and exploit for economic gain, whilst incurring the responsibility to conserve and manage (Herriman, 1997). Thus, the rights conferred by EEZ status are qualified by certain responsibilities. For instance, coastal states must "ensure through proper conservation and management measures that the maintenance of the living resources in the exclusive economic zone is not endangered by over exploitation" (United Nations, 1983).

Given full declaration of EEZs, 35% of former high seas will fall under the jurisdiction of coastal states. The declaration of an EEZ can substantially increase the resource base of individual nations. For example, when India declared an EEZ it gained an area of 2.2 million km^2, with an estimated fishing yield of 4.5 million tons (PTI, 1996). The location of EEZ boundaries can also have significant economic implications given the value of marine resources, and coastal states commonly attempt to maximise the area under their jurisdiction.

Declaring an EEZ is not a pre-cursor to the sustainable use of resources. In declaring an EEZ, the United States sought to "allow coastal states to gain the economic benefits of exploiting marine resources therein" in the hope that "resources would be managed more carefully if a particular manager were identified (Van Dyke, 1995). However, the coastal fisheries of the US have been dramatically over fished during the past decade largely as a result of ineffective and sectoral management, indicating that coherent EEZ management is required.

EEZ Management

The importance of the World's oceans was highlighted by the World Commission on the Environment and Development (1987) when it stated that "looking into the next century, the Commission is convinced that sustainable development, if not survival itself, depends on significant advances in the management of the oceans". The Commission recommended that "coastal governments should launch an urgent review of the legal and institutional requirements for integrated management of their Exclusive Economic Zones, and of their roles in arrangements for international co-operation".

The global adoption of integrated marine resource policies was further encouraged during the Earth Summit in Rio 1992 where the coastal states of the world committed themselves to "integrated coastal management and sustainable development of coastal areas and the marine environment under their jurisdiction" (UNCED, 1992).

The declaration of EEZs and their associated powers over large ocean areas provides an opportunity to develop sustainable policies and management practices for marine resources. The need for management has two main underpinnings:

- **to maintain the ecological balance of the marine areas**

- **to allow the sustainable exploitation of marine resources and in so doing, promoting the economic well-being of the nation involved**

EEZ management seeks to manage EEZ resources and associated land areas, in an integrated, strategic and sustainable manner. Two fundamental requirements for effective EEZ management are co-operation between states with bordering EEZs and good information upon which to make management decisions.

Co-operation Amongst Coastal Nations

The interconnected nature of the oceans and coastal land means that specific areas are somewhat difficult to manage in isolation. The designation of EEZs will require that maritime nations co-operate, especially where states designating EEZs are geographically close and maritime boundaries coincide. The geographic spread of marine resources tends not to reflect maritime boundaries, therefore in order to exploit such resources sustainably, regional policy integration is required. This has led Gubbay (1993) to comment that "the management of EEZs and the high seas undoubtedly needs international collaboration".

The 18 states bordering the Gineau Gulf of West Africa demonstrate the need for strategic management. Gineau states have a combined coastline of 6500km and EEZs totalling 2340 million km2 (Levy, 1987). Severe economic hardship resulting from petroleum shortages, localised political conflict and their position in the global economy, mean these nations are increasingly seeking to exploit their marine resources. The complex nature of the marine boundaries resulting from the proximity of Guinea states has led Bidi (1993) to consider how the collective EEZ resource base will be exploited, asking "must their nationalism lead them to lock it up within the straight and unproductive boundaries of the marine frontiers,...or will they have to... co-operate in order to harmonise their efforts for a united exploitation?". The economic benefits sought by these nations may not be obtainable if exploited on an EEZ specific basis, therefore sustainable development will depend upon a co-ordinated and strategic approach.

Whilst there is considerable potential for conflicting international resource policies a complicating factor occurs where sub-national units have differing views on ocean resource use, in such cases, this "can result in the multiple and possibly conflicting claims by one nation in the international arena" (Hershman, 1996). The need for co-operation in the harmonious use of resources is therefore also an intra-national consideration.

The act of designation of an EEZ can itself lead to conflict, as valuable resources are subject to competing demands. This is demonstrated in the conflict involving the oil rich Spratly Islands in the South China Sea. The archipelago of 60 islands located 100 miles from the Philippine coast (within the declared Philippine EEZ), but are subject to a traditional territorial claim primarily by China, Brunei, Malaysia, Taiwan and Vietnam. The conflict revolves around gaining a share of the resources in that marine area and has led to continuing military operations in the area (Gittings, 1996).

The eagerness of China to expand its zone of maritime jurisdiction reflects a continuing theme in China where the exploitation of the coastal and ocean resources have made a leading contribution to economic development. However, there have been associated resource conflicts, including conflicts between mariculture and shipping development, offshore oil development and fisheries, tidal land reclamation and wetland resource uses and waste disposal and the protection of human health (Yu, 1994). Thus, conflicts may involve a series of resource uses, producing a variety of impacts at a variety of scales.

Through co-operation, resource conflicts and territorial claims can be identified and addressed at an early stage. The availability of and effective use of accurate information will greatly improve the likelihood of co-operation between coastal states.

Information Requirements

A significant obstacle to integrated management is information on which to make management decisions. Marine and coastal data is often patchy, of variable quality, in a range of formats and held by a diffuse set of organisations. The problems associated with the collection of coastal and marine data are well documented elsewhere (Bartlett, 1993; DOE, 1995); they include issues of data collection, compatibility, storage and data gaps.

An extensive and accurate database relating to the physical, biological, social and economic characteristics of the EEZ is necessary to make informed decisions for effective and sustainable management of EEZ resources. The efficient use and correlation of information is of equal importance, as poor information use can lead to misunderstanding, the inefficient use of marine resources and sub-optimal economic benefit. The effective use of sound information is therefore an essential prerequisite for successful management (Russell and Upton, 1997).

Bidi (1992) supports the need for good quality information, and identifies information exchange and co-operation also as being of key importance, describing them as "the only way to ensure... countries derive benefit from the delimitation of their EEZ". This represents a need to integrate data to reflect trans-EEZ processes. An efficient mechanism to manage and display EEZ related information would produce substantial benefits relating to the sustainable management of EEZ resources; Geographic Information Systems are such a system.

The Role of GIS in EEZ Management

GIS have traditionally been considered as terrestrial applications. However, increasing applications are being developed relating to coastal and marine environments, yet the potential for GIS to be used for large scale marine resource management is somewhat under-explored. The key characteristics of GIS of data storage, manipulation, analysis and display suggest that GIS is able to facilitate effective EEZ management by providing:

- **information in a timely manner**
- **information tailored to needs**
- **a cost effective storage medium for data**
- **a national resource and repository of data relating to coastal and marine areas**
- **a means of identifying of data gaps**
- **rationalisation of data collection and research**

The complexity and dynamic nature of the marine environment requires a system which is able to model processes reflecting that complexity. Therefore, to manage the datasets relating to the EEZ ultimately requires a system which adopts the analytical capacity of a GIS and the real time modelling capability of a marine dynamic model.

In the US, the National Oceanic and Atmospheric Administration (NOAA) has sought to create a GIS suitable for coastal and ocean governance out to the edge of the EEZ in the southeast US. The initiative stems from a realisation that US ocean management policies tend to be fragmented and often conflicting, which results in "contentious decision making that often fails to address impacts on ocean resources" (Fowler and Gore, 1997). NOAA considers 3 datasets as fundamental to ocean management:

- **accurate shoreline position**
- **bathymetry**
- **maritime boundaries**

These core datasets can then be combined with datasets relating to physical resources, living resources, historic/cultural resources, economic resources, and foreign materials as appropriate to the issue being examined. Through identifying data requirements for decision making, GIS can feed into the wider process of data gathering and assist in prioritising data needs.

Using GIS to manage data relating to EEZ resources is likely to promote the establishment of partnerships, networking and joint working through the need for sectoral, sub-national and international bodies to co-operate in data collection, management and interpretation. The use of GIS also serves to identify conflict and focus resources on specific coastal and ocean issues which may have previously been overlooked. However, the role that GIS plays in EEZ management will ultimately relate to how GIS output is presented to decision makers and subsequently interpreted.

Conclusion

The management of EEZ resources to facilitate sustainable economic development requires a strategy based on sound information and co-operation between coastal states. GIS has been identified as having a significant role in achieving effective EEZ management through the provision of spatial resource information appropriate for decision making, and through facilitating co-operation and partnership amongst coastal states. Whilst the usefulness of GIS output ultimately depends on the ability of decision makers to interpret and respond to issues identified by GIS analysis, this chapter has highlighted GIS as a key tool in EEZ management and consequently to the economic development of coastal states.

References

Bartlett, D. (1993). *GIS and the Coastal Zone: Past, Present and Future*. Association for Geographic Information.
Bidi, J.T. (1993). Geographical Approach to the Exclusive Economic Zone in Guinea Gulf. *Ocean and Coastal Management.* Vol. 19, pp.137-155.
Champ, M. A. and Ostenso, N.A. (1985). Future Uses and Research Needs in the EEZ. *Oceanus.* Vol. 27, No. 4, p. 62.
DOE (1995). *Earth Science Information Needs for Sustainable Coastal Planning and Management.* DOE.
Fowler, C. and Gore, J. (1997). Creating a Geographic Information System for ocean planning and governance in the Southeastern United States. *http://www.csc.noaa.gov/internal/cfowler/*
Gittings, J. (1996). Alarm as China charts an ocean takeover. *The Guardian.* 17 May 1997, p. 16.
Gubbay, S. (1993). Coastal Zone Management and the North Sea Ministerial Conferences. A *Report* to the World Wide Fund For Nature from the Marine Conservation Society.
Haines, S. (1995). *UN Convention on the Law of the Sea.* Seaways (Supplement) February 1995.
Herriman, M. (1997). Rights and Responsibilities in the EEZ. Living on the edge - the maritime estate *Second Joint Conference: Institute of Australian Geographers and New Zealand Geographical Society.* University of Tasmania, Hobart. 28-31 January 1997.
Hershman, M.J. (1996). Ocean management policy development in subnational units of government: examples from the United States. *Ocean and Coastal Management.* Vol. 31(1), pp. 25-40.

Levy, J.P. (1987). The development of marine resources and the new Law of the Sea: a challenge for developing countries. *Paper* presented at Abidjan Symposium, May 1987.

Prescott, J.R.V. (1985). *The Maritime Political Boundaries of the World.* Methuen London and New York.

PTI (1996). Experts express concern over deep sea fishing policy. *Asia Pacific News Summary.* October 11, 1996.

Russell, I.C., and Upton, A.C. (1997). The contribution of national mapping and marine data collection agencies to resource evaluation and planning in the marine estate. Coastal and Ocean Space Utilization, Singapore. *Conference Proceedings.* Vol. 2 pp. 107-121.

UNCED (1992). Agenda 21. *Chapter 17 Protection of the oceans.* United Nations.

United Nations (1983). *The Law of the Sea: The United Nations Convention on the Law of the Sea.* United Nations.

Van Dyke, J.M. (1996). The Rio principles and our responsibilities of ocean stewardship. *Ocean and Coastal Management.* Vol. 31(1), pp. 1-23.

Yu, H. (1994). China's coastal ocean uses: Conflicts and impacts. *Ocean and Coastal Management.* Vol. 25, pp. 161-178.

CHAPTER 8

The Management Plan of the Wadden Sea and its Visualisation

M. A. Damoiseaux

ABSRACT: Twelve different authorities are responsible for the use and protection of the Wadden Sea, an internationally recognised wetland. They have for the first time compiled an integrated Management Plan. It covers the following eight sectors: environmental protection, coastal management, public works, transport, fishery, energy, recreation, and military activities. GIS has played an important role in the preparation and in the final visualisation of the plan. The Management Plan is visualised in three ways: as a set of 15 maps in the so-called sector notes, meant as background documents for the specialist; a set of 16 maps for the Management Plan itself, meant for wide distribution; a simple PC-application using the original GIS data, named the "Wadden Viewer" and meant for the authorities. A fourth form of visualisation is now in preparation by making the data available via the Internet.

Introduction

The policy for the Wadden Sea is laid down by the Dutch government in the so called "Planologische Kernbeslissing Waddenzee" (Wadden Sea Memorandum) and by the provinces in the "Interprovincial Beleidsplan Waddenzeegebied" (Interprovincial Management Plan Wadden Sea Area). The main objective of these documents for the Wadden Sea is: *the sustainable protection and development of the Wadden Sea as a nature reserve.* Human activities with an economic and/or recreational function are in principle allowed, provided that they do not conflict with the main objective.

Twelve different authorities are responsible for the management of the Wadden Sea. Each of them used to have its own management plan. After the revision of the Wadden Sea Memorandum in 1994, each of these separate management plans had to be revised also. In the Wadden Sea Memorandum there are directives to compile the existing plans into one new plan. In order to promote the implementation of the plan, an "action programme" had to be included. By doing so, the planning structure for the Wadden Sea was greatly simplified.

The compilation of the new management plan and action programme was conducted by a steering committee and eight sector study groups, in which all the concerned authorities were represented. The whole was co-ordinated by Rijkswaterstaat (Directorate-General of Public Works and Water Management), Directorate North-Netherlands. The Survey Department supported the project for the aspects concerning geo-information.

The Management Plan of the Wadden Sea

The Management Plan of the Wadden Sea focuses on the management of the ecosystem and the regulation of the human use of the Wadden Sea. The latter is particularly focused on the prevention or at least restriction of disturbance, contamination and damage to the natural environment in the Wadden Sea, in connection with sustainable use.

The plan was compiled by eight study groups, for the sectors:

- **environmental protection;**
- **coastal management;**
- **public works;**
- **transport;**
- **fishery;**
- **energy;**
- **recreation;**
- **military activities**

Chaired by the responsible authority these sector study groups have described their sector management in short "sector notes". These sector notes provide a good foundation for sector management as well as an explanation of it. On the basis of these sector notes the steering committee has written an integral Management Plan. In this plan attention is also given to the use of statutory regulations, the co-ordination of inspections and upholds the role of information and education, the co-ordination of monitoring, research and data-management of the Wadden Sea.

Use of GIS

For the compilation of a management plan, a lot of spatial information is needed. Because GIS can be of great benefit with respect to this, the use of GIS for the Management Plan of the Wadden Sea was already considered at an early stage. In a project plan the advantages and preconditions of the use of GIS were pointed out and presented to the steering committee and the sector study groups. Emphasis was placed on the fact that GIS was not only of benefit during the compilation of the management plan, but also for the actual management afterwards. Additionally, the Directorate of the North-Netherlands already had an ArcInfo system at their disposal.

The spatial information had to be suitable for visualisation as well for analyses. Therefore, agreements were made on the co-ordinate system, the level of details of the data, the description of the data, the use of a metadata system and a data exchange format. Next a data model was set up for the GIS.

The Sector Notes

For the first draft of the data model it was assumed that there would be a total of thirty-six themes or entities, spread over eight different sectors, plus the topography. Later on a few more themes were added.

The national rectangular RD co-ordinate system was chosen, based on the stereographic projection and with elevation data (in metres) based on Amsterdam Zero (NAP). The same systems are used on topographic maps (scale 1:10.000 to 1:100.000) in the Netherlands.

For the construction of the GIS database a deliberate choice was made in favour of the 1:250.000 scale. This scale makes it possible to also produce maps at a scale 1:100.000 or 1:500.000. In the latter case the Dutch Wadden Sea fits on an A3-size sheet. The initial wish was for a scale of 1:50.000 because several themes were already available at that scale. But, most of the information needed was not yet digitally available and it would have been too great an effort to collect these at such a detailed scale; and, for policy purposes that level of detail is not necessary. For a policy-maker it is sufficient to use maps, for example, with only waterways marked. Only a maintenance officer (and a sailor of course) needs a map or chart with the exact location of the various buoys and beacons. It was also intended to update the database, to support the mandatory annual evaluation and five-year revision of the Management Plan. A GIS database at the scale of 1:50.000 would be too large and too time consuming to maintain for that purpose. For the description of the meta-information a programme called GeoKey was chosen. At that time a prototype was available but meanwhile GeoKey has became a standard within Rijkswaterstaat.

During the compilation of the sector notes the text and the maps advanced together. This is quite unique, since normally attention to maps and illustrations is only given after the text is completed. All 15 different maps were plotted straight from ArcInfo at a scale of 1:250.000 on a HP DesignJet inkjet printer and discussed at several meetings. Every adaptation was immediately carried out in the GIS and the revised maps were presented on the next meeting. In this way, the so-called "discussion maps" grew together with the sector notes. Finally these maps were reduced to a scale of 1:500.000 to fit on an A3-size sheet as a supplement for the final printed sector notes. These serve as background-documents for the specialist. For each sector map some 350 copies were printed (see Fig. 1) on a Xeikon digital printing press, straight from an ArcInfo postscript-file without making final films or printing plates. This turned out to be the most attractive option, considering quality, cost and delivery time.

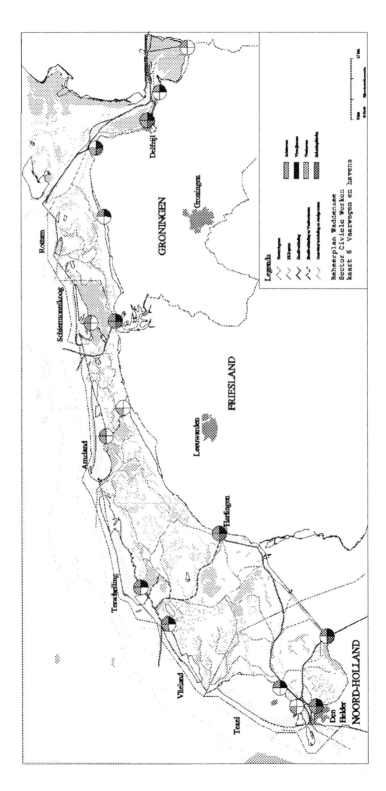

Fig.1. Reduced sector note Map 6: Traffic and transport

The importance of GIS was not only proven through the rapid way in which updated maps could be made, but also by drawing, comparing and fine-tuning e.g. buffer-zones around restricted areas, silence areas, flying routes, shipping routes, and such like. Once more it became clear that for most of the time the intention of the participants was the same, but the delivered files or paper maps were not identical and had to be fine-tuned to get a coherent image.

The Management Plan

Based on the sector notes the integral version of the Management Plan for the Wadden Sea was written by the steering committee provided with 16 maps at the A4-size. For these maps, the ArcInfo coverages were converted to a cartographic system (an Apple Macintosh computer with Freehand-software) and made up again from scratch. This procedure was foreseen since a reduction to A4-size (scale about 1:750.000) is not possible without a certain amount of generalisation. At an earlier stage, a series at the A4-size was produced with ArcView 2 for internal use. The quality was unsatisfactory, due to lack of cartographic functionalities in ArcView 2, especially regarding generalisation, the design of point- and line-symbols and the colour choice.

To present the Management Plan for comment to all the authorities concerned, a small edition of the Freehand-maps was produced on a Canon colour copier at the end of 1995. The comments received were processed in ArcInfo and then plotted again at the original scale. These plots served as revision models and all changes were then drawn by hand (using a mouse) in Freehand. So, no ArcInfo coverage was converted again. This old-fashion-way proved to be a much faster, in spite of the extra integrity checks. Next, 2000 copies of these A4-maps (see Fig. 2) were printed, together with the text of the Management Plan in offset printing.

Information Management for the Benefit of the Wadden Sea

In the Wadden Sea Memorandum it is specified that decision-making on new or altered activities must take place on the basis of the best available information on the expected consequences, particularly with respect to the natural environment of the Wadden Sea.

This information will be collected by means of a research and monitoring programme, carried out by the different authorities concerned. These activities will be essential for a careful and integral tuning and weighting of human use of the Wadden Sea. With the help of adequate gathering, storage and presentation of data and knowledge, the decision process can be based on the best available information. To an increasing degree, attention must be given to the availability and actuality of data and knowledge, which are needed to evaluate the effectiveness of the management actions. By mutual agreement all authorities concerned are going to aim at an adequate and uniform storage and availability of management information in the Wadden Sea. For this, a GIS is indispensable.

RECREATIE

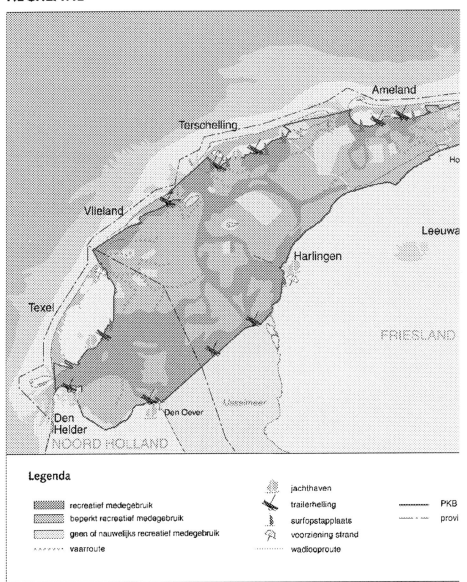

Fig. 2. Part of the Management Plan Map 14: Recreation

The foundation for this has been laid with the Management Plan of the Wadden Sea. In this framework, initiatives have been taken to keep the collected geo-information up-to-date by and in favour of all the authors of the Management Plan. This means, among other things, that it should be possible to access these GIS data on a PC without specialist knowledge. For this purpose it was chosen to use an ArcView application for distribution. This is an easily accessible GIS programme and is also now freeware.

The GIS data are divided into nine different "views" (see Fig. 3), complete with symbols and a short manual distributed under the name "Wadden Viewer" among the authors of the Management Plan of the Wadden Sea. A wider distribution is for not possible yet for reasons of copyright.

The Wadden Viewer-data can also be used in ArcView 2 or 3.0, by those who have that at their disposal. Changing the data, however, is not possible in ArcView 1. So, the users have to pass their analogue revisions on to the North Netherlands Directorate, where these changes will be processed and distributed again. In this way the maintenance of the GIS files seems to be assured.

New Initiatives

Every three years a Trilateral Governmental Conference is held by Denmark, Germany and the Netherlands on the protection of the Wadden Sea. The National Institute for Coastal and Marine Management (RIKZ) of Rijkswaterstaat has now reworked the Wadden Viewer-data for use on Internet.

Acknowledgements

Many people co-operated in the process of visualisation of the Management Plan Wadden Sea. Among them there are three that I want to make special mention to here, in order of appearance: ing. Y.J. Zijlstra, project leader of the Management Plan; P.G. Beukema, GIS specialist and H.A. van Wees, cartographer.

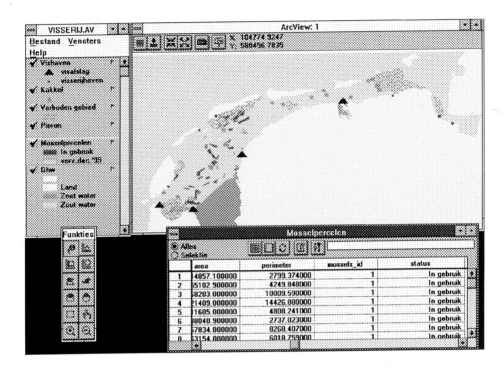

Fig. 3. Screen dump ("view") from the Wadden Viewer: Fishery

CHAPTER 9

Using GIS for Siting Artificial Reefs - Data Issues, Problems and Solutions: 'Real World' to 'Real World'

D.R. Green and S.T. Ray

ABSTRACT: As GIS becomes an increasingly more 'user-friendly' tool, and more people, not necessarily GIS specialists, recognise the benefits of the technology in their work environment, inevitably greater use will be made of the technology for a wide range of applications. The ease with which a GIS can now be used often overshadows the complexity underlying this technology and the potential difficulties that can (and do) arise when this is not fully understood. In practice, there are two routes open to the application of GIS technology for environmental studies. The first is to commission new data for the research. The second is to make use of existing datasets. Data is at the heart of any GIS application. A detailed knowledge about the data sources, the method of collection, capture, scale and sampling strategy, especially if the data are to be used in any analysis, modelling or simulation studies, is fundamental to any application. Unfortunately, information about data (metadata) is seldom available, especially for archival datasets. Furthermore, although it is now relatively easy to acquire digital data, to input, store, manipulate and display this data, and to output the results of any GIS analysis in the practical sense, little consideration is given to the problems associated with data quality and how this will ultimately affect present and future analyses and use of the output for planning and decision-making. The need to raise awareness about data quality for applications is set in the context of the development of an environmental database for the Moray Firth, North East Scotland, and more specifically the use of selected datasets from the database to aid in the proposed siting of an artificial reef. Using this example, the chapter explores the problems associated with the use of both existing analog and digital datasets as the basis for environmental applications, the problems of data acquisition, data quality, data standards, error and how these can affect the operational use of the data in GIS analyses. The solution to such problems appears to lie with improved error assessment and reporting. The outcome of this chapter is an attempt to offer guidance and solutions to researchers and applications specialists undertaking similar studies, by suggesting to what extent studies, such as the artificial reef siting, can safely make use of existing datasets without risking the problems associated with judgements based on inadequate information, and generated or inherent error.

Introduction

Until quite recently most applications of Geographical Information Systems (GIS) have focused on land-based studies rather than marine and coastal environments. A monograph by Bartlett (1994) covers some of the wide range of applications undertaken to date, whilst other papers by e.g. Fairfield, 1987; Davis and Davis, 1988; Riddell, 1992; Deakin and Diment, 1994; and Green, 1994a,b,c; 1995) illustrate some of the specific ways in which GIS has been used to date. With increasing interest now being shown in our marine and coastal environments, more GIS-based applications, particularly those developing Spatial Decision Support Systems (SDSS) (see, for example, Raal and Davids, 1995; Canessa and Keller, 1997) are being developed. Demand from end-users has subsequently placed greater pressure on commercial software developers to provide more user-friendly front-ends to GIS systems, e.g. ArcView for ArcInfo, with the aim of assisting, e.g. coastal zone managers, to use this technology in their work environment. Interest in the benefits of communications and networking technology e.g. the Internet and the World Wide Web (WWW), has recently been examined by a number of studies (e.g. Green, 1996; 1997).

As GIS becomes a more 'user-friendly' tool, and more people, not necessarily GIS specialists, recognise the benefits of the technology in their work environment, greater use will undoubtedly be made of GIS. Whilst this is good in itself, the ease with which a GIS can now be used tends to overshadow the complexity of the technology and the potential difficulties that can (and do) arise when the technology is not fully understood. It is, for example, relatively easy to acquire digital data, to input, store, manipulate and display this data, and the results of any analysis in a practical sense. The problem is that there is seldom any mention of data quality associated with GIS analyses.

In practice there are two routes open to the application of GIS technology in environmental studies. The first is to collect entirely new data for the research. The second is to make use of existing available datasets, both analog and digital. In an ideal world 'starting from scratch' is probably the best approach to any research, except where historical trends are an important component. Acquiring and subsequently using the 'right data' for the 'right job' is then almost a certainty. The appropriate scale, spatial sampling, and areal boundaries are then collected for the specific application in mind. More often than not, however, there are many datasets already available in both analogue and digital format which could potentially be used for other applications, but as an aid to the problem, rather than for quantitative analysis. Unfortunately whilst some datasets may be of use for an environmental application, many may not in fact be appropriate, and may have limited use in practice.

A GIS is relatively simple to develop in theory and the data is easy to acquire, whether it is archival or new, for a wide variety of applications. But, in practice, there are many fundamental considerations to be taken into account when undertaking a GIS project which relate directly to the data available for use in the study. Knowledge of the source of the data, the method of data collection, capture, scale and sampling strategy are all important if they are to be used in any analyses, modelling or simulation studies. Without such information, it is very difficult to make use of the datasets, not in terms of their practical usage, but in a legitimate way. Unfortunately information about the data is not always recorded at the time of collection or forthcoming when multiple datasets from multiple organisations are involved. Furthermore, there are currently few requirements to collect spatial data to any UK standard, and no requirement to document (metadata) the data, although bodies such as the AGI in the UK for example have produced documents advocating the need to consider standards and data quality (AGI, 1996). This means that without having such information to hand when using 'second and third-hand' data, it becomes very difficult to justify and to defend the use of individual and combined datasets for an application, particularly where any analysis is planned. In a practical sense the data can be used. However, there is considerable potential for the generation of error. Without knowledge of the individual and cumulative error contributed at each stage in an analysis it becomes very difficult for end-users to make decisions based on the outcome of an analysis. In some cases the error inherent in a dataset or generated through analysis may not be significant, and may even be minimised. In other cases the output may contain sufficient error that could affect the usefulness of the results, and even decisions made.

This chapter seeks to address some of the problems associated with these data issues and problems, and to offer some practical solutions and guidance for the practitioner. It is set in the context of the development of an environmental database for the Moray Firth, North East Scotland, and more specifically the use of selected datasets from the database for the proposed siting of an artificial reef. Using this example, the chapter explores the problems associated with the use of both existing analog and digital datasets as the basis for an environmental application, the problems of data acquisition, quality, standards, and error, and how these can affect the operational use of the data in GIS analyses. There is also the question of how one makes use of historical or archival data e.g. in sediment movement studies, temporal changes where old datasets have to be used out of necessity. The solution to such problems would appear to lie with improved error assessment and reporting. The outcome of this chapter is an attempt to offer guidance and solutions to researchers and applications specialists undertaking similar studies. It suggests to what extent studies, such as the artificial reef example, can safely make use of existing datasets without risking the problems associated with judgements based on inadequate information, and generated or inherent error.

Artificial Reef Siting

To date it would appear that there is relatively little literature on the siting of artificial reefs utilising GIS technology (see for example, Nakamuna, 1985; Bohnsack and Sutherland, 1985; Matthews, 1986; Grove et al., 1991; Bruno et al., 1996; Herrington et al., 1997; additional reading list of websites).

TRADITIONAL APPROACH

In 1995, AURIS Environmental (Heaps, 1995) undertook a detailed feasibility study of the siting of artificial reefs in the Moray Firth, Scotland, UK. This investigation employed many different data sources and consulted with numerous expert groups. The local fishermen's association was particularly influential and was eventually responsible for helping to short-list the potential sites for the proposed reef. Paper-based maps were used in the analysis but were only small-scale and contained generalised information. The entire selection process employed and described in the AURIS report was essentially empirical and is based on the criteria summarised in Table 1.

A technique referred to as 'Exclusion Mapping', a type of sieve mapping, was employed to eliminate areas with the most obvious constraints, thus helping to narrow down the most suitable location zones. Although this procedure is not described in detail the AURIS Report, the use of conventional maps is normally difficult to undertake and is error-prone.

Whilst this 'traditional' manual approach has been well-tested for many similar tasks and has been deemed reliable in the past, the fact that it is not computer-based means that it also really only provides a one-off solution. Since all the focus is on one task, further analyses of the datasets collected are unlikely to be undertaken, largely because they are so time consuming. GIS

A GIS approach to the problem of locating the optimum site for an artificial reef offers a number of distinct advantages over the traditional approach adopted by AURIS outlined earlier. The most significant of these is the creation of a 'general' environmental database, the datasets of which can potentially be used for numerous other applications, the artificial reef being just one of the many possibilities. The primary advantage is that the source data, held digitally within a GIS is more flexible and accessible and can also be analysed directly using the system's functionality.

Moray Firth Project

The Moray Firth Project undertaken by Ray et al. (1996) followed on from the Moray Firth Review carried out by Scottish Natural Heritage (SNH) (Harding-Hill, 1993). The investigations carried out in this 'Pilot Study' aimed to develop the use of GIS for environmental management, with particular reference to the necessary background research, acquisition of data and the conversion of data into a widely used GIS format (ArcInfo). The purposes of this study were: to ascertain what datasets existed for the study area; to catalogue and document these datasets; to create a digital environmental database; to select those datasets from the total that were deemed to match criteria for siting an artificial reef; and to use those selected as the basis for optimising the siting of an artificial reef using a GIS. Methodologies were also proposed for the continued development of this pilot by developing ideas for a particular siting application; that is, a demonstration system highlighting the potential advantages of utilising GIS technology in these and other areas of coastal zone and environmental management applications.

The first part of the study examined the project life-cycle of developing a Pilot GIS. This included the initial background research, a scoping study, the use of questionnaires (see Appendix), followed by an examination of the problems associated with the acquisition of numerous disparate datasets from many different sources. These datasets were prepared and converted into a common format before being co-registered for compatibility in the spatial database for analysis.

The final part of the study proposed a model, using a GIS approach, for the siting of decommissioned oil and gas structures, for use as steel artificial reefs in the Moray Firth (Figs. 1-4).

The potential use of GIS as a general environmental management tool for applications both in the Moray Firth and other areas was then reviewed.

DATASETS

The main problems found to be associated with the environmental datasets gathered was firstly, their initial acquisition and secondly, problems with some of the datasets. Some of these identified problem areas are summarised in Table 2.

DATA ACQUISITION

There are basically two routes open to any new GIS project application:

- **to make use of existing available datasets.**
- **to start from scratch.**

In most cases the latter is not an economically justifiable approach because of the costs involved in terms of time and cost associated with, e.g. digitising, scanning and labour. In addition, it is seldom practical, may duplicate the work of others, and can be an extremely time-consuming exercise. The former, however, requires that you know what datasets are available; whether they can be used; what costs are involved; and above all whether they are of the right spatial scale for the intended application. Depending upon the goals of the project, it may be that many of the datasets available are indeed unsuitable, bearing in mind that it unlikely that they were originally collected for the intended purpose, in this case artificial reef siting.

TABLE 1
The Selection Process

- Outwith the main trawling areas
- Outwith the main shipping lanes
- With boats and facilities able to exploit a small reef
- Minimal impact on the other local users
- Minimal impact on the environment
- Firm, sandy substrate and
- Accessible to local facilities

More specifically:

- Within a reasonable distance from the shore for sport angling activities (less than 16km)
- With a fairly level topography (for stability and future expansion)
- With optimal hydrography
- At an optimal distance from areas of biological activity
- A steady current through a reef to prevent stagnation and to supply the reef fauna with food
- 30-40m adequate depth (Matthews, 1986)
- firm sand (in particular firm and hard packed)
- should not be placed on an already highly productive habitat
- on the fish path (i.e. for fish congregation) and intercept of the currents; siting planned to intercept the current most effectively with the longitudinal axis orientated to intersect the currents at right angles (Grove et al., 1985)
- sited close to the intended user-group
- placed in less than 50 metres depth (to facilitate diving surveys and ensure the reef is within the photic zone)
- Greater than 20 metres clearance, i.e. ample clearance to navigation
- The reef should be less than half the length of the storm waves

TABLE 2
Potential Problem Areas with Datasets

- Copyright
- Standards
- Cost
- Ownership
- Documentation (Metadata)
- Multiple Users
- Currency
- Format
- Size
- Scale
- Source
- Original Use/Project
- Method of Data Capture
- Updating
- Geographical Extent
- Interpolation
- Georeferencing
- Analysis
- Confidentiality

AVAILABILITY

Although much of the data available can be obtained free-of-charge for an educational research project such as the one carried out here, there were still many datasets which were not easily accessible, despite the non-commercial nature of the project. Some data could only be released for specific lengths of time and for others, deletion of the files was required at the end of the project. Finally, certain data was spatially referenced to a lower resolution than was required or ideal.

CONFIDENTIALITY

Despite the considerable potential value of some datasets to the project, which included the names and addresses of individuals, access is controlled by confidentiality regulations and still others, with defence implications, for example, are also restricted.

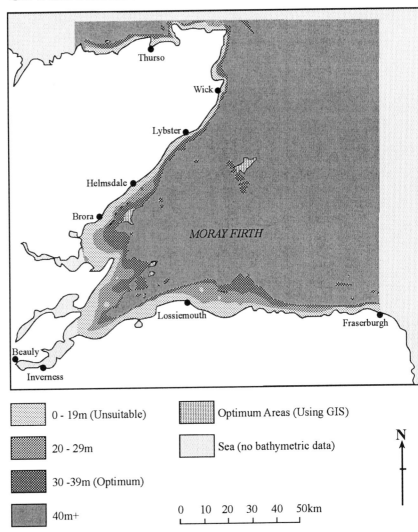

Fig. 1. Location of Optimal and Unsuitable Bathymetry
(map drawn in ArcInfo by Stephen T. Ray)

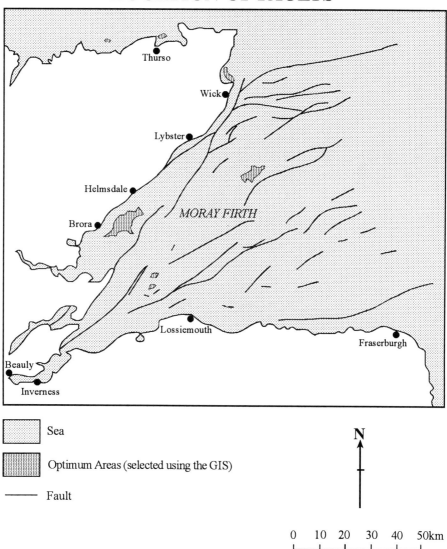

Fig. 2. Location of Faults
(map drawn in ArcInfo by Stephen T. Ray)

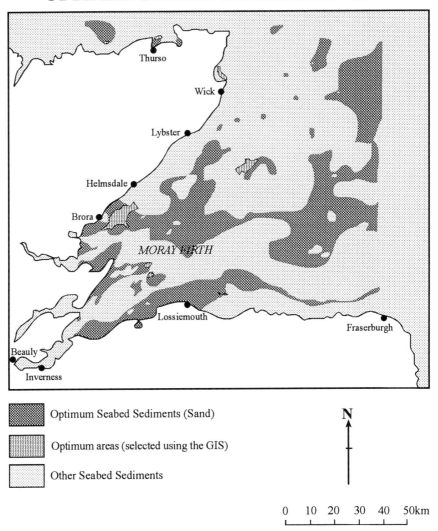

Fig. 3. Location of Optimal Seabed Sediments - Sand
(map drawn in ArcInfo by Stephen T. Ray)

OPTIMAL ARTIFICIAL REEF SITES

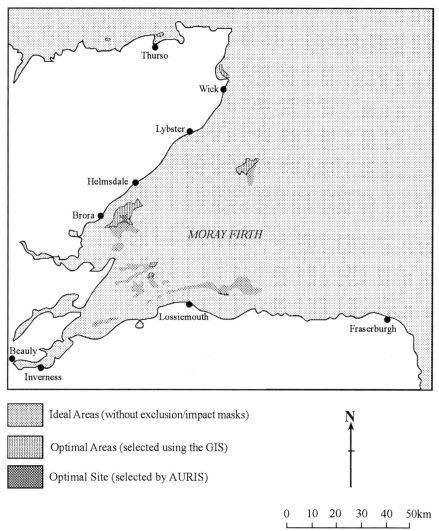

Fig. 4. Location of Optimal Artificial Reef Sites
(map drawn in ArcInfo by Stephen T. Ray)

COPYRIGHT

Copyright varies, with some of the official organisations who are potential suppliers of data, being more restrictive than others. Where project funds are limited then the cost of buying data and then paying royalties may be too great to justify their use, even though they might be considered vital to the application.

LICENCE

Licence agreements may have to be signed for essential data, e.g. geological and related bathymetric data in this project. The conditions attached, although strict, were reasonable in this case and applied to a term of three years, the data being restricted to the Moray Firth Project.

COSTS

Only one licence had to be paid for in this project but due to convenience (in suitable digital form) and the importance to the project, finance in this case was provided for its acquisition. Some important datasets (e.g. for meteorological data and topographic data) would have been quite costly but these were not regarded as essential for the current stage of this GIS pilot application.

GIS COMPETITION

GIS is increasingly becoming part of the work environment of many organisations. Some have a directive to set up a system whilst others are still involved in digitising data with a view to using it 'in-house'. For this and other reasons they are reluctant to pass it on for use within other projects. Of the organisations contacted for this project at least five were creating datasets with a view to setting up GIS for coastal and marine purposes.

Datasets

STANDARDS

Very few of the datasets provided by organisations had any sort of standard attached to them. This had some important repercussions on the possibilities open to the reporting of potential errors and on significance in the decision-making process. The more datasets that are collected to a common standard the higher will be the likely quality especially for use in analyses.

MULTIPLE USERS

The final GIS database, once in working order, may have to enable multiple users, all using the same data. Most GIS can enable this at the present time. However, this brings related problems of data security, overloading the system, IT support, as well as complications with the licence agreements.

SIZE

The size of the data files is always important in any GIS project. However, it is more important in a small organisation with limited disc storage space. The requirement for more frequent backups and data management therefore becomes necessary. Large raster datasets, e.g. satellite imagery, take up a great deal of disk space, and require powerful computer systems to both display and analyse the datasets.

ORIGINAL USE/PROJECT AIMS

In most cases, datasets that are already available usually form the basis for many new projects and applications. One argument used to justify this is that they have multiple potential uses, are already available, do not require additional expenditure, and avoid unnecessary duplication of effort. However, many environmental datasets are in fact acquired for a specific purpose (small projects and are not part of a larger data acquisition strategy), and not the one that individuals find a new use for. Whilst some applications can make use of these, others may find that they are incompatible with other datasets, are of the wrong scale, and so on. Due care and attention must therefore be exercised in the use of archival data and those not acquired specifically for a new project.

METHOD OF DATA CAPTURE

Knowing something about the method of data capture is very important, particularly for a pilot application. Data capture might have been undertaken by trained professionals carrying out surveys to specific requirements, or by amateurs with no real sense of accuracy for fieldwork. The quality of the data may therefore be suspect. This problem can also apply to the data loading, with reference to digitising, scanning, and data import using software packages other than those used originally. These methods of data capture will all lead to many errors which then accumulate throughout the import process. In addition, the provision of a record of the entire process (e.g. a log file) is very important in tracing and identifying a weak link, or a potential source of large errors in the final results. Boats, for instance, with satellite tracking devices, although getting better, are possibly less accurate than a surveyed portion of the coastal strip. Awareness of this and consideration in the GIS use and analysis must all be taken into account.

UPDATING

Knowledge about the currency of the data and the frequency of updates is also vital, especially if different datasets acquired on different dates are being used. Provision of this information in the metadata is vital, and knowing when a dataset was collected, how, and when it was updated is often information that is not available.

GEOREFERENCING

Datasets are often provided as either grid coordinates or latitude and longitude. Therefore a first step is to convert all the datasets to a single common coordinate system. If they are all supplied in the same coordinate system, this will enable more efficient data loading and once again there is less room for error to creep in.

MULTIPLE SOURCES

Data in any environmental application may ultimately come from many different supply sources. As a direct result the dataset will contain many variations e.g. some topics or themes will be very well covered with very detailed datasets for many years and locations, whilst other topics will have been recorded for a restricted number of sites and with insufficient temporal resolution. This means that many data coverages will not be directly comparable thereby limiting their potential for procedures such as GIS overlay mapping analysis. Many datasets may indeed contain more or less the same data on overlapping regions, therefore being a duplication of both cost and effort and limiting the potential for sharing of the datasets between organisations.

FORMAT

Another critical factor is the format in which the datasets are provided. Due to the likely different nature of data from different sources, for which there is no requirement for adherence to a standard, there will be many formats. ArcInfo export files tend to be preferred but more typically data comes in the form of formatted ASCII (either geographical - latitude/longitude - or National Grid), paper lists/tables (which have to be inputted by hand), DXF, DBase, and paper maps (requiring digitising). Many datasets are obtained in a format which can only feasibly be converted into one or other of points, arcs or polygons (points, lines, areas). A variety of data conversion procedures therefore have to be adopted with extensive use often being made of software such as Microsoft Excel followed by the use of ArcInfo routines.

FEATURE TYPE/INTERPOLATION

Many coastal datasets are collected at point sources directly on the coast and can not therefore be interpolated or extrapolated into the marine zone if needs be. Equally there are too few primarily marine datasets and this can prove very restrictive for any spatial analysis. Few datasets appear to be acquired in a digital format that can lead to complex and time-consuming data conversion procedures. Due to survey collection methods of original data some material can have very low precision, both in the survey and for the sample locations. There is also very little information about the actual sources of the data, collection techniques, quality, and known errors. Some potentially useful datasets are of restricted value due to their one-off survey characteristics and the limited spatial locations and time-periods for which they were available.

QUALITY

The quality of available data varies considerably. In part, if it is archival data, this is not surprising as methods of data collection and the equipment used have improved considerably over time. Some data may have been surveyed by trained professionals whilst others had been collected by amateur volunteers. Some datasets may themselves have been previously compiled from multiple and unspecified sources in different formats and thus are devoid of standards and are potentially more error-prone. When metadata is sought from data source contacts some are surprisingly vague about the provenance of many of the datasets even in their own databases.

Desirable characteristics of the data for the pilot GIS are summarised in the following list (Table 3):

Discussion

There are, based upon a simple example of collecting together existing datasets in an environmental database for an application, many factors that need to be considered when planning to use archival and other data of uncertain origin. Knowledge of the datasets available, including their source, format, and quality, can provide a far firmer basis upon which to proceed with a project. Where funds are short, it is inevitable that one option that will be considered is the use of the existing data. This may in fact be possible providing that the applications specialist is clearly aware of the limitations and constraints that exist in the use of spatial and temporal data. It is important to take all these factors into account when undertaking a similar project because it ultimately affects what analyses are carried out, the conclusions drawn, and their potential use for planning and decision-making.

TABLE 3.
Desirable Characteristics of the Datasets required for a Pilot GIS

- **covering both environmental and socio-economic data**
- **reasonably large spatial scale of origination**
- **high density of data values, good coverage and geographical distribution**
- **up-to-date information**
- **some historical datasets to provide for temporal trends**
- **predefined data format**
- **low/no cost implications**
- **full availability and access to the data with few restrictions**
- **information to be provided on data collection techniques**
- **full data description (metadata)**
- **quality of the data including an estimate of the errors involved**
- **digital (preferably ArcInfo) format**

Solutions and Recommendations

As discussed at the outset of this chapter, researchers usually have two choices open to them when using GIS for an environmental application. In an ideal world, and given sufficient funding and manpower resources, a study such as the one outlined above would start from scratch to ensure that the right datasets were available for the right job. Unfortunately, it is not an ideal world, and more often than not existing datasets are often utilised (and sometimes blindly) because of the cost and time involved in starting afresh. This is perfectly acceptable providing the uses that the data are put to are limited to e.g. visualisation or very simple analyses. The researcher must be aware of the pitfalls, and the end-user aware of the quality of the datasets used to derive the output. Clearly where information is available about the component datasets then it is possible to utilise them more fully knowing that they were collected at a certain scale, and using a known sampling strategy. At the end of such an analysis it is then possible to provide a measure of support or level of confidence for the end-user e.g. in the form of an error report that is meaningful to the end-user. Any decisions then made using the analysis can utilise the report e.g. placement of the AR is accurate to within +-100m as a measure of certainty. Where datasets are unaccompanied by detailed information then the researcher should proceed with care, using only the datasets that can legitimately be used, and where analysis is undertaken that some doubt be expressed about the results derived from the analysis. Where component datasets are accompanied by relatively little information, the message is that the researcher should exercise a measure of, if not considerable caution concerning the quality of the results of the analysis.

Table 4 is intended as a set of guidelines for individuals wishing to pursue similar environmental projects to the artificial reef siting example, providing some solutions and recommendations. In effect it aims to be a practical reference for researchers, indicating the stages that might be considered when planning or undertaking a project.

TABLE 4
Options, Solutions and Recommendations

- avoid use of the datasets completely (not practical)
- limit use of the datasets e.g. solely for visualisation/illustrative purposes with no quantitative analysis
- provide error estimates for single and composite datasets as an indication of their usability
- create new datasets from scratch

Summary and Conclusions

Growing interest in the use of Geographical Information Systems (GIS) has led to more and more applications being undertaken. All too frequently, however, the applications of such technology by non-GIS specialists ignore the recognised problems associated with geographical datasets.

This chapter has set out to examine a typical environmental project based around the use of existing datasets with the intention of highlighting some of the possible areas in which there are potential pitfalls for the unwary. It is clear that there are many, particularly for the applications specialist who relies, out of necessity, on archival datasets.

The conclusions of this chapter are that providing the researcher is made aware of and recognises the potential pitfalls and attempts to take account of them in the study, limiting the study to legitimate uses of the data, as outlined, then the existing geographical datasets can be useful. However, if no account is taken of the limitations of the data, then it becomes very difficult to have any confidence in the outcome of any data analysis that makes use of these datasets.

The fact that there are few guidelines provided for those collecting data, and few if any incentives at present to collect data in a standardised form means that many datasets, which are potentially very valuable and useful will continue to have limited use outside of the project for which they were originally gathered and intended. This means that there will be many digital datasets available which can indeed legitimately become part of an environmental database, effectively little more than a data catalogue, but they will not be of much other use unless well documented, and potentially of little real use in an analytical sense.

In the UK the National Geospatial Database Framework (NGDF) (AGI, 1996; Watts, 1997) has been proposed as one way of securing the collection and archiving of standardised geospatial datasets for more widespread and long-term use. Although little has been said to date about marine and coastal datasets (the Ordnance Survey and the UK Hydrographic Office did collaborate on the development of an experimental coastal zone map series and are now collaborating in 2000/2001 with the British Geological Survey (BGS) on a new digital initiative), it is clear that much stands to be gained by formalising the collection and documentation of spatial datasets (both archival and new) which will ultimately facilitate more correct, widespread and long term usage. The most important issue though is to draw attention to the need to provide detailed documentation on datasets that are available either free or at cost so that awareness of the need to acquire and use quality data is brought to the attention of the GIS practitioner.

References

AGI (1996). NGDF document.
AGI (1996). Guidelines for Geographic Information Content and Quality: For those who use, hold, or need to acquire geographic information. AGI Data Quality Working Group. *AGI Publication No. 1/96.* 69p.
Bartlett, D. (1994). GIS and the Coastal Zone: Past, Present and Future. I&E Committee Publication. *AGI Publication Number 3/94.* 36p.
Bohnsack, J.A., and Sutherland, D.L. (1985). Artificial Reef Research: A Review with Recommendations for Future Priorities. *Bulletin of Marine Science.* Vol. 37(1):11-39.
Bruno, M.S., Herrington, T.O., Rankin, K.L., and Ketteridge, K.E. (1996). Monitoring Study of the Beachsaver Reef at Avalon, New Jersey, *Technical Report* SIT-

DL-9-96-2739, July 1996, The Borough of Avalon, New Jersey, Davidson Laboratory, Stevens Institute of Technology, New Jersey, 35p. plus Fig.s.

Canessa, R. and Keller, C. P. (1997). User Assessment of Coastal Spatial Decision Support Systems. In *Proceedings* of CoastGIS'97. August 29th-31st 1997.

Davis, B.E., and Davis, P.E. (1988). Marine GIS: Concepts and Considerations. GIS/LIS 1988. *Proceedings.* San Antonio. ACSM/ASPRS:159-168.

Deakin, R., and Diment, R.P. (1994). Strategic Planning and Decision Support: The Use of GIS in the Coastal Environment. *Proceedings* of AGI'94, Birmingham, England. pp. 13.3.1-13.3.5.

Fairfield, F.M. (1987). Marine Planning with Geographic Information Systems. Coastal Zone'87. *Proceedings:* 1023-1030.

Green, D.R. (1994). GIS and Marine Information Systems: Some Recent Developments in the United Kingdom. In *Proceedings.* Hydro'94. The Ninth Biennial International Symposium of the Hydrographic Society. September 13th -15th. Special Publication No. 33 Aberdeen, Scotland. 6.1-6.27.

Green, D.R. (1994). GIS and Marine Information Systems: An Overview of Some Recent Developments in the United Kingdom. *Proceedings* of the Sixth Biennial National Ocean Service International Hydrographic Conference. US Hydrographic Conference'94. April 19-23, 1994. Virginia, U.S.A.

Green, D.R. (1994). Using GIS to Construct and Update and Estuary Information System. *Proceedings* of the Conference on Management Techniques in the Coastal Zone. Centre for Coastal Zone Management, University of Portsmouth, October 24-25, 1994.

Green, D.R. (1995). User-Access to Information: A Priority for Estuary Information Systems. Published in the *Proceedings* of COASTGIS'95:International Symposium on GIS and Computer Mapping for Coastal Zone Management. February 3rd-5th 1995, University College, Cork, Ireland.

Green, D. R. (1997). *The AGI Sourcebook for GIS 1997.* (with D. Rix and C. Corbin) (June 1997)

Green, D.R., and Massie, G. (1997). FEF Pilot Information System. In *Proceedings.* CoastGIS'97 - The Next Millennium (August 29th-31st 1997).

Grove, R.S., Sonu, C.J., and Nakamuna, M. (1991). Design and Engineering of Manufactured Habitats for Fisheries Enhancement. pp. 109-152. In, *Artificial Habitats for Marine and Freshwater Fisheries.* Seaman, W., and Spragne, L.M. (eds.) Academic Press Inc.

Harding-Hill, R. (1995). *The Moray Firth Review,* Scottish Natural Heritage, Inverness.

Heaps, L. (1995). Feasibility Study for the Construction of a High Profile Artificial Reef in the Moray Firth. *Final Report,* AURIS Environmental, Aberdeen.

Herrington, T., Bruno, M.S., and Ketteridge. K.E. (1997). Monitoring Study of the Beachsaver Reef at Cape May Point, New Jersey. *Technical Report* SIT-DL-96-9-2751, January 1997, The Borough of Cape May Point, New Jersey, Davidson Laboratory, Stevens Institute of Technology, New Jersey, 22p. plus Fig.s and Tables.

Matthews, H. (1986). *Artificial Reef Sites: Selection and Evaluation.* pp. 50-54.

Nakamuna, M., 1985. Evolution of Artificial Reef Concepts in Japan. *Bulletin of Marine Science.* Vol. 37(1):271-278.

Ray, S., Green, D., Wood, and Wright, R. (1996). Putting Nature in its Place. *Mapping Awareness.* Vol. 10(9):20-23.

Ray, S., Green, D.R., Wood, M., and Wright, R. (1996). Getting the Data for the Job: Planning for a Coastal Marine GIS. *Proceedings* if GIS Conference Brno. GIS Frontiers in Business and Science, Part 2, Brno, Czech Republic, 8pp.

Ray, S., Green, D.R., Wood, M., and Wright, R. (1996). Establishing an Environmental GIS for Siting an Artificial Reef. *Paper* presented at a One-Day Seminar of the Association for Geographic Information Marine and Coastal Zone Management GIS Special Interest Group, University Manchester, July 1996. 12p.

Green, D.R., Ray, S., Wood, M., and Wright, R. (1996). Development of a GIS of the Moray Firth and its Application to Environmental Management. *Final Report.* Aberdeen University.

Raal, P.A., Burns, M.E.R., and Davids, H. (1995). Beyond GIS: Decision support for coastal development, a South African example. In R.A. Furness (ed.), CoastGIS'95: *Proceedings* of the International Symposium on GIS and Computer Mapping for Coastal Zone Management. International Geographical Union Commission on Coastal Systems, Sydney. pp. 273-282.

Riddell, K.J. (1992). Geographical Information and GIS: Keys to Coastal Zone Management. *GIS Europe.* Vol. 1(15):22-25.

Scottish Natural Heritage (SNH) (1994). Focus on Firths. *Leaflet,* Edinburgh.

Watts, P.A.G. (1997). Plans for the Coastal Zone. In *Proceedings* of CoastGIS'97. 29th-31st August 1997.

Additional Reading on Artificial Reefs

http://water.dnr.state.sc.us/marine/pub/seascience/artreef.html
http://www.ncfisheries.net/reefs/
http://bigshipwrecks.com/
http://www.fosusa.org/environ/reef1.htm
http://www.pir.sa.gov.au/pages/fisheries/rec_fishing/rec100.htm:sectID=266&tempID=10
http://www.asrltd.co.nz/reefs.html

CHAPTER 10

Collating the Past for Assessing the Future: Analysis of the Subtidal and Intertidal Data Records Within GIS

C.I.S. Pater

ABSTRACT: This chapter considers the potential for integrating disparate data sources for the analysis of temporal change in the near-shore environment of the Wash embayment. The two principle sources of subtidal and intertidal data are the charts published by the UK Hydrographic Office and the surveys of the Port of Boston. From this data, a qualitative assessment has been conducted of modelled and charted topography. Attention is given to the time scale over which charts may be used and of the quality of the data contained. From this recommendations are given to the way in which both the chart resource and other surveying schemes may be used effectively within GIS and in turn, Shoreline Management Plans.

Introduction

The Wash, opening to the North Sea, is the largest embayment in England covering an area of 66,500 hectares (Fig. 1). The embayment, a location of net sediment accretion and experiences a macrotidal environment with a mean spring tidal range of 7.1 metres. The subtidal morphology of the Wash is complex with a multitude of relatively stable channels running between extensive sand banks, generated from the reworking of Holocene deposits. The maximum depth of the Wash (47 metres) is located at the mouth of the embayment and represents the flooding of a lowland clay vale by eustatic sea level change. The intertidal area covers 29,770 hectares and contains a mixture of highly dynamic, soft sediment environments including mudflats, sand bars and the largest area of saltmarsh (4,230 hectares), in the UK (Davidson, 1995). Saltmarsh is a habitat with a very significant ecology in terms of seasonal use by migratory and resident bird species, as a result statutory and non-statutory conservation designations that are applied here have local, national and international importance.

The enclosure and draining of the upper saltmarsh has produced a soil medium that is highly fertile and following centuries of landclaim by the construction of embankments, a protective barrier around the Wash from Gibraltar Point (Lincolnshire) to Wolferton Creek (Norfolk) has been created. To landward of these defences the terrestrial environment is one of high yield, intensive agriculture including the largest area of grade one agricultural land in the UK. However, land drainage has led to peat shrinkage with the relative height of the farmland now three metres below the level of the mean high water spring tide (WEMP, 1996). Storm driven tidal inundation therefore presents a serious risk in this area and maintenance of the flood defences is paramount.

Concerning other commercial activities in the Wash, the main ports of the embayment Boston and King's Lynn, handle between them four million tons of trade annually (WEMP, 1996). It is a result of this freighter traffic through the Wash that the pilots frequently conduct surveys in the approach routes to Boston and King's Lynn. This information is in turn supplied to the Hydrographic Office for use in chart updates. These often surveyed locations are referred to on the charts as the 'Approaches to Boston' and the 'Approaches to King's Lynn'.

Research Objectives

The primary objective of this research is to demonstrate the discrepancy between Digital Terrain Models produced within a Geographical Information System (GIS) and the representation of topography on a conventional Hydrographic Office chart. Bathymetric data sets have also been procured from the Port of Boston from which the resultant DTMs demonstrate the high level of resolution possible in a qualitative investigation of morphological change. Other issues for examination include:

- **compatibility between the available data sets**
- **spatial and temporal extent of the available data sets**
- **suitability of analysis within a GIS**
- **the value of GIS to Shoreline Management Plans**

The primary bathymetric data source for this exercise has been the 1993 edition of the Hydrographic Office (HO) chart 1200 (*The Wash Ports*). To demonstrate the variable level of subtidal and intertidal data intensity across the embayment, soundings from June 1994 and December 1996 have been supplied by the Hydrographer of the Port of Boston. This data represents part of the Port's routine survey work conducted in the 'Approaches to Boston'; particular interest concerns the growth of channels, deeper than Chart Datum, to the north east of the Welland Cut (at the centre of Fig. 2). Access to the lower shore area in the Wash is difficult and the Environment Agency intertidal surveying strategy is reviewed. Attention is directed at the practicalities of linking analysis of the intertidal area with littoral processes in the immediate subtidal zone.

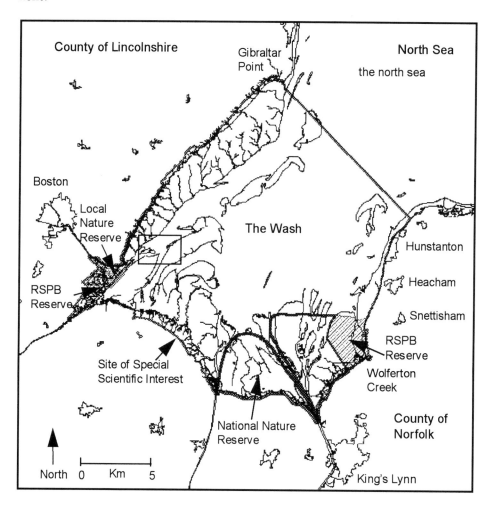

Fig. 1. The Wash Embayment. The boxed area in the western Wash represents the location referred to as the 'Approaches to Boston' (Source: Lee, *pers com* and Paterson, *pers com*)

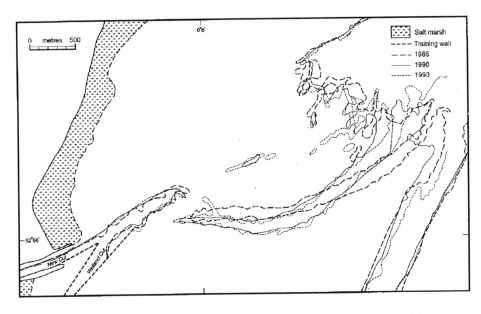

Fig. 2. The 'Approaches to Boston'. Chart Datum from chart 1200 (editions 1985, 1990 and 1993)

The perspective adopted in this study is for a *sea to land* orientation. Research divisions are based on the topographic 'features' of the coastal environment found in the Wash (Table 1) and at the local scale (e.g. an embayment), direct consideration is possible of the relative height levels between the embanked land and the near-shore submarine channels. This permits the adoption of strategies concerning flood defence based on the current physical state of the near-shore and shore area.

TABLE 1.
The topography of the Wash

Topographic Feature	Description
1) Terrestrial environment	The area to landward of the flood defences.
2) The intertidal area	The area between Chart Datum and Mean High Water Springs.
3) The subtidal area	The area below Chart Datum.
4) The embankments	Flood defence.

Shoreline Management Plans

The analysis of both the tidal and the subtidal environment is a central requirement for the production of non statutory Shoreline Management Plans (Ministry of Agriculture Fisheries and Food, 1995). Shoreline Management Plans (SMPs) have the responsibility of promoting the most appropriate means of coastal defence and flood protection in England and Wales within identified shoreline units. These units are based on the concept of littoral cells (Motyka and Brampton, 1993) and they represent a first attempt to classify processes by natural not administrative region. For example, the Wash marks the divide between two littoral cells and within the embayment there is a sub-cell boundary near Snettisham. The sub-cell boundary is defined by the way open shingle beaches are replaced to the south, at Wolferton Creek, by mudflats and saltmarsh (NNSMP, 1996). A key tenet of SMPs is that computational processing should exploit existing technology such as GIS and that the maximum use should be made of existing data sets rather than in the commissioning of new survey work. This research, through the use of a generic GIS, explores the potential of this recommendation to ascertain if this advice is appropriate.

The Selection of Datum

Given the research emphasis of the sea to land perspective, Chart Datum was selected in order to produce a surface with reference to the tidal range not to mean sea level. Chart Datum or the Lowest Astronomical Tide (LAT) is used on HO charts and represents the level below which the tide will not fall under prevalent barometric conditions. This provides an indication of the minimum level of tidal water that can be expected at any one location. However, in coastal and shoreline research, sounding data is frequently converted from a level relative to Chart Datum, to the Ordnance Datum Newlyn (a calculation of mean sea level). By using Ordnance Datum (OD) direct comparisons can be made from a national perspective relieving the difficulties of variables introduced by local tidal regimes. However, it is because of these local variations that at a local scale the use of Chart Datum would seem appropriate. Table 2 provides an indication of the differences in height exhibited between the two datums.

TABLE 2.
Comparison between heights relative to LAT and Ordnance Datum (OD) at Tabs Head. All measurements are in metres. The figures in brackets represent the maximum tidal range in April 1997 (source: HO chart 1200, 1993 edition, the HO Tidal Prediction System and Environment Agency profiles L3A1-L3A3, 1993 to 1996).

Datum	Diff. LAT and OD	MLW Spring	MLW Neap	MLW Neap	MHW Springs	Seaward base embank. (ml)	Top embank. (ml)	Landward base embank. (ml)
LAT	+ 3.7	0.7 (0.1)	2.4	5.6	7.5 (7.89)	7.4	10.1	6.8
OD	- 3.7			1.9	3.8	3.7	6.4	3.1

The adoption of a tidal datum corresponds with the objective of producing a local (embayment scale) spatial model. This datum allows information and data pertaining to the maximum possible extent of the tidal area to be included, thereby allowing consideration of the processes that are inherently responsible for the topography contiguous between the highest and lowest levels of the tide. The charting resource of the HO also represents the only cartographic record of the inshore submerged channels. Within the Wash embayment, qualitative estimation of morphological change using charts was first conducted by Hydraulics Research (HR Wallingford, 1953). Charts produced between 1828 and 1918 were redrawn with a common geo-referencing system that allowed direct comparisons between editions. From this analysis, estimates of change were calculated to ascertain locations of stability or instability. GIS should represent a modern means of continuing this type of analysis.

Digital Representation of Subtidal and Intertidal Topography

During the course of the research both published charts and unpublished collector charts have been digitised spanning the period 1924 to 1997. The data capture was performed using Lites2 software, a component of the Laser-Scan Automated Map Production System (LAMPS). A complete vector coverage was recorded including all spot depths and isolines with attributable values. The data was subsequently used in the production of Digital Terrain Models (DTMs). The purpose of producing DTMs was two-fold, to provide a visualisation of an environment that would not otherwise be readily accessible and for volumetric analysis between sequential chart editions. The DTMs were produced within ArcInfo and preliminary terrain analysis of elevation and declination was conducted through the use of Triangulated Irregular Networks (TINs). The production, by Delaunay triangulation of a 2.5 dimensional or functional representation of the terrain was obtained and viewed first by linear interpretation and secondly via bivariate quintic interpretation (Fig. 3). Linear interpretation provides an artificial angular view of the surface, whereas quintic interpretation has an emollient effect producing a smooth surface, indicating a closer proximity with the actual topography (ESRI, 1991). However, subsequent volumetric analysis of the surface can produce different statistics depending on the surface interpolation adopted. Vertical exaggeration is employed to lend emphasis to the undulation of the intertidal and subtidal topography, an essential function in an area of low relief.

Boston Port Authority Surveys

The Boston Port hydrographic surveys are conducted on an ad hoc basis in two locations in the western Wash, namely the 'Approaches to Boston' (Area 1, Fig. 3) and the Freeman Channel (Area 2, Fig. 3). A good example of the visualisation provided by the DTM can be seen in Fig. 3, Area 1 where a kidney shaped bar, located to the north east of the Welland Cut, can be clearly identified. The routine survey work by the Port of Boston, at this location, is concerned with charting the navigational status of channels in relation to this bar. Surveys in June 1994 and December 1996, show the main approach channel currently runs down the north west side of this bar (Fig. 4 and 5). However, sedimentation is so rapid in this area that on almost a monthly basis, the approach can switch from one side to the other (Davies, Port of Boston, *pers com*).

The sounding location points are produced through the use of a Differential Global Positioning System. The Differential Global Positioning System (DGPS) datum uses the World Geodetic System 1984, which differs from the Ordnance Survey of Great Britain 1936 datum used in plotting HO charts. Therefore, positions of soundings obtained using GPS require a conversion in order to align with the chart. The error expected in the conversion process is in the order of two metres in the horizontal plane. However, given the degree of uncertainty in the original plotting of a transient feature such as an intertidal bar, this estimate would seem acceptable. The precise location and plotting of an amorphous soft sediment bar means that the increasing demand for absolute accuracy has to be balanced against the function of accretion or deposition that can be expected per tide, per season and per year.

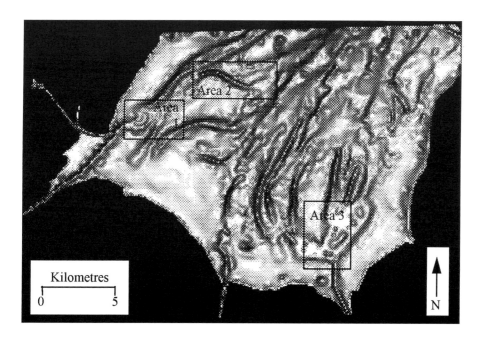

Fig. 3. Bivariate Quintic Digital Terrain Model for the Wash (from chart 1200, 1993 edition). Area 1 encloses part of the *Approaches to Boston*, Area 2 the Freeman Channel and Area 3 the navigation approaches to King's Lynn.

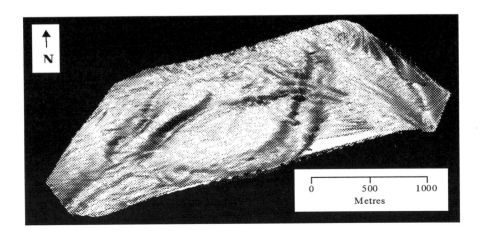

Fig. 4. Bivariate Quintic TIN Digital Terrain Model of the 'Barjune' survey in June 1994. The area of coverage corresponds approximately with Area 1 in Fig. 3. Source: Port of Boston.

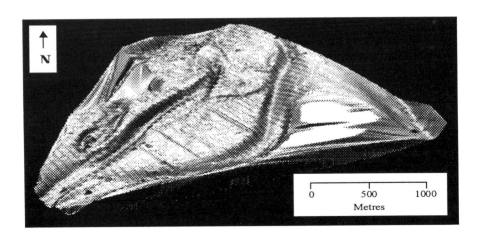

Fig. 5. Bivariate Quintic TIN Digital Terrain Model for the 'TOP612' Survey in December 1996. The area of coverage corresponds approximately with Area 1 in Fig. 3. Source: Port of Boston.

Environment Agency Intertidal Levelling

In regard to other means of determining rates of accretion and erosion, the Environment Agency of England and Wales has conducted biannual surveys around the periphery of the Wash since 1993. The transects, using ground surveying equipment, are at one kilometre intervals and are run from the landward flank of an embankment, across the intertidal area for an average distance of 1270 metres. Additional information on the survey transect includes a description of the ground at each levelling point. For the intertidal area in proximity to the 'Approaches to Boston' three sets of transect data are available (summer 1993, winter 1994 and summer 1994). Ordnance Survey Great Britain 1936 datum is used for the co-ordinate locations and is therefore compatible with Ordnance Survey maps and HO charts. There are however, several issues concerning Environment Agency levelling that will be examination further. For example the optimum distance between transects, the reliability of benchmarks, seasonal variability between surveys and quality assurance for contract survey work. The debate will also be continued between the relative merits of Ordnance Datum and Chart Datum in the maritime environment.

Remote Sensing and Aerial Imagery

The use of remote sensing presents a convenient means to obtain intertidal information, which through the use of ground data can be used to produce a DTM that can be updated rapidly. Synthetic Aperture Radar, which operates independent of meteorological conditions and the sun, provides a viable means of producing topographic details from a satellite platform. Intertidal DTMs in the Wash have recently been produced using this technology (Mason et al., 1995 and Mason, 1997). Aerial photography sorties are also flown over the intertidal zone each year by the Environment Agency as part of a monitoring programme. The survey height accuracy in the intertidal environment needs to be balanced against the resolution that is feasible in the subtidal environment. For nearly the entire intertidal zone, high resolution calibration is possible with the Environment Agency bi-annual levelling data, but this is not available as a matter of course for the subtidal zone. The Port of Boston provides high resolution soundings for particular locations on a seasonal basis, but a pan-embayment subtidal survey is scheduled by the Environment Agency on a five year cycle. Analysis must allow for variable vertical accuracy, variable rates of morphological change and variable survey dates.

Critique of Methodology

There are several stages at which error may be introduced to GIS analysis. These can be highlighted as error into the system, error in the system and errors in interpretation of the GIS output. In the first instant, digitising by hand requires quality control and patience. The interpretation of the chart by the digitiser is also subject to the original interpretation of the data by the hydrographer. In the second error bracket, consideration is needed of the processing technique and interpolation between data points. Finally, training and experience is required to enable accurate interpretation of the models produced from the original data.

For the long term perspective, between 1924 and 1985 datums, projections, scales and area of coverage all vary for the different editions of published charts. Although it may be possible to conduct qualitative estimation over such a temporal scale, the relative merits of this approach must be considered in relation to the quality of the original data sources. Sims *et al* (1995) provides a useful illustration of the complexity of cross-referencing historical cartographic records.

Serious consideration needs to be given to the relative value of this record to computer based modelling, with particular reference to the plotting of the gradient between the intertidal and subtidal areas. With a minimal gradient, contours are inappropriate in the intertidal area and would not, in any case, be included as a prime requisite. However, the level of LAT is an important isoline and problems of DTM construction are encountered if there is minimal data content within the isoline to provide sufficient z value (node) points for a dense pattern of triangulation, the result is a 'flat triangle' error. Flat triangulation is the process in which a planar triangle surface is produced by the presence of three identical data values at digitised node points. For a digitised chart this would occur between isolines of the same depth that contain no additional spot depths. This does not provide a correct reflection of the actual surface topography and will affect the value of visualisation and volumetric analysis, hence the use of linear interpolation to expose these errors which are particularly acute near the LAT isoline. In order to prevent flat triangulation, the density of samples must be very high to provide adequate detail of the surface. However, because of the homogeneous nature of the profile, limited data may be included on the published chart. In terms of visual analysis of a paper chart, only a few spot depths or drying heights across a consistent surface in a non-navigational area provides adequate information.

The provision of data points on charts that can be used as nodes for TIN generation are therefore insufficient for the generation of viable, functional DTMs. The only viable DTMs that can be produced are based on collector charts. This therefore restricts the area of detailed examination from the entire embayment to locations that have a consistent collector chart survey record, namely the approach channels to the ports of the Wash. For example, the functional surface produced from chart 1200 (1993 edition) fails to produce a 'realistic terrain' for the approaches to King's Lynn (see Area 3 in Fig. 3). The scarcity of soundings included on the chart for this location, at its worst, represents one data point with a radius of 900 metres. When the Chart Datum contour is overlain on the DTM (Fig. 6), a clear difference can be identified between the grey shaded terrain and the white flat areas. Dark spots represent isolated peaks within intertidal polygons that should have produced triangulation to form a complete surface both within and without of the intertidal polygons.

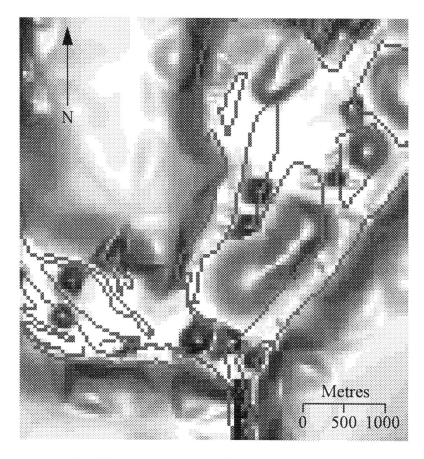

Fig. 6. 'Approach to King's Lynn'. Quintic 2D hillshade with Chart Datum softbreak line overlay (Source: chart 1200, 1993 edition).

Historical and Modern Inconsistencies

From an historical perspective, the system of using local datums for each chart was abandoned in 1968, with the adoption of the LAT national datum and the metrification of survey data. The system of chart projection has also changed from Gnomonic to Transverse Mercator utilising Ordnance Survey Great Britain 1936 datum. On all charts, the relative levels of the neap and spring tides are recorded. Within the Wash the mean high water spring tide varies by 0.5 metres between Tabs Head (7.5 metres in the 'Approaches to Boston') and West Stones (7.0 metres in the 'Approaches to King's Lynn'). The level of Chart Datum to Ordnance Datum also varies around the embayment (3.7 metres below Ordnance Datum at Tabs Head to 3.2 metres below Ordnance Datum at Wisbech Cut), which does complicate the use of Chart Datum. In the absence of intertidal levelling data for the rest of the Wash, the Mean High Water

Spring tide level was adopted as the mean level for the seaward base of the embankments. The alternatives to this is to adopt one spring high tide level for the whole embayment or to apply a vector breakline with z values that vary along its length corresponding to the changing mean spring high tide level. For the 'Approaches to Boston' case study, the Tabs Head level was adopted with converted Environment Agency data providing levels for the embankment.

Concerning the historical chart record, analysis is possible between editions that were produced to a consistent methodology as exemplified by the HR Wallingford analysis in 1953. However a direct link with modern surveys is not possible. Temporal analysis is therefore only possible for the Wash since 1985 with the publication of chart number 1200. However, this edition does not provide a complete coverage of the embayment, nor does the adjacent chart (number 108). The only chart that does cover the whole of the Wash also includes the shoreline as far north as Flamborough Head and is produced at a scale of 1:150,000. Although this provides coverage of a regional scale littoral cell, the data content is minimal and therefore of no use in GIS-based modelling.

Before operational GIS can be implemented in the coastal zone a greater level of consensus, co-ordination and integration of coastal survey work above and below LAT is required. A means must be found for publishing local soundings within national standards that are of real use to near-shore geomorphology. A possible solution could be the assimilation of port survey and intertidal profile within a separate, inshore topographic chart publication to be produced at one scale (e.g. 1:25 000), for the entire UK coastline. This would produce the viable template for an effective GIS. The GIS could then be used to maintain a viable inshore chart record through the direct assimilation of data generated from concurrent monitoring.

Suitability of Analysis within a GIS

The potential contribution of GIS to the examination and monitoring of the interface between the sea and the land is considerable. This can take the form of a system constructed for the spatial analysis of one location (for example an embayment, a sub-cell or cell) or one feature at one location (see Table 1). However, the issue of the survey mosaic is worthy of consideration when evaluating the functionalism of GIS in relation to available data resources. A chart source data diagram on a published chart reveals how it is composed of a survey mosaic that varies both spatially and temporally (Fig. 7). Those areas that fall outside of a frequently surveyed zone will have bathymetric and intertidal data derived from surveys conducted in preceding decades. Given the dynamic nature of this embayment, using the published HO chart record may provide a reasonable level of detail for individual areas, but fail to provide a consistent record of topography over a wider area.

Collating the Past for Assessing the Future 145

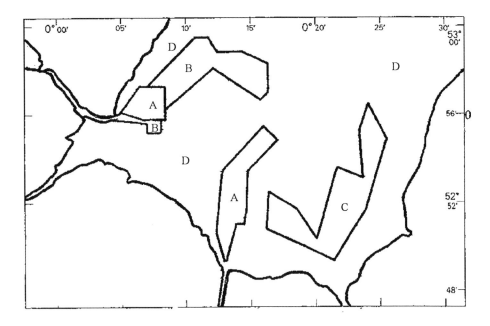

Fig. 7. Source data chart 1200 (1993 edition)

A) Port of Boston Survey   1990-91   1:5000-1:20 000
B) Port of Boston Survey   1989   1:5000
C) King's Lynn Conservancy
   Board Survey   1989   1:37 500
D) Other Surveys: Wimpey Laboratories Ltd   1971   1:25 000
                  Binnie and Partners   1973   1:10 000

For the purposes of this chapter direct quantitative analysis between sequential editions of chart 1200 are inappropriate given this mosaic of data. Until further processing has been conducted between the surveys, the juxtaposition of such disparate data resources inhibits anything more than informed analysis. The relative merits of this approach must be considered in relation to the quality of the original data sources on a HO chart. For example, 1991 subtidal soundings from the Port of Boston compared with intertidal data from 1971.

The Value of GIS to Shoreline Management Plans

An integrated mapping and charting project was conducted between the HO and the Ordnance Survey in 1992 (Bond, 1995), but subsequent development of this project was abandoned. Serious thought then needs to be given to how a spatial computer system spanning the sea and the land can be developed to national standards without the existence of a dedicated coastal spatial database. The justification of such an integrated cartographic resource was confirmed in the subsequent publication of coastal planning and management documents by the Department of the Environment (DoE, 1996) and the Ministry of Agriculture Fisheries and Food (MAFF, 1995). The ideal conditions were created for a review of the UK inshore charting resource, but by failing to invest in this field the effective use of GIS in the preparation and development of Shoreline Management Plans has been seriously hampered. This compounds other pre-existing concerns surrounding GIS such as the cost of hardware, software and data, the usability of an established system and concerning the principle of *ultra vires*, the jurisdiction of its use. The institutional and technical problems identified by Jones (1995) and Collier et al. (1995) remain current and provide an indication as to why in 1997, coastal and marine GIS in the UK seems to have remained at the project stage of development. For an example of how spatial analysis can progress to a working system, the Rijkswaterstaat in The Netherlands provides a demonstration of the effectiveness of an integrated cartographic resource, in terms of both hard copy and digital formats (Damoiseaux, 1995). However, the Dutch are at a considerable financial advantage for the implementation of such a charting record with a coastline only 350 km long, compared to the UK coastline of 15,000 km.

Conclusions

For collating the past from the chart record, the term 'past' must be interpreted as within the last two decades following the publication of chart editions to the same data standards. Concerning the terrain modelling of present conditions through spatial analysis, research is restricted to specific areas where a viable data resource exists. The Wash illustrates, at a local scale, how currently available intertidal and subtidal data (port soundings and Environment Agency profiles) allows quantitative estimation over the whole embayment, but quantitative analysis is possible only in routinely surveyed areas such as the 'Approaches to Boston'. The relevance of this small scale of analysis must also be considered in terms of the dynamics of that one location and in extrapolating conditions to the wider littoral sub-cell and cell. This also applies to the periodicity of the survey strategy and a 'use by' date caveat is needed for the relevance of the data. The composition of survey material within charts also requires a greater deal of co-ordination between concurrent intertidal and subtidal monitoring programmes. The stimulus for integrating current strategies could be the production of a new topographic digital chart to provide a vector template on which a fully functioning, operational level GIS could be constructed. Until this occurs, GIS represents an effective means to display the spatial and temporal survey mosaic of the UK coastal waters. This is only a part of the full contribution that this technology can make, but it demonstrates the increasing difficulty in collating the past and the present and provides a warning for future analysis.

Acknowledgements

The supply of charts (published and unpublished) has been kindly dealt with by the staff of the Hydrographic Office Data Centre, Taunton. The intertidal survey data has been kindly supplied by the staff of the Environmental Agency Flood Defence Department, Peterborough and the subtidal soundings by the hydrographer of the Port of Boston.

References

Bond, B. (1995) Coastal Zone Mapping in the Computer Age: The UK Experience. *Proceedings of CoastGIS '95, International Symposium on GIS and Computer Mapping for Coastal Zone Management*, Cork, Ireland, pp 375-378

Collier, P., Fontana D. and Pearson A. (1995) GIS Mapping of Langstone Harbour for an Integrated Ecological and Archaeological Study. *Proceedings of CoastGIS '95, International Symposium on GIS and Computer Mapping for Coastal Zone Management*, Cork, Ireland, pp 315-327

Damoiseaux, M.A. (1995) From Maps and Atlases to GIS for Coastal Zone Management. *Proceedings of CoastGIS '95, International Symposium on GIS and Computer Mapping for Coastal Zone Management*, Cork, Ireland, pp 165-178

Davidson, N.C. (1995) Chapter 4.1 Estuaries. *In Coasts and seas of the United Kingdom: Flamborough Head to Great Yarmouth.* ed. by J.H. Barne, C.F. Robson, S.S. Kaznowska, J.P. Doody and N.C. Davidson. 63-66. Peterborough, Joint Nature Conservancy Council.

DoE (1996) *Coastal Zone Management - Towards Best Practice.* Department of the Environment. October

ESRI (1991) *Surface Modelling in ArcInfo.* Environmental Systems Research institute. CST 807

HR Wallingford (1953) The Wash. *The Report of the Hydraulics Research Board.* pp 29-37

Jones, A.R. (1995) GIS in Coastal Management: A Progress Review. *Proceedings of CoastGIS '95, International Symposium on GIS and Computer Mapping for Coastal Zone Management*, Cork, Ireland, pp 165-178

MAFF (1995) *Shoreline Management Plans. A guide for coastal defence authorities.* Ministry of Agriculture Fisheries and Food May 1995, publication PB2197

Mason, D.C., Davenport, I.J., Robinson, G.J., Flather, R.A., and McCartney, B.S. (1995) Construction of an intertidal digital elevation model by the 'water-line' method. *Geophysical Research Letters.* Vol. 22, No. 23, pp 3187-3190.

Mason, D.C. (1997) INDUS: Intertidal DEMs Using Satellite Data. University of Reading, NERC Environmental Systems Science Centre. Internet Page (accessed 15/4/97): http://www.nerc-essc.ac.uk/Science/INDUS/INDUS.html

Motyka, J.M. and Brampton, A.H. (1993) *Coastal Management: Mapping of littoral cells.* HR Wallingford, Report SR 328.

NNSMP (1996) *North Norfolk Shoreline Management Plan.* Final draft May 1996, L.G. Mouchel & Partners Limited, West Hall, Parvis Road, West Byfleet, Surrey KT14 6EZ

Sims, P.C., Weaver, R.E. and Redfern, H.M. (1995) Assessing Coastline Change: A GIS Model for Dawlish Warren, Devon, UK. *Proceedings of CoastGIS '95, International Symposium on GIS and Computer Mapping for Coastal Zone Management*, Cork, Ireland, pp 285-301.

Wash Estuary Management Plan (1996) *Wash Strategy Group*, c/o Lincolnshire County Council, Department of Environment Services, Lincoln.

CHAPTER 11

Identifying Sites for Flood Protection - A Case Study from the River Clyde

G. Jones

ABSTRACT: This chapter presents a methodology which utilizes GIS to provide a priority ranking for estuarine sites which require new or additional flood protection. The use of a powerful proprietry spreadsheet for the preparation of the database allows rapid updating of data and for the calculation of different scenarios for flood prediction.

Introduction

Global warming is a widely accepted phenomenon, (DoE, 1996; Wigley *et.al.*, 1992a; 1992b) although the consequences remain controversial. One area of uncertainty concerns the extent to which sea level will rise leading to an increase in the possibility of flooding of coastal areas (Robin, 1986; IPCC, 1990; Warrick and Rahman, 1992).

The occurrence of serious coastal flooding which resulted from severe winter storms, for example on February 26th 1990 at Towyn, North Wales, (Kay and Wilkinson, 1990), and in February, 1991 at Greenock in the Clyde Estuary, served to focus public attention on the possibility of an increase in damage from coastal flooding. It has also caused business and commercial interests to reassess the likely structural, transportation and financial implications of increased flooding (DoE, 1993). In particular, property insurance companies have substantially reassessed the financial liability of coastal and river flooding of low lying areas.

On present evidence it appears unlikely that the predicted rise in sea level around the British coastline due to global warming over the next 30 years would be capable of causing coastal flooding in anything other than specific and localised areas. The susceptible areas are, however, frequently the preferred location of modern day urbanised society in that the gently sloping land adjacent to estuaries provide ideal locations for our cities, industry, transport links and water-based recreation facilities. When expressed as a proportion of the total linear distance of the British coastline the area that is susceptible to coastal flooding due to sea level rise is very small.

The nature of the problem provides an opportunity for identifying the potential areas of impact of coastal flooding. The dynamic nature of the spatial problem combined with the multiplicity of environmental data involved demands a methodology which can manage the data while at the same time allows modelling of the problem to take place under a variety of scenarios.

Main Reasons for Attempting to Predict Flood Risk

There are three reasons for predicting the areas most at risk from the possibility of future flooding:

1) **to allow the identification of the most vulnerable and/or most valuable sections of the coastline;**
2) **to allow for the preparation of a management protection policy which may involve substantial changes in land use over the next 30 years;**
3) **to determine by means of cost-benefit survey the sections of coastline that would justify the massive cost of strengthening the existing flood protection mechanisms.**

This chapter presents a methodology that can help provide information for the first and third points shown on the overhead. The methodology involves the compilation of a spreadsheet comprising minimum elevation values and land use types. Until the availability of the Ordnance Survey (OS) Digital Terrain Model (DTM) data in March 1997, the compilation of elevation data had taken a considerable of amount of time. The OS DTM tiles now available at a scale of 1:10,000 provide either contoured elevations or a 10 metre grid of elevation values. At the time of writing a suitable translator for the OS DTM tiles was not available and instead the methodology used in the chapter will be illustrated by means of the earlier manually-derived elevation values based on a 100 metre grid.

Predictions for Sea Level Change

Sea level has been predicted to rise world wide by about 18 cm by 2030 and, assuming current trends are maintained, by a further 90 cm by 2100 compared to present datum level (Wigley et al., 1992b). These figures are less than previously thought likely, but in combination with a potential increase in storminess (Portney, 1991) will be sufficient to cause serious periodic flooding in low lying areas (Jones, 1994). Stronger winds will result in larger waves which will cause greater erosion along estuary banks. Increased rainfall over the land area will cause a greater run-off of fresh water (Wigley et al,. 1992a). An increase in the discharge of fresh water into estuaries will result in

Identifying Sites for Flood Protection

greater volumes of water accumulating on the surface of estuaries, beneath which will lie the denser salt waters of the sea. The area of accumulation of water will fluctuate depending on the tidal state and on the prevailing weather conditions over the preceding 72 hours and on the ability of the river catchment area to delay the return of precipitation.

Computer Model of the Clyde Estuary

The estuary of the River Clyde displays many of the characteristic land use features associated with managed estuaries in an industrialised country. It is a small river both in terms of its length, 170 km, and its drainage basin, approximately 3836 sq. km (Fig. 1). Despite the industrialisation which occurred along its lower reaches from the early 1800s parts of the estuary still retain a considerable natural beauty. There are three designated Sites of Special Scientific Interest (SSSI), one designated ornithological site and numerous non-designated habitats of local importance. In recognition of its newly recognised role as a provider of recreation, tourism and conservation values, the estuary has been designated a special planning zone by the regional planning authority. A recurrent planning theme of the 1980s was the concept of the Clyde Corridor, a development plan that embraced the estuary and its immediate tributaries. In the 1990s, planning emphasis has shifted towards the construction of a so-called Eco-Plan for the estuary. The scope of the Eco-Plan is wider than its name suggests being involved not only with environmental considerations but also with transport, derelict land and sites with commercial development potential.

The preparation of a database formed the vital first stage of the project. Colour vertical air photographs flown in the summer of 1989 were used for the preparation of a land use base map. The map was digitised and the data stored as polygons with attribute data attached in a separate file. The MAPICS mapping package was used for the preparation of these files. Additional layers of data were digitised, for example, areas of conservation value as designated by Nature Conservancy Council (NCC) (Scotland), now Scottish Natural Heritage (SNH). A digital terrain model (DTM) was also constructed from a raster data set comprising a cell size of 100 metres x 100 metres.

The main purpose of the DTM was to provide the basis for modelling the extent of sea level change within the estuary using the IPCC predictions of 1990 and 1992.

The database for the estuary was compiled using a purpose-designed package called MICROMAPPER. The software permitted entry of data in a spreadsheet format and its subsequent display as a customised two-dimensional colour map. ASCII format files were produced for subsequent export to the UNIRAS package. The data set comprised an area including the estuary and adjacent land on the north and south banks and covered an area equivalent to 5.5 km on a north-south axis and 17.5 km on an east west axis. The lowest elevation point was recorded in each 100m x 100m grid cell. Information was collected from the Ordnance Survey Land Ranger map series, scale 1:50,000. The data base comprising 9625 points was used to construct both two-

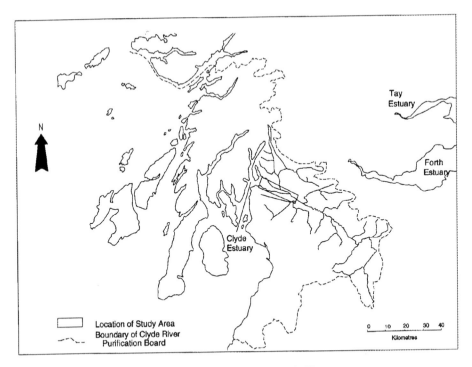

Fig. 1. Location of the River Clyde Estuary

and three-dimensional maps of current high water levels using the UNIMAP graphics package. Other maps were then produced displaying critical water levels above the current high water mark. The current level to which flood protection is provided in the Clyde estuary is 5.0 metres above mean high water level. Maps were produced for a variety of potential flood situations, the example shown in Fig. 2 shows the flood zone that would result from a water level 44 cm above current protection level, this situation represented a freak storm surge condition.

The resolution of detail contained in the initial maps provided a broadly satisfactory indication of areas of land adjacent to the estuary over which water could

Identifying Sites for Flood Protection

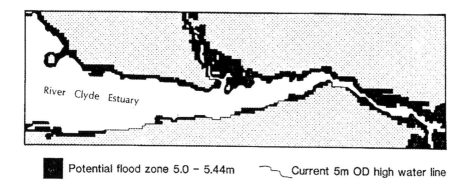

Fig. 2. Preliminary map showing areas flooded by a 44 cm tidal surge above existing flood protection level.

be expected to spill in times of flood. The maps were insufficiently detailed, however, for detailed land use planning purposes and it was recognised that a new vectorised data base was required. For this purpose, four separate sources of data were identified as providing suitable information:

- **Ordnance Survey 1:10,000 scale maps**
- **Admiralty Charts 1:63,360 scale maps**
- **Colour air photographs flown in 1989**
- **Field Survey**

The 1989 air photographs were used to establish the location of high water mark, the existence of mud banks, the location of the main river channels and areas of sedimentation in areas of slack water near areas of conservation value. The field survey provided an opportunity to verify information derived from maps, charts and air photographs. Due to the amount of data compiled for the second data base the study was restricted to an area that extended along the southern shoreline of the estuary for some 15 km.

Three separate data layers were compiled. The first comprised digitised contour lines extracted from the 1:10,000 scale maps for 5m, 10m, 15m, 20m, 25m, 40m and 50m and thereafter at 25m intervals,. The polygonal areas bounded by the contour boundaries were converted to raster data format using IDRISI. Spot height data for land based sites were extracted from the 1:10,000 maps and added to the contour data set. Where necessary, cell height values adjacent to spot height values were adjusted by means of an automated TIN conversion using UNIRAS, (Milne, 1992). Thirdly, water level heights were extracted from the Admiralty Charts and entered as irregular data points prior to UNIRAS being used to generate intermediate values, once again using an automated TIN calculation. The three data sets were collated to provide a final composite data set. Some minor modifications were necessary to incorporate information taken from the air photographs.

UNIMAP was used to produce maps of 1:10,000 scale showing the extent of flooding for situations in which sea level exceeded the current flood protection level of 5.0 metres above high water mark by 0.3m, 0.6m and 1.0m. The UNIMAP mapfiles were exported to IDRISI for conversion from raster to vector line format. The IDRISI vector lines were used as a base for plotting onto OS 1:10,000 maps to produce the final maps showing potential flood areas (Fig. 3).

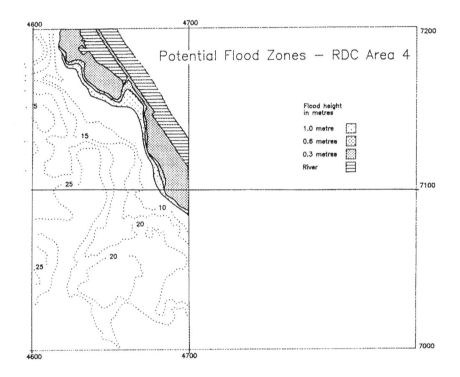

Fig. 3. Example of detailed plot of flood zones for sample area of Clyde Estuary

Interpreting the Results

Examination of the computer generated maps showed that a considerable area of land adjacent to the southern margin of the estuary would be affected by flooding by the year 2030. It must be stressed that the predicted flood areas are those that would result without any new flood protection schemes. The maps cannot provide an indication of the frequency of flooding. They can, however, provide an indication of where additional flood protection measures would result in a reduction to the flood threat or where land use planning policy could be used to relocate land use such that areas most likely to flood would have low value, forfeitable use. In an effort to avoid making value-based judgements from the flood maps an attempt was made to calculate a series of index values in which the flood potential was weighted according to the 'value' of

Identifying Sites for Flood Protection

the land and the depth to which flooding occurred. It was intended that the index values could be used to identify the critical areas where flood management schemes would be of greatest benefit. The matrix for calculating the index values was created using an EXCEL spreadsheet. The use of a spreadsheet allows for rapid and easy changes to be made in any of the parameters used to calculate the flood index values. This method avoids the problem which has plagued some earlier work in which the values used for calculating flood scenarios have been overtaken by new global warming projections.

Calculation of the flood index values required the allocation of weightings that were based on the temporary 'loss' of specific land use during periods of flooding. Five weighting were used. In each grid square the land use with the highest value was applied to the entire cell. The five categories are listed below:

1) land providing services such as main transport links, pipelines and utilities were considered of regional importance and were weighted with a maximum value of 3.0. Severance of any of these services would result in maximum disruption to a population throughout the region. The disbenefit due to flooding would be major and the consequential losses to society great
2) residential property, local transport links and agricultural land of grade 3.i and better were grouped together and weighted with a value of 2.0. This group contained land use types of considerable difference. Although flooding of residential property would bring maximum personal disruption and economic loss to the general public it has not been given the highest weighting because losses can be compensated by insurance cover. Even though personal tragedy levels can be high the loss to society has been judged to be less than that experienced by landuses contained in Category 1. Similarly, local transport was judged to cause minor inconvenience while flooding of the better quality agricultural land was given a weighting of 2
3) land used for industry and designated conservation zones received a weighting of 1.5. The industrial land category may justify further analysis. Industrial types along that part of the Clyde Estuary included in this study were of relatively low value, such as scrap metal yards, boat yards, storage areas, unloading wharves. Installations such as oil and chemical works or electrical, telecommunications or computer plants would justify a higher classification. Designated conservation zones include S.S.S.Is, National Trust property etc. and were included in this category because of their statutory definition
4) areas with no human habitation, no industry, non designated conservation areas, agricultural land lower than class 3.i and communication confined to tracks and paths received a weighting of 1.0 and their inundation was thus considered to have no lasting significance. This category would not normally receive priority in terms of flood protection
5) the final category differed from all the previous categories in that it represented land which could be used as a 'buffer area' and be deliberately allowed to flood. These areas would be sacrificed on either a temporary or

even a permanent basis provided their loss would generate an additional protection to other more valuable areas. In this category might also fall those areas which cannot be economically protected from flooding. This category was weighted at -0.5. Forfeited areas should not be viewed as valueless areas but contribute a valuable protective role and, in addition, contributing important new sites for coastal wet lands to replace those lost by inundation elsewhere in the estuary

Weightings have been allocated on the assumption that the provision of flood protection would be most cost effective if it alleviated the first phase of potential flooding (a flood in excess of 5.0 metres above mean high water spring tides for which flood protection exists). Data has been calculated for flood scenarios at 0.3, 0.6 and 1.0 metre above the critical figure of 5.0 metres above mean high water spring tides high tide. Weightings for 0.6 and 1.0 m scenarios have been allocated proportionately lower weighting values than that for the 0.3 m scenario on the assumption that money spent on raising the flood defence system to restrict the impact of the 0.3 m flood would buy additional time beyond 2030 in which further changes in land use could be assessed in the light of prevailing conditions. It is possible that during this additional time period the inter-governmental actions taken since the Montreal Protocol in 1987 and subsequent meetings would limit the release of greenhouse gases and lead to a reduction in the rate of sea level rise.

The severity of flooding in each grid square was calculated as

- a function of the length of flooded coastline in each grid square multiplied by the weighting factor provided in Table 1
- the area in hectares of flooded land in each grid square multiplied by the appropriate weighting factor

Analysis of Flood Scenario

The results obtained from the flood assessment are presented as Flood Loss Scores (F.L.S.) (see Fig. 4).

Map section 1   Grid squares 1a, 1b, 1c
A flood of 0.3 metres would extend over a narrow flood zone coincident with the raised beach running east-west throughout the area. Although the total amount of flooded land would be small, 152 ha, the potential for disruption is high. It is unlikely that domestic property would be flooded but ground water flooding could result in rising dampness in properties with inadequate damp courses. The egress of water from storm water drains would be prevented thus adding to the accumulation of surface water on roads. A rise in flood water to 0.6 metres would flood an additional 93 ha of the area, confined mainly to railway land. Disruption to road and rail traffic could occur. An additional narrow strip of land would be inundated by the 1.0 meter flood and could

Identifying Sites for Flood Protection

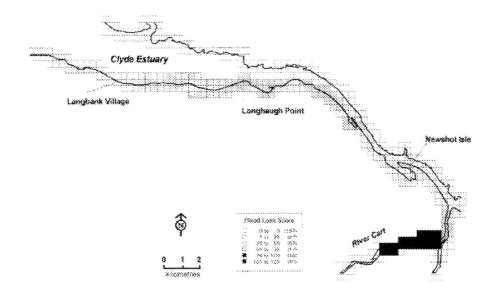

Fig. 4. Flood Loss Scores. Clyde Estuary - South bank

TABLE 1
Weightings used to Calculate a Flood Loss Score (F.L.S.)

	Height above current flood protection level (5.0m above high water mark)		
	0.3m	0.6m	1.0m
Residential development	2.0	1.5	1.0
Pipelines (Water, gas, oil, sewage) Sewage works, reservoirs	3.0	3.0	2.0
Industrial land	1.5	1.0	-1.0
Transport links (motorway, A-roads, main railway lines)	3.0	3.0	2.0
Transport links (all other links)	2.0	1.0	-0.5
Agricultural land (Land use classification Grades 1, 2, 3i)	2.0	1.0	-1.0
Agricultural land (Land Use Classification Grades 3ii, 4, 5)	1.0	-0.5	-2.0
All other land	-0.5	-1.0	-3.0
Designated conservation areas (S.S.S.I., Scottish Wildlife Trust, Scottish National Trust)	1.5	1.0	-1.0
Non-designated conservation areas	1.0	-.1.0	-2.0

cause flooding of domestic property in the village of Langbank. Unless remedial action had been taken at an earlier point in time, flooding of road and rail links would also occur. A section of agricultural land could be sacrificed to sea.

Map section 2   Grid squares 2a, 2b, 2c
A flood of 0.3 metres above current protection levels would result in flooding of the coastal margins of this section of the estuary and would result in substantial disruption to communication links thus bringing the problem of sea level rise to the direct attention of a large sector of the population using road and rail links. The motorway is particularly vulnerable to flooding over a 0.5 km section and currently has only minimal protection from inundation. In addition, the railway between Glasgow and Gourock follows the narrow coastal strip and disruption to rail traffic could be anticipated. The railway connects Glasgow to termini both at Gourock and Wemyss Bay, each of which serve ferry terminals operating to outlying areas. Disruption of transport links would thus have repercussions for travellers over a wide area. In the eastern section of this grid square an earth embankment has already been constructed by the landowner in an attempt to protect agricultural land. This embankment provides protection to a flood height of 5.0 metres above mean high water spring tides. A water level some 30 centimetres above 5.0 metres would allow water to overtop the embankment, and if this occurred, would become trapped behind the bank thus illustrating a specific problem of coastal flood protection. If the protective structure becomes overtopped by flood water, the protection mechanism may then become on obstacle to the easy and rapid removal of flood water. Provision of sluice gates and pumps would be required to drain the land.

A rise in flood level to 0.6 metres would cause inundation to agricultural land and a golf course. Further flooding to 1.0 metre would endanger a short section of class A-road and a substantial area of grade 3.i agricultural land. Based on the criteria used in Table 1 the least cost option would be to forfeit the agricultural land to the sea.

Map section 3   Grid squares 3a, 3b, 3c
A flood of 0.3 metre would inundate an area of land of varying width, including a section of the Erskine Golf Course. Construction of a low concrete embankment has already been made, the purpose of which appears mainly to prevent erosion of land. It would provide minimal defence against flooding. A potentially serious problem could occur at grid reference 4605 7230 where the Erskine Road Bridge crosses the estuary. The rise in sea level could result in inundation of the foundations to the bridge supports, leading to long term damage to the concrete foundations. Up stream of the road bridge, Longhaugh Point and its associated embayment is currently a highly valued bird habitat. A rise in sea level of 0.3 metres would result in changes to the feeding and roosting areas.

The impact of 0.6 and 1.0 metre floods would be to inundate an additional area of valuable fairway and greens of the Erskine Golf Course. Areas of agricultural land could be lost to semi-permanent flooding.

Identifying Sites for Flood Protection

Map section 4   Grid square 4a
A 0.3 metre flood would create losses that were primarily confined to sites of ecological value. The Erskine salting would be permanently waterlogged thus eliminating the current range of plant and animal communities. Two modern commercial development, a large hotel and the Scottish headquarters of the Automobile Association are located on a former river terrace and would suffer some preliminary flood related problems especially an increased possibility of dampness in basements, resulting in increased maintenance costs.

The 0.6 metre flood scenario produces little additional flood problem over the 0.3 metre situation. Dampness to foundations of buildings is likely to increase. Flooding of 1.0 metre would cause problems for both the hotel and the AA headquarters mentioned above. Access roads would be breached. Forfeiting of land becomes a likely option.

Map section 5   Grid squares 5a, 5b
A 0.3 metre flood would cause considerable loss of agricultural land. In addition the loss of two areas of high conservation value would occur - Newshot Island and Newshot inlet. The latter is a salting of considerable ecological value and exhibits a large variety of plant and animal communities. Loss to property would be minimal, although it is possible that a combination of a strong westerly wind and high tide could result in water being driven further inland towards new private residential properties.

The 0.6 and 1.0 metre flood scenario would produce only a small increase in the total amount of flooded land. However, this would extend to the private residential property mentioned above. Some agricultural land could be forfeited to the sea.

Map section 6   Grid squares 6a, 6b, 6c, 6d
A flood of 0.3 m. would result in considerable loss of class 3.i agricultural land and would also threaten a modern sewage treatment site and a length of class A-road. , A part of the Renfrew Golf Course would be jeopardised alongside the east bank of the River White Cart

A progressive increase in flood level to 0.6 and 1.0 metres results in loss of grade 2 agricultural land and an area of horticultural land. Substantial land forfeiting would be necessary.

Conclusion

Application of the methodology described in this chapter has shown that the 0.3 metre flood scenario has shown a substantial proportion of land adjacent to the Clyde estuary as vulnerable to flooding. The direct cost of providing adequate flood protection for the 0.3 metre flood along the entire 15 km section of estuary studied in this chapter would amount to approximately £30 million at 1989 prices. It is suggested that this figure is an under-estimate of the real cost as it takes no account of the inland disruption caused by flooding. By calculating the landuse areas as opposed to linear distance that would be affected by flooding and weighting these figures by the Flood Loss Scores, a nominal disruption figure of £57 million was produced (see Table 4).

The calculation of F.L.S. scores directs attention to the sections of the estuary where flood losses are greatest in terms of extent and financial value. The financial costs of providing improved sea defences along the entire coastline will be prohibitive and accordingly, the previously assumed solution of 'sea-defence-at-all-costs' may no longer be possible in the future. The scenarios made for this chapter have been based on the assumption that the protection of main lines of communication (motorways, A roads and main railway lines) and services essential to the well being of the community (such as sewage plant, electricity generating stations, water pumping stations) will justify protection at all costs as the social disruption caused by their interruption would be too great for society to withstand. In addition, it may be necessary to protect some residential areas although when judged on a thirty year time scale an alternative option may be a phased withdrawal from flood prone areas with owner occupiers receiving compensation from a government-funded 'Flood Defence Fund'.

Even without the threat of sea level rise, Britain would face an expensive reconstruction of its existing sea defences. The life of a sea defence system is estimated at 50 years and many flood protection schemes around the coast of the UK. are now approaching this age, having been strengthened after the 1953 storm surge. Many structures will require major renewal by the first decade of the next century. The cost of this exercise will be immense. The method described in this chapter allows prioritising the most vulnerable key stretches of coastal and estuarine land and indicates where financial investment will be provide a best return. The level of statistical certainty on the likely occurrence of a 0.3 metre sea level rise remains impossible to calculate although the meeting of the Climate Convention in Berlin in April, 1995 provided the strongest scientific support for sea level rise.

The amount of sea level rise beyond the year 2030 will be largely determined by the social response to the threat of global warming in the period up to 2030. Whereas considerable success has been achieved in the reduction of CFCs to the atmosphere the same is not true for the major greenhouse gas, CO_2.

References

Bird, E.C.F. (1993). *Submerging Coasts.* John Wiley and Sons, Chichester.
Chapman, V.J. (1977). Wet Coastal Ecosystems. Chapter 1 in *Ecosystems of the World.* Elsevier Scientific Publishing Co., Amsterdam.
Colophon (1990). *The Delta Project. Preserving the Environment and Securing Zeeland against Flooding.* Ministry of Transport and Public Works. Den Haag, Netherlands
Department of the Environment (1991). *The Potential Effects of Climatic Change in the UK.* UK. Climatic Change Impacts Review Group, H.M.S.O., London
Department of the Environment (1993). *Coastal Planning and Management: A Review.* H.M.S.O., London

Department of the Environment (1996). *Review of the Potential Effects of Climatic Change in the United Kingdom.* H.M.S.O., London.

Gibson, I. (1978). Numbers of Birds on the Clyde Estuary. Chapter 1 *in Nature Conservation Interests on the Clyde Estuary.* Nature Conservation Council, Balloch.

IPCC (1990). *Climatic Change. The IPCC Scientific Assessment.* Report prepared for IPCC by Working Group 1. Edited by J.T.Houghton, G.J.Jenkins and J.J.Ephraums. C.U.P. Cambridge.

Jones, G.E. (1994). Global Warming, Sea Level Change and the Impact on Estuaries. *Marine Pollution Bulletin.* Vol. 28(1), pp. 7-14

Kay, R., and Wilkinson, A. (1990). Lessons from the Towyn Flood. *The Planner, Journal of the Royal Town Planning Institute.* Vol. 76(32), pp.10-12.

Milne, P.H. (1992). Digital Ground Modelling. *Mapping Awareness and GIS Europe.* Vol. 6(3), pp. 33-36

Portney, P.R. (1991). Assessing and Managing the Risks of Climatic Change. pp. 83 - 90, In N.J.Rosenberg, W.E. Easterling, P.R.Crosson and J.Darminster (Eds*.) Greenhouse Warming: Abatement and Adaptation.* Resources for the Future. Washington D.C.

Robin, G. de Q. (1986). Changing the Sea Level. Chapter 7 in *The Greenhouse Effect, Climatic Change and Ecosystems,* Eds. Bolin, B. *et.al.* SCOPE 29 John Wiley and Sons, Chichester.

Stevenson, J.C., Wood, L.G., and Kearney, M.S. (1986). Vertical accretion in marshes with varying rates of sea level rise. In *Estuarine Variability,* ed. D.A.Wolfe Academic Press, N.Y.

UK CCIRG (1991). *The Potential Effects of Climate Change in the United Kingdom.* Department of the Environment. H.M.S.O. London.

Warrick, R.A., and Raham, A.A. (1992). Future Sea Level Rise: Environmental and Socio-Political Considerations. Chapter 7 in *Confronting Climatic Change,* Ed. Mintzer, I.M. Cambridge University Press, Cambridge.

Wigley, T.M.L, Pearman, G.I., and Kelly, P.M. (1992a). Indices and Indicators of Climatic Change: Issues of Detection, Validation and Climatic Sensitivity. Chapter 6 in *Confronting Climatic Change,* Ed. I.M. Mintzer, Cambridge University Press, Cambridge.

Wigley, T.M.L, and Raper, S.C.B. (1992b). Implications for climate and sea level of revised IPCC Emission scenarios. *Nature,* Vol. 357, pp. 293-300.

CHAPTER 12

Arctic Coastal and Marine Environmental Monitoring

H. Goodwin and R. Palerud

ABSTRACT: Akvaplan-niva has participated in Norwegian, and international marine field cruises from the North Sea to the Kara Sea, north west Russia, since 1989. During this time Akvaplan-niva have collected a large amount of environmental data. An environmental database has been designed and built to store this data, which is now linked directly to GIS software ArcView, for retrieval and display. ArcView has been customised for Environmental Impact Assessment studies based on the Pechora Sea. Specific algorithms have been developed for computing risk assessment in the event of environmental pollution and from a variety of human activities.

Background

Since 1992 Akvaplan-niva has participated in a number of joint international scientific cruises to the Barents Sea, Svalbard, and the North West Russian Arctic (Fig. 1). The cruises to the NW Russian Arctic have been in collaboration with Norwegian and Russian partners such as the Norwegian Polar Institute and Murmansk Marine Biological Institute (MMBI). Samples collected include surface sediments, core sediments, fish samples, and water column profiles. The samples were later analysed for various contaminants including a large range of heavy metals, PCBs, and PAHs. These efforts resulted in the collection of a huge amount of data describing the environmental state of parts of the Arctic that had previously been unavailable to western scientists.

It was clear that a well structured environmental database, and geographical information system should be designed and built not only to store these data, but also accommodate the possibility for spatial analysis of the data, and to be readily available for risk assessments and environmental impact assessments in the future.

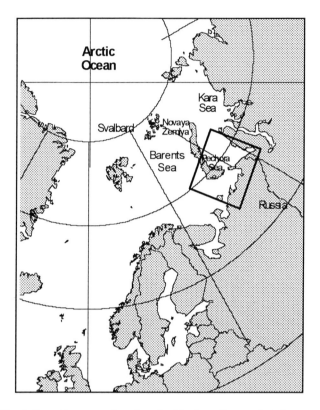

Fig. 1. Location map showing the Pechora Sea and study area.

With the end of the Cold War and the relaxing of travel and business restrictions in the former Soviet Union, possibilities for business and industrial developments for Western and Russian partners have opened up. Of significant interest to the west are the huge reserves of minerals, such as gold and diamonds, and of course the potential for offshore and on-shore oil exploitation. One such area is the Pechora Sea and surrounding mainland administered by the Nenets Autonomous Okrug and the Komi Republic. Considerable development by Russian oil companies has already taken place here, with little regard to the environment, and several international oil companies have already indicated their interest here in the past few years. The Arctic is a fragile and very sensitive environment, as we saw in Alaska after the Exxon Valdez oil spill disaster (Exxon Valdez Oil Spill Symposium, 1993), where large pristine areas were destroyed by crude oil. It is therefore of the utmost importance that we have an industrial strategy defined, and the necessary tools to predict and monitor the effects of oil exploration and exploitation in this remote and fragile environment.

Data Collection

As in any GIS the success of the system depends on the quality and on there being a wide and comprehensive range of environmental data contained in the system. A number of Russian institutes have large amounts of data from these Arctic areas, describing bird, mammal, fish and benthic communities. It is therefore essential that good co-operation between western and Russian partners is maintained. It is also important to note that all data entering the database undergoes rigorous quality assurance and quality control. An environmental atlas has been compiled by INSROP (International Northern Sea Route Program) (Bakken et al., 1995), of which Akvaplan-niva played an important role in data collection, describing the part of the environment along the Northern Sea Route from north-west Russia to Japan. This atlas could provide a comprehensive database for our particular study area. A similar program between Akvaplan-niva and Petrozavodsk State University is developing a digital environmental atlas of the White Sea.

For the data that we have in the system it has been considered important to include metadata describing the data source and quality, references and ownership etc, and where available and relevant, an image is included, showing for example the most dominant species of a benthic community (Fig. 2). By making use of ArcView's Hotlinks function it is possible for the user to retrieve this descriptive information about the data by simply clicking a given polygon in the map. This facility provides readily available information to the user about the suitability of a data set for a particular project or environmental assessment.

Base map data is crucial in any GIS so that we can simply present our spatial data and results from queries in a geographical environment. There is little in the way of digital maps available for this area of the Arctic. The digital base map chosen here was the US Defense Mapping Agency Digital Chart of the World 1:1,000,000 (DCW). There are also a number of British Admiralty Charts, and some Russian maps at various scales showing bathymetry with varying degrees of accuracy in paper form. Since the DCW is an immense collection of geographic and thematic data it was deemed necessary to develop Avenue programs in ArcView to make accessing and displaying areas and particular themes of interest simpler and more efficient. The programs include allowing the user to select from a small scale map the DCW tiles they are interested in displaying, and then select from a list which themes to display (e.g. Political and Oceans, Roads, Rivers etc), and whether to display points, lines or polygons. The selected themes are then added automatically to the active view and classified with predefined legends. This is the first step in developing a user-friendly environment which at this level requires little knowledge of ArcView in displaying maps for an area of interest.

Software Development

Akvaplan-niva has built an environmental database to store data collected from the Arctic on an MS SQL Server using MS Access and ArcView GIS as the interface. In addition several modules have been developed which demonstrate the potential of GIS in the role of environmental monitoring and risk analysis.

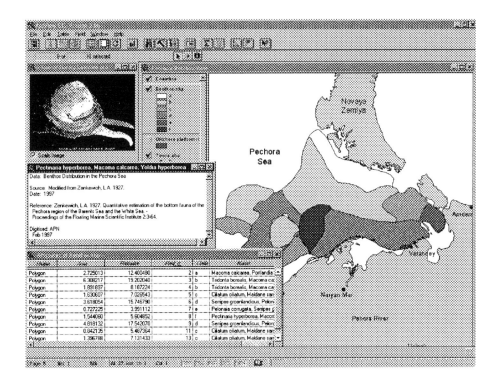

Fig. 2. Example of how ArcView can display resource data (in this case benthic communities) and retrieve not only the underlying attribute data, but also descriptive metadata, and imagery.

Data Input to the Environmental Database

The database was designed primarily to store the vast amounts of data Akvaplan-niva has collected each year from the North Sea environmental surveys around oil installations, and from scientific cruises in the Arctic. Other data includes that collected by scientists on fieldwork, both freshwater and marine samples. An interface built in MS Access provides a simple user interface for input of environmental data collected on cruises, and from fieldwork. The interface guides the user through a series of screens for inputting information about the sampling location, the type of sample taken (e.g. sediment or fish sample), sampling method, type of analysis performed, and the results of analysis returned from the laboratory (e.g. concentration of heavy metals in seabed surface sediments). Some standard queries have been written in MS Access to generate output for standard reports for different environmental surveys, and to retrieve the data and manipulate it into a suitable format for use in statistical analyses.

GIS Interface

The second interface is the GIS interface that offers a wide range of data retrieval, manipulation, analysis, and modelling possibilities, provided by a flexible and user-friendly environment. The software used is the ESRI developed PC ArcView. The work here concentrated on developing a GIS model to promote the potential advantages of using GIS for planning new marine and coastal activities. We did not spend time establishing comprehensive set of resource data for a specific area, nor did we develop user specific applications. It was envisaged that these phases of system implementation would be carried under concrete user requirements to a given task or as part of a co-operation project.

One possibility with ArcView is that you can build interactive links to a number of database platforms. We established a direct link to our SQL database using SQL connectivity and the appropriate ODBC drivers. Programming in Avenue allowed us to develop an interactive system where the user is asked a series of questions designed so that they retrieve only the particular environmental monitoring data they are interested in. Two different approaches were adopted. The first was designed more for our own specific needs where the user can select data collected from a particular field cruise. The second is a more general approach where the user defines the area of interest by using the mouse to draw a box on a map on the computer screen, from which coordinates bounding the area are calculated and returned to the database for retrieving the requested data (Fig. 3). Further questions help direct the user to retrieve specific data types, and their values stored in the database. Appropriate commands are also available for displaying these data on the map view with predefined, or user defined legends.

Modelling with GIS

As mentioned earlier the Pechora Sea is of considerable interest to oil companies for exploring and exploiting new oil resources. There is therefore a need to develop management strategies for controlling activities in this part of the Arctic, and for devising contingency plans for dealing with potential oil spills and other accidents related to increased activity in this area. It is intended to demonstrate how we can use GIS to predict the risk to environmental habitats under both normal operating circumstances and in the case of a potential accident.

A number of avenue programs were developed to show some of the spatial modelling capabilities within ArcView and how we can use the system for decision analysis and environmental impact assessments in the coastal and marine ecosystems. For the purpose of this model a number of fictitious datasets were established, including fish, mammals, birds etc. Each of the polygons defining a particular habitat was assigned a sensitivity index (0-3) representing the importance, or vulnerability of the habitat to environmental change. These values in reality would have to be assigned by a scientific specialist. The INSROP Dynamic Environmental Atlas (Bakken et.al., 1995) identified and recorded Valued Ecosystem Components (VECs) along the Northern Sea Route. This information could be included here.

Some of the modules developed here include calculating influence zones around oil installations and pipelines, demonstrating the potential effects caused by a hypothetical oil spill, assessing the effects of establishing new ports and terminals.

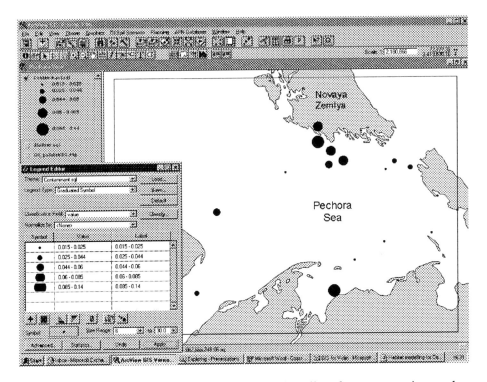

Fig. 3. The ArcView interface has been modified to allow the user to retrieve and display data directly from the SQL database.

1. *Oil platform development :* We have attempted to show how a user, involved in environmental assessment of oil platform location can interactively locate a proposed or existing oil platform on the GIS map, which must be situated at a known oil or gas field. They create an interactive buffer zone around the installation representing the estimated or expected influence area. The program then goes on to calculate the area of VECs for the fictitious ecosystems expected to be affected by such an activity, the number of animals estimated to be in this area. A calculation based on the sensitivity index can indicate whether the VEC will be affected by the installation, and to what extent. As a decision tool this value could be used to say whether such an activity is acceptable under the threats it poses to the environment. In some cases it may simply not be possible to relocate such activities, since oil companies must drill for oil where they find it. However, such knowledge provided by an intelligent GIS could help indicate that certain additional precautions should be taken to reduce the influence zone of an installation.
2. *Oil pipeline development :* This module performs similar calculations to the module described in 1) above. Here the criteria are that a pipeline must start at an oil platform and terminate at a port terminal. A buffer zone is again calculated from a user specified distance, and the program assesses the environmental risk for the pipeline passing through or in the vicinity of a VEC. If the risk is unacceptable the

user is asked if they would like to relocate the pipeline, and the calculation is repeated. These tools calculate relative values of risk to the environment, which can be useful when comparing alternative cases in the planning process to help the manager chose a solution with the least risk to the environment.

3. *Shipping routes :* Shipping routes are often influenced by the sea-ice conditions in the Arctic seas, and can therefore not always take the shortest or most 'environmentally friendly' route to their destination. This module calculates areas of preferred shipping activity, where buffer zones are established around sensitive areas, and average ice conditions are considered. The user can input the time of year they are interested in, and then display a polygon within which are the preferred areas for shipping with respect to VECs and sea-ice, if relevant. In practice ships navigate with the aid of satellite imagery in ice prevalent waters, and the ability to include this information, and update 'preferred' shipping areas in real time are essential.

4. *Oil spill scenario :* This is the most powerful set of tools written for this model, and they demonstrate some of the more advanced sides of GIS. The idea behind this module was to show that it is possible to build some simple modelling directly into a GIS and use the results of the to calculate the risk posed to the VECs. Alternatively it would be possible to import the results generated by true oil dispersal model and use these results to run the risk calculations. This model is not a complex true mathematical oil drift, and so does not model oil drift realistically. However, it does show that the implication of such a model is feasible. The user is requested to input the location of a hypothetical oil spill interactively on the computer screen, and enter the quantity of oil discharged. The next phase is to model the drift of oil based on ocean currents, and optional wind parameters (wind speed and direction) which are user defined, for a user specified time interval. The model then goes onto calculate the area, quantity and density of oil that would come in contact with VECs. Finally a value associated with the environmental risk to each VEC is calculated, along with a figure representing the risk to the affected environment as a whole (Fig. 4). In the situation where an oil spill has occurred, it is obviously not possible to relocate the incident elsewhere. However, the results generated by this model would not only provide real values as to the risk posed to the environment but also present the situation clearly in digital map form, such that an oil spill contingency plan can be quickly drawn up. This may involve taking measures to deflect the oil to other areas, so it would be useful to recalculate the risk to habitats that may then be affected. Other measures may include using dispersants, which could have side effects on particular ecosystems (IMO/IPIECA, 1996). By running a simple query in the GIS, the manager could retrieve metadata files describing whether or not this is an acceptable solution in this particular case.

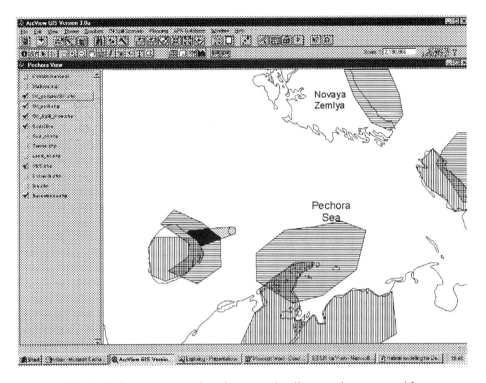

Fig. 4. Risk assessment when for example oil comes in contact with Valued Ecosystem Components

The Future

The future use and development of this system will depend largely on collaboration between interested partners from both the oil industry and Russian local governments. This chapter clearly demonstrates the ability of GIS as a powerful decision support and environmental impact assessment tool for use in the coastal and marine environments. It has been shown that by developing an 'intelligent' system, management decisions can be simplified by calculating relative risk to the different VECs and choosing the solution that is most acceptable. The user interface can be developed in such a way that the user does not have to be a 'GIS specialist' to operate the system, but in contrast should be a scientist or consultant familiar with the geographic region and type of data being analysed. One important advantage of using GIS in the marine environment is that we can work in three dimensions, calculating effects on creatures suspended in the water column, or to the benthic communities living in the surface sediments. However, for GIS to assist in coastal and marine planning the need for detailed, up-to-date, and accurate resource and sensitivity maps for specific areas of interest is essential to provide the backbone to the GIS database. There is relatively little known about the environmental state of the Russian Arctic, so we need to build on the limited knowledge we have, so that we can ensure future industrial developments in this area will not lead to environmental disasters in this fragile environment. In addition it is necessary that strict policies for controlling the exploration and exploitation of these Arctic waters are established and enforced by the Russian Republics.

The main goals of this project have been to promote the benefits of GIS technology for environmental impact assessment within a remote region of the Arctic which, in the near future, will be exposed to ever increasing risk for environmental damage as a result of hydrochemical developments. At the same time Akvaplan-niva have acquired a considerable amount of experience and expertise, and have amassed a large amount of environmental data that could form a base for environmental monitoring surveys in this part of the Arctic. The benefits have been two-fold in that Akvaplan-niva developed a GIS as an interactive interface to their environmental database, and at the same time demonstrated the analysis, and modelling potential of GIS for EIA and as a decision analysis tool.

References

Bakken, V., Brude, O.W., Larsen, L.H., Moe, K.A., Wiig, Ø., Sirenko, B., Gavrilo, M., Belikov, S.Y., and Garner, G.W. (1995). INSROP Dynamic Environmental Atlas. *IST'95 INSROP Symposium Tokyo '95*. pp. 213-221.
IMO/IPIECA (1996). Sensitivity Mapping for Oil Spill Response. *IMO/IPIECA Report Series*. Vol. 1.
Exxon Valdez Oil Spill Symposium (1993). Programs and Abstracts. *Exxon Valdez Oil Spill Symposium*. February 2-5, 1993 Anchorage, AK, USA.

CHAPTER 13

A GIS Application for the Study of Beach Morphodynamics

L. P. Humphries and C. N. Ligdas

ABSTRACT: This research is a continuation of work conducted at the University of Sunderland since 1990 on the impacts of the accumulation of millions of tonnes of colliery spoil on the beaches. Regular surveys of the beaches were undertaken from 1991 spanning a period of about 5 years during which all mining activity ceased on the Durham coast. During mining operations the accumulation of spoil on the beaches affected both the geomorphology and the geochemistry of the beaches. It may also have masked the impact that deep mining of multiple coal seams under the coasts had on the beaches. Once tipping ceased, it was expected that spoil would erode rapidly and that the pattern of erosion would show changes parallel to the shoreline. This was the case on some of the beaches but on others, notably the Dawdon blast beach, anomalies appeared showing localised patches of changes at orientations at variance with the expected pattern. Dawdon Blast beach has been extensively undermined and it is possible that some of these patterns may have been produced by structural changes in the bedrock caused by subsidence. To test this hypothesis, a Geographical Information System (GIS) was used to study changes of the beach surface over the years immediately superseding the cessation of mining, focusing on 3 study areas (Dawdon, Easington and Horden) which present different undermining conditions. It was found that the spatial pattern of significant elevation changes in the Dawdon beach is considerably different from that observed in Easington and Horden. The possible relationship to coastal colliery panels was further tested by obtaining the strain, subsidence and surface heights along the transect from the centre of the worked panels lying east and west of the beach area in Dawdon. The implications for coastal management on this coastline are that areas directly undermined may be suffering subsidence impacts causing increased levels of erosion of beach material. This is an important factor of local significance in assessing coast vulnerability and risk of cliff erosion since changes in bedrock caused by subsidence diminish the role of the beach as a protective influence on coastal defence. The role of the GIS in this research is not only to identify spatial patterns and relationships but also to combine, assess and weight the different factors affecting coastal vulnerability.

Introduction

A long history of coal mining in the Coal Measures seams, underlying the Co. Durham coastline and extending several hundred meters off-shore, has altered the geomorphology of the area and left a legacy of despoiled beaches due to the practice of disposing the colliery spoil directly onto the beaches. At the peak of this activity (before 1983) over two million tonnes of minestone was disposed annually onto the beaches.

The three beaches that we chose to examine are Dawdon, Easington and Horden on the Co. Durham coast of North East England (Fig. 1). All three belong to the same littoral cell, 1c (Motyka and Brampton, 1993), and they exhibit similar geological and geomorphological characteristics; cliffed coastline (Magnesian Limestone cliffs with heavy overburden of boulder clay at the backshore) and sand beaches containing a high proportion of shales and limestones. Both the geomorphology and the geochemistry of the beaches have been largely affected by the colliery waste. Beaches extend beyond the natural coastline and the sand in most of them is heavily polluted by coal. Littoral processes are characterised by a moderate southward drift and tidal currents intensify the southward transport of sand and colliery waste on the lower beach and on the nearshore sea bed.

Using sedimentary and morphological indices obtained from data collected between 1991 and 1995, all three beaches have been classified as *reflective* beaches (Humphries, 1996) based on the beach stage model by Wright and Short (1983). This means rather steep beach slopes which are not dissipative of wave action. Cliff recession in these three beaches seems to proceed at lower rates (0.11 m/yr for headlands and 0.3 m/yr for cliffs backing the bay) than those observed at beaches north of Ryhope during the period 1983-1993 (Anon, 1991; Anon, 1993).

Off-shore underground colliery panels have been excavated in roughly 6 seams, ranging from 340 to 465 m depth, over a period of more than 50 years (Table 2). The most recent ones can be seen near the Dawdon area and the northern part of Easington beach (Fig. 1).

It is probable that subsidence could well have played some part in shaping the beaches studied here. Calculations conducted by a subsidence engineer from British Coal have revealed subsidence on the Northumberland coast in the order of one metre (T. Dixon pers. comm.). On the Co. Durham coastline, the impacts on coastal erosion of subsidence of this order may have been minimised by the accumulation of colliery spoil on the beaches.

The aim of this research was to investigate the morphological changes of the beaches during a 5-year period after the closure of the mines in the area and identify patterns or anomalies that can be associated to the mining activities and in particular

GIS for the Study of Beach Morphodynamics

the existence of the coastal underground mines (Fig. 1). The thick dashed line indicates the cliff trace digitised from 1:10000 OS maps. Boxes indicate the study areas. The "Main" and "Yard" colliery panels close to the Dawdon beach area have been worked during the late 1980's. "High Main" has been worked during the 1960's and all the other panels prior to 1947.

This chapter describes the work that was carried out using a GIS to compile and analyse the data. Results were then compared with theoretical values and areas of surface impact of predicted subsidence in the area.

Data Development and Analysis

a. Survey points

The main body of the field research has been conducted by using a total station (Geodimeter 400) to survey several times a year, three profile lines on each of the main beaches affected by the spoil at Dawdon, Easington and Horden (Fig. 1). Measurements were taken at regular time intervals to cover seasonal variations. Instrument error is estimated to be less than 1 cm. OS maps (1:10000) were used as a backcloth for the cliff line and high/low water marks.

The GIS software that was used was the ArcInfo Workstation version which can handle both vector and raster data in surface analysis.

Survey points from beach profile data were used to investigate the development of beach surface over the study period. From these datasets, the corresponding TINS for every study area were constructed. Lattices representing the elevation maps of the beaches were produced for every survey that was carried out using bivariate quintic interpolation of the related TINS. The smoothing effects of this method were considered appropriate since we were dealing with relatively small TINS and map extents in each beach. The cell size was set to 10 m. This spacing was considered appropriate taking into account the distance between survey points and profile lines. To assess the error in the elevation data, the lattice containing only the measured elevation at the survey points was compared to the lattice containing only the extracted interpolated heights at the same points. The mean difference of the resulting lattice is very close to 0, so the standard deviation can be considered as an estimate of the RMS error. The error of any two resulting maps after overlay operations was calculated as the square root of the sum of squared errors of the two maps. Table 1 shows the errors (m) for the resulting difference maps.

The elevation lattices for all 3 beaches were then used to produce slope and aspect maps, one for each survey. Data collected in March and June 1991 were used as "base" data from which changes in beach elevation for the coming months were examined. The mean values of slope, aspect and "difference" elevation data were also studied over the five year period, taking into account the estimated error, in order to assess any variability of the mean size of these characteristics with time (Figs 2, 3, 4). However, the study concentrated on identifying spatial variations and patterns. Elevation differences were studied as "cut" and "fill" changes of the beach surface. These changes were further analysed by assuming thresholds. Mean "cut" values were

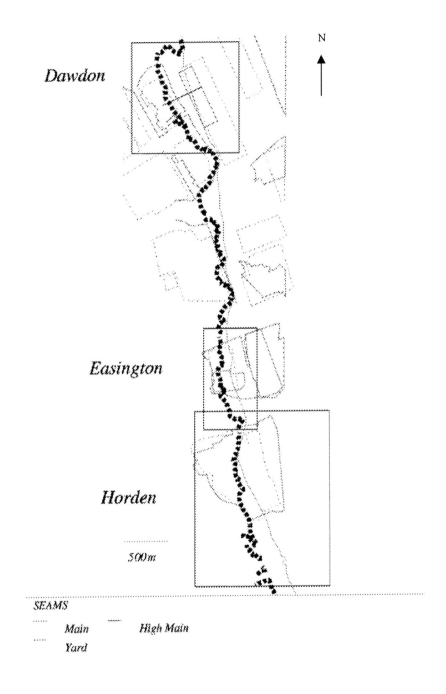

Fig. 1. Location map

"cut" changes calculated this way can be seen in Fig. 5. Unusual negative (cut) and positive (fill) changes, produced by assuming a mean - 2s / mean + 2s threshold, were reclassified using a binary classification. Cumulative maps for unusual "cut" and "fill" changes where then produced by adding the reclassified "cut" and "fill" maps. The resulting maps (2 for each study area) show occurrence of unusual negative and positive changes (Fig. 6a, b, and c).

TABLE 1
The errors (m) for the resulting difference maps

Dawdon		Easington		Horden	
Mar-91	0.38	Aug-91	0.62	Aug-91	0.84
Aug-91	0.35	Dec-91	0.33	Dec-91	0.31
Oct-91	0.32	Feb-92	0.38	Feb-92	0.74
Feb-92	0.30	Apr-92	0.38	Mar-92	0.18
Apr-92	0.34	Jun-92	0.32	Apr-92	0.27
Jun-92	0.84	Sep-92	0.48	Jun-92	0.29
Sep-92	0.60	Dec-92	0.37	Nov-92	0.51
Oct-92	0.29	May-94	0.36	Dec-92	0.40
Feb-93	0.26	Jun-95	0.41	Feb-93	0.82
Mar-93	0.79	Jul-95	0.29	Jun-93	0.44
Sep-94	0.18			May-94	0.59
Oct-94	0.37				
Nov-94	0.76				
Dec-94	0.45				
Jan-95	0.98				
Feb-95	0.52				
Apr-95	0.61				
May-95	0.45				
Apr-96	0.67				

b. Subsidence
Data was obtained from British Coal from underground abandonment coal plans at a scale of 1:10,560 covering the beaches in the study. From these plans, information on the seams worked (John Ellis pers. comm.), and the charts supplied by the Subsidence Engineer's Handbook (Anon, 1975) it was possible to calculate a prediction of maximum subsidence that would affect the beaches. Table 2 shows the 6 main seams in the study areas.

The calculated value gives a theoretical maximum. It is more practical to calculate how this subsidence changes from this maximum over the centre of a worked panel to the limits of subsidence impact at the surface. This value is a function of the distance from the centre of a panel in terms of depth and the width to height (Subsidence Engineers Handbook, 1975). The subsidence and strain profiles are then drawn for each seam.

TABLE 2
The 6 main seams in the study areas

Seam	Date of extraction	Depth of seam (m)	Height of seam (m)	Width of seam (m)	Subsidence (m)
High Main	1960-66	342	2.13	133	0.75
Main	1987-88 (Dawdon)	372	1.17	274	0.78
Yard	1980-82 (Dawdon)	384	1.47	159	0.53
Maudlin	<1947	403	1.58	232	0.87
Low Main	<1947	419	1.22	402	1.00
Hutton	<1947	464	1.83	412	1.42
		Total extracted=	9.40m	Total maximum subsidence =	5.35m

Active subsidence after excavation takes place after the face has advanced. This is almost fully developed after the face has advanced a distance about equal to the draw beyond the surface points to be examined. The angle of draw (which predicts the area on the surface influenced by the subsiding panel) from both panels covers the beach area so it is reasonable to assume that the beach has been affected by subsidence from both panels and probably within a year of cessation of excavation.

Fig. 7 is an example of a subsidence and strain profile calculated for two worked panels underlying the cliffs and the nearshore zone at Dawdon Blast beach. The strain and subsidence profiles were calculated, using charts from the Subsidence Engineer's Handbook, for values from just one seam, Main, for a range of survey data. Main was the seam worked most recently (1988) and the seam nearest the surface underlying the section of beach shown in Fig. 8. Subsidence from multiple seams is added so that for the panels shown (which do not include High Main which is not worked here) the maximum subsidence is 4.6 metres.

Profiles of "cut-fill" difference maps and elevation maps (Fig. 9), from the centre of the cliff panel of the Main seam to the centre of off-shore panel, were drawn from the beach surfaces for all the surveyed months in Dawdon. The direction of these profiles is perpendicular to the shoreline. These profiles were then compared to the strain and subsidence theoretical profiles.

Results

Surface analysis has shown that Dawdon presents a host of different characteristics, in both size and spatial occurrence of the features that were examined, compared to the other 2 study areas in Easington and Horden.

The slope maps for Dawdon generally reveal a different morphological profile in the second half of 1991 and early 1992 from that of later months. A typical example is shown in Fig. 10(a) for October 1991. Two quite distinct sections can be seen with the foreshore part being steeper. Aspect maps (example in Fig. 10(b)) show a similar pattern, with the steeper nearshore part facing east and the backshore part facing west. In the second half of 1992, this pattern changes completely and the overall trend is that the mean slope values increase with time (Fig. 2) while the beach surface becomes a rather undulating surface. This is easily observed in the elevation profiles (Fig. 9).

Slope maps for Easington and Horden do not show the same characteristics as Dawdon and beach orientation is solely towards the east. No increase in the mean slope value is observed either (Fig. 2). Changes in beach elevation appear to be seasonal in most cases. Undulations of mean elevation and mean "cut" values in Fig. 9 highlight the dry-wet season periodicity. In Dawdon, however, certain unusual patterns ("sink holes" and undulations) in the beach surface prompted us to investigate the surface development further. Relatively higher values of "cut" changes seem to be concentrated along a narrow zone at the lower parts of the foreshore in Easington and Horden, but occur in more distinct patches in Dawdon beach. Notably a sink hole is apparent at the north end of Dawdon beach during the spring and early summer months of 1992.

"Cut-fill" changes over the 2s threshold (Fig. 6(a)) indicate that in Dawdon over the study period, losses of beach material occurred in a much larger area than those in Easington and Horden. Taking into account the similar topography, geology and climatological conditions in the 3 study areas, the unusual "cut" patterns in Dawdon indicate a higher degree of vulnerability at this beach.

To further investigate these observations, the profiles of heights and difference values were used to test the hypothesis that the height of the beach can be predicted from a combination of strain values, subsidence and difference features. The plot of cut features (Fig. 9) shows a good relationship with the edge of underground panels. A multiple regression of the parameters difference value, strain and subsidence on height was conducted to test this relationship. In order to be able to use multiple regression the data must be normally distributed. The normal probability plot of the 26 data points used showed a good approximation to a straight line so the test was appropriate. Strain is proportional to subsidence and inversely proportional to depth so that the maximum strain over an extraction is proportional to S/h, where S is the maximum subsidence and h is the depth of the seam. It is appropriate to just calculate the subsidence and strain profiles for the seam nearest the surface as strain decreases with the depth of the seam. Multiple extraction will have an impact however but is not considered here.

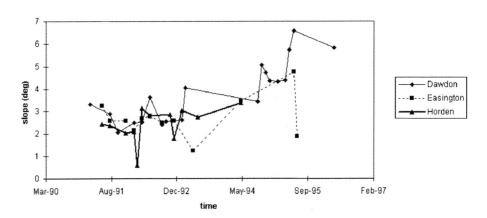

Fig. 2. Slope mean values for the 3 study areas

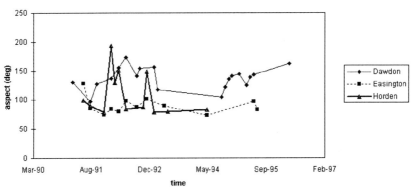

Fig. 3. Aspect mean values for the 3 study areas

GIS for the Study of Beach Morphodynamics

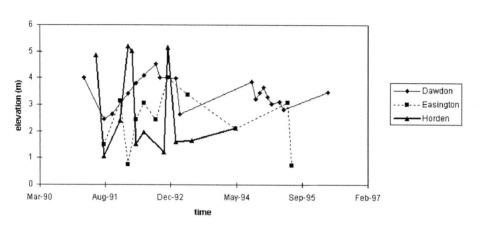

Fig. 4. Mean elevation

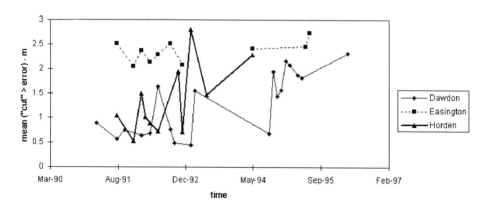

Fig. 5. Mean "cut" values. The mean has been calculated for values above the estimated error as shown in Table 1.

(a) (see next page for caption to Figures 6a-c)

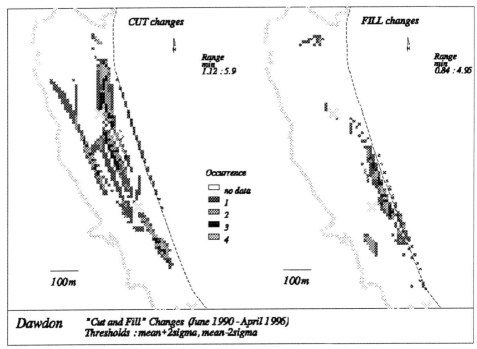

(b)

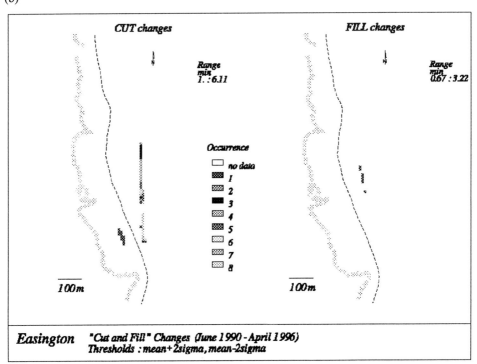

(c)

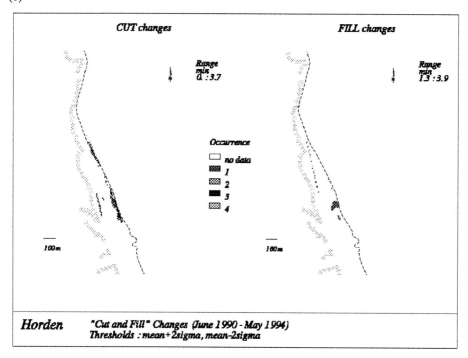

Fig. 6. (a) Cumulative "cut" and "fill" values for Dawdon. These values are changes outside the 2σ threshold. "Min" figures on the graph indicate the range of mean+2σ and mean-2σ values outside which "cut" and "fill" changes were taken into account. Cell values are only an indication of occurrence of a change. Data from a survey carried out by British Coal surveyors in June 1990 was used as the base map to estimate changes. The reason for doing so was the good quality of data in this survey. If March 1991 is assumed as the base map, "cut" and "fill" patterns remain the same and are enhanced. This can probably be attributed to colliery waste still being tipped on Dawdon beach during 1991; (b) Cumulative "cut" and "fill" values for Easington. The base map is March 1991;(c) Cumulative "cut" and "fill" values for Horden. The base map is March 1991.

The results show a good correlation, $R^2 = 0.64$, so 64% of the variation in height on the beach can be predicted by two factors associated with mining in addition to the value of the erosion features. Fig. 11 shows the fit between the actual and predicted values of surface height. These statistics give quantitative support to the relationship highlighted by the GIS plots.

Discussion

These results indicate the extent to which the spoil on the beaches for years has masked the impact of mining activity. With the erosion of the spoil following the cessation of tipping the impact on beach geomorphology has begun to emerge and will do so for some years to come. The impact of mining on other areas not protected to the same extent has been reported to the north of this study area.

At Newbiggin, subsidence from undersea mining is believed to have caused sand to migrate to the centre of the bay and coastal dumping of mine waste north of the town may have been responsible for a change in sea currents so that sand is now no longer replaced on the beach. The loss of sand was blamed on mining subsidence in the bay and in areas of offshore rocks which protected the bay by absorbing wave energy (Tooley, 1989). Tooley (1989) stated that wherever there is significant mining settlement on the foreshore along the north east coast, coastal erosion and coastal protection problems are the result.

Similar results were recorded in the report of an overall survey of the coastline for Castle Morpeth Borough Council (Anon, 1985) which was conducted to calculate the amount of subsidence from coal-mining and how it had affected the coast. The subsidence contours showed a maximum subsidence of 3.6 metres between 200 and 800 metres from the cliff line. Immediately north of the study area a consultant's report for Sunderland Borough Council reported problems with ground control points, notably discrepancies in the vertical plane which could not be accounted for. The general conclusion was that this was due to a general drop in level of between 200 and 300 mm between 1974 and 1988. There does not to have been any subsidence impacts in this area from 1988 onwards.

The results from this research support the link between mining activity and subsidence found elsewhere on this coast and presents a quantitative analysis of the factors influencing such subsidence and the subsequent impact on beach height.

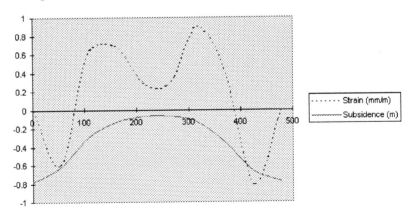

Fig. 7. Strain and subsidence profile for the Main coal seam

GIS for the Study of Beach Morphodynamics 185

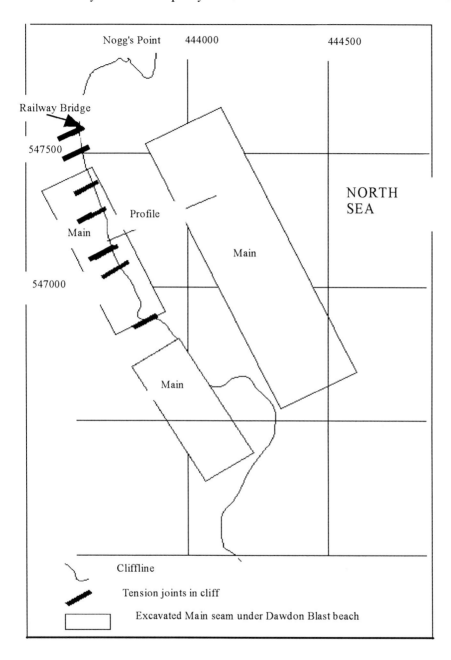

Fig. 8. Location map Dawdon and Main coal seam. The thin line from the centre of the cliff panel to the centre of the off-shore panel indicates the line for which strain, subsidence and elevation profiles are drawn.

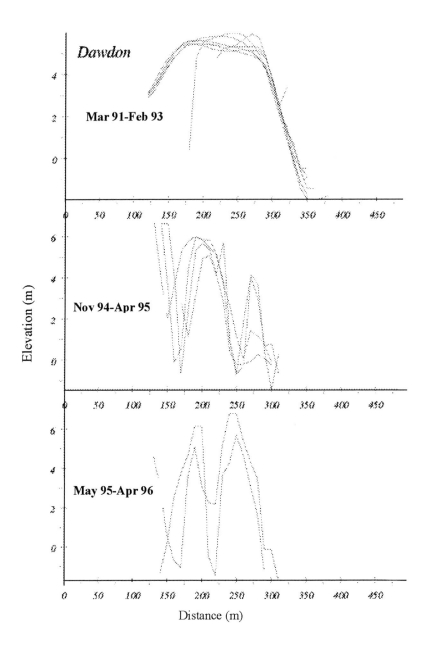

Fig. 9. Profile of "cut-fill" map for all surveys in Dawdon. Positive values indicate "cut" changes and negative values indicate "fill" changes.

GIS for the Study of Beach Morphodynamics 187

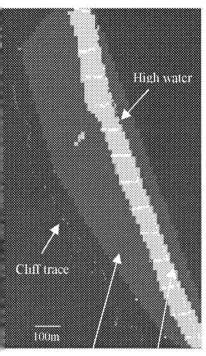 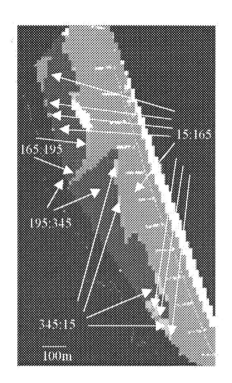

Slope <2.5 degrees   Slope > 2.5 degrees

SLOPE   **ASPECT**

Fig. 10. (a) Dawdon, October 1991. Slope. (b) Dawdon, October 1991 Aspect.

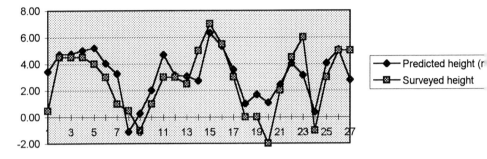

Fig. 11. Predicted and actual heights from the Dawdon Blast beach surveys

A further note on the GIS functionality should be made here. The use of the GIS affected both the survey and data collection approaches. Once the potential for studying spatial relationships of a variety of factors and phenomena associated with coastal vulnerability was realised, the distribution of survey points and type of data to be collected was reviewed. The development of a coastal GIS to study factors affecting cliff erosion (Ligdas, 1996) in the North East coast is now being expanded to include tools that analyse a much wider range of coastal processes. A major weakness in proprietary GIS is the lack of statistical capabilities. This means that either code has to be written to accommodate this or statistical analysis must be carried outside the available GIS functions.

Conclusions

Surface analysis tools of the GIS have allowed the examination of the beaches as a whole and the identification of spatial patterns that are unusual or anomalous with respect to their neighbours. Elevation errors have been assessed and in conjunction with the available spatial resolution they should be taken into account in the resulting images.

Surface analysis identified features and possible areas of high vulnerability that could have not been resolved by a simple profile analysis of the survey data. In Dawdon beach, during the study period, a much larger area is affected by erosion compared to Easington and Horden beaches. A possible interpretation is that subsidence has worked its way to the surface of the beach and is concentrated on weak spots exploiting the nature of the solid geology. The Magnesian Limestone contains abundant joins and possibly underground solution cavities. The "hole" which was developed at the north end of Dawdon could have been caused by a sink hole.

The time period of the study is obviously not adequate to establish log-term patterns and continuous monitoring of the beaches is required. The GIS database however, offers the opportunity to store and organise the data in a uniform and coherent manner. It is important to design the database correctly in order to ensure data continuity and allow frequent updates. Most significantly, the assessment of coastal vulnerability requires the integration of a wide range of factors which are spatially dependent and a GIS, despite certain inherent problems in data analysis, can play that role.

References

Anon (1970). *Colliery waste on the Durham Coast. A Study of the Effect of Tipping Colliery Waste on the Coastal Processes.* HR Wallingford. 26p.

Anon (1975). *Subsidence Engineer's Handbook*, National Coal Board Mining Department.

Anon (1985). An overall survey of the coastline. Prepared for Castle Morpeth Borough Council by Babtie, Shaw and Morton Consulting Engineers.

Anon (1993). *Durham Coast Management Plan : Coastal Processes Report.* Prepared by Posford Duvivier for Durham County Council and Easington District Council.

Anon (1991). *An Investigation into the Recession Rate of the Sea Cliffs between Hendon Sea Wall and Ryhope Outfall.* Prepared by Bullen and Partners for Sunderland Borough Council.

Humphries, L. P. (1996). A coastal morphology classification system for beaches on the Co. Durham coast modified by the addition of colliery spoil. In, *Partnership in Coastal Zone Management,* Eds. Taussik, J., and Mitchell, J., Samara Publishing Ltd.

Ligdas, C. N. (1996). The study of coastal processes in the North East of England using a GIS. In, *Partnership in Coastal Zone Management*, Eds. Taussik, J., and Mitchell, J., Samara Publishing Ltd.

Motyka, J.M., and Brampton A.H. (1993). *Coastal Management, Mapping of Littoral Cells.* HR Wallingford.

Tooley, M. (1989) The Flood Behind the Embankment, *Geographical Magazine,* November 1989.

CHAPTER 14

Determination and Prediction of Sediment Yields from Recession of the Holderness Coast

R. Newsham, P.S. Balson, D.G. Tragheim, and A.M. Denniss

ABSTRACT: The rapidly eroding cliffs and foreshore of the Holderness Coast represent one of the largest sources of sediment discharging in the southern North Sea. For effective coastal management it is important to understand the sediment sources, transport pathways and depositional sinks. The sediment budget is regarded as one of the key information needs for sustainable planning and management. In order to quantify the sediment yield from recession of the Holderness Coast, digital photogrammetry has been used, together with nearshore bathymetric survey data, to produce a single DTM for a length of over 52 kilometres of coastline. Other data including geological sections, sediment lithology, recession rates and the location of coastal defences have been added to the DTM to create a GIS for the Holderness Coast. The GIS has been used to predict the volume and nature of the sediment yield assuming that historical recession rates continue. The GIS can also be used to predict future yields using assumptions of accelerated recession or the implications of arrested retreat due to construction of defences.

Introduction

The Holderness Coast stretches for over 50 kilometres from the Chalk promontory of Flamborough Head in the north to the shingle spit of Spurn Head at the mouth of the Humber Estuary in the south (Fig. 1). The coastline consists dominantly of cliffs up to 38 metres high in unlithified glacial tills and other glacigenic sediments. The coast

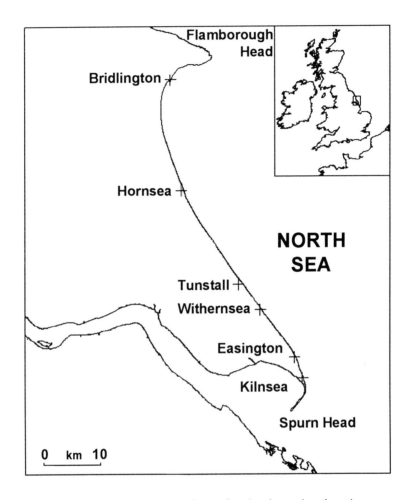

Fig. 1. Map of Holderness Coast showing its regional setting

is largely undefended except at the seaside towns of Bridlington, Hornsea and Withernsea. Coastal erosion of the undefended stretches results in cliff recession rates up to around an average of 2 metres per annum which makes the Holderness coastline one of the most rapidly eroding in the UK. The rapid recession releases large quantities of sediment onto the beach and into the nearshore zone. This sediment consists of a very heterogeneous mixture of particle sizes ranging from boulders to clay. The sands and gravels contribute to the maintenance of the Holderness beaches and the Spurn Head peninsula. Silts and clays, which dominate the sediment yield, are transported in suspension in the nearshore waters. This suspended sediment may have final 'sinks' including the mudflats and saltmarshes of the Humber Estuary and The Wash and other depocentres elsewhere in the southern North Sea. In the past, public concerns have focused on the loss of farmland and cliff-top property caused by retreat of the cliffs that have led to calls for construction of coastal defences. Construction of 'hard' engineered defences will inevitably lead to a reduction in sediment yield from coastal recession, with consequent, but largely unknown, impacts elsewhere. Although there are many published estimates of sediment yield for the Holderness coast (e.g. Redman, 1869; Pickwell, 1878; Dossor, 1955; Valentin, 1971) these are based on coastal length and assumptions of average cliff height, cliff recession rate and cliff composition. Consequently estimates have varied by up to an order of magnitude. This paper describes work currently in progress to improve on the quantification of sediment yield by using digital photogrammetry and GIS data analysis techniques. The establishment of a sediment budget is regarded as one of the key information needs for sustainable planning and management(Department of the Environment, 1995).

The work was carried out at the British Geological Survey (BGS) as part of the NERC Land-Ocean Interaction Study (LOIS).

The Holderness GIS

The method used in the study involved the creation of a DTM using digital photogrammetry derived from scanned vertical aerial photographs and combined with other datasets in a GIS environment.

The project was carried out using a suite of Intergraph software packages. The initial photogrammetry was undertaken using Intergraph ImageStation. The DTM derived from the photogrammetry was transferred to a Windows NT workstation and manipulated using Intergraph Terrain Modeler. Terrain Modeler is a specialized package developed for analysis, manipulation, modification and graphical display of elevation modelling data and is built on Microstation, which is a computer-aided drawing package.

The datasets can be summarised as follows:

3D DTM - from aerial photography
Colour aerial photography at 1:6000 scale was acquired by the NERC aircraft during low-tide conditions (around 10:15 am) on 14 April 1995 along the Holderness coastline. At this scale, the photographs generally covered 75% land, 5-10% beach and the remainder sea. In order to use the photographs photogrammetrically, a GPS ground control survey was commissioned to provide 80 points for forty photographs. These were generally chosen to occur within the triple overlap areas of adjoining stereomodels. The accuracy specified for the GPS ground control was better than ± 0.1 m in Easting, Northing and Height. The actual mean transformed coordinates of the 11 stations produced root mean square error (RMSE) values of E=0.033 m, N=0.020 m and H=0.004 m. Colour diapositives were made from the negatives and scanned at a resolution of 20 microns to produce 3 band, red-green-blue digital images, each measuring 432 Mb in size. The images were subsequently compressed using the JPEG method and six overviews added to each, resulting in a final colour image size of about 132 Mb.

The photogrammetry was undertaken using an Intergraph ImageStation, a dual-function image analysis and digital photogrammetric workstation. It is estimated that individual DTM points may have an RMSE height accuracy better than ± 0.20 m. A ground-truth survey using a Nikon Total Station is currently being undertaken and should provide more quantitative information about co-ordinate accuracy. Lines representing the cliff-top and bottom were digitised manually into the file from the image. These features are placed into a Microstation file. The x,y,z position of the features are loaded into Terrain Analyst to create a digital terrain model of the area. As each different feature is loaded into Terrain Analyst it is assigned an attribute relating to the type of feature it is. For example, the lines representing the cliff-top and base are loaded as a breakline; this prevents the model from rounding off the cliff-top and base. Further details of the photogrammetric techniques used can be found in Balson *et al.* (1996).

3D DTM - from bathymetry
Water depth data was used to construct a DTM for the nearshore area. Data was captured for an area of sea floor out to approximately 2.5 kilometres from the low water mark where a prominent break in slope is assumed to represent the depth beyond which shoreface erosion becomes insignificant. Spot heights were loaded into Microstation and the elevations were corrected to Ordnance Datum to enable the final model to be merged with the onshore DTM to create a single DTM of the eroding cliffs and shoreface (Figure 2).

Cliff Geology
Horizontal sections which show the lateral and vertical distribution of the geological formations exposed in the Holderness cliffs were drawn by Bisat (Catt and Madgett, 1981). The original hand-drawn sections, housed in the archives at BGS, were digitised using Microstation. The section was rescaled in segments, between points with known Eastings and Northings, to reduce the effect of horizontal scale errors in the original drafting. Each geological formation was captured as an individual polygon. The vertical axis was converted to heights relative to Ordnance Datum. The digitised section had a vertical exaggeration of x10, to assist in the visualisation of the polygons. Each polygon was marked with a code to identify the specific geological formation. Bisat's sections show the formations exposed only above beach level. For the area below beach level and on the shoreface the geology must be assumed. For simplicity it has been assumed that only Basement Till is exposed below beach level. In the future it is hoped that further study will improve knowledge of the distribution of Quaternary formations of the Holderness Coast.

To assess the accuracy of the cliff-top topography in the original horizontal section produced by Bisat, a line was placed in Microstation to follow the current cliff-top on the DTM. Terrain Modeler created a profile across the model and placed it graphically into the Microstation file. The original profile was then overlain and registered using known points on each profile. Slight discrepancies between the original profile and the profile generated from the DTM are probably due to the difference in the age of the two profiles. In some areas the height of the cliff has reduced as the cliff has eroded back.

Geological Boundaries

Boundaries between the geological formations on the land area in plan view were digitised from BGS 1:10,000 geological maps using Microstation. These lines can be draped onto the DTM in Terrain Analyst.

Erosion Rates

Valentin (1954; 1971) determined the amount of cliff-top recession along the Holderness Coast by comparison of distances from fixed points depicted on 1:10,560 Ordnance Survey maps published in 1852 and field measurements made in 1952. His measurements were spaced at approximately 200 metre intervals along the entire coast. The measurements thus represent average rates of recession averaged over a 100 year time period. Recession since 1952 may have exceeded 50 metres at some locations. The grid reference of each measurement given by Valentin represents the intercept between the cliff-top of 1952 and a line at right angles to it. Each measurement was considered to represent a value for an orthogonal line from the modern coastline through the grid references given by Valentin.

Additional orthogonal lines were constructed within Microstation at 10 metre intervals along the cliff section. These can be assigned values which are intermediate and in approximate ratio to the values entered from Valentin's data on either side. By using the values at this closer spacing, and assuming that the average rate remained constant with time, a new cliff-top after a given number of years of recession can be predicted.

Lithological data

Lithological data for each geological formation have been taken from published data (e.g. Madgett and Catt, 1978) and from field observations. For this study each of the identified geological formations was given a single average composition. In the future it is hoped to be able to take account of the lateral variability of the lithological composition within individual formations by laterally subdividing formations and assigning different values.

Sea defences

The location of existing sea defences, such as the sea walls, which prevent cliff erosion at the towns of Withernsea, Hornsea and Bridlington, were digitised in Microstation.

Calculation of Sediment Yield

To determine the total volume of sediment potentially yielded by recession of the coast, the DTM was migrated horizontally landwards by a given amount and the volume between the DTM and the migrated DTM calculated. The amount of migration can be, for instance, 1 metre, 10 metres or a distance equivalent to 10 years at the present long-term average rate. Similarly an assumed accelerated or decelerated rate can be used in order to assess the implications of future change. Erosion yields can also be hindcast but it is necessary then to assume that the present cliff-top elevation profile is representative of the past profile. The length of coastline can be divided into sections with different migration values if necessary.

After migrating all the features in the Microstation file landward by a given amount at right angles to the cliff top a second DTM was generated. For the migrated DTM a polygon was placed in the Microstation file along the cliff-top and around the model to seaward to exclude the land area from the volume calculation. Areas protected by sea defences are removed from the calculation at this stage. This is achieved by placing two polygons in the Microstation file each side of the sea defences. Shoreface yield can be taken into account by joining the polygons at the base of the sea defences.

Terrain Analyst calculated the difference between the two DTMs within the polygon. An ASCII report file was generated which contains the value for the volume between the two DTMs (Fig. 3). The ASCII file contains two sets of values, the first being the volume of the model where the migrated surface is below the original surface. This value represents the amount of sediment potentially released by erosion for the selected distance. The second value is the volume of the model where the migrated surface is above the original surface. This second value can be ignored, as there would not have been sediment present in these areas and therefore would not affect the yield. Thickness maps can also be generated during the volume calculation. This enables visualisation of the distribution and intensity of the erosion along the eroding shoreface.

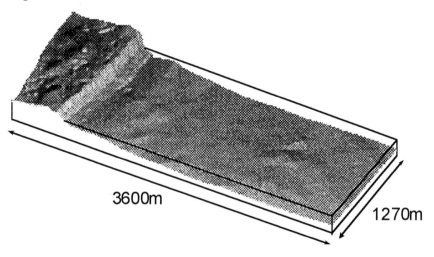

Fig. 2. Shaded perspective view of DTM for a section of the coast near Easington

Determination and Prediction of Sediment Yields

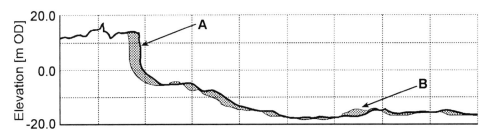

A = The original surface is above the migrated surface. Gives correct volume calculation.

B = The migrated surface is above the original surface. Sediment previously removed these areas.

Fig. 3. Horizontal section through original DTM. The change after lateral migration is indicated by the cross-hatched area, which represents the volume of sediment yielded

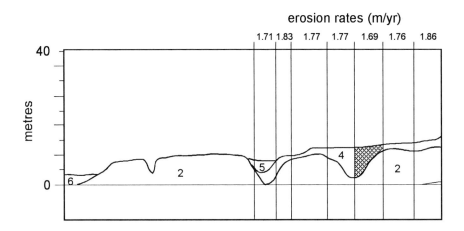

e.g. Mud yield =
Area of geological formation (cross hatched area)
X erosion rate X percentage mud

Fig. 4. Horizontal geological section used to calculate the sediment yield by lithology

Calculation of Sediment Yields by Lithology

The subsequent behaviour of the sediments after erosion from the cliff and shoreface is to a large part dependent on its mineralogy and grain size. Sand and gravel contributes to the beach and is moved southward by longshore drift. Mud is transported southwards mainly in suspension in the nearshore waters. It is necessary to determine the relative contribution of different size categories from the eroding shoreline in order to properly assess the impacts of any future changes in the sediment budget.

The horizontal geological sections have been used to determine these contributions. For this study it has been assumed that each formation is vertically and laterally consistent enough that a single average composition can be assigned to each. The section has been divided into 'cells' bounded by the locations of the recession rate values given by Valentin (1954, 1971) (Fig. 4). The area of each formation polygon in m^2 were multiplied by the recession rates to obtain the volume contribution in m^3 for each formation. These values were multiplied by the values for the average composition and summed to give the total yield of gravel, sand and mud for any given stretch of coastline. Further values can be calculated using different scenarios of future cliff recession rates and using the original DTM to ascertain the new height of the cliff after a specified period of recession.

Future

Although the calculations of sediment yield by lithology can be calculated indirectly, it is hoped that in the near future the volume by lithology will be calculated directly from the model. It is hoped that the boundaries between each lithological unit (digitised from the cliff section) will be directly loaded into the model as surfaces.

Summary

A GIS based on Intergraph software and hardware has been created for the Holderness Coast. The GIS is able to use 3D DTMs derived from digital photogrammetry and bathymetric surveys to calculate the volume of sediments which would be yielded by incremental landward recession. Data on the thickness and distribution of geological formations is included which can be used to calculate the contributions made by mud-, sand-, and gravel-sized fractions to the total volume. The amount of recession can be varied to determine yields for any predicted rate of retreat. This enables the impact of future changes in recession rate to be assessed. Different rates can be used for different stretches of coast to take account lateral variations, as for instance may occur after the construction of coastal defences.

Acknowledgements

Published with the permission of the Director, British Geological Survey (NERC). This is LOIS Publication No. 411 of the LOIS Community Research Programme of the Natural Environment Research Council.

References

Balson, P.S., Tragheim, D.G., Newsham, R., and Denniss, A.M. (1997) Predicting sediment yield from the recession of the Holderness coast, UK. Coastal Zone 97. Boston, Massachusetts. July 1997.

Catt, J.A., and Madgett, P.A. (1981). The work of W.S. Bisat, F.R.S. on the Yorkshire Coast. In: *The Quaternary of Britain*. Eds: Neale, J. and Flenley, J.Pergamon Press. pp. 119-136.

Department of the Environment (1995). *The Investigation and Management of Erosion, Deposition and Flooding in Great Britain*. HMSO.

Dossor, J. (1955). The coast of Holderness: the problem of erosion. *Proceedings of the Yorkshire Geological Society*. Vol. 30:133-145.

Madgett, P.A., and Catt, J.A. (1978). Petrography, Stratigraphy and Weathering of Late Pleistocene tills in East Yorkshire, Lincolnshire and North Norfolk. *Proceedings of the Yorkshire Geological Society*. Vol. 42:55-108

Pickwell, R. (1878). The encroachments of the sea from Spurn Point to Flamborough Head, and the works executed to prevent the loss of land. *Minutes of the Proceedings of the Institution of Civil Engineers*. Vol. 51:191-212

Redman, J. B. (1869). pp. 493-497 in discussion of a paper by W. Shelford, On the outfall of the River Humber. *Minutes of Proceedings of the Institution of Civil Engineers*. Vol. 28:472-516.

Valentin, H. (1954). Der landverlust in Holderness, Ostengland, von 1852 bis 1952. *Die Erde*. Vol. 6:296-315.

Valentin, H. (1971). Land loss at Holderness. pp. 116-137 in: Steers, J.A. (ed.) *Applied Coastal Geomorphology*. Macmillan: London.

CHAPTER 15

Tracing the Recent Evolution of the Littoral Spit at El Rompido, Huelva (Spain) Using Remote Sensing and GIS

J. Ojeda Zújar, E. Parrilla, J.M. Pérez, and J. Loder

ABSTRACT: The El Rompido spit at the mouth of the River Piedras in SW Spain currently forms a notable example of a right angled estuary. This spectacular geomorphological unit, and the rapid rate of longitudinal progradation that have characterized its more recent evolution, has drawn the attention of many investigators. This chapter summarizes some of the cartographic results of a study of the spit's recent evolution using various different data and information sources including maps, charts, aerial photographs, and satellite images analysed in a GIS.

Introduction

The spit El Rompido at the mouth of the River Piedras in SW Spain currently forms a notable example of a right angled estuary. The 12km. coastal spit that we see today has evolved from a system of barrier islands, still active until the late nineteenth century, which extended from the mouth of the Guadiana river on the Portuguese border to the Rio Piedras. At the end of the nineteenth century the majority of the tidal inlets, which separated the barrier islands, began to close, thereby contributing to the formation of a continuous ridge of coastal dunes which now ends in extreme distal point of the El Rompido spit. Situated on a very dynamic coast which is subjected to a strong eastward littoral drift, recently modelled mathematically at up to 260000 m^3/yr. (Medina, 1991), the spit is geomorphologically and ecologically significant and it has been declared a Natural Protected Area by the Regional Government of Andalucia.

This spectacular geomorphological unit and the rapid rate of longitudinal progradation that have characterized its more recent evolution have drawn the attention of many investigators (Borrego et al., 1992; Dabrio, 1982; Medina, 1991).

The chapter summarizes some of the cartographic results of a study of the spit's recent evolution using various sources - maps, charts, air-photos, satellite images - treated digitally in a GIS context. IDRISI was used for satellite images, Desktop Mapping System for digital photogrammetry and stereoplotting, and ArcInfo - ArcView 3.0 for general mapping output and spatial analysis.

Time Scales

Analysis of the spit's recent evolution has been handled at different time scales:-

i) **long term change**

Evidence from old maps and charts make it quite clear that the change from a system of barrier islands to an estuary-spit occurred at the end of the nineteenth century.

ii) **Decadal time scale**

Since its establishment the spit has been subject to rapid longitudinal progradation. Historical cartography, current maps, and the earliest air photos integrated geometrically and consistently into a GIS have formed the basis for quantification. The temporal rates of longitudinal progradation have been very significant (40-60m/yr.) throughout the period. Progradation has been by the accretion of beach ridges at the spit's distal end, and as interpreted from satellite images and air photos these beach ridges reveal a clear evolution from an alignment perpendicular to the coast towards a more parallel alignment. This progressive change in orientation has been interpreted in terms of weakening tidal currents and also in terms of greater and lesser availability of sediment supply.

Nautical charts with detailed point bathymetry have made it possible to evaluate the sedimentary balance of the submarine beach (shoreface) between the reference dates of 1943 and 1981. This was done by generating separate DEMs (TIN Data Structure) and exporting them to a grid format for analysis of spatial and volumetric changes (chart differencing).

The volumetric difference between the reference dates gives an annual rate of 119,000 m^3/yr., which, given the theoretical rate of 260,000 m^3/yr. of sediment transport, suggests that the spit incorporates some 45% of the theoretical sediment transport mobilized by littoral drift. Finally the three dimensional graphs and the sedimentary balance image show how the volumetric changes reflect, from a spatial perspective, the change from an estuary model with a single main tidal channel and a clear associated submarine delta to two tidal channels model and a tidal delta which extends parallel to the coast.

iii) Annual temporal scale

Analysis has concentrated on the years 1984 - 1996 in an attempt to trace the evolution of the coastline (backshore/foreshore) in the emerged spit and its associated beaches (using air photographs and digital steroplotting in Desktop Mapping System 4.0) and in the sandy intertidal sediments, in effect tidal deltas, (using satellite images and digital image processing in IDRISI).

The results are very interesting in that the evolution of the photo-interpreted coastline shows a change in behaviour. The emerged spit undergoes minimal morphological change and practically ceases its longitudinal progradation whilst the adjacent beaches, which previously moved in step with the spit's prolongation continue their eastward migration (90 m/yr.), although this is no longer directly related to the progradation of the emerged spit.

The westward migration of the beaches is now explained in terms of the intertidal sediments. The chapter covers the evolution of these intertidal sand bodies from 1989-1996 in a set of enhanced TM Landsat images that present two conclusions.

i) the intertidal sediments have continued their eastward progradation at rates similar to the adjacent beaches providing shelter from the dominant south-westerly wave action and augmenting the beach accumulation processes

ii) there is a clear cyclicity in progress

* In the dry years (1989-1994) with low storm evidence and low river discharges, the intertidal sediments move eastward, progressively deflecting the main channels until they are parallel to the coast and only one main tidal channel operates in the mouth.

* During the wet phase 1994-1996 with high storm incidence and high river discharges the intertidal sediments are broken up and a second tidal channel is established which becomes hydraulically more efficient than the old channel which is gradually reduced. In this way, part of the intertidal sediment is passed beyond the estuary mouth and continues its eastward migration driven by the littoral drift.

Key points for further investigation

- what happened at the end of the 19th century?

- **what is the relationship between the evolution of the spit and the changing orientation of the coast at its current distal point?**

References

Borrego, J., Morales, J.A., and Pendon, J.G (1992). Efectos derivados de las actuaciones antropicas sobre los ritmos de crecimiento de la flecha litoral de El Rompido. *Geogaceta*, 11: 89-92.

Dabrio, C.J. (1982). Sedimentary structures generated on the foreshore by migrating ridge and runnel systems on microtidal and mesotidal coast of Spain. *Sedimentary Geology*, 32:141-151.

Medina, J.M. (1991). La Flecha de El Rompido en la dinamica litoral de la costa onubense. *Ingenieria Civil*, 80:105-110.

CHAPTER 16

Littoral and Shoreline Processes in Large Man-Made Lakes

A. Sh. Khabidov

ABSTRACT: The generation of GIS and their application for environmentally sound management of the coastal zone are impossible without knowledge of environment itself. Traditionally, the emphasis is given to oceans, marginal and enclosed seas or natural lakes in studies of the coastal environment. Evidence from the coastal environment of man-made lakes is limited and mechanisms of its changes are not generally agreed upon. Because of this, the findings of investigations of the sedimentary environment of large man-made lakes and the associated littoral and shoreline processes will be discussed in this chapter.

Introduction

No less than 30,000 man-made lakes can be counted around the World. In response to the creation of reservoirs landscape changes took place over an area of 1.5 m. km^2, the most marked changes occuring at the boundaries between water bodies and adjacent terrestrial patches. This has generated a need for the development of coastal zone management of man-made lakes (Avakyan et al., 1987; Voropaev and Vendrov, 1979).

Most, if not all, of the responsible authorities dealing with the coastal resources of oceans, marginal and enclosed seas, natural and man-made lakes recognize the interrelationships between the natural coastal system and the fabric of society. This leads to management in which both are taken into account or, in other words, to environmentally sound management. Thus, coastal zone management seeks a harmonious interaction between the socio-economic system and the relevant constituents of natural systems.

Two thirds of mankind have chosen to live, work and recreate on land and cities near the coast. However no less than 60-70 per cent of all the sedimentary coasts of oceans, seas, natural and man-made lakes appear to be receding as the effect of natural processes as well as human intervention. Because of this, investigation of littoral and shoreline processes provided a basis for integrated coastal planning and management.

The man-made lakes differ in nature not only from the oceans, marginal and enclosed seas, but from the natural lakes as well. There are good grounds to believe that such lakes are not unified depositional sedimentary environments and properly represent a complex of different ones (Khabidov, 1985; Khabidov et al., 1998; Zhindarev et al., 1998). Determination of the main environments of man-made lakes and inherent to them features of coastal processes is the subject of the present work. This chapter is based on the evidence from reservoirs of plain regions, since, as it is stated by Avakyan et al. (1987), that these man-made lakes are the most widespread ones.

Main Peculiar Features of Man-Made Lakes

GENERAL OVERVIEW

A reservoir is a man-controlled and time-varying water body for storage, which is created within both natural and artificial basins. Since the majority of the lakes are created by virtue of damming rivers, so far as in general their main features are associated with river valley morphology and geological peculiarities, 'inherited' hydrological factors, and seasonal and/or long-term water level fluctuations due to the river discharge changes and human activities.

Since the morphology of river valleys has been much studied and discussed in the literature (e.g. Easterbook, 1993; Ritter, 1978), only the features of direct concern here will be listed. Firstly, the bulk of river valleys have an asymmetric shape. Secondly, complex-shaped river plains and diversified terraces are common features of the valleys. Lastly, the relative depth and width of valleys usually increases downstream in the river because of the base level stepping down and lateral erosion. However, on numerous occasions the river valley width is the subject of wide variations due to local geological distinctions.

Among the inherited hydrological factors are channel flows. These flows (Matarzin et al., 1977) result from the lengthwise gradient of water surface which, in its turn is produced by the differences in the river discharge at the entrance section of a man-made lake and at the dam. The gradient peaks at the area of river and water body junction. Thereupon it decreases approaching zero at the dam. The mean velocity of channel flows changes in response to changing of water surface gradient. Exceptions to this rule occur within the cascade of reservoirs where they arise from uneven discharge at the adjoining hydropower plants, and yet in either case the highest channel flow velocity is observed in the flooded river channel with lesser flows in the nearshore areas of the reservoirs.

Seasonal and/or long-term fluctuations occur in the water level of man-made lakes because of the annual changes of discharge and effective water storage in the reservoirs. The relation between them may range from 8% to 20% under the seasonal and from 20% to 50% under the long-term control, respectively, exclusive of the tropical and equatorial areas (Voropaev and Vendrov, 1979). The water level fluctuations vary in amplitude as the relation. Besides, it changes along a man-made lake fetch: the magnitude of fluctuations reaches a maximum value at the vicinity of dam and is negligible at the junction of a river and reservoir.

Wind and waves are further important factors affecting the depositional sedimentary environment of man-made lakes. This effect arises by virtue of the fact that during the reservoir filling, as its water level rises, the lake fetch length, area, and mean depth increase. Clearly the wave climate of a man-made lake is associated not only with the frequency of winds and their velocities, but with the morphology of the lake basin and the water fluctuation pattern as well. For these reasons, the wave height, all other factors being equal, increases away from the junction of river and reservoir at all instances.

It seems reasonable to say that, taken together, the factors outlined above are responsible for differentiating amongst depositional sedimentary environments of man-made lake basins. Let us consider this as in interplay between geomorphic, geological and physical factors by the example of Novosibirsk Reservoir. This man-made lake is well suited for the explanation of such mutual relations, because of the isolated location, simple plane shape, classical morphology of the basin, and uniform composition of sedimentary coasts.

THE PATTERN OF PHYSICAL PROCESSES IN MAN-MADE LAKE

The Novosibirsk Reservoir was formed in 1956-1959 in Western Siberia, Russia when the Ob River was dammed ca. 20 km upstream from the city of Novosibirsk (the dam: latitude 55°N, longitude 83°E). The man-made lake reaches 220 km southwest and has the following morphometrical features: maximum, mean, and minimal widths are 22, 10, and 2 km, respectively, maximum and mean depths are 25 and 9 m, respectively, water surface area is 1070 km^2, total and effective water storage are 8.8 km^3 and 4.4 km^3, respectively, and total shoreline length is 550 km.

The lake's basin inherited main features of morphology and geology of the Ob River valley. The valley had a simple plane shape and an asymmetric shape in section, That is why the deepest area of the man-made lake basin is shifted towards the right shore. The river plain and three terraces of sand and sandy loam origin were presented in the original relief (Beirom and Shirokov, 1968).

Seasonal changes in the water level of the Novosibirsk Reservoir include: (i) qater level rise with duration ca. 50 days at a rate of ca. 0.1 m per day, (ii) the level of stabilization at the project mark of high water (mean duration ca. 120 days), and (iii) water level subsidence with duration ca. 195 days at a rate of ca. 0.02 m per day. The fluctuations of ca. 5 m per year at a distance of 135-140 km from the dam remain practically the same and then a gradual decrease is observed; as this takes place, wide fluctuations are fixed now and then. Notice that effective water storage of Novosibirsk Reservoir (4.4 km^3) is well below the Ob River discharge (more than 54 km^3 per year), with the result that the lake is filled annually and has a relatively stable water level fluctuation regime.

At a latitude of 55^0N, the man-made lake freezes in winter. The ice season lasts about 180-190 days from November to April or May. The ice thickness can reach ca. 1.0-1.2 m. The water level fluctuation during the iceless season is ca. 3.5 m. In keeping with the general rule, the highest velocities of channel flows in the man-made lake (up to 1.5 - 1.7 m/sec) during the iceless season occur in the flooded Ob River channel with lesser flows at the depth of closure of the man-made lake (Table 1).

Based upon evidence derived from the observations, the storms of west-south-westerly, south-westerly, south-south-westerly, coinciding with or similar to the fetch of Novosibirsk Reservoir have the largest strength, duration, and frequency. Table 1 gives the values of significant wave heights at the depth of closure of the lake which were measured under the west-south-westerly wind with velocity of 20 m/s. As it may be seen, the wave climate of the Novosibirsk Reservoir is truly dissimilar within different parts of the man-made lake. During the iceless season, in the upper shallow area of 65 km long, wave height does not exceed 0.5-0.7 m even under severe storms. Outside this area, in the deepest places in the lake the wave height reaches 3-3.5 m. Here waves of 1m in height are observed approximately for 74 days per season, waves of 1-2 m height - more than 9 days per season, waves of 2-3 m - about 0.8 days per season, and waves of 3 m and more - up to 0.06 days per season.

Surface Processes in the Coastal Zone

DEPOSITIONAL SEDIMENTARY ENVIRONMENTS OF MAN-MADE LAKES

Judging from the environmental data, within the basin of a model man-made lake sequential substitution of relief-forming and depositional sedimentary environments along its fetch are observed. As this takes place, the mainly fluvial sub-environments are changed by predominantly wave ones. These changes are caused by inherited morphology and geological structure, peculiarities of water level changes, water circulation in the lake and wave climate typical for it. Three main types of the environments can be distinguished (Fig. 1):

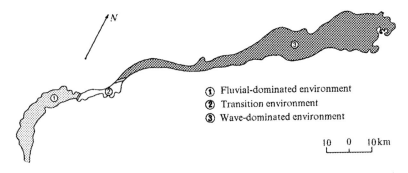

Fig. 1. Depositional sedimentary environments of Novosibirsk Reservoir

1) Fluvial-dominated environment in the site of 60-65 km long (220-160 km away from the dam). River deltas, in particular, constructive deltas, are similar to the environment mentioned since fluvial processes play a key role in formation of both
2) The transition environment of 15-20 km long on the lefthand coast and 30-35 km long at the right one, where both the waves and channel flows are responsible for geomorphic processes
3) Wave-dominated environment, where the wave-induced surface processes play a key role. Total length of this area reaches 130-140 km

TABLE 1
Distinctive channel flows velocity at 0.5 m from the bed and changes of the significant wave height lengthwise the Novosibirsk Reservoir under the predominant wind of 20 m/s.

Distance From the dam, km	Water level rise Flow velocity, m/sec			Water level stabilization Flow velocity, m/sec			Water level subsidence Flow velocity, m/sec		
	Flooded River channel	Depth of closure Right shore	Depth of closure Left shore	Flooded River channel	Depth of closure Right shore	Depth of closure Left shore	Flooded river channel	Depth of closure Right shore	Depth of closure Left shore
220	1.40	*	*	0.60	*	*	0.45	*	*
200	1.50	*	*	0.58	*	*	0.32	*	*
180	1.10	*	*	0.45	*	*	0.26	*	*
160	0.75	0.28	0.16	0.38	0.20	0.12	0.21	0.15	0.07
140	0.46	0.20	0.11	0.29	0.16	0.07	0.17	0.09	0.05
120	0.36	0.10	0.05	0.25	0.05	-	0.15	0.04	-
100	0.30	0.04	-	0.19	-	-	0.12	-	-
80	0.24	0.02	-	0.14	-	-	0.11	-	-
60	0.18	-	-	0.10	-	-	0.09	-	-
40	0.08	-	-	0.06	-	-	0.05	-	-
20	0.04	-	-	0.03	-	-	0.02	-	-
1	0.02	-	-	0.02	-	-	0.02	-	-

Distance From the dam, km	Significant wave height at the depth of closure, m	
	Right shore	Left shore
220	*	*
200	*	*
180	*	*
160	0.51	0.38
140	0.84	0.95
120	*	*
100	1.21	1.25
80	1.44	1.42
60	1.57	1.62
40	1.88	*
20	2.00	2.05
1	2.50	2.15

* measurements were not made

It seems likely that the relief-forming and depositional sedimentary environments, revealed at the Novosibirsk reservoir, are common types of environment for man-made lakes of plain areas with similar features. Small differences in hydrodynamics, the configuration of the basin, the original relief and geological structure between one lake and another may simplify or complicate the scheme presented. Nevertheless such changes do not cause serious repercussions. Furthermore, based on the data provided by the studies of Reineck and Singh (1973), Matarzin et al. (1977), Allen and Collinson (1990) and others, a large body of remote sensing data, and field surveys, Khabidov et al. (1998) suggested that certain environments are present within the basins of piedmont man-made lakes as well.

LITTORAL AND SHORELINE PROCESSES WITHIN MAIN ENVIRONMENTS

At the entrance section of Novosibirsk Reservoir, annual normal sediment flux ranges up to 6.9 mln.m^3 per year. The major part of the sediment, mainly sand, is deposited over the fluvial-dominated area, because of the low wave energy of the lake basin. The erosion is most often observed in the channels and at the terrestrial patches adjacent to them, with the total erosion rate not more than 4.7 mln.m^3 per year (Lysenko, 1967). This being so, within the fluvial-dominated environment of the lake numerous aggradative isles, operating, cut-off and overgrown river channels separated by drained or semi-drained terrestrial patches as well as shallow subenvironments (depths of less than 2.5-3 m) with quiet or even stagnant state are typical features. The latter ones are most often presented by a diversity of complex bays, river plains, small lakes, and swamps.

The front of the delta-like area is marked by geomorphic changes and the distinct replacement of coarse deposits by fine-grained ones on the bed; as this takes place, it is more advanced in the dam direction of the righthand side of the coast of the man-made lake. The shallow sub-environments of the lake basin are completely substituted for deeper environments downward from the front.

The geomorphic changes are most pronounced in coastal morphology and dynamics. The shoreline of the transition zone of Novosibirsk Reservoir presents the continuous lines of cliffs or steep slopes caused by erosion. Typically, a sandy beach with a narrow berm (not more than 1.5-2 m) is adjacent to an escarpment. The bottom at the nearshore zone has a slope from 2 –3° up to 6-8°. At the sedimentary shores, the modern erosion rate reaches maximum (3-5 m/yr) in circumstances where waves interact with channel flows.

Here, the longshore sediment transport due to combined channel flows and wave-induced currents is dominant in the nearshore zone. However, judging from the replacement of combined current and wave ripples by wave ripples (Khabidov et al., 1998), the role of channel flows is progressively decreasing outward from the front of delta-like area and, in contrast, the role of wave-induced currents is increasing. The cross-shore sediment transport rate increases concurrently because of the undertow occurrence. As a consequence different accretive landforms at the windward side of natural and artificial coastal barriers are common within this area. Again, bay-mouth bars, single tombolos and spits are fixed here too and in doing so many bars and spits extend in the direction opposite to the general direction of channel flows. The shore-normal sediment transport gives rise to more or less stretched sites of the accretive terrace bordering the nearshore zone; the terrace face typically has a slope of 9-10° exceeding 2.5-3 times the averaged slope of the bottom within the nearshore.

During the initial filling of the Novosibirsk Reservoir and just after its formation, the rate of coastal erosion within the wave-dominated area was the highest and on some sites it was as great as 25-30 m/yr. Terrigenous material entering the nearshore zone as a result of bluff erosion was transported mainly in the offshore direction, because of rather high steepness of original slopes. As the nearshore zone width increases and the beach slope decreases due to cliff recession and sediment accretion at the periphery of the offshore area, the erosion rate began to reduce. At present the eroded shores recede at the mean rate of 4.5 m/yr and, in places, the erosion has been terminated.

Within 15-20 years the position of brow of the accretive terrace bordering the nearshore zone was relatively stabilized though casual fluctuations of the brow take place even nowadays. It marked the change of prevailing tendencies in sediment transport, that is the longshore sediment transport began to prevail over the cross-shore one. Similar to the wave-dominated shores of seas, natural and other man-made lakes, the shoreline of Novosibirsk Reservoir at the wave-dominated area represents a series of bow-shaped terrestrial patches and therefore the longshore sediment transport usually occurs as two-way migrations, the rate of which may be as great as 60,000-70,000 m^3/yr.

Owing to the combination of the processes of erosion, sediment transport and accretion, diverse landforms are being formed in the coastal zone. Among these features the wave ripples, megaripples, antidunes, both longshore and transverse bars, cliffs, benches and accretive terraces bordering the nearshore zone, barrier beaches, spits, baymouth bars, cuspate bars, both single and double tombolos, etc., should be mentioned.

Conclusions

Any one of large man-made lakes shows similarities to the type of lakes studied and it is not a unified depositional sedimentary environment. These reservoirs represent a complicated complex of different environments, such as fluvial-dominated and wave-dominated ones as well as a transition environment, where both wave-induced and fluvial processes take place. It is felt that certain of the environments are present within the basins of most man-made and natural lakes. Probably, the exception are lakes marked by a significant magnitude of the water level fluctuations and/or absence of conditions for windy wave formation.

Within the basins of man-made lakes, along the lake fetch, a sequential substitution of the fluvial-dominated sedimentary environments by predominantly wave ones is observed. The substitution of environments is caused by diversified geomorphic, geological and physical factors, and in turn drives the littoral and shoreline processes. Clearly the strategy of environmentally sound coastal zone management for man-made lakes should take into account all these factors and the interactions with each other. Therefore, for instance, the most efficient techniques of coastal stabilization have (Silvester and Hsu, 1993) natural prototypes.

Acknowledgment

This work was supported by Russian Foundation for Basic Research under projects # 93-05-12057 and 96-05-64448.

References

Allen, P.A. and Collinson, J.D. (1990). The lakes. In: *Sedimentary environments and facies*, Mir/World Publisher's House, Moscow: 85-122.
Avakyan, A.B., Saltankin, V.P. and Sharapov, V.A. (1987). The man-made lakes, Mysl Publisher's House, Moscow: 9-107.
Beirom, S.G. and Shirokov, V.M. (Eds) (1968). *Novosibirsk Reservoir' coastal zone development*, Nauka/Science Publisher's House, Novosibirsk, 1968: 5-140.
Easterbook, D.J. (1993). Surface processes and landforms, Macmillan Publishing Company, New York: 93-184.
Khabidov, A. Sh. (1985). *Coastal morphology and dynamics: a comparative analysis of seas, natural and man-made lakes*. Ph.D. thesis, Moscow State Univ., Moscow: 121.
Khabidov, A. Sh., Zhindarev, L. A., and Trizno, A. K. (1998). *Relief-forming and depositional sedimentary environments of large man-made lakes*, Nauka/Science Publisher's House, Novosibirsk, 1998.
Lysenko, V. (1967).
Matarzin, Yu.M., Bogoslovsky, B.B., and Mazkevich, I.K. (1977). *Hydrological processes in man-made lakes*, Perm State Univ., Perm: 24-81.
Reineck, H.E., and Singh, I.B. (1973). *Depositional sedimentary environments with reference to terrigenous clastics*, Springer-Verlag, Berlin: 212-354.
Ritter, D.F. (1978). *Process Geomorphology*, William.C.Brown Publishers, Dubuque, Iowa: 205-302.
Silvester, R., and Hsu, J.R.C. (1993). *Coastal stabilization: innovative concepts*. A Simon & Schuster Company, Englewood Cliffs, New Jersey: 107-540.
Voropaev, and Vendrov, S.L (1979). *World reservoir*, Nauka/Science Publisher's House, Moscow: 9.
Zhindarev, L. A., Khabidov, A. Sh., and Trizno, A.K. (1998). *Coastal dynamics of seas, natural and man-made lakes*, Nauka/Science Publisher's House, Novosibirsk, 1998.

CHAPTER 17

Coastal Zone Management: The Case of Castellón

I. Rodríguez, A. Lloret, and J. M. de la Peña

ABSTRACT: The aim of this chapter is to present the methodology used for the stretch of coast between the ports of Castellón and Sagunto, in order to introduce and compare all the factors related to that coast for its Integral Management. The methodology has been developed so that it can be generalized to any other stretches of coast, given the modifications of the input parameters which depend on the characteristics of the particular stretch, as well as on the possible alternative action steps taken to facilitate and speed up communications between the various administrations involved, which in turn maximizes the resulting Integral Management. In this sense, the Geographic Information System (GIS) is seen as an indispensable tool to handle, combine and manage the various data, as well as to allow the planning of strategies for the best management of coastal areas with special problems, keeping in mind all the aspects of the affected littoral and its interrelationships.

Introduction

The concern on behalf of the General Directorate for Spanish Coasts to approach every coastal operation from all possible aspects, through multidisciplinary studies, led its staff to ask the Center for Studies of Ports and Coasts of the CEDEX to elaborate a methodology that would allow for the integral management of a stretch of coast. The basic content would be to develop a tool for the evaluation of alternative coastal action proposal considered by the General Coastal Directorate taking into account the demands and endeavours from entities involved or having competencies in coastal management.

It was decided that the first trials would be done on a pilot area, and the most adequate was thought to be that which lies between the ports of Castellón and Sagunto, since it is one of the stretches of coast that most concerns the General Coastal Directorate at present. That concern is not only due to the regressive state in which the area is found, but also because it was the object of a major study at the beginning of this year, ordered by the General Coastal Directorate and undertaken by the Center for Studies of Ports and Coasts dealing with the littoral dynamics of the Castellón coast.

The general scheme of the methodology was shaped in the meetings that followed, being agreed that the evaluation of the alternative action steps would take place according to four different aspects, which were:

1. **Beach quality**
2. **Financial considerations**
3. **Social considerations**
4. **Ecological impact**

After specifying the possible alternative actions for the General Coastal Directorate in the stretch of coast between the ports of Castellón and Sagunto, the evaluation parameters were defined, taking into account the alternative proposals and the criteria for their assessment.

A scale between 0 and 2 was introduced in order to carry out the evaluation, as well as the need to give each parameter under consideration its adequate weight or importance. The methodology thus created would allow for the modification of those weighted scores according to the requests forwarded by the authorities involved in the election of the actions. This could be done following prior discussions with experts in the particular fields.

Two significant aspects are worthy of comment: One is that the methodology will be applicable to any other stretch of coast, after previous modification of the starting parameters that will depend on the characteristics of said stretch, as well as the possible alternative action steps. That way, the proposed method facilitates and speeds up discussions between the different entities involved, so as to maximise the Integrated Management of a coastal stretch.

The second aspect is that the system set forth here is not closed, allowing for any necessary modifications that may come up as the result of conversations between the involved entities. That way, it is simple to obtain new comparative results through the GIS.

Location and Problem Aspects of the Castellón Coast

For the most part, the Castellón coast, between the harbours of Castellón and Sagunto, is made up of very old marshes and dried-up coastal lagoons, whose configuration is that of a plain behind the beach, of low altitude over the sea level, and a levee of sediments, in many cases dune chains, of higher altitude that separates the land from the sea. The constant erosion, and in some cases, construction, have caused such levees to disappear, exposing the low-lying land in the plain to flooding from the sea. Higher sea levels coincide with the storms in the area.

Coastal Zone Management: The Case of Castellón

The solutions given to this problem have always been specific and drastic: the construction of rigid structures, building artificial levees from rubble blocks, or groyne fields in some areas by filling them with gravel or boulders.

In the present situation, it would seem necessary to have a global approach to the coast with continuous monitoring in certain characteristic points along those profiles, which would reach the maximum sea-depth, and would encroach about 300 to 400 meters inland.

Methodological Aspects

DEVELOPMENT AND KNOWLEDGE OF THE ENVIRONMENT

The work plan for the Integral Management of a coastal stretch should be divided in a series of action steps to be analyzed and studied (Fig. 1). The first consideration should be given to the creation of guidelines and organization that would bring together all the involved entities to take the final decision. As the first step in the study it is important to know, on one hand, the *medium* objective of the study, divided up as follows:

1) Physical
2) Chemical-biological
3) Ecological

and, on the other hand, the *agents* that act on that medium, which can be divided into:

1) Physical
2) Human

The action of the agents on the medium generates the *processes*, that can be classified into:

1) Littoral
2) Environmental
3) Socio-economic

Once, the medium, the acting agents and the processes are known we can deduce the consequences that any action steps would generate on a stretch of coast, as well as the sensitivity in each state to natural risk. We could then select the alternative action in a first approach basis, after the problems are detected and premises are in hand.

STUDY OF THE ALTERNATIVE ACTIONS

To begin with, there were five general alternative action scenarios chosen:

1) Alternative 0 (no steps taken)
2) Build up of dune profiles and bypassing along the existing harbours
3) Closed, rigidized beaches in the midst of urban stretches and no actions taken elsewhere
4) Beach nourishment with periodic renourishment
5) Pocket urban beaches and rubble mound longitudinal breakwater in other areas

EVALUATION PARAMETERS

The evaluation of alternative action steps would be carried out considering four differentiated parameter groups, to wit:

1) Quality of beach (shape, dimensions, quality of sediments and water, climate, services, accesses, etc.)
2) Financial assessment (direct and indirect costs, direct and indirect benefits)
3) Social assessment (social response, landscape, compatibility with other policies, unemployment, etc.)
4) Ecological considerations (natural spaces, humid areas, other habitats)

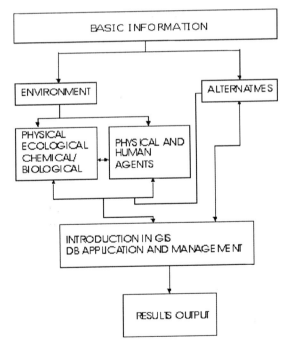

Fig. 1. action steps to be analyzed and studied for the Integral Management of a coastal stretch

In Fig. 1. a scheme is presented that shows the parameters and their relations as given above.

METHOD OF PARAMETER EVALUATION

The parameters will be evaluated according to:

$V_i = P_i \cdot i$ = the value of the parameter.

Taking the following into consideration:

P_i = Weight = Importance of the parameter to characterize the quality of the beach (from 0 to 1);
i_i=*index*=relative value of the parameter (0=bad, 1=regular, 2=good).

In order to evaluate the indexes, absolute values are taken into consideration for *quality of beach*, and relative values (alteration produced according to the alternative taken) for *social evaluation and ecological impact*. The *monetary evaluation* will not be done by indexes, but by monetary units.
      The weights will be determined according to one of the following methods, based on consultations:

1) *Ordering by ranking*:

$$P_i = \frac{\sum_{j=1}^{m} R_{ij}}{\sum_{i=1}^{n}\sum_{j=1}^{m} R_{ij}} \qquad (1)$$

m = number of individuals consulted
n = number of parameters
R_{ij} = rank that individual i attributes to parameter j.

2) *Classification by scalar degrees*:
where the weight that individual j assigns parameter i is taken from the following expression:

$$P_i = \frac{\sum_{j=1}^{m} P_{ij}}{\sum_{j=1}^{m}\sum_{i=1}^{n} P_{ij}} \qquad (2)$$

$$P_{ij} = \frac{E_{ij}}{\sum_{i=1}^{} E_{ij}} \tag{3}$$

where E_{ij} is the value of the scale that individual j assigns to parameter i.

INTEGRATION IN THE GIS

Since most of the data are of a spatial nature, it is possible to integrate them in a GIS, which offers us the possibilities of storing, modifying and relating any type of spatial information. In this case, the GIS will work as a database, whose information has been previously geo-referenced, and from which all the thematic layers of interest will be generated, and which together with other cartographic or statistical data, will be used to facilitate a more accurate evaluation of the study area.

In addition to decomposing reality into different topics: relief, lithology, floors, rivers, settlements, roads, etc., the GIS offers us another great advantage because it can make for an interrelation between the different layers, which gives these systems a surprising capacity of analysis. The maps stored in the computer can handle very complex requests, or even be combined algebraically in order to produce derived maps that show real or hypothetical situations.

The data to be integrated in the GIS come from various sources:

. **Satellite images from the study area**
. **Aerial photographs in greater scale**
. **Various kinds of cartography (evolution, topographic and geomorphological maps, nautical charts,...)**
. **Field data (points taken with GPS, water samples, sediments, etc.)**
. **Data charts (sediments, flora and fauna characteristics, ...)**
. **Other sources**

Each and every piece of data introduced in the GIS has to necessarily pass through a cartographic standardization process, which will allow us to convert such data to a common "datum" by means of cartographic processing which creates a Cartographic Data Base.

Most of the data with spatial information are easily transformed into points, lines and areas with labelled attributes for each entity. Other data, such as satellite images, aerial photography, and oblique photographs, will be included in the raster format after going through the geo-referencing process.

Once the needed and indispensable data have been introduced, the next step is the management and operation of data, for which the GIS basic tools are used (superimposing, comparison, longitudinal and area calculations, buffers, alternative search, spatial analysis). The final result is through graphs, cartography or data charts, which will later serve as a discussion document when it comes to choosing an action alternative in the coastal stretch under study. Fig.2 shows the scheme for the strategic and operative design in the use of GIS.

In essence, the GIS provides, in the case of the Integral Management of the Castellón Coast, a coherent storage system for the spatial information, in such a way that its manipulation and/or updating can be done with the least effort. In addition, it can provide cartographic models using the transformation or combination of different variables; coincidence tables between maps; superimpose two or more layers of information, and facilitates the graphic presentation of results because it allows the access to different peripherals controlled by the same system. But what really makes it the tool capable of handling projects of this nature is its capacity for simulation making it possible to generate forecasts prior to any decision-taking.

In the case of Castellón, because the project is just under way, we are not totally aware of the real problems that this undertaking involves. We do have a general idea thanks to the broad literature that exists on that subject, but until we finish collecting the information, and the integration and operation phase is initiated, we will not truly know the capabilities of GIS in Coastal Environment Management.

Bibliography

Bartlett, D. (1995). Coastal GIS Database as Models of Reality. Cork-Irlanda: *Mast Advanced Study Course on GIS.*
Bosque S.J. (1992). Sistemas de Información Geográfica. Madrid: *Editorial* Rialp (451 pp.).
Chuvieco, E. (1990). Fundamentos de Teledetección Espacial. Madrid: *Editorial* Rialp (453 pp.).
Clark, J.R. (1996). *Coastal Zone Management.* Handbook. Lewis Publishers CRC Press (694 pp.).
Clark, M.J. (1995). Designing Information Approaches for Coastal Zone Management. Cork-Irlanda: *Mast Advanced Study Course on GIS.*
Clark, M.J. (1995) Coastal Zone GIS. Cork-Irlanda: *Mast Advanced Study Course on GIS.*
Clark, M.J. (1995). Issues in Coastal GIS. Cork-Irlanda: *Mast Advanced Study Course on GIS.*
Clark, M.J. (1995). Planning Support and Decision-making. Cork-Irlanda: *Mast Advanced Study Course on GIS.*
Committee on Coastal Erosion Zone Management (1990). *Managing Coastal Erosion.* National Academy Press (182 pp.).
Gómez Orea, D. (1992). Evaluación de Impacto Ambiental. Madrid: *Editorial* Agrícola Española S.A. (223 pp.).
Graham, D.M. (1995). *Defining Eenviromental Monitoring.* Sea Technology.
Gutiérrez Puebla, J., and Gould, M. (1994). SIG: Sistemas de Información Geográfica. Madrid: *Editorial* Síntesis (223 pp.).
Hiland, M. J., Byrnes, M.R., McBride. R.A., and Jones, F.W. (1993). Change

Analysis an Spatial Information Management for Coastal Enviroments. *MicroStation Manager* (pp. 58 - 61).

Jimeno Almeida, R. (1993). Por un modelo de gestión del litoral. Gijón: II Jornadas Españolas de Costas y Puertos (Vol. II, pp. 229 - 246).

Peña Olivas, José Manuel de la. (1996). Estudio Evolutivo de la Costa de Castellón (puerto de Castellón-puerto de Sagunto). Madrid: Centro de Estudios de Puertos y Costas del CEDEX *(Technical Report* to D.G. de Costas).

Peuquet D.J. (1994). It's about Time: A Conceptual Framework for the Representation of Temporal Dynamics in Geographic Information Systems. *Annals of the Association of American Geographers* n° 84, Vol. 3, pp. 441 - 461.

Ricketts, P.J. (1992). Current approaches in GIS for coastal management. *Marine Pollution Bulletin,* Vol. 25, 1-4, pp. 82-87.

Rodríguez Santalla, I. (1996). Los SIG en Estudios de Evolución Costera: Ejemplo el Delta del Ebro. Madrid: II Jornadas sobre el delta del Ebro, Centro de Estudios de Puertos y Costas del CEDEX.

Townend, I.H. (1993). Coastal Management. Londres: M.B. Abbot & W.A. Price *(Coastal Estuarial and Harbour Engineers Reference Book).*

Wallace, J. (1995). Data Types and Potential Sources. Cork-Irlanda: *Mast Advanced Study Course on GIS.*

CHAPTER 18

Evaluating the Coastal Environment for Marine Birds

S. Wanless, P.J. Bacon, M.P. Harris, and A.D. Webb

ABSTRACT: The marine environment is being increasingly exploited by fisheries and the oil and gas industry. Conservationists urgently need the ability to identify the processes that determine patterns of abundance of marine species. We describe a preliminary Geographic Information System (GIS) in which spatial data on environmental variables (seabird colony locations, sea depth and seabed sediments) are integrated with realistic energy constraints faced by marine birds during the breeding season. A simple foraging model predicts the spatial variation in the quality of given locations as potential feeding sites under different feeding conditions and stages of the breeding cycle. We show how the approach can be used to help managers identify key marine areas and assess the impacts of environmental change or damage.

Introduction

With the increasing exploitation of Scottish waters both for fisheries and oil and gas production, the pressure on marine communities has intensified. From a conservation point of view the need to identify the factors and processes that determine patterns of abundance in marine species has never been more critical. In this chapter we use a Geographical Information System (GIS) to integrate the results of recent studies on seabird feeding behaviour, collected using radio telemetry techniques, with spatial information on three environmental variables: namely, distance from the breeding colony, water depth and seabed sediments. In any given area this approach allows us to investigate how environmental characteristics potentially influence the feeding performance of marine species under varying levels of prey availability (Wanless et al., 1998).

The species we selected as a model was the shag *Phalacrocorax aristotelis* which is an important avian predator in the inshore marine community. It is endemic to the northeast Atlantic and Mediterranean and more than 20% of the world population breeds in Britain and Ireland (Lloyd et al., 1991). The shag is a foot-propelled pursuit diver, that relies heavily on the lesser sandeel *Ammodytes marinus* as a food supply (Harris and Wanless, 1991). It is predominantly a benthic feeder (Wanless et al., 1991a; but see Grémillet et al., in press). In a recent paper we have described the development of this integrated approach in detail (Wanless et al., 1998), our aim here is to demonstrate the potential of the technique as a management tool in coastal systems. Specifically we show how it can be used to: a) help delineate marine Special Protected Areas (SPAs) b) assess the potential impact of an oil spill; and c) estimate the effects of a reduction in food availability, such as might be brought about by overfishing.

Study Area

To demonstrate the GIS approach we selected a study area that encompassed the Firths of Forth and Tay in southeast Scotland (Fig. 1), an area which holds important concentrations of shags. The main breeding location is on the Isle of May, with smaller colonies on many of other islands in the Firth of Forth (Joint Nature Conservation Committee/Seabird Group Seabird Colony Register). From a management point of view the area is of particular interest because of its proximity to both the major sandeel fishing grounds centred on the Wee Bankie (c. 30 km east of the Isle of May) and the high level of tanker traffic en route to and from the refinery at Grangemouth further up the Forth estuary. Information on the bathymetry and seabed sediments for the area was obtained from British Geological Survey maps (BGS, 1986). These data, together with those on the sizes and locations of seabird breeding colonies, were digitised and stored in a GIS. The bathymetry (bands of 0-10, 11-20, 21-30 m etc.) and sediment type polygons (Fig. 2) were intersected and then superimposed on a 1x1 km grid of the area to produce small (<1 km^2) polygons of known depth, sediment type and location. The physical environment of the area was relatively complex with a variety of sediment types and water depths generally being less than 60 m.

The Foraging Model

Full details of the model are given in Wanless et al. (1998). Unlike many marine birds the plumage of shags is not fully waterproof, and after feeding they must return to the coast to dry their feathers. In the nesting season this requires breeding adults to return to the colony to attend the eggs or chicks. Thus a fundamental environmental constraint is the distance of a feeding location from a breeding colony and hence the time costs incurred by a bird in flying to and from the feeding area. The average flight speed of a shag is 13.2 m sec^{-1} (Pennycuick, 1987). Birds typically make one feeding trip per day during incubation, whilst during chick rearing the modal number is three per day (Wanless and Harris, 1992). One adult is always present at the nest, and nest

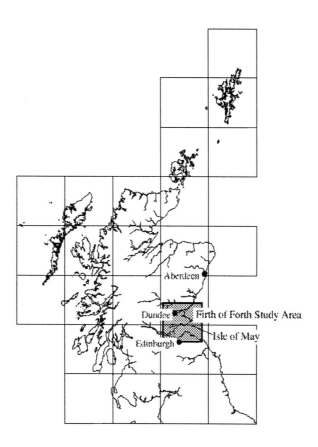

Fig. 1. The location of the Firth of Forth study area on the east coast of Scotland.

duties are shared fairly equally, so each bird has only a maximum of half the daylight hours for foraging (about 7 to 8 hours each in summer at this latitude).

The second fundamental constraint is depth of water at each potential feeding location. We used the bathymetry data in the GIS together with an equation relating diving efficiency (the proportion of the total dive cycle spent foraging on the seabed during the 'bottom' phase of each dive (bottom-time)) and dive depth (given in Wanless et al, in press) to estimate the predicted diving efficiency of a shag for any given polygon in the study area. Results from the model indicated that most of the area was potentially suitable feeding habitat for shags. However, because diving efficiency declined with depth and depth changes with distance from the coast, polygons with the highest diving efficiencies tended to be located close inshore, while offshore areas tended to have relatively low efficiencies.

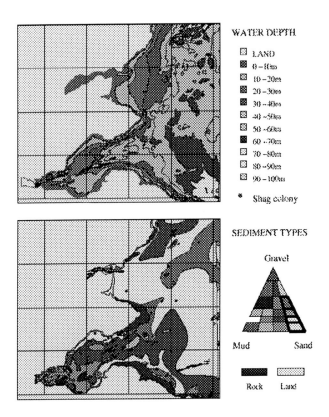

Fig. 2. Sea depths and seabed sediments types in the Firth of Forth. Data reproduced with permission of the British Geological Survey and the UK Hydrographic Office. In the Key for the sediment types, the apex of the triangle represents gravel, the left hand side of the base mud, and the right hand side of the base sand: the thick black border at the bottom right of the triangle denotes those sediments considered suitable for sandeels.

The third stage of the model development was to incorporate information on the likely distribution of potential food for shags. This was based on the observation that shags feed predominantly on lesser sandeels (Harris and Wanless, 1991) and that the most important factor influencing the distribution of these fish is the seabed substrate: sandeels are more likely to be found in sandy sediments (Reay, 1986). Accordingly, we used the GIS to select out areas where the seabed sediments were classified as sand or gravelly sand, as a first approximation to potentially suitable feeding areas for shags. Suitable habitat was defined as having a sand to mud ratio of greater than 9:1, containing less than 30% gravel, and having particle sizes varying from 625 mm - 2 mm (British Geological Survey, 1994). The Firths of Forth and Tay contain large areas of sandy sediments and there was no evidence that such sediments were associated with any particular depth bands. Predicted diving efficiencies within potentially suitable feeding habitats were highest in the inner Firth of Forth and around the entrance to the Tay estuary.

Total daily energy requirements of shags were estimated from the basal metabolic rate (BMR = 739 kJd^{-1}, Bryant and Furness, 1995) and a field metabolic rate equivalent to 3 x BMR during incubation and 4 x BMR for an adult provisioning a brood of three chicks at the time of their peak energy demand. Peak daily metabolised energy for a chick was estimated from equations given in Weathers (1992). Assimilation efficiency was set at 0.85 (Dunn, 1975) and the energy density of 1 group sandeels was taken to be 6.7 kJg^{-1} (Hislop et al., 1991). Thus during incubation we estimated that a shag had an average daily requirement of 389 g of sandeels and an adult provisioning three chicks required 920 gd^{-1}. To estimate the time needed to meet these energy demands we also needed information on typical feeding rates of shags. Such data are difficult to obtain but results from recent work on the Isle of May suggest that, on average, an adult shag feeding on 1 group (fish hatched in the previous calendar year, typically more than 10 cm in length) sandeels over a sandy substrate obtains 0.2 g sec^{-1} of bottom-time foraging (max 0.6 g sec^{-1} foraging, Wanless et al., 1998).

Having now described both the energetic and environmental parameters used to estimate the daily feeding times of a marine bird, we now demonstrate how such an approach can provide a useful tool for coastal zone management. We consider three management aspects: the identification of sea areas important to marine birds at particular colonies; possible indirect effects of oil pollution; potential consequences of reduced stocks of prey species.

Identification of Sea Areas of Importance to Marine Birds

With the introduction of the EC Habitats and Species Directive there is an increasing requirement for the UK to give legal protection to marine wildlife and marine habitats. The EC Directive calls for the establishment of a European network of protected areas to be made up of Special Areas of Conservation and Special Protection Areas (as defined in the Wild Birds Directive). In the context of identifying key marine areas, information on the distribution of seabirds at sea can be used to identify areas that appear to be consistently important (Carter et al., 1993; Skov et al., 1995). However, the use of a GIS-based modelling approach can not only indicate which sea areas are likely to be important but can also be used to explore the consequences of changing patterns of human activity and management on marine species. Fig. 3a shows the spatial pattern of predicted daily feeding times for shags breeding on the Isle of May, a National Nature Reserve (NNR), firstly during incubation and secondly during chick rearing under average feeding conditions. The most favourable and areas which would require more than the maximum available eight hours per bird per day in order to obtain adequate food are shown. An SPA centred on the Isle of May would clearly be inappropriate for shags: the great majority of the most favourable (< 3 hours) foraging areas are both inshore and distant from the island, while most of the sea further offshore is inappropriate habitat or cannot be utilised (> 8 hours). Assessment of the extent and location of suitable feeding habitat is also greatly altered by the stage of the nesting cycle. Fig. 3a contrasts the low-demands of the incubation period with the higher demands of chick rearing. The extensive areas requiring less than 3 hours feeding during incubation disappear completely during chick rearing, when require 5 to 8 hours feeding are required to supply the larger food demand.

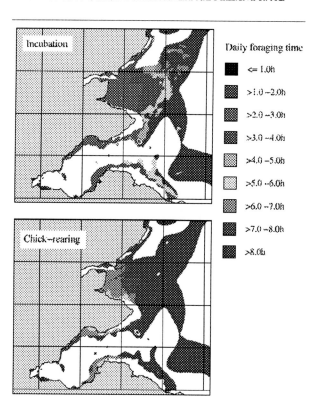

Fig. 3a. Estimated total daily feeding times for shags breeding only on the Isle of May to feed anywhere in the area shown. The feeding costs are given in hourly bands from 1 to 8 and above 8, and shaded. The time calculations assume an average prey-capture rate of 0.2 g of fish sec^{-1} foraging time on the bottom, and are given for two phases of the breeding cycle: the egg incubation period (top) and chick-rearing (bottom).

The system can also be used to contrast the importance of sea areas with regard to a single colony (protect the nature reserve) compared with their importance to shags at all the colonies in a region (protect the regional seabird community). The energetic calculations of Fig. 3a are based on flight costs to the Isle of May NNR only, although there are some other shag colonies in the Firth of Forth. Fig. 3b presents a similar analysis, but uses flight distance energetic data from each sea-polygon to the nearest colony. The shores of the Firth of Tay and the northern shore of the Forth do not have shag colonies (due to a lack of suitable cliffs), so the main changes in this region are along the southern shore of the Firth of Forth and in Largo Bay on its northern shore. Here, during chick rearing, appreciable areas that would require 6 to more than 8 hour feeding trips by shags from the Isle of May can be exploited in only 3 to 6 hours by shags from other colonies. In other circumstances, where colonies were more widely distributed, the effects could be more extreme.

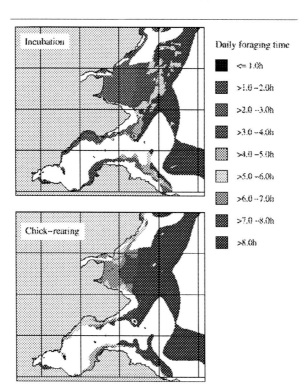

Fig. 3b. As for Fig.3a, but with the travel costs to and from each part of the sea area now referred to the nearest shag colony, not always to the Isle of May

Indirect Effects of a Notional Oil-Spill

We envisage a notional accident where an oil-spill occurs near Fife Ness (the eastern promontory between the Firths of Forth and Tay), and is moved along the northern shore of the Firth of Forth by wind and currents (Fig. 4). We assume that dispersed oil affecting the sea-bed and its fauna makes the 'slick zone' appreciably less suitable for shags to feed in during the chick rearing season: we emphasise that this is an illustrative example only. Current information on the likely effects of oil on coastal marine fish do show toxic and pathological effects, but are rather unclear about effects on fish densities (e.g. Exxon Valdez Oil Spill Symposium, 1993). The inset of Fig. 4 shows the affected slick zone and its daily foraging time cost bands for feeding shags, while Table 1 lists the areas of the various time-cost bands affected, and their proportions of the total areas of those same time-cost bands available to shags on the Isle of May, which is the nearest colony. Thus a simple program in the GIS system allows an actual, or potential slick's impact area to be assessed in relation its likely impact on the chick-rearing of nearby colonies. In circumstances where shags from several colonies might concentrate in a small area (sand over shallow water) the effects of a small oil spill in the 'wrong' area could be dramatic and widespread, and such GIS approaches could identify such important areas in advance.

TABLE 1
Different foraging efficiency (time costs) bands for shags, showing: the total areas in those bands available to shags breeding at the Isle of May, the total areas lost to a notional oil slick (see Figure. 4) and the percentage of area in the time band lost. Areas (ha) with respect to Isle of May Shags.

Areas (ha) with respect to Isle of May Shags

Time Class	Total May Area	Slick Area	Percent May area lost
<1	0	0	0
>1 - 2	0	0	0
>2 - 3	54	0	0
>3 - 4	3,040	0	0
>4 - 5	7,299	811	11
>5 - 6	11,189	959	8
>6 - 7	17,870	607	3
>7 - 8	17,820	614	3
>8	127,569	819	1

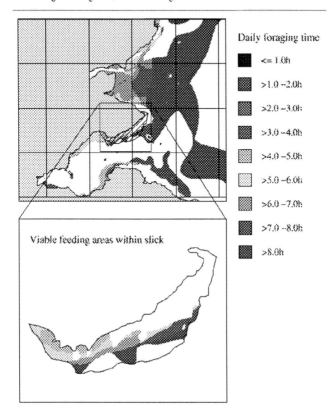

Fig. 4. Effects of a notional oil-spill on feeding areas available to shags. The top map represents feeding costs, (for all colonies, as for Fig. 3b), for average prey-capture rates during chick rearing, and shows the location effects of the notional oil-slick. The inset shows details of the feeding zones within the notional slick area. See Table 1 for the areas involved when the Isle of May is the nearest colony

Notional Effects of Reduced Food Supply

During the shag breeding season sandeels migrate inshore from the deeper waters of the North Sea into shallower regions that are more profitable for shags to exploit (Wright, 1996). Natural variation in the numbers and timing of these movements will presumably result in widely differing prey-capture rates for shags breeding at the Isle of May both within and between years. Following Wanless et al. (1998) we illustrate the effects of a plausible range of feeding rates on predicted daily feeding times. The top map in Figure. 5 corresponds to a prey capture rate of 0.1 g sec^{-1} (mass of prey captured per second of bottom-time spent foraging), typical of conditions for a year of poor food supply; the bottom map represents favourable conditions, for a prey capture rate of 0.6 g sec^{-1}. These may be compared to the average conditions for chick rearing, which are illustrated as the lower map of Figure. 3b. The contrast between the situations of Figure 5 is dramatic. At the high capture rate around half the area can be successfully exploited for under five hours total foraging effort per bird per day: at the low capture rate nowhere can be exploited successfully for less than six hours per day, and over nine-tenths of the area readily exploited (<= 5 hours) at the high feeding rate has become unexploitable (requiring more than 8 hours).

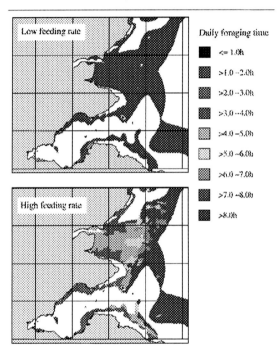

Fig. 5. Notional effects of reduced food supply. Estimated total daily foraging times for breeding shags during the peak energy demands of chick-rearing. The top map assumes a prey capture rate of 0.1 g of fish sec^{-1} foraging time on the bottom, the bottom a capture rate of 0.6 g sec^{-1}. See Fig. 3b (lower) for average rate of 0.2 g sec^{-1}

These results should be considered in the context of the recent commercial exploitation of sandeels on the Wee Bankie (ICES, 1995), and the possibility that overfishing could reduce the capture rates of shags below those used here, perhaps resulting in much of the Firth of Forth becoming much less suitable as a feeding habitat able to support shags rearing a normal sized brood.

Discussion

Surveys 'At Sea' of fish and seabirds are difficult, expensive and often confined to small areas. Our chapter illustrates how crucial ecological data on feeding behaviour can be integrated with simple but extensive GIS data on the marine environment to predict those areas most likely to be important for supporting shag populations. We discuss elsewhere (Wanless et al., 1998) how these predictions could be checked against seabird survey data, and the problems introduced by census counts that include non-breeding shags that are not confined to colonies but can return elsewhere on the coast after feeding.

We have used the shag to illustrate this approach, as its bottom-feeding habits facilitate estimating the constraints on its feeding. However, collaborative work in progress as part of a European Union funded study of the effects of large industrial fisheries on non-target species aims to examine in detail the feeding behaviour of mid-water and surface feeding seabirds (particularly the Common Guillemot (*Uria aalge*) and Kittiwake (*Rissa tridactyla*)) as well as the spatio-temporal distribution (including the diurnal vertical distribution in the water-column) of the fish shoals on which they feed. Given such data, similar simple models may allow us to identify other areas of the sea likely to be of particular importance to other top avian predators.

Acknowledgements

This work was funded partly by an award from the Leslie and Dorothy Blond Trust and the European Commission study contract 95/c 76/15, 'Effects of large scale fisheries on non-target species'. We thank the British Geological Survey for allowing a GIS version of their data to be used in the project. Some data on bathymetry are reproduced by permission of the Controller of Her Majesty's Stationery Office and the UK Hydrographic Office.

References

British Geological Survey (1986). 1:250 000 series, sea bed sediments and quaternary geology. *Tay-Forth sheet*. British Geological Survey, Keyworth.
Bryant, D.M., and Furness, R.W. (1995). Basal metabolic rates of North Atlantic seabirds. *Ibis* 137: 219-226.
Carter, I.C., Williams, J.M., Webb, A., and Tasker, M.L. (1993). *Seabird Concentrations in the North Sea: An Atlas of Vulnerability to Surface Pollutants*. Joint Nature Conservation Committee, Aberdeen.
Dunn, E.H. (1975). Calorific intake of nestling double-crested cormorants. *Auk* 92: 553-565.

Exxon Valdez Oil Spill Symposium (1993). *Program and Abstracts*. Exxon Valdez Oil Spill Trustee Committee, Anchorage, Alaska: pp 231-267.

Grimillet, D., Argentin, G., Schulte, B., and Culik, B.M. (in press). Flexible foraging techniques in breeding cormorants Phalacrocorax carbo and shags Phalacrocorax aristotelis: benthic or pelagic feeding. *Ibis,* Vol. 40(1).

Harris, M.P., and Wanless, S. (1991). The importance of the lesser sandeel Ammodytes marinus in the diet of the shag Phalacrocorax aristotelis. *Ornis Scandinavica* 22: 375-383.

Hislop, J.R.G., Harris, M.P., and Smith, J.G.M. (1991). Variation in the calorific value and total energy content of the lesser sandeel (Ammodytes marinus) and other fish preyed on by seabirds. *Journal of Zoology* (London) 224: 501-517.

ICES (1995). *Report* of the Working Group on the Assessment of Norway Pout and Sandeel. ICES CM 1995/Assess: 5.

Lloyd, C., Tasker, M.L., and Partridge, K. (1991). *The Status of Seabirds in Britain and Ireland.* T. & A.D. Poyser, London.

Pennycuick, C.J. (1987). Flight of auks (Alcidae) and other northern seabirds compared with southern procellariiformes: ornithodolite observations. *Journal of Experimental Biology,* 128: 335-347.

Skov, H., Durinck, J., Leopold, M.F., and Tasker, M.L. (1995). Important bird Areas for Seabirds in the North Sea. *Birdlife International,* Cambridge.

Wanless, S., and Harris, M.P. (1992). At-sea activity budgets of a pursuit-diving seabird monitored by radiotelemetry. In *Wildlife Telemetry: Remote Monitoring and Tracking of Animals,* pp. 591-598. Ed. by I.G. Priede and S.M. Swift. Ellis Horwood, New York and London.

Wanless, S., Burger, A.E., and Harris, M.P. (1991). Diving depths of shags Phalacrocorax aristotelis breeding on the Isle of May. *Ibis* 133: 37-42.

Wanless, S., Bacon, P.J., Harris, M.P., Webb, A.D., Greenstreet, S.P.R., and Webb, A. (1998). Modelling environmental and energetic effects on feeding performance and distribution of shags *Phalacrocorax aristotelis*: integrating telemetry, geographic information systems and modelling techniques. *ICES Journal of Marine Science,* Vol. 54, No. 4, Aug 1997, pp. 524-544

Wanless, S., Harris, M.P., Burger, A.E., and Buckland, S.T. (in press b). The use of time-at-depth recorders for estimating depth ulilization and diving performance of European Shags. *Journal of Field Ornithology.*

Weathers, W.W. (1992). Scaling nestling energy requirements. *Ibis* 134: 142-153.

Wright, P.J. (1996). Is there conflict between sandeel fisheries and seabirds? A case study in Shetland. In: *Aquatic Predators and their Prey,* Eds. Greenstreet, S.P.R. & Tasker, M.L., Blackwell Scientific Publications, Oxford UK. 208p.

CHAPTER 19

Initial Attempts to Assess the Importance of the Distribution of Saltmarsh Communities on the Sediment Budget of the North Norfolk Coast

N.J. Brown, R.Cox, R.Pakeman, A.G.Thomson, R.A.Wadsworth, and M.Yates

ABSTRACT: Very high resolution imagery from an airborne multi-spectral scanner has been used to estimate the distribution of different saltmarsh communities along a thirty kilometre stretch of the North Norfolk coast. Field observations have been used to develop a mathematical relationship between the vegetation, physical environment and sediment accumulation. This relationship has been used to produce provisional sediment accretion maps for the North Norfolk coast.

Introduction

The inter-tidal zone is very dynamic and changes occur on every tide. In addition to these frequent but gradual changes, the occasional extreme event can significantly alter the shape and character of the 'landscape'. This dynamism is part of what makes the inter-tidal zone such an interesting area of study, but it also generates problems in quantifying change. It is important to remember that the inter-tidal zone provides a number of valuable 'environmental services', in particular saltmarshes protect land and housing from flooding, act as a nursery for fish and shellfish and are a place of recreation.

Work on this chapter arose from the activities of the BIOTA (Biological Influences On inter-Tidal Areas) programme within LOIS (Land Ocean Interaction Study). A description of the aims and objectives of the BIOTA programme and especially the development of a prototype spatial decision support system (SDSS) can be found in Brown et al. (1997). In this chapter we concentrate on the development of a spatially articulate model to fit within the SDSS framework to allow users to investigate and describe the accumulation of sediment on the vegetated parts of a salt marsh.

Method

STRUCTURE OF THE MODEL

A mathematical model based on analysis of field data has been constructed to estimate the rate at which sediment accumulates on a saltmarsh. The model has been designed to fit within the SDSS framework to allow a variety of users to investigate the sensitivity of the assumptions made and to investigate the possible effects of sea level rise. Analysis of field observations was carried out using stepwise linear regression for a variety of vegetation types independently. In each case the depth of sediment accumulated on a single tide was found to be significantly positively correlated with the length of time an area is submerged and the amount of sediment in the water and negatively correlated with distance from the nearest creek and the density of the sediment. These individual equations were then combined to give the form shown below. Considering accumulation per unit time the vegetation coefficient (v) can be shown to have units of area. This implies that, all other things being equal, species with a large or complex cross sectional area trap sediment more efficiently than smaller or less complex species. The final form of the model is:

$$\Delta h = \frac{v * t * s}{d * \rho} \tag{1}$$

where Δh = the depth of sediment accumulated (m), v = is a constant dependent on the type of vegetation, t = time submerged (s), s = concentration of sediment in the water column (kg m^{-3}), d = distance from the nearest creek (m), ρ = sediment density (kg m^{-3}).

Assembling the necessary data for the model requires the integration of remotely sensed and ground based surveys with predictions of tide height.

SOURCES OF DATA

Vegetation type
Estimating the vegetation cover on the inter-tidal zone, has been carried out using classified data from the CASI (compact airborne spectrometer instrument). Data were collected with an approximate resolution of 5 meters. The CASI image used here is shown in Fig. 1. Additional flight lines cover the remaining parts of the Norfolk coast as well as the whole of the Wash and Lincolnshire coast as far as the Humber Estuary, however, these have not yet been classified.

The Effect of Saltmarshes on Sediment Deposition 235

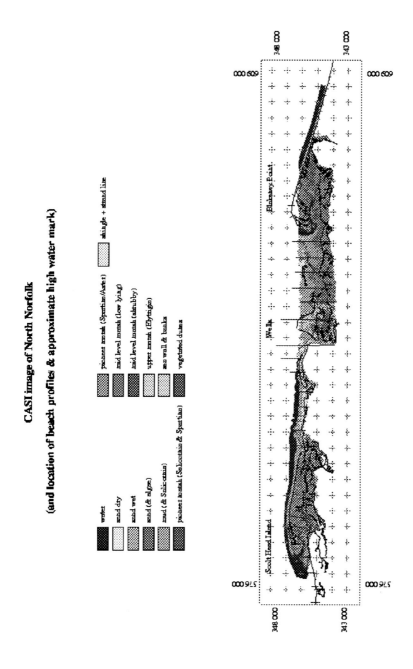

Fig.1. CASI image of North Norfolk
(and location of beach profiles and approximate high water mark)

Classification was performed using a supervised clustering technique to produce thirteen classes; classification and verification is described in Thompson et al. (1996). The abundance of the different classes is shown in Table 1.

TABLE 1
Classes defined from the CASI imagery

Class ID	Description	Area (ha)	Area (%)
1	Water	755	15.7
2	dry sand	180	3.7
3	wet sand	1078	22.5
4	sand & *algae*	391	8.1
5	mud & *Salicornia*	0	0
6	pioneer marsh (*Salicornia* & *Spartina*)	227	4.7
7	pioneer marsh (*Spartina* & *Aster*)	731	15.2
8	mid level marsh (low lying vegetation)	672	14.0
9	mid level marsh (shrubby vegetation)	379	7.9
10	upper marsh (*Elytrigia*)	266	5.5
11	sea walls & banks	14	0.3
12	vegetated dunes	63	1.3
13	shingle & strand line	49	1.0
	Total	4805	

The problems with geocorrecting imagery collected from scanning instruments mounted on aircraft should not be underestimated as it is necessary to know the exact location and orientation of the instrument every 50th of a second. To confound matters there is very little "fixed" detail in the inter-tidal zone which can be used for ground control. The image used in this study has a root mean square error of approximately 25 m.

Duration of submergence
As the predicted rate of sedimentation is directly related to the length of time an area is under water it is important to be able to estimate the elevation of any point. Other groups within LOIS are employing a subtle method to develop an inter-tidal DTM (digital terrain model) based on remotely sensed radar information and a hydrodynamic model of the sea (see Mason et al. (1997) for further details). Unfortunately this subtle DTM is not yet available and recourse has been made to the beach profiles collected by the NRA (National Rivers Authority, see acknowledgements). These have been collected at intervals of one kilometre along the entire coast of the East Anglian region (from Lincolnshire to Essex). By combining the

profile information with the CASI imagery the relationship between the shape and stability of the beach and the vegetation can be revealed.

Fig. 2 shows a profile across Scolt Head Island. Note the dunes and shingle ridges protecting the salt marsh, and the stability of the profile over a three year period. Fig. 3 shows a more exposed profile one kilometre to the east of the town of Wells where a single minor ridge is sufficient to protect stable salt marsh vegetation this is, however, conspicuously different from that at Scolt Head Island. Fig. 4 shows a profile mid way between Wells and Blakeney Point where the vegetation is very similar to the Scolt Head profile, however, the bare sediment is much more dynamic than the in the other two profiles.

It is interesting to observe the very narrow and consistent range of elevations within which the different species exist and this has been used to refine a DTM based on the NRA profiles. Construction of the DTM followed a two stage process. First the profiles were "stitched" together to form a surface using a triangular irregular network approach. Obviously with profiles so far apart any form of interpolation will miss many major topographical features (rivers, sand banks and so on). The initial surface was then refined by exploiting the relationship between ground cover and elevation along the observed profiles. For each cover type the mean and standard deviation of the elevation was used to generate a range of 'expected' values to use as an 'adaptive filter'. Each point in the DTM was examined in turn and where the interpolated surface was much higher or lower than might be expected (given the cover type), the elevation was adjusted. Where the interpolated elevation was within the expected bounds it was not altered.

The level of the tide has been calculated using the Admiralty harmonic method. One secondary port, Wells, is within the study area, but the predicted tidal range is very low due to the convoluted nature of the channel to the harbour. Fortunately two other secondary ports, Hunstanton and Cromer are respectively to the west and east of the region. Tidal heights from these two ports are used to interpolate values in between.

By combining information about the elevation of each point with the predictions as to the height of the tide, it is possible to estimate the duration of submergence at any point.

Distance from the nearest creek

The distance from the nearest creek has been estimated by reclassifying the image to try and identify creeks and then apply a buffer around them. Note, however, that even a resolution of five meters is rather coarse for the complex topography of the typical salt marsh, this may explain why the mean distance to a creek (80 meters) appears rather high by this method.

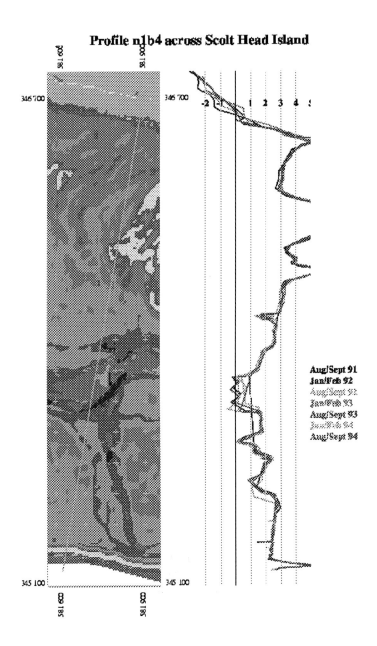

Fig. 2. Profile n1b4 across Scolt Head Island

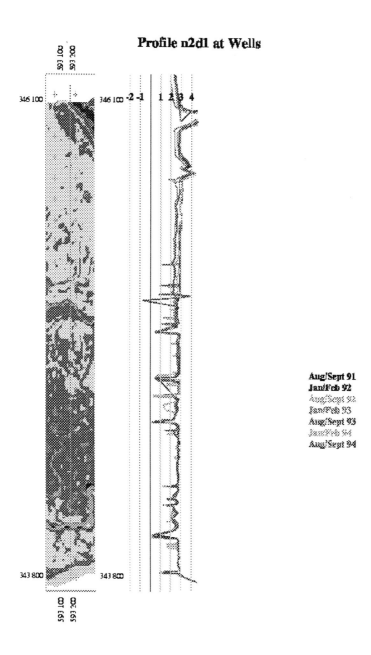

Fig. 3. Profile n2d1 at Wells

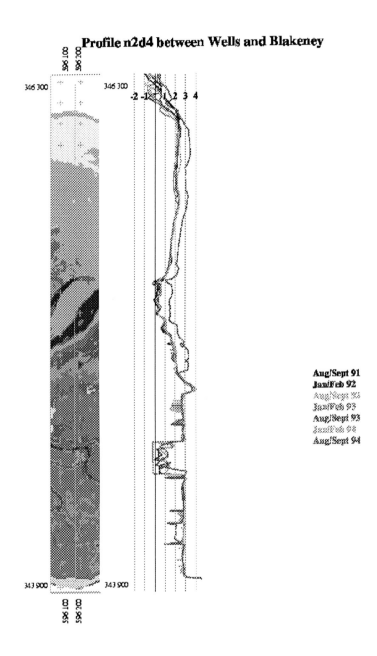

Fig. 4. Profile n2d4 between Wells and Blakeney

Sediment trapping by vegetation
Sediment accumulation by different species is based on a series of short term (single tide) observations. Filter papers are fixed horizontally within the different types of vegetation before the tide approaches. After the tide recedes the filter papers are collected, dried and reweighed to estimate the amount of sediment trapped. Environmental data around each filter paper, such as the elevation and distance to the nearest creek, is measured as well as the sediment concentration on the incoming tide.

Results

Fig. 5 shows the predicted accumulation of sediment for 1995 using a sediment concentration in the water of 1 kg/m^3 and a sediment density of 400 kg/m^3. Users of the model can select and alter these Fig.s according to their own knowledge and experience. The average accumulation rates are relatively low, averaging only about three millimetres per year over the vegetated part of the marsh. This still represents a significant volume of sediment (roughly 48 thousand cubic meters). Note that the Fig.s for accumulation are rather lower than the predicted rate of sea-level rise, which by some estimates (Warick and Oerlemans, 1990) may exceed 10 mm/year over the next hundred years.

Rigorous validation of the model is still being carried out, but, initial results can be compared to rates reported in Steers (1960) along three transects at Scolt Head Island. Accretion was measured three times over a twenty two year period. Unfortunately, there is no information about the vegetation and only the only elevation data is for the first station on each transect. Rates of accretion varied from zero to 12 mm/year. This was measured in relation to patches of sand spread on the marsh surface. Current practice is to use fine white clay that may more safely be assumed to be neutrally buoyant in the surrounding sediment. Little systematic variation can be seen along individual transects or over time. Fig. 6 shows the data from the lowest transect, Missel Marsh, for the 17 points (out of the original 37) which could be found in all years. Whether the missing sites represent erosion or other changes is unknown, so that conclusions about the net accretion over the whole marsh are difficult to draw. Fig. 7 shows the data from all three transects plotted against the elevation of the first station on each transect it demonstrates how relatively minor differences in elevation can have a major influence on the rate of accretion.

Discussion

The results of this analysis should be taken as indicative of what is possible. It should be noted that the results presented in this chapter are "typical" and are not expected to reflect the best possible fit with observed values. There are a number of deficiencies with the approach which are currently under active review. At the moment comparisons between the model and long term observations appear to indicate that the model over-predicts at low elevation and under-predicts at high elevations.

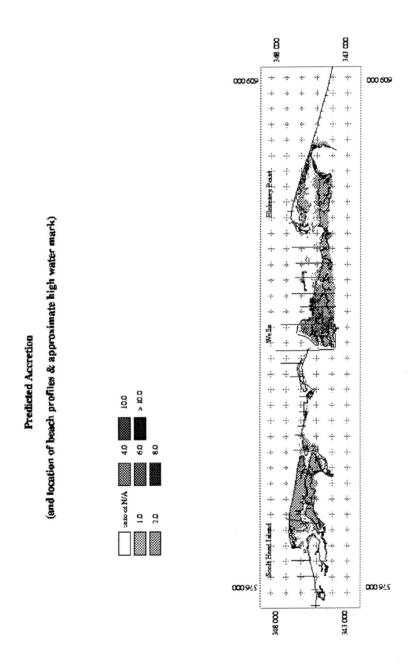

Fig. 5. Predicted accretion
(and location of beach profiles and approximate high water mark)

The Effect of Saltmarshes on Sediment Deposition 243

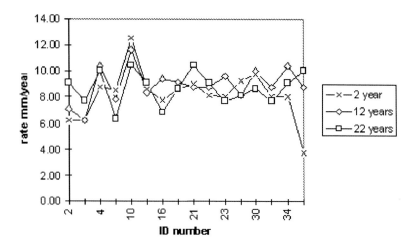

Fig. 6. Annual accretion rate Missel Marsh between 1935 and 1957 Stn. 1 at 2.28m AODN (source Steers, 1960)

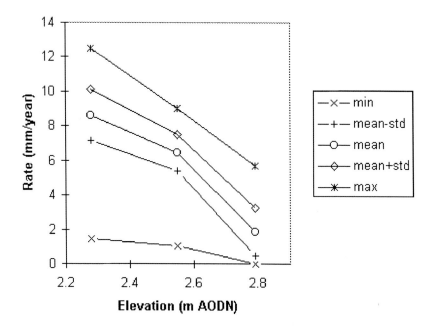

Fig. 7. Averaged accretion rates at Scolt Head 1935 to 1957 (source Steers, 1960)

Constructing a DTM from filtering an interpolated surface by ground cover provides a certain amount of circularity in the argument. In addition, problems with geo-referencing the image and the way in which the profiles record every creek mean that there is an unknown uncertainty with the final predicted elevations. The method for identifying creeks is possibly deficient and will lead to underestimates of deposition away from the front edge of the marsh.

One of the most difficult parameters to estimate is the concentration of sediment in the water column, and how much of it represents "new" sediment and how much is being transported around on the local scale.

The Admiralty method of predicting tidal levels is designed for particular ports, the physical characteristics of many parts of the coast are likely to lead to rather different tidal behaviour. Use of a hydrodynamic model which takes into account the bathymetry would be preferable.

Basing long term predictions from short term observations are likely to lead to compounded errors. In particular the importance of occasional rare events is impossible to capture in such an approach.

Conclusions

This chapter reports initial attempts to estimate accretion rates across a large area of vegetated saltmarsh. Although there are significant uncertainties about the approach, there appear to be few practical alternatives. To rely on digging up markers as in the experiment reported by Steers (1960), introduces problems with extrapolating net accretion due to the steady loss of sites. Attempts to use ground survey to describe the shape of the surface to the required degree of accuracy (to within a millimetre) implies very high costs as well as the practical problem of deciding exactly where the surface of wet mud is. These initial findings will help to quantify the flux of materials between the land and the ocean and help to assess the impacts of pressing environmental problems such as sea-level rise.

Acknowledgements

This work was funded under the NERC Land Ocean Interaction Study. We wish to thank the National Rivers Authority (now the Environment Agency) for access to their transect information. We would also like to thank: Jim Eastwood , Robin Fuller & Tim Sparks for help with analysis of the imagery and Laurie Boorman & Selwyn McGrorty for help with collecting data on the ground (or rather mud!).

References

Brown, N.J., Cox R., Pakeman, R., Thomson, A.G., Wadsworth, R.A., and Yates, M. (1997). Development of a spatial decision support system for the BIOTA project within LOIS. Proceedings *CoastGIS'97* Aberdeen 29th-31st August 1997

Mason, D.C., Davenport, I.J., Flather, R.A., and Gurney, C. (1998). A digital elevation model of the inter-tidal areas of the Wash produced by the waterline method. *International Journal of Remote Sensing.* Vol. 19(8): 1455-1460.

Steers, J.A. (1960). Physiography and Evolution. In, Steers, J.A. (Ed.) *Scolt Head Island.* W.Heffer & Sons Cambridge UK.

Thompson, A.G., Eastwood, J.A., Fuller, R.M., Yates, M.G., and Sparks, T.H. (1996). Remote sensing the intertidal zone in the LOIS project. RSS96 Remote Sensing Science & Industry. *Proceedings* of the 22nd Annual Conference of the Remote Sensing Society. 11-14th September 1996 University of Durham. 298-303.

Warick, R.A., and Oerlemans, H. (1990). Sea Level Rise. In, Houghton J.T., Jenkins, G.J., and Ephraums, J.J. (Eds.) *Climate Change: The IPCC Scientific Assessment,* Cambridge UP, UK, 257-282.

CHAPTER 20

Quantifying Landscape / Ecological Succession in a Coastal Dune System Using Sequential Aerial Photography and GIS

S. Shanmugam and M. Barnsley

ABSTRACT: This chapter presents an attempt to measure the path of habitat and vegetation succession in a coastal dune system (Kenfig NNR, south Wales) using remote sensing and GIS. The loss of slack habitats associated with the continuing stabilization of this dune system is a major cause for concern. These habitats support a range of plant species, including the rare fen orchid, *Liparis loeselii,* as well as other hydrophytes. A decrease in their areal extent implies a reduction in biodiversity. To quantify the overall rate and spatial dimension of these changes, a series of aerial photographs dating from 1962 to 1994 were digitised and analysed in an image processing system. The resultant maps, transferred to a vector-based GIS, were used to derive a transition matrix for the dune system over this period of time. The results indicate that there has been a marked reduction in the total area of bare sand (19.6% of the dune system in 1962, but only 1.48% in 1994) and a decline in both the areal extent and the number of dune slacks. Over the same period of time, there has been an increase in *Salix repens* dominated habitats, at the expense of pioneer species. Analysis of the habitat maps, together with hydrological data, within the GIS suggests that even the dry slacks have the potential for further greening and to support invasive species. In terms of habitat management, however, there is still scope to restore many of the slacks to their original state. It is estimated that at least 24% of the area occupied by partially and moderately vegetated slacks could be rehabilitated.

Introduction

The coastlines of the world, over 440,000 km in length, represent both a dynamic natural environment and an important context in which a diverse range of human activities, as well as geomorphological and biological processes, interact. Apart from their economic and recreational importance, coastal zones assume significance for the following reasons: a) there is often high diversity - both biodiversity and landscape diversity - in a very small area; b) they are active in both geological and geomorphological terms; and, c) they offer an important - quite often unique - habitat for animals and plants. Owing to the increasing pressures on coastal areas in the form of recreation, pollution and mineral extraction, the need for integrated coastal zone management (ICZM) is increasingly evident. One of the main aims of ICZM is to resolve the issues and conflicts relating to the various pressures outlined above, while considering the requirements for nature and landscape conservation.

Issues of biodiversity and nature conservation are perhaps most pronounced for dune systems, which form approximately 20% of the area occupied by world's coastal landforms and which are especially rich in species of plants and animals. Coastal dune systems also offer particularly suitable sites to study the ecological significance of the life cycles and growth form of plants. This is because, unlike many other terrestrial habitats, they frequently provide sites that are in a state of succession: thus, they combine the special interests of a successional sequence and, because the process of dune formation is often continuous, they may contain the earliest phase of succession as a permanent feature of the area. These features not only make them areas of special research interest, but have also led to some of them being categorized as Protected Areas. Biodiversity, however, the key to ecological equilibrium, can only be maintained in coastal dune systems where dune and vegetation succession are active and ongoing. The over-stabilization of coastal dunes - often as much a problem as erosion - leads to a loss of biodiversity

Accurate vegetation and habitat maps are essential prerequisites to an improved understanding of the problems associated with dunal landscapes, to monitor vegetation succession therein and, hence, to plan and institute effective conservation and management programmes. They can form the basis of, and produce the spatial dimension to, resource information systems and change-detection techniques. Conventional surveying techniques and *in situ* measurements clearly have an important role in producing such maps, but they are time-consuming, manpower-intensive and, hence, expensive C particularly in the context of long-term monitoring programmes. For this reason, attention is increasingly being focussed on the use of airborne and satellite remote sensing data, combined with the spatial analytical capabilities of modern Geographical Information System (GIS) technology. Hartog et al. (1992), for example, combined aerial photography and GIS to derive transition matrices in a study of the succession of dune vegetation structure resulting from changes in the level of ground-water in the Amsterdam Waterworks dunes. They present ideas as to how these data can be used to analyse spatial patterns on the dune surface and to model landscape succession. Similarly, Davis et al. (1994) provide an account of the use of remotely-sensed images (Landsat-TM data and aerial photographs) and GIS technology to characterise vegetation communities in south western California. The authors demonstrate how a vector-based GIS, combined with remotely-sensed data, can be used to produce improved landscape-ecological maps compared to those generated using traditional mapping and manual cartographic procedures.

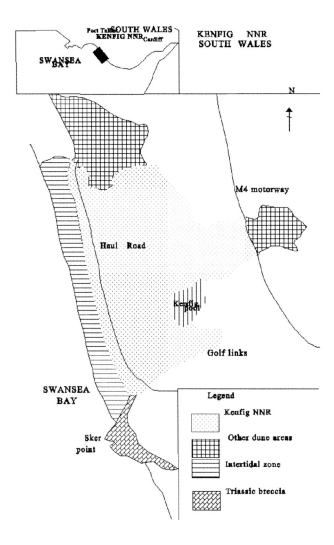

Fig. 1. The Kenfig Burrows National Nature Reserve (Kenfig NNR) in south Wales

Study Area

The study area examined in this chapter is the Kenfig Burrows National Nature Reserve (Kenfig NNR) in south Wales (Fig. 1).

Kenfig NNR is a remnant of a much larger sand dune system that once stretched along the coast of south Wales from the River Ogmore to the Gower peninsula C was established as a Site of Special Scientific Interest (SSSI) in 1977 and, at 600 hectares, is one of the largest of the 50 NNRs in Wales. It contains 575 species of flora, of which 550 are native to the U.K. Of these, 50% are native Welsh flora and 23 are nationally scarce species. Kenfig is, however, experiencing a reduction in biodiversity due to the continuing stabilization of the dune system. The three main consequences of the stabilization process are that: a) there has been accelerated succession towards a dune-heathland climax, due to the expansion of vegetated areas at the expense of areas of mobile sand; b) sand mobility has almost stopped; and c) the surviving pockets of original dune habitat are increasingly fragmented and isolated. The significance of the loss of biodiversity associated with over-stabilization is highlighted by reference to a number of important plant species.

- **Kenfig is the stronghold of the rare and declining fen orchid, *Liparis loeselii*, supporting more than 95% of the total British population. This species is the only higher plant occurring in Britain to be listed as a priority species in Annex II of the original EC Habitats and Species Directive. It is also listed as requiring protection under the 1992 Bern Convention and is on Schedule 8 of the Wildlife and Countryside Act, 1992. Despite this, it is still in decline in Britain. At Kenfig, it occurs in wet dune slacks which are being invaded by *Salix repens*, *Phragmites australis* and *Hippophae rhamnoides* (Jones 1992)**
- **form *Ammophilia arenaria* (marram grass) in the foredune areas is declining due to competition from other plant species**
- **form *Rumex rupestin* (shore dock), and *Lomosella subulata* (welsh mud wort), both pioneer species, are now almost lost**

To help conserve biodiversity and to preserve the unique nature of the Kenfig NNR, there is a pressing need to monitor and understand vegetation and dune succession in the area, including the processes that have led to the present succession towards a dune-heathland climax status.

Aerial Photography and Aerial Photo-Ecology

The need for plant ecologists to study the vegetation relationships of entire regions was emphasised as early as 1926 (Howard, 1970). Today, this regional approach is encouraged by applying aerial photographs and satellite imagery to the study of both vegetation and terrain. It is at the level of macro-ecology that aerial photographs are most valuable where they help to correct the imbalance between plant geography and plant ecology. Aerial photographs can be used to identify and delineate plant assemblages at the physiognomic or formation levels (Howard, 1970).

The concept of a *photo-community,* refers to a distinct assemblage of plant species discernible in (stereo-pairs of) aerial photographs at a specified scale. This concept, however, has its limitations, since the photo-community varies not only with scale but also with several other factors such as photo texture, tone *etc.* In this study, the photo-community concept has been used to a limited extent, but the emphasis has been more towards the delineation of habitats and vegetation densities. The vegetal association of the habitats has been inferred with the assistance of recent vegetation maps and ground checks.

Methodology

The methodology adopted in this study is similar to that used by Hartog et al. (1992). Thus, we have visually interpreted the aerial photographs (Table 1) based on the keys developed by Hartog et al. (1992), although these will not be described here. Figure 2 shows aerial photographs of the northern part of the study area acquired during 1962 and 1994.

Clearly, there are limitations to the ability to identify and map the many types of vegetation species and communities in Kenfig NNR using panchromatic or colour aerial photographs, the most important ones being the spatial scale and spectral limitations of these media.

As mentioned earlier, the focus of studies of ecological diversity and nature conservation has shifted from species to habitats, with this change in emphasis accelerating in recent years. This general observation is also true for Kenfig NNR, where loss of habitat has been the major cause for concern among the nature reserve managers. Consequently our aim has been to map habitats based on a visual assessment of vegetation density and structure in aerial photographs. Taking into account the nature of the study, the scale of photographs used, and the structure of the vegetation at Kenfig NNR, a simplified classification of habitats was produced (Table 2).

The habitat maps generated from the aerial photographs were digitised using a semi-automatic approach involving scanning, conversion to grids, and then semi-automatic digitising. The vector coverages generated in this way were subsequently geo-referenced, co-registered and rectified with reference to control points derived from Ordnance Survey maps of 1:10000 scale.

The Haul road which runs parallel to the coast along the western margin of the NNR was laid after 1962 and has had an influence on the morphology of the dunes, vegetation and hydrology of the adjoining areas (Jones, 1993). To have a common area of study before and after 1962, the Haul road has been used as the western boundary and has been superimposed on the habitat map of 1962. A similar procedure was

TABLE 1
Details of aerial photographs used in the study.

Year	Type	Scale
1962	B/W Panchromatic	1:10,000
1971	B/W Panchromatic	1:5,000
1991	Colour	1:10,000
1994	Colour	1:25,000

TABLE 2
Classification key of habitats used in this study and their equivalent NVC types.

Habitat classification (present study)	Corresponding NVC types
Open sand	Mobile dunes SD4-SD6
Partially vegetated dunes	Semi-fixed dunes SD7a-SD7g
Moderately vegetated dunes	Dune grassland SD8, SD9 and SD12
Densely vegetated dunes	
Partially vegetated slacks Moderately vegetated slacks Densely vegetated slacks	Dune slacks SD13-SD17
Woodland/Scrub	Scrub and Woodland SD18 and W

Quantifying Landscape / Ecological Succession 253

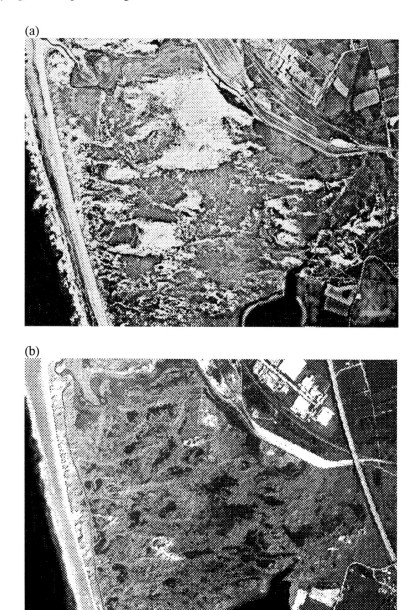

Fig. 2. Aerial photographs of Kenfig NNR acquired during (a) 1962 and (b) 1994. Note the presence of the Haul Road in the west and the motorway (M4) in the east in the 1994 photograph

adopted using the M4 motorway as the eastern boundary which is present in the 1991 and 1994 photographs but not in the 1962 and 1971 photographs. Each parcel of a given habitat is thus a closed polygon in a coverage. The vector coverage for a given year comprises a number of different layers, where each layer contains polygons representing the parcels of a given habitat. The attributes of each polygon in any given coverage include its area, perimeter, habitat type and hydrotype (Fig. 3).

Vector GIS and Dune-Landscape Succession Analysis

Overlays were produced of the individual habitat maps for each year considered in this study. The resultant change vectors present a vivid picture describing the type of successional changes that took place between 1962 and 1994. The changes are reflected not only in terms of an increase or decrease in the areal extent of individual habitats, but also in terms of partial or complete change of each habitat parcel into another stage in the successional sequence.

To understand the nature of landscape/habitat succession, each habitat was compiled as a separate layer (*i.e.*, vector coverage) and overlays were constructed between corresponding layers from different years. A total of 156 vector coverages, representing all the habitat / landscape types for the four years in consideration, were generated. Figure 4 is an example of a vector coverage containing the eight habitat/ landscape types. The areal extent of each of the habitats / landscape was computed for each of the years studied. The results confirm that there has been a general greening of the dune system since 1962. In other words, there has not only been a substantial decrease in the areal extent of open sand / mobile dune, but also a reduction in the generation of dune slacks and an increase in the biomass on existing slacks. The most likely explanation of this is that there has been a decrease in the mobility of the dunes, as a consequence of over-stabilisation.

A quantification of the change brought about in the dunal system over the years, described qualitatively above, is given in Figure 5. This shows that about 20% (approximately 82 hectares) of the study area within the NNR was covered by mobile sand during the early sixties, but that this had reduced to a mere 1.5% (6.5 hectares) by 1994. Similarly, the areal extent of the partially-vegetated sand/dunes (PVD) reduced from 26% in 1962 to 2.7% by 1994. On the other hand, the area of moderately vegetated dunes (MVD) increased from 18.7% to 56.8% over the same period of time.

In the case of the dune slacks, the partially vegetated slacks have reduced in extent from ~12% to 1.6%: this is a result of their transition into more densely vegetated slacks (0.24% in 1962 to 5.65% in 1994). The slacks also have a tendency to transform into woodlands, the area of which has increased eight-fold. Table 3 provides

Quantifying Landscape / Ecological Succession

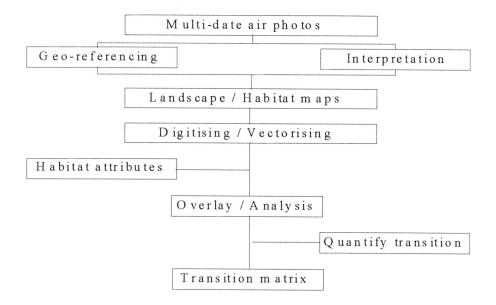

Fig. 3. Schematic representation of research methodology

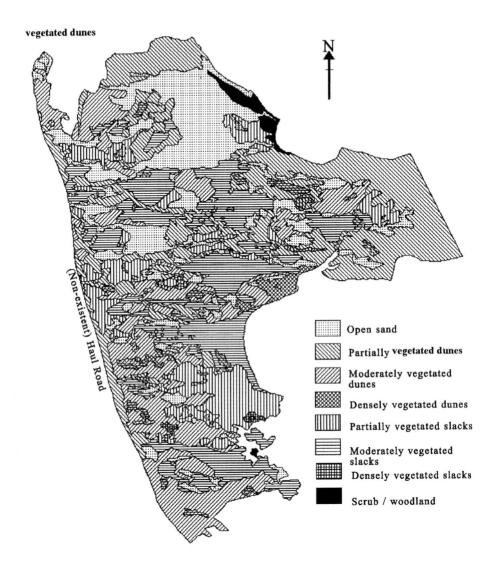

Fig. 4. An example of a digital vector coverage of the landscape / habitat map of Kenfig NNR derived from aerial photographs acquired in 1962

Quantifying Landscape / Ecological Succession

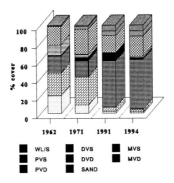

Fig. 5. Transition diagram showing changes in the type and the areal extent of landscape / habitat units between 1962 and 1994

an overview of the changes in the spatial structure of vegetation in the dune and slack habitats. A more detailed analysis of the successional changes for specific habitats, inferred from Table 3, is as follows. In 1994, the open sand (OS) cover had reduced to 5.6% of its areal extent in 1962. This was due to growth of vegetation on the dunes.

Thus, 8.8% of the open sand habitat had been transformed into partially vegetated dune habitat, 64.5 % transformed into moderately vegetated dune, and 0.6 % to densely vegetated dune by 1994. While this general trend indicates the greening and stabilisation of the dunes, there is also evidence of some erosion/dune mobility which has resulted in the conversion of 1.17% of the partially vegetated dunes into open sand. Analyses of the change vectors between 1971 and 1991 show that the geomorphological process of erosion/dune mobility seems to have almost ceased sometime after 1971.

A similar analysis for the dune slacks shows that significant erosional/dune mobility processes were in operation during 1962 and for some years later, resulting in certain areas of the slacks being converted to dunal habitats. By 1994, however, 24.7% of the partially-vegetated slacks (PVS) had been converted into moderately vegetated slacks (MVS). This conversion was principally along the periphery of the slacks. Comparison of vegetation growth on the dunes and in the slacks indicates that the dunes were greening at a faster rate than the slacks. In the dunal habitat, the areal extent of PVD conversion to MVD is higher than any other type of habitat conversion. The rate of succession towards densely vegetated dunes (DVD) and scrubland is not as effective as the succession towards moderately vegetated dunes (MVD). Field observation showed that MVD habitats could easily be restored to OS status by mowing or removing the vegetation cover. Restoration of DVD into OS is more difficult as it involves intensive, repeated mowing and complete removal of seabuck thorn (including the roots) by manual cutting and injection of weedicide into the roots.

The interest of the managers of the NNR lies more with the slacks, as they are the habitats that support the rare species, such as *Liparis loeselii*, and that are being invaded by *Hippophae rhamnoides* and *Salix repens*. In this context, 71.5 % of the area under PVS in 1962 had been transformed to MVS, DVS (densely vegetated slacks) or WL/S (woodland / scrub) by 1994. The areas of PVS that existed in 1994 were the ones created due to erosion during or soon after 1962. Similarly, 70.6% of MVS has been transformed to DVS and WL/S, while only 29.9% has remained as MVS. This study has brought to light the fact that there was some kind of erosional and dune mobility

TABLE 3

Transition matrix showing habitat changes for part of Kenfig NNR between 1962 and 1994. Key: OS = open sand; PVD = partially vegetated dune; MVD = moderately vegetated dune; DVD = densely vegetated dune; PVS = partially vegetated slack; MVS = moderately vegetated slack; DVS = densely vegetated slack; WL/S = woodland/scrub. Figures in italics = area in hectares. All other elements of the matrix = percentage of area in 1962.

			1994						
			OS	PVD	MVD	DVD	PVS	MVS	DVS	WL/S
			6.17	*11.20*	*189.80*	*49.50*	*6.76*	*92.09*	*23.55*	*33.07*
1962	OS	*76.64*	5.36	8.80	64.50	0.62	1.58	14.25	1.18	2.63
	PVD	*104.48*	1.17	1.98	73.50	0.40	0.89	15.02	0.15	7.41
	MVD	*75.20*	0.90	1.15	55.10	35.20	1.65	6.10	-	0.20
	DVD	*3.70*	4.10	-	-	93.80	-	-	-	2.10
	PVS	*47.63*	0.001	1.21	24.70	2.51	0.95	41.60	23.20	5.98
	MVS	*83.03*	-	0.52	-	-	-	29.90	55.60	15.00
	DVS	*0.98*	-	-	-	-	-	-	50.86	49.20
	WL/S	*3.98*	-	-	-	-	-	-	-	100.00

process going on after 1962, as a small percentage of vegetated dunes and slacks have been converted into open sand or less vegetated landscapes.

Slack Hydrology as an Attribute

Overlays of the habitat maps of different years has helped to quantify the direction, amount and rate of transition/succession, described above. As the hydrological regime plays an important role in determining vegetation growth and the subsequent successional trend in the dunal system, an attempt was made to demarcate slack areas that are undergoing rapid transition due to hydrological factors. The dune slacks at Kenfig NNR were classified into one of four types by Jones (1993) based on their hydrological status:

- **Type 1** slacks are typified by very shallow winter-flooding. During periods of flooding the water table displays a subdued response even to prolonged periods of heavy rainfall.
- **Type 2** slacks are similar to the above, but flood to a greater depth. The depth of flooding rarely exceeds 30cm, however, and for most dip-wells in the group was found to lie between 10cm and 25cm;
- **Type 3** are characterised by deep winter-flooding; and
- **Type 4** slacks are characterised by extremely deep winter-flooding.

A vector (point) coverage consisting of 192 dip-well locations (Jones 1993) C together with their corresponding hydrological characteristics (hydro-type/attribute) C was combined with the habitat change vector maps. This indicated that: (i) slack types 2 and 3 are the most vulnerable to transition to denser vegetation types; (ii) faster vegetation growth occurred in areas originally characterised by slack types 2 and 3; and (iii) after becoming moderately vegetated, hydrological factors have little impact on the transition into dense slack and woodland.

Limitations of this Study

The most difficult (and perhaps error-prone) part of studies such as this is rectification of the aerial photographs to some reference map projection. Although the utmost care was taken in geo-referencing the photographs and rectifying them, the scale of the photographs was different for each date, so that distortion and subsequent error in matching may have resulted in a small percentage of error creeping in the geometry of the vectors and eventually in the change statistics. A second consideration is that colour photographs, had they been available for all the years under consideration, would have provided more reliable information on vegetation type and density.

Conclusions

The general conclusion that can be drawn from the integration of aerial images and GIS used in this study is that, while the processes of erosion and dune mobility were active at Kenfig NNR at the start of the period considered (1962-1994), the dunes have become increasingly stabilised and there has been a cessation of dune-slack generation. It was also noted that the greening of the dunes has been slower than that of the slacks, and that at least 25% of the slacks could be restored to the conditions that support the survival of rare and pioneer species given appropriate management practices.

Habitat monitoring is an important prerequisite to the management of coastal dune systems, which are highly dynamic in nature. Aerial photographs provide a reasonably good source of information about the temporal changes and transitions that have taken place in recent decades. Their value in this context derives from the comparatively long time-series of data that they provide (*c.f.* the multispectral images produced by digital remote sensing devices). GIS is an ideal tool with which to examine both the amount and the rate of transition of habitats, as well as to demarcate the areas that need urgent attention in terms of management and conservation. In combining these technologies, this study has helped to highlight those areas and habitats that could be rehabilitated under appropriate management programmes and, ultimately, to ensure that the biodiversity of this coastal dune system is maintained.

Acknowledgements

The authors would like to thank the Commonwealth Scholarship Commission for providing the opportunity for S.S. to carry out research in the UK., the officials at Kenfig NNR for their suggestions and help, Dr. Peter Jones (Countryside Council for Wales, Bangor) for having provided the baseline vegetation map and hydrological data and Paul Pan (University of Wales Cardiff) for many helpful discussions and advice.

References

Dargie, T.C.D. (1995). Sand dune vegetation survey of Great Britain. A national inventory. Part 3: Wales. *Report of the JNCC.* Scotland.

Davis, F.W., Stine, P.A., and Stoms, D.M., (1994). Distribution and conservation status of coastal sage scrub in southern California. *Journal of Vegetation Science.* Vol. 5:743-756.

Hartog, M., van der Meulen, F., and Jongejans, J. (1992). Dune landscape development and changing groundwater regime: Quantitative landscape succession with help of a GIS. *Coastal dunes. Geomorphology, Ecology and Management for Conservation.* Balkema. Rotterdam.

Howard, J. A. (1970). *Aerial Photo-Ecology.* Faber and Faber, London.

Hurford, C. (1992). A survey to monitor the fen orchid *Liparis loeselii* in dune slacks ND6 at Kenfig NNR. *Countryside Council for Wales report.* Bangor. Wales.

Jones, P.S. (1992). Autoecological studies on the rare orchid *Liparis loeselli* and their application to the management of dune slack ecosystems in South Wales. *Coastal dunes. Geomorphology, Ecology and Management for Conservation.* Balkema. Rotterdam.

Jones, P.S. (1993). Ecological and hydrological studies of dune slack vegetation at Kenfig NNR, Mid Glamorgan. *Ph.D Thesis.* University College of Wales Cardiff.

CHAPTER 21

Geomatics for the Management of Oyster Culture Leases and Production

J. Populus, L. Loubersac, J. Prou, M. Kerdreux, and O. Lemoine

ABSTRACT: The coastal zone of Charente-Maritime in France has a leading position in oyster culture production. Yet it is being jeopardized by a number of problems such as overstocking, inadequate distribution of culture types, wild stock in excess, and strong siltation processes. Some of these problems result from a lack of communication and realization within the professional community, others due to obsolete management tools. The introduction of geomatics is believed to help provide some answers, besides alleviating the classical task of lease management. This chapter shows how the system was built, the spatial data geo-referenced and linked to administrative data. Examples of queries on the spatial data base reveal some otherwise invisible facts and allow prioritization of management decisions. The system also acts as a communication tool between the actors, i.e. the professionals, the local authorities and the research sector. Perpectives on further uses are given.

Introduction

Heral (1986) described the relationship between annual productions and oyster biomass over one century. He concluded that the observed oyster growth decrease was due to overstocking that limited the carrying capacity of the bay. Moreover, large tidal flats do offer big opportunities for shellfish cultivation but suffer from high biodeposition rates (Sornin, 1983). On this basis, spatial management of oyster culture seems to be a key factor for the future « sustainable development of the oyster industry ». In relation to the administration in charge of this management (Affaires Maritimes), IFREMER developed a GIS in order to provide a useful tool for integration of different spatial features such as bathymetry, leasing ground location, sediment types, regulations, phytoplankton and hydrodynamic model simulations, etc... Cross analysis of these layers should provide adequate advice for the better use of both environment and space.

Material and Methods

THE GIS PLATFORM

ArcView was selected as the GIS platform for the project, whereas Autocad was selected for the digitization of cadastre maps. During data implementation, the formats .dwg and .dxf were successively used, then transformed into Arcinfo coverages and ArcView « shapefiles ». The shapefile is an ESRI proprietary format that, unlike a coverage, does not store the topology, which makes it a lot easier to manipulate. When the topology is needed, it is computed locally. Tabular data remain unchanged from the coverage format. Raster data may be imported into ArcView and stored either as a « grid » or as an image. Grids, which keep pixel values inside a table, can be manipulated using the ArcView extension « Spatial Analyst » through various functions : slopes, isolines, enhancement, area summarizing etc...

SCRIPT DEVELOPMENT

When clean data sets are ready for use in an ArcView project, it is desirable for the end-user to perform his spatial analyses using only resident functions. The software offers a number of standard facilities, plus a wealth of scripts that only need to be copied from the library and run for specific purposes. For instance, to make the union of several identical shape files, a specific script has to be invoked, are not available in the interface toolkit. When the script has been loaded, if recurrent use is needed, it may easily be either iconised in the ArcView interface or made resident in the project. However, as little code as possible was written in the course of the project.

THE REFERENTIAL

In coastal zone mapping the Mercator projection system is the current one used for the purposes of navigation. Inland, the French IGN (Institut Géographique National) uses the Lambert conformal conical projection. Considering this, and also its links with other data sources and its potential users, the latter land-based reference was more appropriate than a maritime one. The Lambert projection was also adopted by the local services of the Ministry of Transport in charge of dredging and harbour maintenance, in particular for depth surveys.

THE INTERPOLATIONS

Interpolation is required when values of a given variable are needed at locations without measurements. This is the case here for the output of depth soundings surveys, where point samples have to be made into either a grid file or isolines. It may also concern biological data such as samples of oyster biomass. Geostatistics help us to understand the spatial structure of a phenomena (a variable) by building its experimental variogram and applying the model of the structure to make estimates in unknown places. After building the experimental variogram and interpreting it, a model of the structure and a neighbourhood are chosen. The calculation of the value at the unknown grid location is the weighted sum of all samples points contained in the neighbourhood, the weights being provided by the model. Kriging was performed here with Isatis, software developed and distributed by Géovariances.

The Lease Cadastral Maps

DESCRIPTION

One hundred and seventy cadastral paper maps cover the tidal area of Marennes-Oléron (Fig. 1), containing 22000 leases for a total surface area of 2900 hectares.

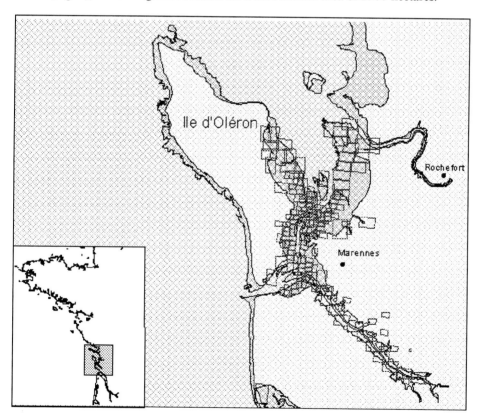

Fig. 1. Location map of Bassin de Marennes-Oléron in Charente-Maritime, France.

Their attribution and management is run by Affaires Maritimes, a body within the Ministry of Transport. Until recently, this job was still performed by hand, i.e. graphical corrections on paper maps. The whole system was traditionally geo-referenced to a local coordinate system centered on the spire of the church of the town of Marennes. Although most sheets overlap correctly, there has never been any control of the consistency of the whole structure. The system was « paperwise » viable. With the advent of its digitization and the prospects thereby raised, the question of an accurate geo-referencing could no more be avoided.

The digitization of the 22000 parcels was contracted to a chartered surveyor, who delivered the work under the dwg (Autocad) format containing the following layers : parcels, channels, «îlots » (banks), coastline and miscellaneous. The sheet and parcel numbers were attached to each parcel label point. Later, an ArcInfo routine had to be written in order to transform the polylines and polygons into a topological structure and to add to each parcel its unique number (a concatenation of the above two). A comprehensive work of polygon cleaning, dangle correction and other trouble shooting had to be carried out to end up with a reliable cadastre structure under ArcView.

GEO-REFERENCING

In order to build a transform polynomial, some conspicuous points known in both coordinates systems had to be found. The first operation consisted of assessing the positional accuracy of various church spires, and conspicuous marks surrounding the basin, which were likely to have been used for positioning. Discrepancies of up to 150 metres were found to exist between distances computed in local coordinates and in the Lambert projection. As more points were necessary, a field survey was organized in order to dGPS the few additional marks appearing on some sheets. An accuracy of 2 metres RMS was obtained, typical of a base station not farther than 20 km. The basin was split in 4 zones where different polynomials were applied using surrounding tie-points. Table 1 shows a set of five points in the NE zone which exhibits an average shift of 18.5 metres. The dispersion of 5.4 metres shows that the positions within the local referential are not consistent. This will obviously generate a distortion on the final positions computed by the Helmert polynomial.

However, this is probably the best achievable result using global polynomials based on only a few points. The alternative would be to work on a sheet basis, in which case 3 points per sheet would be necessary, a great deal of work. The lack of conspicuous points (the corner of a lease on the map having no physical existence in the field) would in all cases limit the accuracy.

BANK MAPPING

Parcels in the field are grouped into entities referred to as « banks », which are both physical units and management units, i.e. under the supervision of a bank committee. It is therefore quite a relevant geographical entity to be stored in the database. The way to extract banks is as follows. First of all, a query is made of the field « bank name » within the administrative data, after a logical join has been

TABLE 1
Positional accuracy of 5 marks in N.E. zone of Marennes-Oléron basin. The discrepancy is between distances in the Lambert and local systems.

Mark name	Distance discrepancy (metres)
Barques	10.9
Encluz	17.3
Les Anses	19.0
Piedmont	15.8
Refuge	29.5
Average/Std. dev.	18.5 / 5.4

performed with the parcel graphical file. All parcels of a given bank are highlighted. Spatial merging of all parcels may then be performed which results in one (or several) global polygon (s). Completion is achieved using the interactive graphical editing tools of ArcView in order to end up with a polygon, which is labelled after the bank name. All separate banks are then unioned to form a new shapefile.

More attributes may be appended to the bank entity. Apart from bank committee data for instance, some biological data are available per bank : this is the case of the biomass of wild stock for instance.

THE ADMINISTRATIVE DATA BASE

The « parcel administrative data base » contains 32 items of information relevant to each lease, i.e. its unique number, the owner number, the bank it belongs to, its specific rearing activity, various dates (last movement, begin and end-dates) and so forth. Another file is the « owner data base », with various civil status data, which may be linked to the previous one through the owner number.

These data are mostly for administrative use, although some of it may be of interest for management purposes. Previous experience of a similar data base construction and management had been gained by the authors with the development of a GIS for pearl oyster culture in French Polynesia (Chenon, 1990).

This whole semantic data base is centrally maintained by « Affaires Maritimes » for the whole of France and it undergoes frequent updates. Therefore, in order to maintain the data integrity, it is never modified in the project and is exploited through logical joints with the lease graphical file.

The Depth Chart

THE INTERPOLATION METHOD

The whole area covered by oyster and mussel culture inside the Bassin is currently surveyed by DDE (Harbour Maritime Service), in charge of coastal civil engineering. Their main concern is a siltation trend that has been occurring for many years and that is suspected to be the result of extensive oyster breeding on «tables», i.e. metal structures that stick out two feet off the ground. Depth surveys were conducted in 1970, 1985 and recently over the period 1994-96. Soundings were performed every ten metres on transects 100 metres apart and corrected for tidal effects, to provide a mapping on scale of 1/10 000. Tidal constraints limited transects ashore to an altitude of approximately 4 metres above chart datum.

The data are delivered as (X,Y,Z) ascii files, X and Y being Lambert coordinates in metres and Z in centimetres positive for depths above the chart datum. The variograms generated by Isatis (de Chambure, 1991; Constantin, 1996) in two orthogonal directions, i. e. NNW and ENE, both local (range of 500 metres) and egional (range of 1500 metres) are shown in Fig. 2.

The fitting using a Gaussian function is made good up to 400 metres. A slight « nugget effect » of a few centimetres is adopted, in order to avoid matrix inversion problems. The NNW regional fitting (up to 900 metres) is made to be better than the ENE one, as sounding profiles are run East-West. The neighbourhood is an oval one, with 100 and 250 metres as respective horizontal and vertical semi-axis values. The interpolation is performed on a 20 metre mesh size grid encompassing the whole basin. A section of the the depth digital terrain model (DTM) is shown on Fig. 3, with depth varying between -12 and +4 metres.

MERGING DEPTH TO LEASES

The aim of this operation is to provide the average depth of each lease. It is a classical problem of raster/vector conversion (Durand, 1994), the depth DTM and the leases being respectively under these forms. The choice of the method depends on the relative size of the objects concerned. If the parcel size and the DTM pixel are of comparable

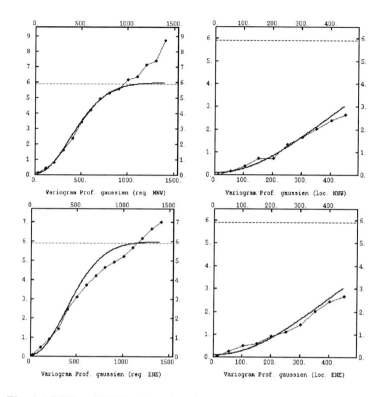

Fig. 2. NNW and ENE-oriented regional and local variograms of depth soundings adjusted by a gaussian fit with slight nugget effect

size, it is recommended to convert pixels into polygons, perform an intersection and compute the weighted sum of intersected depth polygons contained in each cadastre polygon. In this present case, with a parcel size of around 40 metres and pixel size of 20 metres, this method is inappropriate, as the intersection of both polygons coverages would yield over 100000 polygons. It is more reasonable (and much less time consuming!) to use depth pixel centroïds as representative of the pixel value and compute the average of all centroïds contained in a given parcel. This operation has the advantage of being readily available in ArcView Spatial Analyst (« summarize »). It ends up with the attribute « Elevation » appended to the lease file as shown in table 2 (Elevation in metres is positive above chart datum). Fig. 6 illustrates the case of Banc de Ronce, with parcels colour-coded according to their elevation.

TABLE 2

Lease file attribute table. « Site » is the bank name, Cm1nat is a code for culture type

Shape		Cm1nat	Site	N parcel	
Polygon	940	40	PERQUIS	24305517	2.8
Polygon	915	40	RONCE	25208051	0.7
Polygon	1388	31	PERQUIS	24302315	2.5
Polygon	1127	31	PERQUIS	24303816	2.6
Polygon	914	40	SABLE DE RONCE	24301591	2.6
Polygon	985	31	PERQUIS	24301814	1.8
Polygon	577	40	RONCE	25207560	0.2
Polygon	904	31	SABLE DE RONCE	24205382	3.3
Polygon	906	40	SABLE DE RONCE	24304132	3.3
Polygon	831	31	SABLE DE RONCE	24205493	7.8

The Tidal Model

Shellfish leasing grounds are located on large tidal mudflats and therefore are submitted to a variable immersion duration from sea to shore. Due to their filtering activity, oyster growth depends directly on this parameter. A tidal model is written under the Avenue ^{TM} programming language. The script running on ArcView provides water level estimates at given times. The calculation is made by using formulae given by the EPSHOM (Hydrographic Survey) manual. The results were compared to tables given by EPSHOM and show very little discrepancy. However, this calculation is not available when the distance with the reference harbour is too large.

This model is used to calculate the immersion duration at each depth. Water level calculations are recorded in a ASCII file with a six minute time lag. By reading this file between two given dates (t_1 and t_2), another script permits to generate a DBF file where the frequencies of water level values at a 0,1 meter accuracy (F_z) are recorded. The immersion durations (DT_z) for each depth Z are calculated by summing the frequencies over the values deeper than Z.

$$DT_z = 6 * \sum_{max}^{z} F_z \qquad (1)$$

The DBF file is joined to every depth table to provide maps of immersion duration. «Scope for growth» of oysters is a isometric function of Dtz ($SFGz = DT_Z * X$) where X is a complex function for determination of growth per time unit. SFGz represents the gain in flesh dry weight for an oyster. In this way, Fig. 4 illustrates the effect of a theoretical siltation of 0,5 metres on the spatial distribution of immersion duration, in per cent of loss of immersion time. This calculation provides direct coefficient (DT_{z1}/DT_{z2}) for calculation of « differential scope for growth » at two different depths.

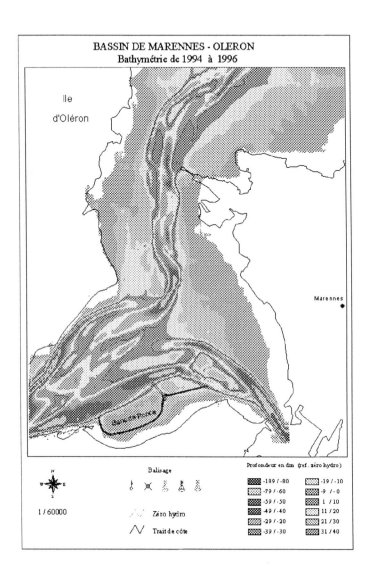

Fig. 3. Digital Terrain Model (DTM) of the tidal zone (southern basin).

$$SFG_{z2}=(DT_{z1}/DT_{z2}) * SFG_{z2} \quad (2)$$

Moreover, a ratio (PDTz) could be calculated by dividing for each depth the immersion duration by the total duration.

$$PDT_z = DT_z / (t_2-t_1) \quad (3)$$

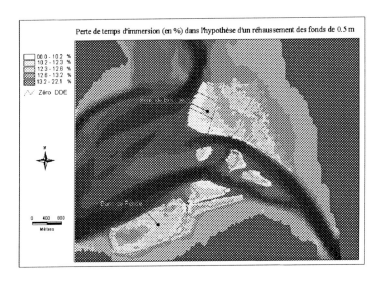

Fig. 4. Immersion duration loss (in %) for a theoretical siltation rate of 50 centimetres on oyster leasing grounds in the south of Marennes-Oléron basin.

Application to Dredging Plans

THE SEDIMENTOLOGICAL BALANCE OF THE BASIN

Obviously, the sedimentological balance is positive in the southern par of the basin. Siltation is particularly strong in « off the ground » culture areas where two feet high metallic structures produce heavy sediment deposition. However, the quantification of this phenomenon remains very approximate and tools are needed to prioritize dredging and cleaning operations.

The 1985 depth survey was processed with geostatistics exactly the same way as described above for the 1994-96 survey. By subtracting the two situations, a difference map was obtained and transformed into a grid file in ArcView. Using the « summarize » operation applied to the bank file, it is possible to display the average sediment balance per bank. Fig. 5 denotes an almost homogeneous accumulation of over 30 cm on the banc de Bourgeois, whereas deposition is more variable across banc de Ronce with values between zero and 30 cm.

THE DREDGING OPERATION OF PERQUIS/RONCE

In this area prone to siltation, the idea is to improve the flush effect of the ebb tide through channels surrounding some of the banks. The Conseil Général of Charente Maritime is currently funding dredging operations carried out by contractors under supervision of the local Harbour Maritime Service. Two operations are programmed for impending execution in several phases, the Etier de Perquis in the East (2 phases) and the Coursière des Lézards around the Ronce bank (Fig. 6).

The site width is 50 metres, which means all parcels intersecting this corridor have to be expropriated and their owners granted equivalent leases in the vicinity. The dredging corridor width itself is only 20 metres. The objective is to compute the sediment volume between the present elevation and the projected one Z_{final}.

The volume computation makes use of the depth grid file. The elevation to be reached is 1.0 metre above chart datum. After keeping only the pixels above 1 metre in the grid file (by a thresholding query on the grid table). An averaging of all pixels values contained in the polygon is then performed (using the above-mentioned « summarize » tool) which yields its average depth $Z_{average}$. The volume is computed as :

$$(Z_{average} - Z_{final}) * area \qquad (4)$$

In Etier de Perquis, the first phase concerns a surface to be dredged of 22000 m² on an average sediment height of 0.7 metres, that is a volume of 15500 m³. In Coursière de Ronce, 50000 m² are concerned on a height of 1.6 metres, i.e. a volume of 81000 m³.

Future Prospects

A number of further uses of the cadastre data base are foreseen.

- In biology, stock assessment is a major issue in this basin. In order to assess the global biomass, the delineation of homogeneous zones of culture type is needed. As regulations may not be respected, the only way to approach reality is to interpret airphotos previously registered to the cadastre file. Every couple of years, B&W airphoto campaigns are undertaken with simultaneous oyster biomass sampling
- Productivity, mortality, epizooties are also of highest concern. Ways are sought to couple results of trophic and sedimentological models with field observations in test parcels, together with the incidence of rainfall sewage outfall
- In socio-economy, there are prospects of a better understanding of the lease exchange system by computing a « theoretical » lease value based on major bio-

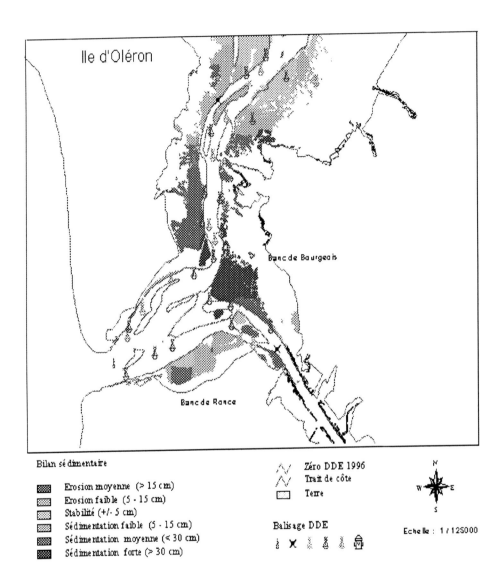

Fig. 5. Sedimentary balance of Bassin de Marennes-Oléron averaged on a bank basis

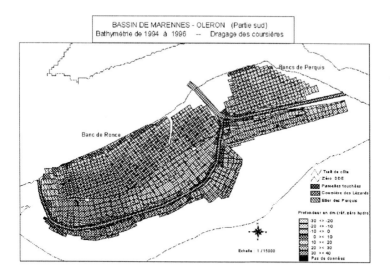

Fig. 6 The leases concerned by the operation. In Etier de Perquis, 8 parcels amount to a total surficial area of 14.8 hectares, whereas in Coursière de Ronce 67 parcels amount to 79 hectares.

physical parametres (essentially emerging time, position in terms of productivity, etc...). As market values are known, the comparison of the two might allow to refine the notion of anticipated parcel value

Conclusion

The shell culture cadastre clearly is an essential element of coastal management in France, a) for the economic value of shell production, b) for its impact on coastal environment. Its digitization is underway within Affaires Maritimes, the Administration in charge, with technical assistance by Ifremer. For lease management itself, the transition from paper maps to graphical files associated to a semantic data base will not only alleviate the traditional task of Affaires Maritimes, but also open new possibilities, namely the spatial illustration of queries made on the data base, which was otherwise unfeasible.

For wider purposes, several bodies are interested in possessing the digitized cadastre. Among them, Ifremer is in charge of dealing with biological and environmental aspects and providing advice to all possible actors in the coastal zone. The cadastre is a fundamental layer of anticipated GIS's to be set up in its various coastal stations. If only a couple of applications of data mixing were shown in this paper, indeed a host of them are expected in the near future, with a view to improving decision-making.

References

Chenon, F., Varet, H., Loubersac, L., Grand, S., Hauti, A. (1990). « SIGMA POE RAVA », a Geographic Information System of the Fisheries and Aquaculture Territorial Department. A tool for a better monitoring of public marine ownerships and pearl oyster culture. *Actes du colloque international « Pix'îles 90, Télédétection et milieux insulaires du Pacifique, approche intégrées ».* IFREMER, Nouméa-Tahiti, novembre 1990, pages 561-572.

Constantin, V. (1996). Cartographie du Golfe du Lion. Modèle numérique de terrain bathymétrique par les logiciels Isatis et Trismus, *rapport IFREMER - DRO/GM/96.34*, 26 p.

de Chambure, L. (1991) Traitement par numérisation d'une carte bathymétrique et calcul d'un MNT, *Rapport Géovariances*, 38 av. Franklin Roosevelt, BP91, 77212 AVON, France. FRA/033/149/91, 20p.

Durand, H., Guillaumont, B., Loarer, R., Loubersac, L., Prou, J., Heral, M. (1994). "An example of GIS potentiality for coastal zone management : pre-selection of submerged oyster culture areas near Marennes Oléron (France)". *EARSEL Workshop on Remote Sensing and GIS for Coastal Zone Management*. Delft, The Netherlands, 24 - 26 oct. 1994, 10 p.

Héral, M., Deslous-Paoli, J.M., Prou, J. (1986). Dynamiques des productions et des biomasses des huîtres creuses cultivées (crassostrea angulata et crassostrea gigas) dans le Bassin de Marennes-Oleron depuis un siecle. *CIEM C.M. 1986/F:41*.

Sornin, J.M., Feuillet, M., Héral, M., Deslous-Paoli, J.M. (1983). Effets des biodépôts de l'huître Crassostrea gigas (Thunberg) sur l'accumulation de matières organiques dans les parcs du Bassin de Marennes-Oléron. *J.Moll.Stud.*,Suppt. 12A, 185-197.

CHAPTER 22

GIS and Aquaculture: Soft-Shell Clam Site Assessment

A. Simms

ABSTRACT: The 1992 collapse of the Newfoundland, Canada northern cod fishery and the subsequent closure of a majority of the inshore ground fishery has placed a focus on the development of aquaculture within the province. In May 1995 Innovative Fisheries Inc. of St. John's, Newfoundland conducted field studies to evaluate the soft-shell clam resources on three sand flats near Burgeo, Newfoundland. GIS can be used to examine issues regarding the development and management of the soft-shell clam beds. GIS can also be applied to examine the issue of "competing uses" for the proposed soft-shell clam aquaculture site. The information presented in this study indicates that GIS is an important tool for the aquaculture industry. These systems can be used to monitor, quantify and evaluate the soft-shell beds near Burgeo. Management issues such as water quality, resource sustainability as well as the economic viability of the clam resource can be assessed within a GIS environment. The results of the analysis in this study suggest potential problems with faecal coliform contamination from local cottages. Finally, data collection for aquaculture site assessment is required if a resource is to be managed effectively. GIS applications provide insights into the quality of the physical environment as well as the sustainability of a resource. However, it is the aquaculture operators who ultimately make the final decisions.

Introduction

The 1992 collapse of the Newfoundland, Canada northern cod fishery and the subsequent closure of a majority of the inshore ground fishery have placed a focus on the development of aquaculture within the province. Aquaculture development is also encouraged by government policy and funding. Traditionally, commercial aquaculture within Newfoundland coastal areas was limited and produced either cultured mussels or trout. Recently, salmon, scallop and cod farming have been attempted with limited success. Therefore, a majority of aquaculture farms are based on "seeded" rather than naturally occurring resources. Given the government incentives to expand and diversify aquaculture, developers are also examining the potential of harvesting naturally occurring species such as the under utilized soft-shell clams (*Mya arenaria*).

In May 1995 Innovative Fisheries Inc. of St. John's, Newfoundland conducted field studies to evaluate the soft-shell clam resources on three sand flats near Burgeo, Newfoundland (Fig. 1). The field studies involved the collection of data on:

1) **clam biology**
2) **hydrology**
3) **water quality**
4) **land-use**

within the study area. This database will provide the information required for evaluating the economic viability and the management of the resource. This data also fulfils a requirement on government regulations for licensing aquaculture operations. The field data are spatial and the non-spatial attributes, relating to resource assessment and management issues, can be integrated, analyzed and mapped using geographical information systems (GIS).

GIS and Aquaculture

GIS has been used in aquaculture studies for at least 10 years. Kapetsky et al. (1988), Manjarrez and Ross (1995) used GIS to evaluate the suitability of coastal areas for fish farming activities. In these studies, the aquaculture potential of a coastal area was determined by factors such as bathymetry, water quality, exposure, landuse and proximity to other facilities. Aquaculture management issues such as the multiple uses of estuarine waters, the impact of water quality on shellfish leases, aquaculture and habitat availability and conflict issues between aquaculture operations and marine waterfowl habitats have been addressed by Clarke (1990), Legault (1992), and Simms (1994), respectively.

The development of spatial databases for soft-shell aquaculture sites will permit the use of GIS as a decision support tool which can help both the aquaculture developers and the government regulator in assessing and managing the clam beds near Burgeo. The advantages of using a GIS as a part of the decision making process are:

1) **GIS provides the capability to integrate, scale, organize and manipulate spatial data from many different sources**

2) **data can be maintained, updated, extracted and mapped efficiently**

and

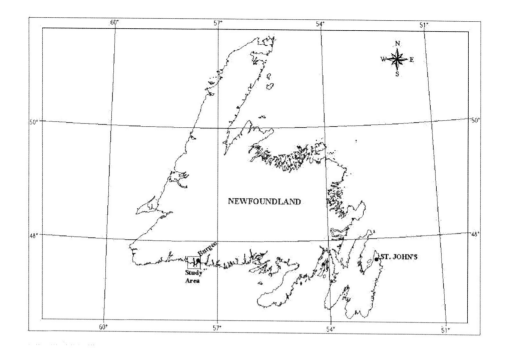

Fig. 1. Study area

3) GIS permits quick and repeated testing of models which could be used to aid the decision making process

These characteristics of a GIS can be used to examine issues regarding the development and management of the soft-shell clam beds. Development issues include water quality and avoidance of competing landuses. The primary management issue is the sustainability of the resource in the study area. This can be evaluated by the continual monitoring of the clam size distributions and water quality whereby the spatial databases can be updated for analysis and mapping in a GIS.

GIS can also be applied to examine the issue of "competing uses" for the proposed soft-shell clam aquaculture site. In this situation the decision-makers can take an active part in structuring a possible solution to a problem. Since many resource management problems affect more than one group the decision making process needs to incorporate the participation of the local inhabitants, aquaculture industry and the appropriate government agencies. Thus any issues regarding "competing uses" can be evaluated by mapping the various scenarios of conflicts or environmental impacts that a "use" may have on the soft-shell clam environment.

Spatial Data Structures

A GIS should be capable of:

1) creating digital abstracts of the real world
2) effectively handling of these data
3) providing new insights into the relationships of or among spatial variables

and

4) creating summaries of these relationships (Berry, 1993)

The data structure used to store spatial data will determine how the data can be encoded, analyzed and displayed. There are two basic data structures raster and vector. Vector data structures use discrete (x, y coordinates) points or line segments to represent spatial objects such as points, lines or polygons. Raster structures subdivides the area into a regular grid of cells (rows and columns), and these cells are used to represent point, lines or polygons (Berry, 1993). The location of vector objects are viewed as precise where each object is given an explicit X and Y coordinate. The grid cells of a raster structure provide approximate or implied location of an object. For example, if the grid cells of a raster map is 25 x 25m then the location accuracy of an object is said to be ± 25 m.

A more important factor when selecting a vector or raster data structure is the implications on the encoding, storage, analysis and display. Generally, vector data structures are used for inventory, descriptive queries, spatial database management, and computer mapping. Raster structures are used for prescriptive analysis, spatial statistics and modeling (Berry, 1993). GIS applications for aquaculture require both structures, and ideally the system used for analysis would provide the advantages of both vector and raster data structures. The Burgeo field data are stored initially as vector data and converted to raster as needed. In this study, vector and raster GIS capabilities are provided by ArcView^{TM} 3.0 and its Spatial Analyst 1.0 module (ESRI, 1996).

Data Collection

The data used in this study can be divided into primary and ancillary databases. Primary data include all field data such as clam biology, hydrology, water quality and landuse (Table 1). The 1:50,000 digital topographic maps and scanned aerial photographs are considered ancillary data.

TABLE 1
Spatial and Non-Spatial Data Used in the Burgeo Soft-shell Clam Study

Data Group	Spatial Objects	Attributes (Non-Spatial)
Clam Biology	Point	Length & Width (mm), Weight (grams)
100 x 100m Grid	Polygon	Linked to all data by GEOCODE
Hydrology	Point & Polygon	Water Depth (m), Current Speed & Direction, Bottom Substrate, Tidal Data
Water Quality	Point & Polygon	Faecal Coliform, Salinity, Temperature, Rainfall (24, 48 & 78 hr. period)
Landuse	Point	Cottage Locations (Coded as non-pollution, potential or definite pollution source.)
1:50,000 digital topographic map	Point, Line & Polygon	Coastline, Contours, Lakes, Rivers, Cultural Features and Vegetation Cover
Registered and Edge-Matched 1:12,500 Area Photographs (1989)	32 bit Raster	Used for interpretation and visualisation of site characteristics.

The storage of any spatial data within a GIS requires that location of the objects be defined as either projection (metric X and Y) or degree (latitude and longitude) coordinates, and the data are represented as points, lines or polygons. For this study, ground control points (GCP) were used to generate a 100 X 100m grid on the two estuaries (Indian Hole and Little Baraswav) and the sand flats located in Big Baraswav (Fig. 2). The GCP(s) are marked in the field with surveying poles. These grids are stored as polygons in an ArcView™ vector shape file. A unique numeric code, named GEOCODE, is used to link the non-spatial information database (e.g. water depth, bottom type, etc.) with the geo-referenced hectare polygons. In addition, the unique code is also used for reference in the field. The map coordinates and the unique code of each hectare grid centroid are stored as way-points in a global positioning system (GPS). The GPS is used to location the grids for data collection and harvesting. Therefore, because of the need to produce operational field maps, where the unique numbers must have a logical base for referencing in the field, an arbitrary number cannot be used for coding the grids. The numbers are based on the concept of a quadrant whereby NE, NW, SW and SE quadrants are assigned numeric values of 1,2,3, and 4 respectively. Thus, if a 1 hectare grid is the first cell to the Northeast of the GCP it is given the code 111 which means that the cell is located in the Northeast and is the first row and column to the Northeast of the GCP. When the grid is located in the second row to the north of the GCP and 20 columns away from the GCP in the Northeast quadrant it would be code 1220. The first digit is the quadrant code while the following digits represent the rows and columns.

The soft-shell clam biology, water quality, and landuse data are stored as points. The clam biology database represents clam samples taken at the center of each grid. Individual clams were measured (length and width) and weight. For each clam, the data was stored as a single record in the point database. This meant that a single location has more than one record. However, the points represent a sampled area of approximately 0.25m sq. However, before any descriptive analysis or mapping can be performed on this point data the information must be converted to densities or counts for each polygon in the 100 x 100m grid. The water quality points are permanent sample stations (Fig. 2) that will be used for environmental monitoring. These data are stored as x, y coordinates and as an ArcView^{TM} polygon vector shape file. Raster Thiessen polygons are formed by a proximal mapping function in Spatial Analyst 1.0 (ESRI, 1996), and the shape of the polygon is determined by the distribution of the points. A polygon is formed around a point by constructing boundaries halfway between a point and its nearest neighbor. The result is a series of irregular shape polygons (converted to vector shape files) that can be used to map water quality within a vector data structure (refer to Fig. 4). The reason for this approach is twofold; first there are some observations missing from the database that would make the point distribution too sparse for contouring by spatial interpolation algorithms. Second, if the database is transferred to a system without the capability to build Thiessen polygons or create a continuous surface by interpolation the existing polygons could be used for mapping water quality attributes.

A GIS Approach to Aquaculture Site Assessment

The data collected on clam biology, hydrology, landuse, and water quality as well as the inclusion of aerial photography and a digital topographic map provides the information required for an aquaculture site assessment. This evaluation should address issues on the suitability of the environment, potential competing landuse, and the sustainability of the soft-shell clam resource. The following sections present a preliminary analysis of the field data whereby a GIS approach is used to map and analyze the data in terms of:

1) **the physical site characteristics and landuse**
2) **water quality**

and

3) **the soft-shell clam resource**

The clam resource will be evaluated in terms of spatial variability in the density of various size class distributions and population characteristics.

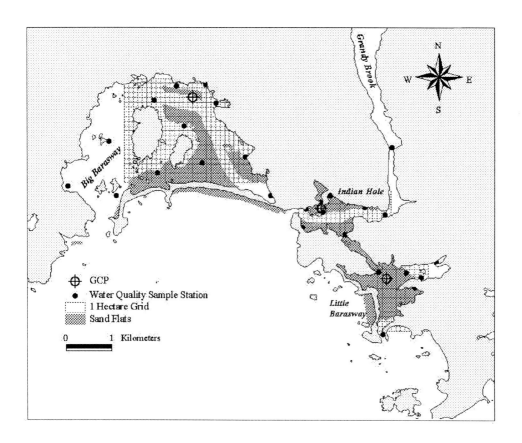

Fig. 2. Study area - survey data grids and points

Site Characteristics

The soft-shell clam inhabits bays and estuaries, intertidally and subtidally in fairly shallow water to depths of approximately 9m (Newell and Hidu, 1994). The potential aquaculture sites near Burgeo, Newfoundland consists of two active estuaries (Indian Hole and Little Barasway), and one former estuary (Big Barasway) of the Grandy Brook (Fig. 3). The three sites are sheltered from prevailing southerly and westerly winds by sand and gravel spits. The sand flats for Big Barasway and Indian Hole (indicated by a solid white line on Fig. 3) are in fairly shallow water. The water depths at median tide in these two areas range from 0.30 to 2.16 m. Little Barasway has a delta forming (see Fig. 3) and the median tide depth ranges from 0.19 to 0.88 m. The substrate in Big and Little Barasway is predominantly sand while Indian Hole is more diverse with a combination of sand and weeds. All three areas have one or two hectares compose of mostly gravel or weed substrate.

The hills surrounding the potential sites are sparsely vegetated with exposed bedrock. Given the lack of vegetation cover and the distribution of lakes, streams and rivers in the region there is a possibility of high fresh water runoff during sustained periods of rainfall. Since a majority of the cottages are a known pollution source the freshwater runoff may cause water quality problems (Fig. 3). There are also cottages located upstream on the Grandy Brook. The seaside and river cottages are usually occupied frequently from May to October. This coincides with the summer spawning and harvesting periods and could present a potential contamination problem. Finally, field observations indicate that a majority of the sand flats in Big Barasway is subject to predation by gulls and seals while the areas in Indian Hole and Little Barasway appear to experience much less predation.

Water Quality

The location of cottages in the study area (Fig. 3) introduces the possibility of contamination from faecal coliform. This may affect the commercial viability of a clam bed because there are government regulations that control the closing or opening of a contaminated area for harvesting clams. In Canada, if the faecal coliform density exceeds 13.99 MPN/100 ml the area is closed to harvesting until the density falls below this value. During the field surveythe water quality monitoring stations were sampled at regular intervals from May 28 to Sept. 13, 1995. During this period fecal coliform densities ranged from 1.9 to 920 MPN/100 ml. During a majority of the time the densities were at 1.9 MPN/100 ml, which is below the standard set by the Department of Fisheries and Oceans. However, during extended rainfall events (e.g. 52.4 mm over 72 hours from Sept. 17 to 19, 1995) the fecal coliform densities range from 17 to 920 MPN/100 ml. This event is presented in Fig. 4.

Prior to the rainfall event the fecal coliform densities were at acceptable levels, however by Sept.19, 1995 all areas were highly contaminated (Fig. 4). Two days later (Sept. 21, 1995), most of Big Barasway had returned to acceptable levels except for an area in the north, however, Indian Hole and Little Barasway remained highly contaminated. A survey of the water quality stations on Sept. 21 indicated that fecal coliform levels at Big and Little Barasway had returned to acceptable levels, while Indian Hole still had areas that were contaminated (Fig. 4). The time series graph in Fig. 4 suggests that the maximum contamination occurred on Sept. 19. Big and Little Barasway appear to flush completely within 5 days while Indian Hole may take longer because of contamination from upstream sources.

GIS and Aquaculture: Soft-Shell Clam Site Assessment

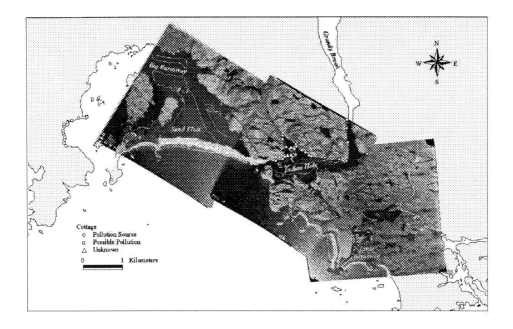

Fig. 3. Aquaculture site characteristics and cottage location

To date the long term impact of fecal coliform contamination on clams is not known. Studies have demonstrated that clams can rid themselves of contaminates if they are exposed to clean seawater for approximately 48 hours. This process is only viable for marginally contaminated clams (Supan and Cake, 1982). On going field surveys are examining the impacts of these type of runoff events on the clam beds. Furthermore, this is the first water quality study conducted on the estuaries and the residents were not aware of the fecal coliform contamination problem. Another solution to the problem is to present the information to the cottage owners and inform them of potential health risks, since the contamination affects the recreational use of these sand flats. This is a situation where the pollution can be controlled at the source.

Water qualities such as salinity and temperature affect the growth, mortality and the production of the soft-shell clam. For example, clams need salinities of at least 5 parts per thousand (ppt) to survive, but thrive in salinities from 25 to 35 ppt. Studies have

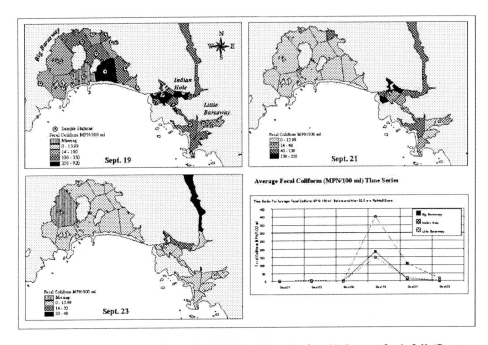

Fig. 4. Fecal Coliform (MPN/100ml) levels after 52.5 mm of rainfall (Sept. 19th-23rd, 1995)

demonstrated that, during excessive freshwater runoff clams may experience 90 percent mortality in areas where salinities reduced to less than 5 ppt (Newell and Hidu, 1994). During the runoff event from Sept. 19 to 23, 1995 the salinities at Indian Hole and Little Barasway dropped below 5 ppt for several days, e.g. salinities ranged from 2 to 30 ppt, but returned to acceptable levels within 5 days (Fig. 5). The short-term impacts of low salinities on clam mortality have not been studied in this area. However, calms can survive low salinities if water temperatures are low (Allen and Garrett, 1971). From Sept. 19 to Sept. 23, 1995 the water temperature ranged from 11 to 14.5 degrees Celsius and there were no high clam mortality recorded during this period.

When the water quality attributes fecal coliform and salinity are mapped as Thiessen polygons in Figs. 4 and 5 the maps reveal information about potential flushing rates and current dynamics. The maps provide the aquaculture operator with insights to possible problem areas. Given that Big Barasway flushes faster and did not record salinities below 5 ppt, this area could be seeded with clams from Indian Hole and Little Barasway.

Mapping and Analysis of Soft-Shell Clams

GIS is a tool that is capable of quantifying and visualizing the characteristics of the of the natural clam resource. The clam biology database is used to examine the spatial density of the resource, estimate the market biomass, and develop a potential harvesting plan. These issues are address by descriptive and prescriptive queries within ArcView™ and the Spatial Analysis module (ESRI, 1996).

Preparation of the clam biology data for analysis included the calculation of clam densities as well as an estimate of the total biomass for each area. The attribute clam length was subdivided into 4 classes and they are:

1) **Juvenile - length < 35 mm (LT 35 mm)**
2) **Pre-Recruits - length >= 35 and < 50 mm (GE 35 to LT 50 mm)**
3) **Recruits - length >= 50 mm and < 63.5 mm (GE 50 to LT 63.5 mm)**
4) **Brooding Stock – length >= 63.5 mm (GE 63.5 mm)**

A Query Summarize function in ArcView™ (Fig. 6) was used to summarize by GEOCODE, the total number of clams for each size class. Thus each 100 x 100 m grid in the recruit

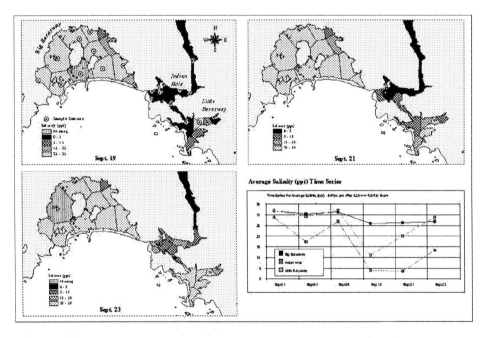

Fig. 5. Salinity (ppt) levels after 52.5mm of rainfall (Sept. 19th-23rd 1995)

group represents the market clam population. The procedure for calculating clam densities involved the addition of the four clam size class fields (attributes) to the clam biology point databases. For each record a value of 1 was assigned to a size class attribute if the clam length value was within the specified class interval otherwise the value was set to 0. The study area is assigned a total count for each clam size class. Given that the sampling occurs in a 0.25/m sq. area at the center of each grid, the density for each size class was calculated as: n/area, where n is the total number of clams in a grid. An estimate of the market clam biomass was calculated using the ArcView™ Query Builder and Field

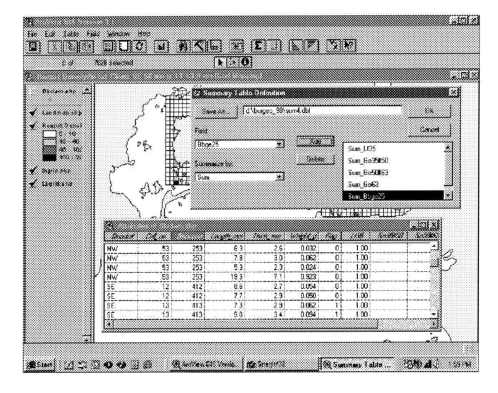

Fig. 6. ArcView Descriptive Query Summary

Calculator. A market biomass is calculated by estimating the total number of clams in a hectare grid. The estimation of the biomass is based on several assumptions and they are:

1) **The clam densities are homogeneous within a 100 X 100 m grid.**
2) **It takes 1500 recruit size clams to make a bushel**
3) **One bushel of recruit size clams weights approximately 27 kilograms (Robert, 1981)**

The biomass estimate for each grid is calculated as follows:

$$((\text{clam density/m. sq.} * \text{grid area}) / 1500) * 27$$
(1)

The result is a clam shell weight in kilograms for each 100 x 100 grid. A grid is economically viable only if the recruit clam density is greater than or equal to 10/m sq. This criteria can be used with the ArcView™ Query Builder (Fig. 7) to select grids for the biomass calculation. After the grids are selected a Field Statistics function can be used to calculate the biomass. For example, Fig. 8 illustrates how the biomass was calculated for Big Barasway. Note that the sum value, in the Statistics for BB_biomass field, is 236345 kilograms. This number represents the predicted biomass for Big Barasway.

The 100 x 100 grids are used for descriptive analysis such as the calculation of densities. The polygons can also be used for mapping attributes. Fig. 9 (A) illustrates the use of the grids to map the distribution of mature clams densities/m sq. (e.g. clams >= 25 mm). Mature clam densities of 161 to 269/m sq. are acceptable for good growth while densities higher than 269/m sq. can impede growth because of competition for food and space (Newell and Hidu, 1994). From an aquaculture management perspective areas with excessive mature clam densities, depending on the size class distribution, need to be culled or harvested. Thus mapping of the mature clam densities can identify areas that require culling or harvesting as well as areas of low densities that require seeding. A possible solution to the excessive density problem is to move the culled juvenile or pre-recruit clams to areas of low densities.

Mapping clam densities with vector polygon data will reveal trends, but if the data are to be used in any prescriptive analysis the spatial data must be converted to a raster format. The visualization and analysis of clam densities in a GIS can be applied to a raster version of the original grid. However, the clam density distribution is a continuous spatial variable and should be mapped as a continuous surface where possible. Continuous surfaces are created in a raster GIS by using spatial interpolation algorithms. ArcView™ Spatial Analyst provides an inverse distance squared (IDW) and a spline interpolation routine. The process of converting the clam density data from polygon to point coordinates for interpolation involved several steps. Firstly, stratified random samples of coordinates were generated for each of the three areas. The sampling density ensured that 25m spacing occurred between the points. Secondly, a point-in-polygon procedure was used to attach the grid density size class data to the points database. An inverse distance squared algorithm

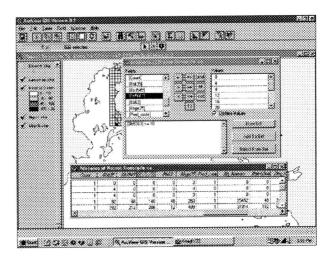

Fig. 7. ArcView Query Builder - "Select Recruit Density GE 10/m. sq."

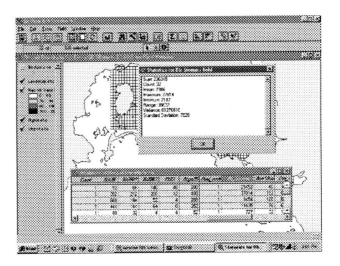

Fig. 8. ArcView Field Statistics for Biomass - Based on Query "Select Recruit Clam Density GE 10/m.sq."

GIS and Aquaculture: Soft-Shell Clam Site Assessment

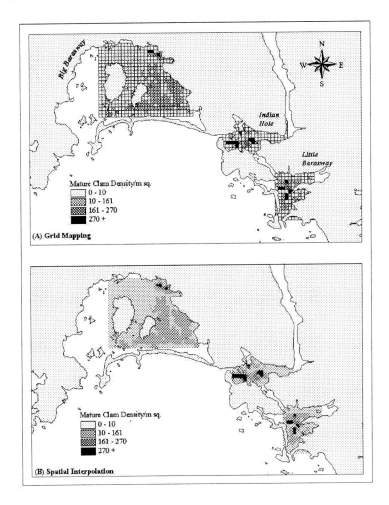

Fig. 9. Comparison of (A) Grid and (B) Interpolation Mapping - Mature Clam Density/m.sq.

was used to create continuous surfaces for all clam size density classes. Fig. 9 (B) illustrates the interpolation result for mature clam density. A comparison of Figs. 9 (A) and (B) indicate that some spatial averaging occurs, (e.g. the maximum density is reduced from 488 to 473/m sq.) however, the interpolation preserves the original trends in the data. Fig. 9 (B) indicates that, in some areas, there is a need to cull the mature clam population, especially at Indian Hole.

According to Newell and Hidu (1994) the expected spatial distribution of clams in an area ranges from highly dispersed in the juvenile size class to a highly aggregated pattern in the recruit and brooding classes. Mapping the size class densities as continuous surfaces indicates that a similar pattern occurs in Big Barasway (Fig. 10) and to a lesser degree in Indian Hole and Little Barasway. Fig. 11 confirms the increasing spatial aggregation from juvenile to brooding stock classes whereby an inverse relationship exists between the larger clam size classes and percent area occupied. For example, in Big Barasway juvenile clams (LT 35 mm) are found in 95 percent of the surveyed area while brooding stock clams (GE 63.5 m) are found in only 17 per cent of the area.

Fig. 11 also presents information on the clam population characteristics. In Big Barasway there is a large decrease in clam density from the juvenile to pre-recruit class (Fig. 11 (B)). The average density drops from a high of 124/m sq. for the juvenile class to 30/m sq. for the pre-recruit class. There is also a dramatic change in densities at Indian Hole where the average density drops from 125/m sq. for the pre-recruit class to 53/m sq. for the recruit class. Little Barasway shows a more gradual change where the average densities are 57, 55, 43/m sq. for the juvenile, pre-recruit and recruit classes respectively. Indian Hole has the highest densities in all classes while Big Barasway has the second highest juvenile density but the lowest densities in all other classes. The frequency size class information presented in Fig. 11 (C) indicates that 80 percent of the sampled clam population in Big Barasway is juvenile and the recruit population accounts for only 10 percent of the sampled data. The trends at Indian Hole and Little Barasway are similar. The juvenile classes at these locations account for 47 and 42 percent of the clams sampled, however, Little Barasway has a higher percentage of recruits (23 percent) than Indian Hole with 14 percent. Although Indian Hole has a higher density of clams/m sq. Little Barasway has the highest percentage of recruit clams suggesting a possible higher survival rate for clams at this location. The low percentages associated with the larger clams at Big Barasway needs further study in order to understand the dramatic changes in the observed densities. Although the area appears to be experiencing high predation, an analysis of the data did not reveal any correlation between the presence or absence of predators and the size class distribution.

Aquaculture Management and GIS

GIS can be used to manage and organize operations at an aquaculture site. The harvest plan presented in this chapter is an example that demonstrates how a GIS can be used to help aquaculture operators map and evaluate their proposed activities. Through GIS the operator

GIS and Aquaculture: Soft-Shell Clam Site Assessment

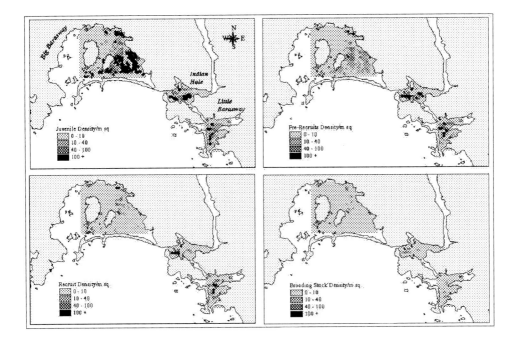

Fig. 10. Clam Density/m.sq. by Class Size

can identify and map the areas to cull or harvest as well as evaluate the expected yield along
a proposed harvest path. These applications can be called prescriptive because the results of the analysis can be use as rules or guidelines for harvesting clams.

The use of a GIS to develop aquaculture operation scenarios requires the definition of criterion or rules for a particular activity. In this study the following rules are used to identify areas for harvesting or culling and they are:

1) **The density of recruit clams must be greater than or equal to 10/m sq.**
2) **When the density of mature clams exceed 269/m sq. the area must be culled.**

The Map Query tool in ArcView^{TM} Spatial Analyst is used to define areas where the mature clam density is greater than 269/m sq. and where recruit densities are equal to or greater than 10/m sq. The result of this analysis is two binary maps where 0 and 1 indicates that the condition is false and true respectively. The Map Calculator tool in Spatial Analyst is used to combine the maps to identify areas of no harvesting, harvesting only as well as sites

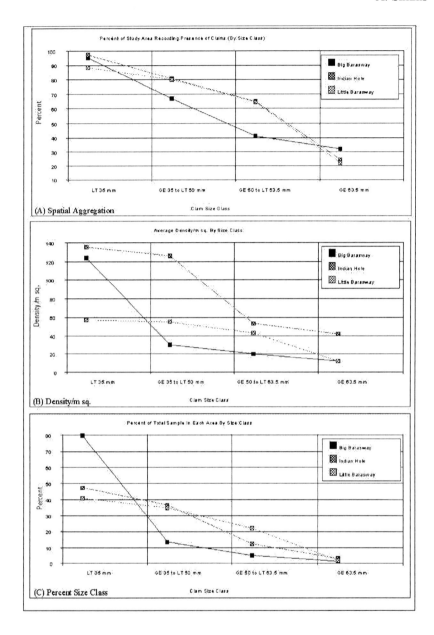

Fig. 11. (A) Spatial Aggregation, (B) Density and (C) Percent Size Class Trends

where both culling and harvesting is required. The binary maps are used to calculate the harvest plan map where:

[Map of Harvest Areas] + ([Map of Cull Areas]* 2).

GIS and Aquaculture: Soft-Shell Clam Site Assessment

This expression will produce a map with four values. The no harvest, harvest only, cull only and the combined cull and harvest areas will be coded 0,1,2,3 respectively. The harvest plan map produce by this analysis is presented in Fig. 12 and the harvest only or cull and harvest areas are easily identified. A profile analysis was performed at Indian Hole on the estimated metric ton/hectare map. A proposed harvest track was digitized on the Harvest Map and a profile was generated (Fig. 12). The profile result indicates a low expected yield of about 1 metric ton/hectare at the start and a high of 26 tons/hectare is available at about 300 m from the start. However, at about 1 kilometre from the start the expected yield drops to 0 tons/hectare. The area where the expected yield drops to zero is at the beginning of a no harvesting zone. These areas may consist of either juvenile or pre-recruit clam beds that should not be disturbed. Thus the harvest plan map could be improved by adding other layers such as maps that identify areas that maybe important for juvenile or pre-recruit clams. The profile analysis can also be used to evaluate proposed harvesting tracks and select only those tracks that offer the highest yield/hectare.

The calculation of harvesting area and the expected yield/hectare provides an estimate of the total market biomass for all three areas. Big Barasway has the highest market biomass at 236 metric tons followed Indian Hole and Little Barasway with 223 and 216 metric tons respectively (Fig. 12). An examination of the average yields/hectare indicates that Indian Hole is the most productive at 13.94 metric tons/hectare followed closely by Little Barasway at 11.37 tons/hectare. Finally, Big Barasway is the least productive area at 7.37 tons/hectare.

Conclusion

The information presented in this study indicates that GIS is an important tool for the aquaculture industry. These systems can be used to monitor, quantify and evaluate the soft-shell beds near Burgeo. Management issues such as water quality, resource sustainability as well as the economic viability of the clam resource can be assessed within a GIS environment. The results of the analysis in this study suggest potential problems with faecal coliform contamination from local cottages. Furthermore, at Big Barasway the relatively low counts and densities at the recruit and brooding size classes raise questions about clam mortality. This trend is also supported by the fact that the Big Barasway average yield/hectare is approximately half of the average yield calculated for Indian Hole although the area is twice as large (Fig. 12).

Finally, data collection for aquaculture site assessment is required if a resource is to be managed effectively. GIS applications provide insights into the quality of the physical environment as well as the sustainability of a resource. However, it is the aquaculture operators who ultimately make the final decisions.

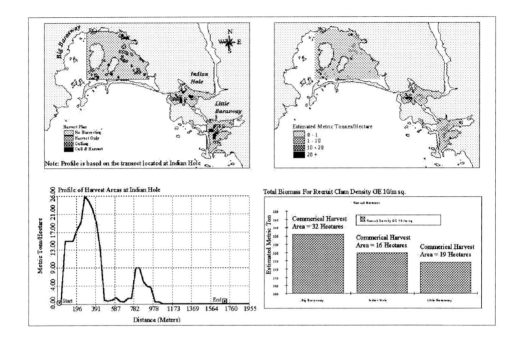

Fig. 12. Harvest Plan for Aquaculture Sites

References

Allen, J.A. and Garret, M.R. (1971). *The Excretion of Ammonia and Urea by Myra Arenaria. Comp. Biochem. Physiol.* 39A: pp 633-642.

Berry, J. K. (1993). *Beyond Mapping - Concepts Algorithms and Issues in GIS.* GIS World Inc. USA.

Burrough, P.A. (1986). *Principles of Geographical Information Systems for Land Resource Assessment.* Clarendon Press, Oxford.

Clark, W.F. (1990). *North Carolina's Estuaries: A Pilot Study for managing Multiple Uses in the States Public Trust Waters.* Abermarle-Pamlico Study Report 90-10.

Demers, M.N. (1997). *Fundamentals of Geographical Information Systems.* John Wiley and Sons Inc., USA.

ESRI (1996). *ArcView™ 3.0 GIS and Spatial Analyst 1.0 software.* Environmental Systems Research Institute Inc., Redlands, California.

Kapetsky, J.M., McGregor, L., and Nanne, E. (1988). *A GIS to Assess Opportunities for Aquaculture Development: A FAO-UNEP Grid study in Costa Rica.* FAO Fisheries Technical Paper 287. United Nations, Rome.

Legault, J.A. (1992). Using A Geographical Information System to Evaluate the Effects of Shellfish Closure Zones on Shellfish leases, Aquaculture and Habitat Availability. *Can. Tech. Rep. Fish. Aquat. Sci.* 1882E: iv+10.

Manjarrez, J. A. and Ross, L.G. (1995). GIS Enhances Aquaculture Development. *GIS World*, Vol. 8, No. 3, March. pp. 52-56.

Newell, C.R., and Hidu, H. (1994). *Soft-Shell Clam.* Fish and Wildlife Services, U.S. Department of Interior. Biological Report 82 (11.53).

Robert, G. (1981). *Stock Assessment of Soft-Shell Clams in Open Shellfish growing areas of Prince Edward Island.* Fisheries and Oceans Canada, Halifax, N.S.

Simms, A. (1994). *Report on Geographical Information Systems Mapping of Aquaculture Information, Waterfowl Concerns, Water Quality and Other Pertinent Parameters.* Technical Report – Canadian Wildlife Service, Dartmouth, N.S.

Supan, J.E. and Cake, E.W. (1982). Containerized-Relaying of Polluted Oysters in Mississippi Sound Using Suspension, Rack, and on Bottom-Longline Techniques. *Journal of Shellfish Research*, Vol. 2, No. 2, pp 141-151.

CHAPTER 23

Evaluation of Ecological Effects of the North Sea Industrial Fishing Industry on the Availability of Human Consumption Species Using Geographical Distribution Resource Data

J. Robertson, J. McGlade, and I. Leaver

ABSTRACT: This report presents an analysis of the ecological impacts of the industrial fisheries in the North Sea. The geographical distribution of fish stocks was used to demonstrate overlap areas where feeding strategies occur. Potential consumption levels are calculated using specific diet requirements of human consumption stocks based on consumption rates and trophic efficiencies. Impacts on human consumption stock sizes indicate a transfer efficiency problem related to food deficiencies caused by the industrial fishery. The effects of geographical scaling procedures on the projected consumption levels with different future stock sizes are investigated to establish management procedures. Recommendations are made for the development of GIS systems more appropriate for the handling of fisheries data.

Introduction

The North Sea stocks of human consumption fish species have been seriously depleted by a combination of overfishing and poor or inappropriate capture practise. The Food and Agriculture Organisation of the United Nations (FAO, 1994) has recently warned of the high exploitation levels of North Sea cod, haddock and whiting stocks by the commercial fishing fleet and the International Council for the Exploration of the Sea (ICES) has been giving clear and regular warnings about the depleted state of the cod stock for many years (Anon, 1995), and has also posted warnings on other stocks. The current status for North Sea stocks as given by ICES (see Appendix 1) indicates that many are either overexploited or just at the safe limit. Management advice extends from reduction of effort to the setting of precautionary TACs based on recent catches.

Clearly, we do not enjoy prolific fish stocks in European waters, yet fishing effort is not significantly decreasing. As a result the stability of fish stocks in the North Sea for human consumption is now a matter of concern both within the fishing sector itself as well as amongst European Union (EU) member states.

A number of factors have influenced the changes in observed fishing patterns in the North Sea, not least of which have been the Common Fisheries Policy (CFP) with its associated technical instruments, the continued extraction of fish resources for oil and meal and the increased globalisation of fish products and processing (see Fig. 1 for North Sea industrial, demersal and pelagic landings from 1970 to 1994). What has become more apparent, however, is that the integrity of the ecosystem itself may also be at risk not only because of overfishing but also through pollution, environmental changes and ecological forcing via the effects of a) the industrial fisheries b) current fishing practises in the North Sea and c) other extractive processes. For example, cod, haddock, whiting, saithe, mackerel, herring, and flatfish are generally caught for human consumption using large mesh trawls. A large mesh ostensibly allows juveniles to escape, but recent studies have indicated that there is a significant amount of discarding of juveniles at sea.

In 1994, total landings from the North Sea industrial fishery in ICES Area IV were 1307k mt, slightly below the 1974-93 mean of 1439k mt; the targeted by-catch of human consumption species i.e., haddock, whiting, cod and saithe, was 24k mt. This is now known to be an underestimate for the Norway pout fishery. Given that much of this by-catch (principally and particularly from the pout fishery) comes from recruiting age classes, any ecological overlap between the industrial and human consumption fisheries will be likely to have an effect on the productivity of human consumption species. However, a recent European Union study (Commission of the European Communities, 1992) concluded that whilst a 40% reduction in the industrial fisheries gave modest increases in whiting and haddock, there was only a minimal change to the catches of cod, Norway pout, sandeel, saithe, herring and sprat because of increased discarding.

Much work has been undertaken by ICES member countries on the dynamics of all the key commercial species. Annual assessments provide managers with an analysis of the level of exploitation with respect to two key biological stock reference points i.e., growth and reproductive potential. Where there are sufficient data, the Advisory Committee on Fisheries Management (ACFM) has defined a minimum biologically acceptable spawning stock level (MBAL). Generally this equates to the level of stock below which there is evidence of recruitment failure, or the lowest level of spawning recorded. ACFM has identified three categories of stocks; those below MBAL or expected to become so (i.e. outside safe biological limits), those above MBAL and those whose state is indeterminate.

Environmental and ecosystem effects as well as fishing can distort the value of MBAL. The ICES Multispecies Assessment Working Group has responded to this by developing a multispecies virtual population analysis (MSVPA), involving predator-prey interactions and preferences. To add information and validate the model, a number of stomach content surveys and feeding studies have been undertaken. But the enormous data requirements and wide range of analytical issues arising within the MSVPA model make it an inappropriate tool for annual assessments and day-to-day management. On the other hand, the data generated for MSVPA can be used in ecosystem models such as ECOPATH (Christensen and Pauly, 1992; Christensen, 1995) to examine a number of key ecosystem properties such as impacts and the flow of biomass.

In this study, MSVPA, ECOPATH and other models were used to try to obtain a clearer picture of the effects of different fisheries on ecosystem dynamics than is currently available from the single species assessments alone. GIS distributional resource data are used for individual species to ascertain species interactions. For example, the industrial species comprise a significant part of the diet of cod, haddock, whiting, mackerel and saithe. Thus, one potential explanation for the decline in human consumption stocks over and above the effect of directed fisheries, is the alteration in food availability caused by extraction of large amounts of fish biomass. One of the aims of this chapter is to investigate whether the food demands of the human consumption species are being met. We also need to know whether there would be enough food for

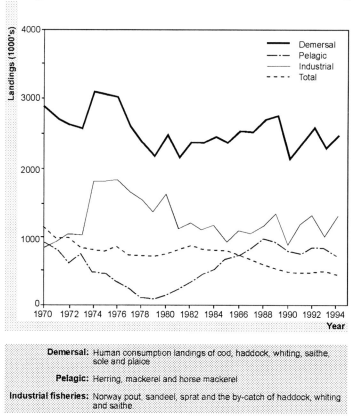

Fig. 1. Landings in the North Sea 1970-1994

the diet requirements of future generations of increased stock sizes. There is little point in having protracted discussions on effort limitation of fishing power, technical measures or other measures to protect the stocks if those fish stocks have no, or not enough, food to sustain them in the long run. In this chapter we investigate the impact of the North Sea industrial fishery on the food available to human consumption species, and attempt to answer the question by analysis of existing data sets, recourse to the literature and by calculation of the weight of fish consumed in the North Sea to further investigate changes in trophic level transfer efficiencies.

We therefore approach the problem from three angles to examine the partitioning of direct food consumption levels of the human consumption species within the North Sea. We use the consumption levels from the 1981 ICES stomach sampling exercise and base total consumption on total stock biomass estimates from 1983 to 1985. Then, using MSVPA consumption inputs based on the 1981 and 1991 ICES stomach sampling exercises, we analyse various predator-prey interactions using the ECOPATH model and compare consumption levels between the years 1977 to 1986 and 1987 to 1994, to see whether there are any changes in consumption and whether they are due to diminishing stocks of human consumption species or alterations in prey selection.

In terms of explicit controls on fisheries, these are mainly restricted to regulating fishing mortality via overall catch levels and technical measures such as mesh size, minimum landing sizes, and area and seasonal closures. The dynamics of fishing mortality, however, are heavily influenced not only by ecological interactions and the types of fishing vessels and gears used etc., but also by market forces. In the North Sea there is widespread evidence that over quota landings, by-catches and discards represent a significant proportion of the tonnage of human consumption species extracted from the sea. To help estimate the level of tonnage involved, identify the reasons behind these practices and estimate their potential role in the future, interviews with the fisheries sector in the UK were carried out.

The North Sea Industrial Fishery and its Fish

The industrial fisheries target sandeels, Norway pout, sprat and herring using small mesh trawls. The landings are broken, pulverised and pressed to extract meal and oil that are principally used to feed terrestrial and aquatic farmed animals. Some oil is added to human food such as biscuits and margarine.

The main commercial industrial fishery is for sandeel, of the *Ammodytidae* family and for Norway pout, *Trisopterus esmarki*. Other smaller, but no less important, industrial fisheries exist for sprat, *Sprattus sprattus;* horse mackerel, *Trachurus trachurus.* Smaller amounts of blue whiting, *Micromesistius poutassou* are also taken.

Sandeel and pout constitute the two largest tonnages of industrial species taken. Sandeel comprise five species in the North Sea which are the common sandeel *Ammodytes tobianus,* Raitts sandeel *ammodytes marinus,* the smooth sandeel *Gymnammodytes semisquamatus,* the greater sandeel *Hyperoplus lanceolatus,* Corbins sandeel *Hyperoplus immaculatus* (Knijn et al., 1993). The lesser sandeels, common sandeel, Raitts sandeel and smooth sandeel are on average up to 25cm long and comprise the majority of commercial catches. The much larger Corbins and greater sandeel grow to 35cm and can be a smaller part of the commercial catch.

Sandeel are shoaling pelagic fish. They prefer sandy or sand/gravel mixtures on the seabed to bury in. Sandeel are distributed widely throughout the North Sea. However, the main commercial fishery is conducted to the west of Jutland in Denmark, and on the Dogger Bank in the central southern North Sea and off the east coast of Scotland from the river Forth estuary northwards. A small fishery exists in Shetland waters.

Sandeel feed mainly on planktonic copepods and crustacean larvae. Some polychaete worms, amphipods and euphausids are also eaten. The greater sandeel also feeds on fish including other sandeels. In general the species live for about five years. Sandeels are believed to bury in the sand during winter but many are found in the stomachs of predators during this period which may suggest that they are available for food for periods of the winter or that predators dig them from the sand. They apparently spawn during winter and produce demersal eggs which adhere to the seabed. sandeels are a major food source for a variety of fish, seabirds and marine mammals. They are captured by demersal trawls constructed with very small meshed netting.

Norway pout is found in the northern and central areas of the North Sea. It is usually found at 80 to 200m water depths but can also be found at up to 450m in the Norwegian Deep. It is a small (10-20cm) species rarely attaining four years old. Mature fish diet consists of crustaceans and small fish. It is generally found near the seabed and the commercial fishery is carried out mainly by demersal trawls. Like sandeels, Norway pout are an important food for other fish species within the North Sea. Like sandeels the pout are captured within very small meshed demersal trawls.

Fig. 2 shows the landings by year since 1953 for sandeel and from 1959 for pout (Anon, 1996). The sandeel fishery enjoyed a steady increase in catches up to 1977 from when a concerted effort began to increase catches to a high of just over one million tonnes in 1989. Some 775,000 tonnes have been taken on average by the fishery, each year over the last ten years. Landings from the pout fishery rose steadily to 1994 from when a decrease has occurred due partly to the imposition of a Total Allowable Catch (TAC) on the fishery and the establishment of the "pout box" in 1985 which reduced the area for pout fishing. Interestingly, the pout box was set up as an exclusion zone for pout fishing to protect human consumption species which had previously been captured as by-catch in large quantities within the small meshed pout trawls. Elsewhere in the North Sea by-catches of human consumption species are still taken within the pout fishery.

The sprat is mainly concentrated in the southern section of the North Sea and in the case of summer juveniles, particularly in the German and Southern Bights. Some are found in the Moray Firth where there was a large fishery in the nineteen sixties and seventies. It does not occur to any great extent in the northern or central

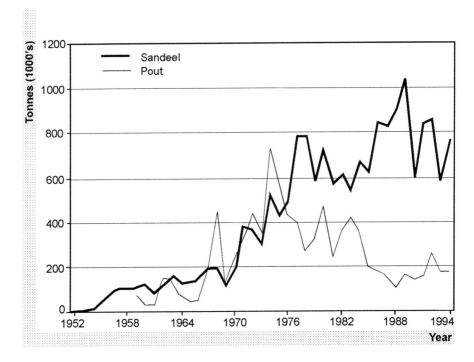

Fig. 2. Total landings of sandeeel and pout by the industrial fishery (1952-1994)

part of the North Sea. Sprat ranges between 5 and 15cm in size and seldom live longer than five years.

The Regulatory Framework

The current regulations in the North Sea industrial fishery within the Common Fishery Policy are designed to deal with two main issues; by-catch and control of the total tonnage allowed to be captured. There is, however, a glaring omission in the sandeel fishery where there is no limit on the quantity which may be caught each year. Controls on the industrial fisheries are by the following main measures.

• **a ban on the capture of herring for industrial purposes within EU waters and a ban on the landing in the EU of herring caught for industrial purposes (R[EEC] 2115/77)**

• **a 10% by-catch upper limit of human consumption species which may be taken in any of the small mesh industrial fisheries (R[EEC] 3094/86)**

• **a 15% by-catch limit of which not more than 5% may be cod and haddock within the Norway pout fishery (Article 14 of R[EEC] 3882/91)**

- a ban on processing operations which may be used to control the levels of apparent by-catch on board the vessel (Article 10a of R[EEC] 3094/86

- a prohibition on fishing for Norway pout with small mesh nets in a particular area of the North Sea called the "Norway pout box" (Annexe 1 of R[EEC] 3094/86

- Total Allowable Catches for the Norway pout, sprat, blue whiting and horse mackerel in Area IV.

Apart from the EU regulations each EU member country may have separate and additional measures to control the fishery. For example, Denmark in 1996 implemented a loss of fishing licence scheme for over by-catch limits within the sprat fishery. Some other measures are in force but are not listed here.

Feeding Interactions and Stomach Content Data

In describing the various interactions which will ultimately have an effect on the production and yield of fish from the North Sea, it is important to look at variability in feeding patterns. One method is to obtain stomach samples from as wide a range of areas, and time as possible, to see if any general patterns emerge which can be used to establish and/or validate different models. These data can be augmented with information from feeding studies.

To date, two "Year of the Stomach" surveys have been undertaken, one in 1981 and 1991; additional sampling has also been undertaken on research surveys in various quarters of the year in 1980, 1982, 1985 and 1987. In this analysis the results from both surveys have been used in the calculations of total biomass consumed by predators; the 1981 data were used for the period 1983-1985 and for the 1977-1986 period and the 1991 data, as revised by the 1995 Multispecies Assessment Working Group (ICES CM 1996/ Assess:3) for the period 1987-1994. Reference was also made to the studies on cod feeding (Daan, 1973), whiting (Hislop et al., 1983), and mackerel (Mehl and Westergard, 1983).

In the MSVPA model, the food composition of a particular predator age group is predicted from the relative abundance of the various prey age groups weighted by their relative suitability as prey. A basic assumption is that suitabilities are constant for each predator age prey combination over time. The suitabilities or preferences are proportional to the probability of encounter between the prey and predator multiplied by the probability of the predator eating the prey once encountered.

For suitabilities to remain constant, the product of these two probabilities needs to stay the same. Over the past two decades there have been large changes both in absolute abundance in species such as herring and sandeel, as well as their spatial occurrence. For example, 28% of the total sandeel biomass was distributed in the northern area in 1981, while in 1985, 86 and 87 it went from 15% to 20% to 35% respectively. Thus there are very likely to have been changes in suitabilities due to alterations in spatial overlap.

Changes in suitability can also occur as a result of prey switching i.e. the probability of a predator eating a prey item altering with respect to local and global prey biomass levels. In the North Sea little is known about such effects. There are also potential changes in suitabilities caused by non-linearities in the model itself. For example, when prey switching is included in the MSVPA model, the equations can exhibit one stable and two unstable solutions (Hildén, 1988). In this analysis prey switching *per se* was not included, although the effects were noted in the output of the ECOPATH model.

Method for Estimating Predator Consumption Rates

Various methods have been used to estimate rations or consumption rates; historically these have involved gastric evacuation rate models (Daan, 1973; Elliott and Persson 1978; Jobling, 1986; Persson, 1986; dos Santos, 1990; Bromley, 1991). Estimates are based on the volume of the stomach contents or in-situ gut fullness and gastric evacuation rates. The ICES standard estimates are derived from the model of Daan (ibid.)

A more recent study by Hansson et al. (1996) has compared the results obtained from a selection of these gastric evacuation models with a bioenergetics model developed by Hewett and Johnson (1992). The model simply examines consumption (C) as:

Consumption = Metabolic loss + Waste loss + Growth

<u>where:</u> **metabolic loss = respiration + specific dynamic action; waste loss = egestion + excretion; and growth = somatic + gonadic growth.**

The results show that the gastric evacuation models currently used by ICES for cod consistently produces low rates especially for age 1 fish; the Bromley model predicts annual consumption 300% higher than the standard ICES model; the dos Santos model gives consumption rates for age 5 fish that are twice those of the other models; and the bioenergetics model gives rates intermediate to these. As the authors point out, the North Sea fish community has changed considerably over the past few decades, with a nineteen-fold change in herring biomass over the period 1974 - 1989 (Anon., 1991). Such significant changes in prey availability will not necessarily be reflected linearly in the growth rates of predators, because of the trade-offs in prey type (e.g. lean or fat) and the non-linear relationship between food consumption and growth. This presents a significant problem in the MSVPA, because few experimental data sets exist to be able to establish and validate a bioenergetic model for the main predator species other than cod.

In this study the bioenergetics results were used for cod; for the remaining species the daily food consumption rates (g/day) were calculated after Greenstreet (1996). For haddock, saithe and whiting the mean stomach weight at age in each quarter (w) was taken from Daan (1989), and Jones' (1974) digestion model applied.

Water temperature in the North Sea was assumed to be 6°C in quarter 1, 7°C in quarter 2, 10°C in quarter 3 and 8°C in quarter 4. A digestion coefficient (Q) of 0.2 g/hr was assumed for all prey at a reference temperature of 12°C. For plaice and sole a similar approach was used, and reference made to the daily consumption rates of Basimi and Grove (1985). The function to relate water temperature to the instantaneous rate of digestion was $k = 0.025*W^{0.068}*e^{0.086}T$. For the mackerels the digestion model of Mehl & Westergard (1983) was used. For Norway pout, reference was made to dietary data from Albert (1991) and prey conversion rates from Rumohr et al. (1987); the Jones' digestion model was used. For herring, dietary data were derived from Last (1989) and Daan's digestion (1973) model; in the absence of any further information the herring daily consumption rates were applied to sprat and sandeels. Given the results in Hannsson et al. (ibid) it is likely that these overall ration estimates are highly conservative.

Diet Content and Total Consumption 1983-95

FOOD TYPES

The prey species found in the stomachs are divided into classes (after Greenstreet, 1996) *plankton* - copepods, euphausids, fish larvae, and eggs; *benthos* - worms, crustacea, bivalves; *pelagic planktivore* - sandeels, herring, sprat, norway pout ; *pelagic piscivore* - mackerel; *demersal benthivore* - plaice, rays; *demersal piscivore* - cod, haddock, whiting, saithe, dogfish, gurnard.

Each predator and its main prey species are expressed as a percentage weight (tonnes) of total diet (see Fig. 3). Table 1 sets out the equivalent numerical results.

In general terms, the diet of whiting, saithe and mackerel consists of between 37%-60% of pelagic planktivores of which pout and sandeel are the two species mainly consumed. Saithe and mackerel consume plankton to the extent of 35% - 50% of total diet respectively whilst the proportion of plankton in the diet of cod, haddock and whiting accounts for between 5% - 11%. Cod and haddock consume fairly large proportions of benthic material (ie; 34% - 42% of total diet) whilst the other species consume less than 1% - 14%.

SPECIFIC COMPONENTS OF THE TOTAL DIET FOR EACH SPECIES

Cod diet: The largest component of cod diet is benthic species (34%). Demersal piscivores which include prey such as cod itself (cannibalism), haddock, whiting, saithe and a mixture of other roundfish amount to 22.6% of diet. This is followed closely by pelagic planktivores (including pout and sandeel) at 21.2% of total diet. Pelagic piscivores account for an insignificant <1% of the diet. The mixture of diet components would appear to remain similar between quarters. The proportion of benthos might be expected to increase during winter months when sandeel are reported

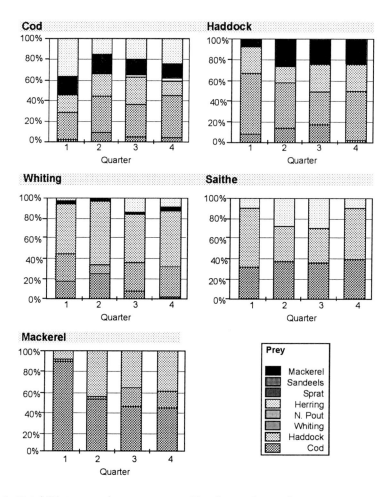

Fig. 3. Total Diet: prey classes consumed by the predators, by quarter, expressed as a percentage of total weight (tonnes)

to be buried in the sand for long periods but this does not appear to be the case probably because cod dig for sandeels in the sand.

Haddock diet: Haddock consume 42% of benthic food whilst pelagic planktivores (mainly sandeel and pout) account for 25% of diet and demersal piscivores some 21%. A reasonable quantity of plankton material (11%) is also eaten.

TABLE 1
Total Diet: percentage of all prey consumed per day for each quarter and over the whole year for period 1983-85 (after Greenstreet, 1996)

Prey	Cod % predator					Haddock % predator					Whiting % predator				
	Q1	Q2	Q3	Q4	Year	Q1	Q2	Q3	Q4	year	Q1	Q2	Q3	Q4	year
Plankton	1.53	8.56	4.95	3.58	4.85	7.76	13.38	17.78	2.66	11.18	16.79	25.18	7.39	2.23	8.30
Benthos	27.29	36.18	31.53	42.13	34.02	59.62	44.57	31.06	46.25	42.21	27.62	9.02	28.49	29.86	13.66
Pelagic planktivores	18.02	22.38	27.72	13.47	21.2	24.74	14.54	27.88	27.86	24.64	49.71	62.44	48.18	54.72	59.87
Pelagic piscivores	0.00	0.00	1.24	2.81	0.97	0.00	0.00	0.00	0.00	0.00	0.00	0.02	0.02	0.00	0.01
Demersal benthivores	16.69	18.22	15.47	15.15	16.39	7.88	27.50	23.09	23.13	21.87	2.92	0.91	1.23	4.02	3.39
Demersal piscivores	36.47	14.67	19.11	22.87	22.57	0.00	0.00	0.19	0.11	0.10	2.95	2.43	14.69	9.17	14.77

Prey	Saithe % predator					Mackerel % predator				
	Q1	Q2	Q3	Q4	year	Q1	Q2	Q3	Q4	year
Plankton	30.52	36.07	35.55	39.02	35.40	90.39	52.85	45.07	44.17	50.24
Benthos	0.14	0.32	0.20	0.08	0.18	2.35	2.15	18.69	16.33	12.14
Pelagic planktivores	60.03	35.93	34.23	51.98	44.85	7.25	45.00	33.88	39.48	37.47
Pelagic piscivores	0.00	0.00	0.00	0.00	0.00	0.00	0.00	0.00	0.00	0.00
Demersal benthivores	0.14	0.00	0.00	0.09	0.05	0.00	0.00	0.03	0.00	0.01
Demersal piscivores	9.18	27.67	30.02	8.83	19.51	0.00	0.00	0.33	0.00	0.13

Whiting diet: Some 60% of whiting diet comes from the pelagic plaktivores (and most of this consists of sandeel). Benthic material and demersal piscivores each account for about 14% of diet. Plankton accounts for 8% of diet with about 3% of demersal benthivores comprising the remainder. Sandeel are a major component of whiting diet with equal amounts eaten across all four quarters.

Saithe diet: Plankton and pelagic planktivores account for 79% of total saithe diet. Demersal piscivores account for 19% of diet and the remainder of insignificant amounts of benthos and demersal benthivores.

Mackerel diet: Half of mackerel diet consists of plankton whilst pelagic planktivores (mainly sandeel) account for 37%. Benthic material is also taken (12%).

CONSUMPTION OF FISH PREY SPECIES BY FISH PREDATORS

Fig. 4 sets out the daily tonnages of fish prey eaten by the fish predators. The data are displayed by quarter. Table 2 illustrates the consumption per day of each prey species by each predator expressed as a percentage of the total weight consumed per quarter and year. Fig. 5 sets out the percentage consumption of prey species by quarter.

Predator cod: There is no single fish prey which dominates the diet of cod. It is a generalised feeder and there is no dominant species taken at any particular period of the year (except for sandeel during the second quarter which amounts to 40% of diet). However, cannibalism in cod is 8.5% of its yearly fish diet. Cod also eat extensive quantities of whiting (20.7%) and haddock (21.2%). Sandeels account for 16.4% over the year and pout for 17.2% of cod yearly diet. Incidental amounts of herring (9%), sprat (4.7%) and mackerel (2.2%) are taken.

Predator haddock: Norway pout (49.7%) and sandeel (45.9%) dominate the diet of haddock. No cod or mackerel appear in the diet of haddock whilst very small quantities of haddock (0.6%) whiting (0.2%) herring (0.8%) and sprat (2.8%) are taken. The Norway pout component of the diet is mainly devoured during the first two quarters although haddock do consume significant amounts of pout during the second two quarters. The main prey species during the second half of the year is sandeel with small amounts taken during the first quarter and none during second.

Predator whiting: Sandeel (44.9%), sprat (15.7%), pout (14.9%) and herring (11.1%) dominate the diet of whiting. Haddock (8.3%) and whiting (4.9% cannibalism) are lesser targets for the whiting. An insignificant proportion of cod and mackerel are taken (<1%). The dominance of sandeel in the diet reaches its peak during the second quarter. This may be associated with post spawning heavy feeding to build body weight.

TABLE 2
Fish Diet: Average daily percentage consumption of fish per quarter for period 1983-85 (after Greenstreet, 1996)

Prey	Cod % predator					Haddock % predator					Whiting % predator				
	Q1	Q2	Q3	Q4	Year	Q1	Q2	Q3	Q4	year	Q1	Q2	Q3	Q4	year
Cod	10.07	10.89	5.44	9.19	8.55	0.00	0.00	0.00	0.00	0.00	0.21	0.71	0.01	26.00	0.27
Haddock	19.38	11.39	23.94	30.59	21.22	0.00	0.00	1.21	0.23	0.57	2.14	1.55	14.02	9.80	8.25
Whiting	37.49	17.31	10.36	18.63	20.68	0.00	0.00	0.12	0.43	0.20	2.43	1.25	8.67	4.71	4.94
N. Pout	6.65	8.87	31.43	16.68	17.18	76.23	87.13	48.35	32.17	49.74	1.20	10.02	9.87	23.59	14.88
Herring	6.16	6.35	16.04	4.27	9.12	2.7	0.00	0.00	0.36	0.79	13.77	0.76	20.37	6.94	11.06
Sprat	9.16	5.17	2.46	1.67	4.67	9.56	12.87	0.75	0.00	2.79	39.75	3.64	10.11	21.06	15.69
Sandeels	11.12	40.02	7.74	11.80	16.37	11.41	0.00	49.47	66.81	45.90	23.50	82.05	36.94	33.64	4.89
Mackerel	0.00	0.00	2.59	7.16	2.17	0.00	0.00	0.00	0.00	0.00	0.00	0.02	0.02	0.00	0.01

Prey	Cod % predator					Haddock % predator				
	Q1	Q2	Q3	Q4	year	Q1	Q2	Q3	Q4	year
Cod	0.00	0.84	0.93	0.00	0.41	0.00	0.00	0.55	0.00	0.21
Haddock	9.15	29.63	35.65	10.99	20.76	0.00	0.00	0.27	0.00	0.11
Whiting	0.65	0.39	0.29	0.46	0.45	0.00	0.00	0.14	0.00	0.05
N. Pout	76.27	46.69	36.17	71.30	58.84	3.85	0.54	20.38	67.49	23.45
Herring	1.39	0.96	0.95	1.11	1.11	0.00	1.03	19.44	17.83	11.94
Sprat	0.71	15.00	0.22	1.84	0.76	3.85	3.33	17.81	10.04	10.45
Sandeels	11.83	19.35	25.79	14.31	17.67	92.31	94.56	41.41	4.64	53.58
Mackerel	0.00	0.00	0.00	0.00	0.00	0.00	0.55	0.00	0.00	0.20

Predator saithe: Norway pout (58.8%), haddock (20.8%) and sandeel (17.7%) dominate the fish diet of saithe. Herring (1%), sprat (<1%), whiting (<1%) and cod (<1%) account for less than one percent each of the diet. No mackerel are taken.

Predator mackerel: Sandeel accounts for 53.6% and pout 23.5% of mackerel fish diet. Smaller proportions of herring (11.9%) and sprat (10.5%) are taken. Insignificant quantities of cod (<1%), haddock (<1%), whiting (<1%) and mackerel (<1%) are taken.

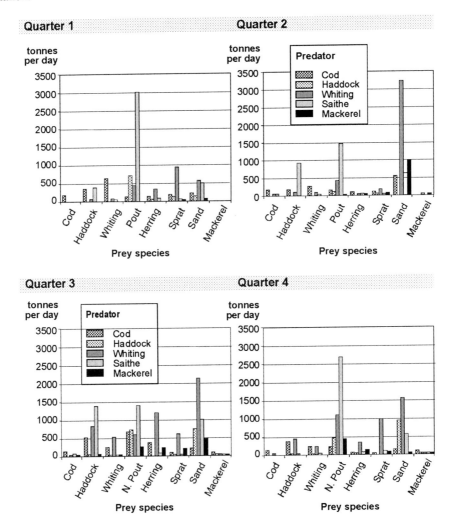

Fig. 4. Daily consumption of fish prey by fish predators (Cod, Haddock, Whiting, Saithe, Mackerel)

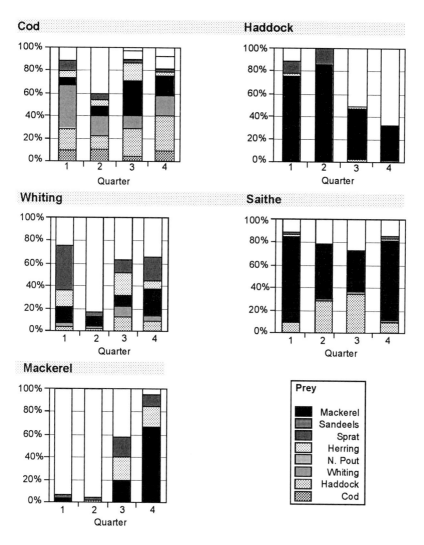

Fig. 5. Percentage of fish prey consumed per quarter by the five main fish predators

TOTAL ANNUAL TONNAGES OF PREY FISH EATEN

These results are combined to illustrate the annual percentage of fish prey taken by each predator (see Fig. 6). Sandeel and Norway Pout dominate the diets of mackerel, saithe, whiting and haddock and the annual average tonnage of each prey species which are eaten by all the predators combined are illustrated in Fig. 7. The corresponding numerical data in Table 3 illustrates the consumption per quarter and total annual consumption.

SANDEEL RESULTS:

Total annual predation on sandeels by the five main human consumption fish predators is 1,276,004 tonnes. This represents 32 per cent of the total tonnage of fish prey consumed. During the peak consumption of sandeels in the second quarter of the year some 56 percent of all fish diet consists of sandeels. The total consumption is a underestimate as apparently the 1981 ICES stomach data did not pick up all the sandeel consumption due to a computer coding error (ICES, 1997). The revised factor (which is not used in this calculation because it is not available) will need to be taken account of in any re-calculation of the data.

The second quarter coincides with the requirement for post spawning human consumption fish to concentrate on building body weight and therefore food intake is high. Sandeels are available in high numbers at this same period and the probability of a predator encountering them is much higher than during winter months. Competition

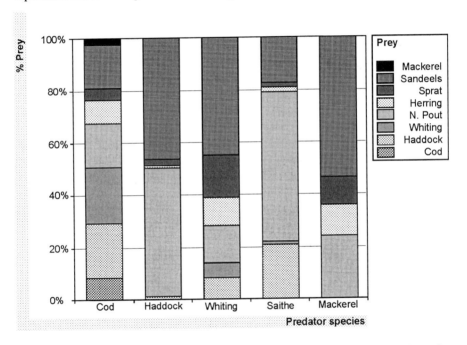

Fig. 6. Overall percentage of each fish prey species taken by each predator (source: Greenstreet, 1996)

with commercial fisheries is also high during this period. Sandeel landings for 1994 of the industrial fishery from each ICES statistical square (30 by 30 mile squares) are given in Fig. 8. The data for each square are presented as a percentage of the total annual landings by Denmark, Norway and United Kingdom. Although sandeel catches are spread over the entire North Sea the industrial fishery has favoured areas where aggregates are encountered to give high and sustained daily catch rates which are

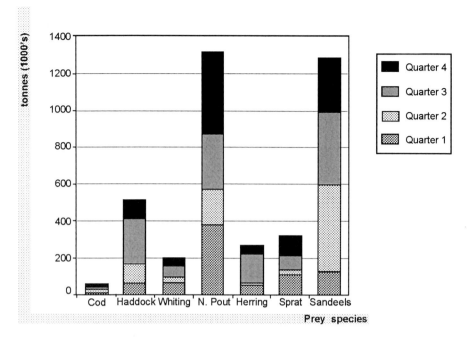

Fig. 7. Annual total consumption of the main fish prey species (quarterly consumption shown) (source: Greenstreet, 1996)

necessary for the economic viability of the vessel. The main sandeel fishing areas are illustrated in Fig. 9 for 1993 (landings by Denmark and Norway) and Fig. 10 for 1994 (landings by Denmark, Norway and United Kingdom). The plots use a geostatistical gridding method (Journel, 1989; Cressie, 1990) to express nodes around which concentrated fishing takes place. Hot spots occur on grounds to the west of Denmark which produces the greatest quantities and this coincides with the need for quality fish to the factories; the closer to home the shorter the steaming time to the factory the fresher the fish will be. Dogger Bank, Wee Bankie off the river Forth Estuary in Scotland (from which landings have only recently developed although a Russian fishery previously existed here in the 1960s) and the Buchan coast off Scotland derive additional catches which in some years can match the tonnages taken from Danish/Norwegian waters.

Haddock consume large quantities of sandeel. Haddock are mainly concentrated from the middle to the west side of the North Sea south of 56.5°N latitude and across the entire North Sea north of this latitude. Heavy concentrations of haddock are associated with masses of sandeels at the Northumberland coast, Wee Bankie and northwards along the Scottish coast including Orkney and Shetland waters.

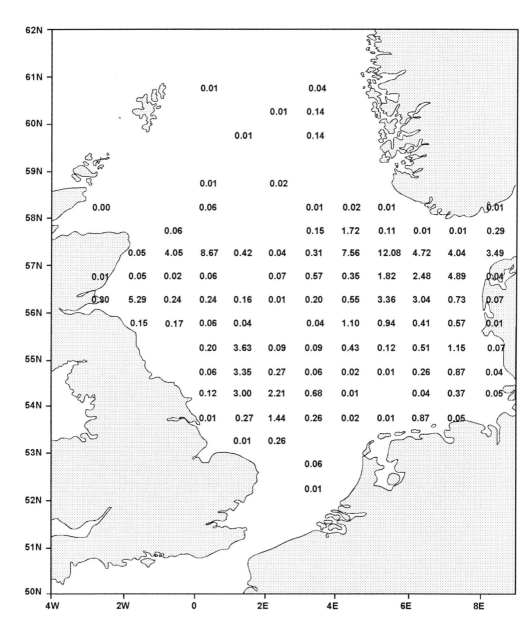

Fig. 8. Sandeel landings by statistical square as a percentage of the total catch in 1994 (source: ICES ACFM)

Nearly half of whiting fish diet consists of sandeel. Whiting of greater than 20 cm congregate heavily in the southern North Sea during summer when sandeel are

Evaluation of Ecological Effects of the North Sea 315

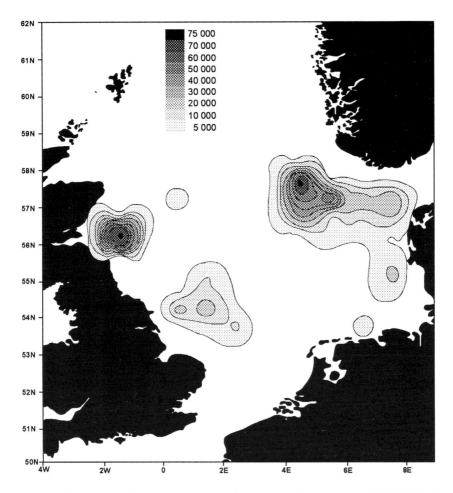

Fig. 9. Sandeel landings in 1993 - catch concentrations (source: ICES ACFM)

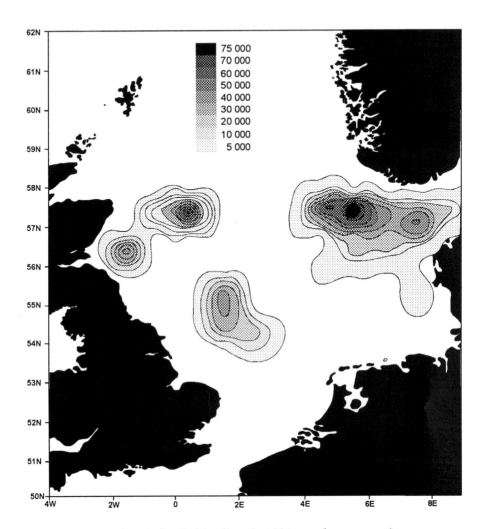

Fig. 10. Sandeel landings in 1994 - catch concentrations
(source: ICES ACFM)

available on the western side along the English and Scottish coasts, on the Dogger Bank and across the Dutch coast and northwards along the offshore Danish coast.

Whiting associate with the entire north-south length of the North Sea during winter. They may well be taking advantage of the pout in northern parts during winter months as well as sandeel, sprat and juvenile herring in southern parts over the whole year.

TOTAL SANDEEL CONSUMPTION BY THE ANIMAL KINGDOM

Adding seabird consumption of, on average, 200,000 tonnes of sandeel per annum, mammal outtake of 45,000 tonnes (both these estimates are taken from Anon, 1994; the mammal consumption does not include that taken by the estimated 380,000 porpoise in the North Sea) and the commercial catch (as indicated in ICES ACFM reports) to the consumption by fish predators gives a total consumption set out in Fig. 11. This is compared with the Spawning Stock Biomass and Total Stock Biomass calculated by ICES for each year from 1976 to 1992. Apparently, total consumption of sandeels equates to the total stock available. ICES do not place confidence limits to their sandeel stock assessment and this may explain why our takeout plot lies higher than the total stock for some years. Even though ICES MSVPA estimates of sandeel consumption by fish predators are lower than our estimates, the reduced total consumption would still lie perilously close to the total stock assessment. It would appear that the food requirements of the North Sea animal kingdom are not being met (at least in respect of sandeel).

POUT RESULTS:

Some 1.3 million tonnes of pout are consumed by North Sea human consumption fish predators. It is interesting to note that the main predators of pout are haddock and saithe of which the older fish of those species are concentrated in the northern part of the North Sea where the main pout concentrations are found.

Haddock and pout are both widespread throughout the northern section of the North Sea, north of 58°N latitude. Older cod are also mainly distributed in northern waters but their intake of pout amounts to a smaller overall proportion of their diet.

The total stock biomass of pout in 1994 (this is the most recent year for which a total is available) as calculated by ICES (Anon, 1996) is 2,590,065 tonnes in quarter four of the year. The commercial industrial fishery accounts for 172,100 tonnes (ACFM, 1994). In total, some 1,483,153 tonnes are taken for food by fish predators and by the commercial fishery. This amounts to 57.3% of the total stock. It does not include natural deaths (such as old age, disease, parasitism and accidental destruction by trawl gear nor any consumption by mammals).

Biomass Consumption Comparison 1977-1986 and 1987-1994

ESTIMATING PREDATOR CONSUMPTION RATES 1977-86:

Total consumption of each species for the 10 year period 1977-1986 were again derived using the values given in Greenstreet (1996); numbers and mean weight at age

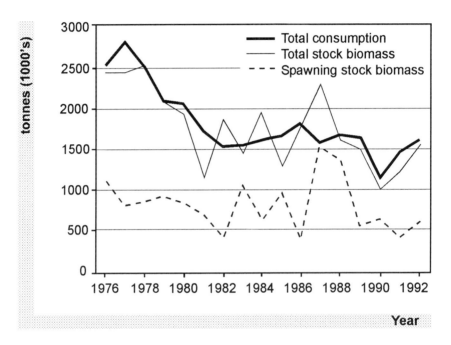

Fig. 11. Total non-human consumption of sandeels from the North Sea compared to spawning stock biomass (SSB) and total stock biomass (TSB) Note: total consumption includes predation by fish, birds and mammals added to the industrial fishery catch (SSB & TSB source - ICES ACFM)

for each species in each quarter averaged over the years were taken from the MSVPA (MSA Working Group, 1990) "key run"; the stomach composition was taken from the 1981 stomach data. Consumption by quarter between the key species over this longer period is shown in network diagrams (see Figs. 12a, b, c and d).

ESTIMATING PREDATOR CONSUMPTION RATES 1987-94:

For the period 1987-94 an MSVPA was run using the revised 1991 stomach data (see ICES CM 1996/Assess:6). Mean weight and numbers at age were taken from the single species assessments where possible or an average was used from the MSVPA key run. Data for 0 group whiting and saithe were taken from Daan (1973) and length-weight relationships where needed from Coull et al. (1989). Consumption by quarter between the key species is again shown in network diagrams (see Figs. 13a, b, c and d).

The annual average biomass consumed by individual predators within the 1977-86 and 1987-94 periods is given in Table 4.

Ecopath Model

TRANSFER EFFICIENCIES:

Using an ECOPATH model (see Christensen and Pauly, 1992), the structure and functioning of the North Sea ecosystem for the periods 1977-1986 and 1987-1994 were compared. The ECOPATH model used in this analysis differs from that given in ICES CM 1992/Assess:16 in that only the key species plus detritus and primary productivity were defined rather than the 22 groups plus detritus of the earlier model. The category "*others*" was primarily based on flatfish. Estimates of detritus and primary productivity for the whole period were taken from Jones (1984), Franz and Gieskes (1994), Hannon and Joiris (1987), ICES CM 1992/Assess:16 and the North Sea Study survey data stored at the British Oceanographic Data Centre, NERC-Proudman Oceanographic Laboratory. Input data were derived from MSVPA runs using the 1981 and 1991 stomach data for the two periods respectively; information about diet compositions for the key species was taken from Daan (1989), Daan et al. (1990) and Greenstreet (1996); consumption rates were derived as for Table 4. The values were then used to derive an annual set of statistics for each period. The key results are given in Table 5.

Eight trophic levels were identified in the analysis. The allocation of species to trophic level and relative flows by trophic level were: I phytoplankton and detritus; II other species; III Norway pout, herring, sprat, sandeels, plaice and sole; IV cod (52%), haddock, whiting (90%), saithe (78%); V cod (46%), haddock (2%), whiting (9%) and saithe (21%); VI virtually no human consumption species at this level; VII cod (1%), whiting (1%) and saithe (1%) and VIII cod (1%).

The transfer efficiencies of the North Sea altered significantly between the two periods; in particular there was a large decline in efficiencies in trophic levels IV-VIII in the latter period, together with a loss of several trophic interactions. It would appear that these changes have been largely brought about by alterations in the top predators and the numbers of species consumed by them. This would suggest that a trophic "trickle" has appeared in the ecosystem, in which prey switching has diminished, and the suitabilities of predatory species altered.

POTENTIAL FOR RECOVERY OF OVEREXPLOITED FISH STOCKS:

ACFM has adopted two precautionary biological criteria to ensure the sustainability of stocks; one relates to growth, the other to reproduction. Whilst per capita growth is an

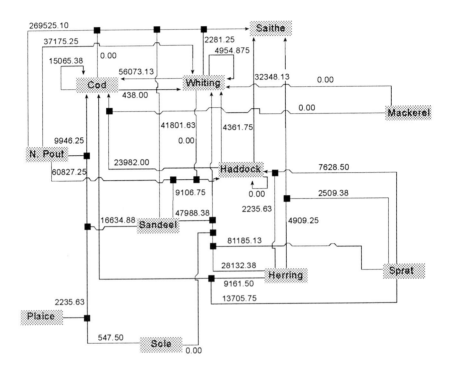

Fig. 12a. 1977-86 mean consumption (tonnes) for 1st quarter

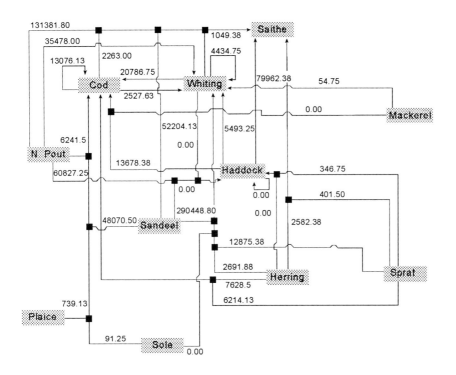

Fig. 12b. 1977-86 mean consumption (tonnes) for 2nd quarter

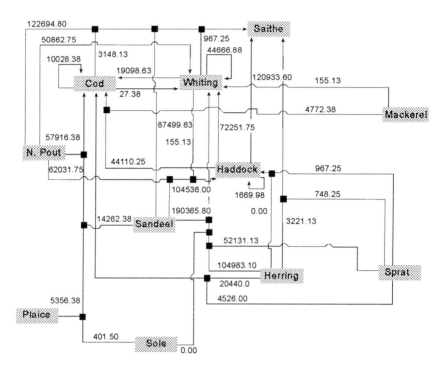

Fig. 12c. 1977-86 mean consumption (tonnes) for 3rd quarter

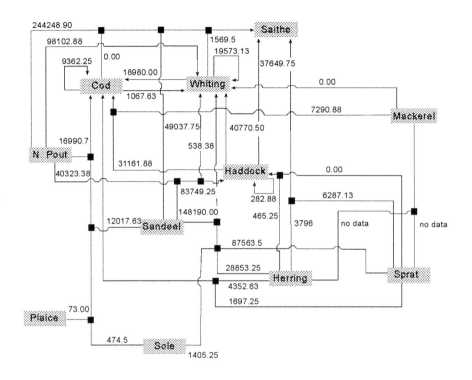

Fig. 12d. 1977-86 mean consumption (tonnes) for 3rd quarter

obvious criterion to monitor and manage, perhaps the most fundamental issue in fisheries management is the relationship between spawner abundance and recruitment. The importance of this relationship has been further underscored by the phenomenon, otherwise known as depensation, inverse density-dependence or the Allee effect, in which reduced reproductive success coincides with low population size. Populations with multiple equilibria may suddenly shift to a regime in which a reduction in fishing mortality is insufficient to induce stock recovery.

To date, a lack of conclusive evidence has led some scientists to state that recruitment overfishing is impossible. But, in two recent studies, Myers and Cowcadens (Myers *et al.*, 1994) (pers. comm. R. Myers, DFO, Newfoundland Canada) have examined 200 populations and 129 species respectively, including the most commercially important North Sea species, to see a) whether fish recruitment is related to spawner abundance and b) whether depensatory effects are present. They concluded that when there is a sufficient range in spawner abundance the answer to the question is yes at all levels but that depensatory effects could only be detected in Icelandic herring and Pacific salmon. In an overall sense then this should not be an issue in current discussions about the North Sea.

From a practical point of view it is important that if fishing mortality thresholds are used in management then they should hold true with observations on spawner abundance and reproductive success. If as Mace (1994) has suggested poor recruitment occurs at one half of the maximum of the stock recruitment (S-R) curve, then the shape of the S-R curve is very important. Only in this way will it be possible to tell whether the threshold estimate is sufficiently conservative without needlessly restricting the harvest. The S-R curves for the key North Sea species were thus examined following Myers et al. (1994) to determine the degree of risk associated with current fishing levels. Where possible the threshold values used were MBAL otherwise 50% of maximal recruitment was used. The S-R data were divided into those above and below the threshold, and then regressions performed on the log-transformed data. The S-R plots for the key species are shown in Fig. 14 (for cod, haddock, saithe, Norway pout, whiting, plaice and sole respectively). From the Myers approach cod, haddock, plaice, whiting and sole all appear to be in a category where the threshold currently set is reasonable; for saithe, where there is a negative relationship between stock and recruitment, the threshold value is difficult to apply.

Calculating the ratio of recruitment above and below such thresholds is important in the argument that reducing a stock below the estimated level will have an impact on productivity. In the absence of extensive data, the threshold must be seen as a precautionary approach. However, it does suggest that within the current management regime, biomass thresholds to protect recruitment could be used to define both the biological and economic risks associated with higher levels of fishing (McGlade and Shepherd, 1992). In the current regime however, as most of the stocks are above the threshold values, the argument has to be made that not only are there biological losses accruing but also long-term economic ones as a result of the fishing practices in place.

Evaluation of Ecological Effects of the North Sea 325

North Sea Fleet Dynamics

Throughout Europe, fishing is an important but relatively minor activity in terms of employment and output. Of the four countries primarily concerned in the North Sea fisheries, fishing is more important in Denmark and the Netherlands than in the UK and Germany. But, the industry is very location specific, so recent trends have had

TABLE 5
Transfer efficiencies (%) for the North Sea in the periods 1977-86 and 1986-94

Period	Trophic Levels					
	III	IV	V	VI	VII	VIII
1977-86	13.5	32.8	21.1	24.7	22.8	26.9
1986-94	14.0	27.0	12.8	14.0	13.7	13.9

TABLE 6
Summary of responses from interviews held with participants in UK fisheries

Number of Responses	Issue	Range of Estimates (%)
10	Black/brown fish estimates: Landed in Scottish mainland ports by UK vessels	40-45%
	Landed in Isles ports by UK vessels	60-70%
	Landed in UK ports by UK vessels	40-50%
10	By-catch of human consumption species by UK vessels	38-45%
5	By-catch of species in the industrial fishery	35-38%
5	By-catch of human consumption species by Spanish vessels	0-5%

TABLE 7
Transfer efficiencies (%) for the North Sea for the periods 1977-86 and 1986-94 (as for Table 5); with 30% increase in landings of the MSVPA run of human consumption species (30%); with the industrial fishery species catches and by-catch cut by 50% for the current year (NI94) and using the earlier period as a basleine (NI77-86)

Period	Trophic Levels					
	III	IV	V	VI	VII	VIII
1977-86	13.5	32.8	21.1	24.7	22.8	26.9
1986-94	14.0	27.0	12.8	14.0	13.7	13.9
30%	14.0	28.9	15.9	17.6	17.0	18.1
NI94	12.2	28.9	15.9	17.6	17.0	18.1
NI77-86	11.9	32.8	21.1	24.7	22.8	26.9

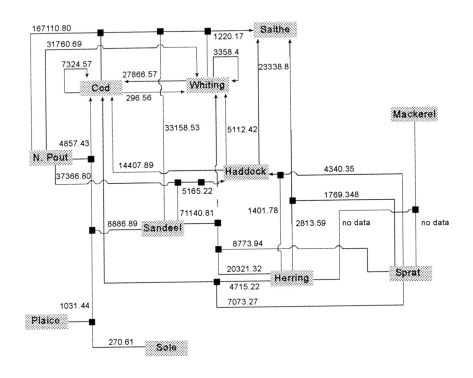

Fig. 13a. 1987-94 mean consumption (tonnes) for 1st quarter

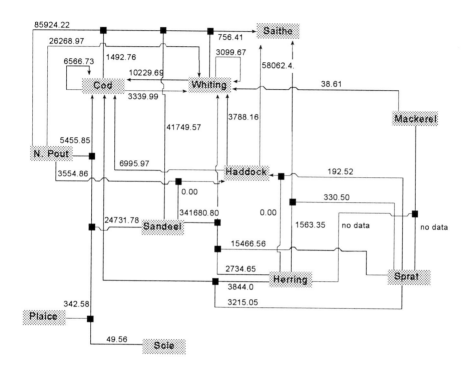

Fig. 13b. 1987-94 mean consumption (tonnes) for 2nd quarter

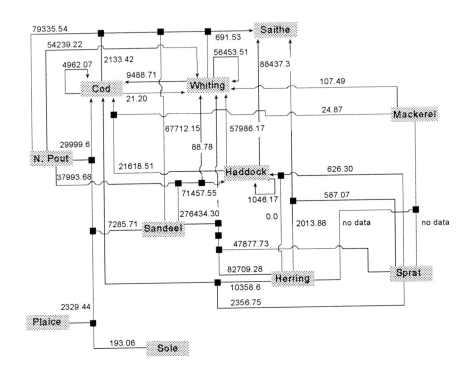

Fig. 13c. 1987-94 mean consumption (tonnes) for 2nd quarter

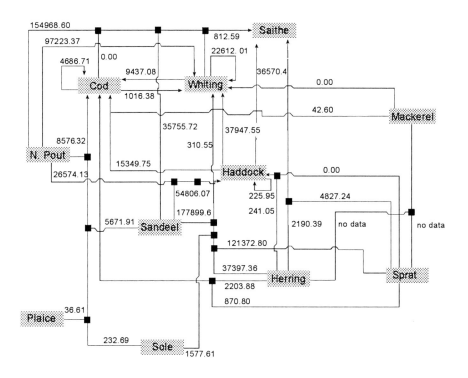

Fig. 13d. 1987-94 mean consumption (tonnes) for 4th quarter

equally adverse effects throughout the region. The UK has the largest number of fishermen of the four countries, more than half the estimated 30,000 fishermen in the UK, Netherlands, Germany and Denmark. However, as a proportion of national employment fishing in Denmark is four times that in the Netherlands or the UK.

In terms of vessels, the UK fleet is the largest of the four followed by the Danish. In terms of capacity (gross registered tonnage), the UK fleet is also the largest followed by Denmark and the Netherlands, although actual registered tonnage by British vessels is somewhat distorted by vessels that have reflagged in the UK. The average capacity per vessel (kW) is much larger in the Netherlands (800kW/vessel) than the other countries. However around 70% of the UK fleet still consists of small vessels (10m) and that these generate much part-time employment.

The Danish fleet is the largest in terms of catch of the four countries represented in the northern seaboard area, followed by the UK, the Netherlands and Germany. Approximately 70% of the Danish catch by volume (25% by value) is for reduction purposes - the production of fish meal and oil. The other countries are more oriented to fish for human consumption. National preferences and the composition of the fleets cause catch composition to vary; for example in the UK mackerel, herring, cod and haddock are important; whereas in Germany it is cod and in Denmark it is cod and herring. It is thus a general feature of these countries, with the exception of the Netherlands, that their fisheries depend heavily on the state of the cod and other roundfish stocks.

While the bulk of fish landings are destined for consumption within Europe, usually in added-valued processed products, there is an increasing level of imports from outside the community. This is particularly the case in Denmark and reflects in part its need to maintain a very extensive processing industry. Denmark's fish and fish products account for over 5% of total Danish exports by value; Dutch exports are about 60% of those in Denmark. The UK and Germany are net importers of fish and fish products.

Extensive surveys of the fishing industry in various member states, have shown however that declining catches have led to economic and social problems not only for the fishing communities but also for processors. The measures for conservation and control, introduced under the CFP, have tended to exacerbate these problems through the creation of others, including illegal fishing; increased efficiency of vessels encouraged indirectly by the quota system; high levels of discards; and flags of convenience. The processing industry also suffers from a shortfall of local raw materials and is increasingly dependent on imports. As both species and quality of imports continue to diverge from those afforded through local supplies, the industry will need to develop new products or processing techniques.

To gain a clearer idea of the impacts of illegal fishing and bycatches on the North Sea ecosystem, a series of confidential interviews with the industry were undertaken. The results, summarised below in Table 6, starkly underline the extent of blackfish (over quota illegally landed fish) in the market place and the levels of bycatch on UK vessels. This is in sharp contrast with the bycatch levels on Spanish vessels for

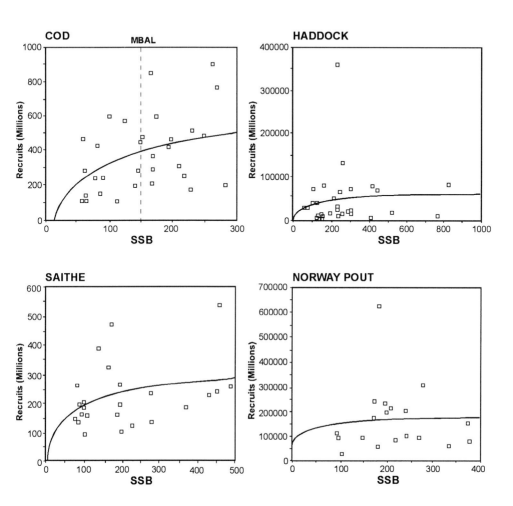

Fig. 14. Stock (000mt) versus recruit (millions age 1 for whiting, plaice and sole)

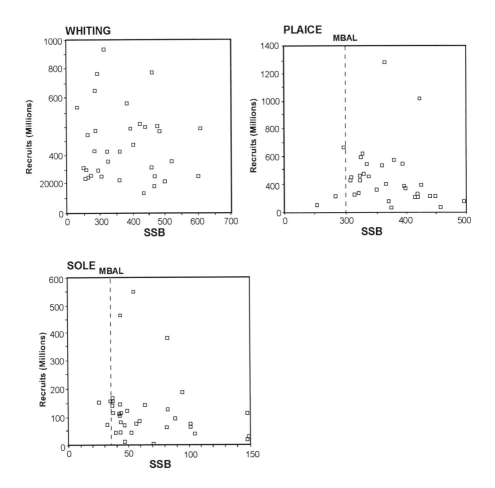

Fig. 14. *Continued* stock (000mt) versus recruit (millions age 1 for whiting, plaice and sole) example, where virtually everything is landed and used (although much of this is illegal undersize fish).

The availability of illegally landed fish, increased imports and cheaper substitutes has led to a stabilisation of fish prices, and as a result, the long-term economic forecasts for high quality, human consumption species are generally optimistic in the harvesting sector, despite ICES warnings of overfishing and uncertainties in fish production. This suggests that the current levels of bycatch and illegal fishing are unlikely to be reduced over the next few years unless there is a radical change in the management of North Sea fisheries or the ecosystem itself.

The other major aspect that the interviews looked at was the impacts of the industrial fisheries and the overall adherence to regulations. Regarding the latter, all interviewees indicated that excessive regulations meant that most fishermen considered themselves to be operating outside the legal framework for much of the time that they were at sea. Overall, there was little appreciation of the potential effects of the industrial fisheries other than it was a Danish issue. Rather the main perception was that quality was now of paramount importance, and that crews and POs (Producer Organisations) alike were having to improve fishing and handling practices to maintain fish quality throughout the market chain.

The Biological Effect of the North Sea Fisheries

MODELLING THE EFFECTS OF CHANGES IN THE ECOSYSTEM AND FISHERIES:

As a result of using the 1981 stomach data for the earlier period (1977-86) and the revised 1991 stomach data for the period 1987-1994, the biomass consumed by the individual predators from this analysis (see Table 4) are larger by as much as 30% in comparison with the published results from the ICES MSA Working Group. However, these differences are small in comparison to the potential losses through illegal landings and discards to the overall tonnage extracted from the North Sea.

Using the estimates of bycatch plus illegal landings of human consumption species derived from the interviews above, a new MSVPA was run for the period 1987-1994 with a 30% increase in the catches of cod, haddock, whiting and saithe. The data were then used to run an ECOPATH model to examine the transfer efficiencies and the trophic impacts. A second MSVPA run was then undertaken using this revised set of population figures and with the industrial fishery on Norway pout, herring and sandeel reduced nominally by 50%. A third run was also undertaken to examine the effects of using the earlier period as a starting condition for reduction of the industrial fishery. The results were averaged over a five year projection and put into an ECOPATH model. The transfer efficiencies and impacts of these 3 models are given in Table 7 and Figs. 15, 16 and 17.

As can be seen in Table 7, the inclusion of 50% more in the landings of human consumption species in the MSVPA generates a noticeable increase of 2 - 4% in the trophic transfer efficiencies from levels IV - VIII. This shows that the roundfish fishery, operating as it does on older age classes, can have a profound effect on the actual efficiency of the whole system. What is more, the trophic transfer efficiencies lie between the results of the two decadel averages, suggesting that biomass landed either illegally or lost through discards is higher than 30%.

Looking at the effects of a reduction in the industrial fishery, only the level III transfer efficiencies change: the 1.8% reduction is probably indicative of the fact that any benefits coming from an increase in biomass of industrial species are dissipated within the trophic level itself. However, it is the comparison of the baseline runs that gives some indication of the problem. It would appear from the transfer efficiencies that under the current ecological regime there is a strong possibility that gains in the lower trophic levels will not lead to changes in the transfer efficiencies in the upper levels in line with those observed in the 1977-86 period, but that inclusion of illegal landings etc., brings the estimates together.

The impact diagrams (Figs, 15, 16 and 17) show that the effects of those changes on pairwise interactions are small but that cumulatively the negative impacts of fisheries in the earlier decade were spread across all the trophic levels and have now become more concentrated.

Discussion with Conclusions and Recommendations

(a) Fisheries are experiencing reduced levels of cod, haddock, whiting, mackerel and herring. This means that fish imports are required for human consumption within the European Community to offset the shortfall of fishery products from our own seas.

(b) The food preferences and requirements of the fish within the North Sea have noticeably altered during the period 1987 to 1994.

(c) Human consumption species eat mostly sandeel and Norway pout in the fish component of their diet.

(d) Breeding failures in bird colonies are evident as a consequence of reduced quantities of fish high in lipids to feed the young (see for example Anon, 1994). Probabilities of mammals encountering suitable prey food are diminishing.

As expected, the introduction of the illegal landings and discards into the biomass models has a significant effect on the upper trophic levels in terms of their transfer efficiencies; the reduction of the industrial fishery only effects level III. The impact diagrams show only small changes in the interactions and impacts between the predator and prey species given the 1994 situation, compared to the effects that removal of the industrial fishery would have had in the earlier period.

It would appear that the high levels of illegal landings and discards from the human consumption fishery mitigate against the full effect of a reduction in the

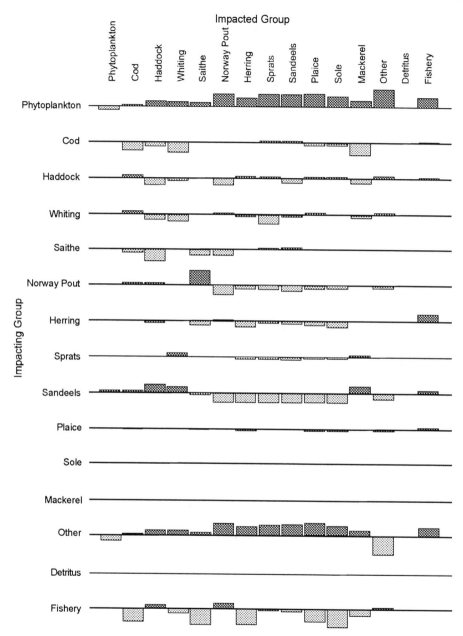

Fig. 15. 30% run with 30% increase in historical landings of human consumption species 1987-94

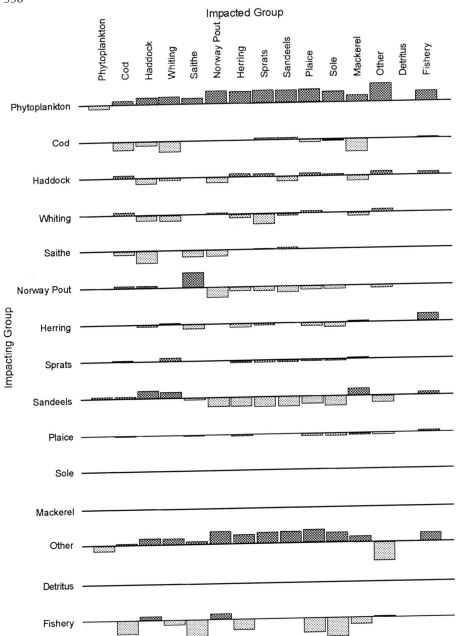

Fig. 16. NI 94 run with 50% decrease in the industrial fishery

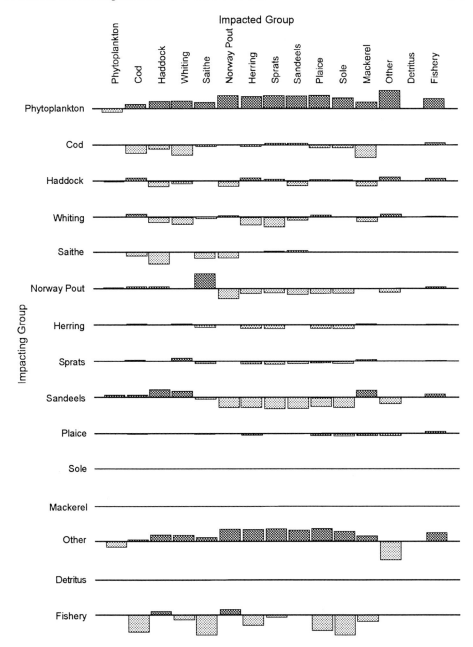

Fig. 17. NI 77-86 run with 50% decrease in the industrial fishery during 1977-86

reported industrial fishery. However, given that illegal landings are also a key element in some of the industrial fisheries it is clear that some reduction should be undertaken as soon as possible. Alternatively, methods by which the by-catch species may be excluded from the trawls should be investigated. There is little effort being made by the fishing industry to reduce discards in the human consumption fishery (except for Scottish, English, Irish and Swedish fishermen who use enhanced selective devices in prawn trawls, Robertson (1993) and by Danish fishermen in cod trawls). The fishing industry should stop discarding of juvenile fish from the human consumption fishery; stop the practise of by-catch within the industrial fisheries for pout and sprat; limit the outtake of sandeels and stop the practise of "blackfish" landings. However, economic as well as legal sanctions will clearly be needed to achieve this. The result could be an enhanced fishery for the industry.

Alterations in the ecosystem, which have occurred as a result of the high levels of exploitation on species in the higher trophic levels, have now given rise to a situation in which the suitabilities and prey preferences have changed. Given the fact that many of the human consumption species are also potentially at stock levels where recruitment and growth overfishing are already occurring, it would seem that the only rapid way to remediate the situation is to combine the two processes of reduction in the industrial fishery as well as making steps towards controlling or eliminating the high levels of illegally landed fish. This is a course already well known to the EU member states, but what is certainly needed to create the appropriate management structure is a full accounting of the illegal landings associated with all North Sea fisheries.

Greater emphasis must be placed on fish species interactions by way of who eats who and under what conditions. The consumption of fish by birds and mammals must also be taken account of more positively and fully within the ecosystem. We can no longer afford the luxury of paying no heed to the food needs of the creatures within the North Sea system. The food requirements are important elements controlling the life of the stocks or the ability for the stocks to grow. All other stock parameters will follow naturally if enough food is available. Continuous monitoring of the food requirements of fish should be undertaken and some form of calibration of future food availability versus increasing stock size is needed. Immediate action is needed if food availability reduces to levels that cannot sustain the stocks. This is the correct basis on which catching levels on the various stocks should be controlled.

Based on recent stomach sampling exercises it appears that sandeel and pout are the two most important food items for the human consumption fish species of the North Sea. Greater levels of protection of these stocks should be afforded to ensure enhanced food supplies within the system to allow the stocks to grow. The calculations contained in this chapter suggest that a reduction in the sandeel fishery would allow more food to become available to the predators who rely heavily on the sandeel for sustenance and allow increased spatial overlap between food and predators.

There is clearly a question as to whether or not enough food exists in the North Sea ecosystem to feed the fish. As Gislason and Kirkegard (1996) suggest "If the stock sizes of the fish predators should increase, for example due to an overall decrease in fishing effort in the human consumption fishery in the North Sea, it may be necessary to decrease the fishing mortality of sandeel as well". Their concerns about lack of food for fish are mirrored by Greenstreet et al. (1997) who state "The food requirements of piscivorous demersal fish also do not appear to be adequately supplied from within the North Sea, but, in this case, immigration is not though to provide the shortfall". Our analysis supports the view of the above authors that ecosystem difficulties are occurring in the North Sea.

Given that there is a lack of food and given that sandeel constitutes the main element of the diet of the most important human consumption species, the precautionary principle should be applied to the North Sea sandeel fishery to limit the catches. This will ensure that food is made more readily available to other fish, birds and mammals within the system.

This could be achieved by various means i.e.

1) **the closure of the Dogger Bank, Wee Bankie and Buchan areas of the North Sea**
2) **Realising the importance of the whole of the western side of the North Sea to major world class breeding colonies of seabirds and of mammals and not least to human consumption fish stocks, the imposition of a "sandeel median line" along the 4 degree east longitude line westwards of which all fishing for sandeel should be banned may be a useful measure**
3) **Further being mindful of the very great importance of the sea west of Jutland, Denmark for the rearing of juvenile cod which feed on sandeel, although apparently to a lesser extent than older cod, it is suggested that the sandeel catch be limited in this area**
4) **A precautionary TAC set at an agreed limit lower than the recent catch average**

Such possible precautionary approaches to the problem fall in line with FAO's recent guidelines on the precautionary approach to capture fisheries (FAO, 1995).

The sandeel stock assessment requires greater accuracy. Only then can more robust conclusions be made on the percentage outtake of sandeel by predators. Until such time, the precautionary principle should be applied to limit the catch.

It would appear that a new thinking is required as to how we consider the North Sea fish stocks. The portioning of food requirements of the human consumption species must include the forward calculation which accounts for the future food needs of an increased stock size. It is essential that larger stock sizes do not eat themselves down or suffer lack of food and thus negate their increase through improved management structures. The only method by which this may be achieved is by ensuring

1) **that the high lipid species (i.e. sandeel, Norway pout, herring and sprat) are available in as high quantities as possible as food for the human consumption species**

2) **levels of illegal catches and by-catches are properly estimated and included in the calculations of fish stock status, and**

3) **a precautionary principle is applied at the ecosystem level**

The management of fisheries data for the North Sea is unsatisfactory. Whilst efforts are being made to integrate oceanographic and ecological relationships, there is no real investigation of food requirements of fish (other than Greenstreet et al, ibid) and integration of this with all other data. GIS presents great opportunities to develop an overview of the North Sea in micro and macro detail with all known data meshed with various interactive programmes to give true pictures of the resource both as a historical review in helping to model how and why it was, with up to date geostatistical comparisons between the animal kingdom to calculate future resource strength (see for example Meaden and Chi (1996) who make some suggestions). An information system *per se* is not what is intended here, but rather a fully integrated data holding system at the first tier with predictive and analytic programmes allowing data manipulation above this with clear visualisation of the area for resource management purposes. Much data exists which is impossible to access due mainly to bureaucratic creep or national political interest protection. We need a more open and truthful system of data sharing to allow a greater range of experts to analyse the data and this may open better possibilities for understanding the consequences of fishing mortality. Satellite monitoring of fishing vessels within EU waters will start in 1998 with over 25m industrial trawlers. Such information will eventually allow real time assessment for resource management within a GIS system.

We should like to recommend that i) a workshop to examine the widespread consequences of changes in the fishing practises in the North Sea should be held in the near future and made open to industry, EU Member State scientists and policy makers and other interested parties particularly fishery stakeholders and ii) a fishing industry consultation workshop be established quickly to consider and formulate a Code of Conduct for Responsible Fishing Operations in the North Sea. The Code would provide an opportunity to address the controversial and difficult issues (i.e. over quota catches and juvenile fish discards amongst others) by involving the industry and fishers from each of the Member States represented. Long term, cost effective practises would be established replacing existing practises with those that meet goals of sustainable fishing and which are genuinely acceptable to those whose livelihood is in fishing. iii) a workshop to explore the possibility of setting up a fully integrated GIS for the North Sea fisheries be undertaken soon.

Acknowledgements

We warmly appreciate the help and assistance and discussions with a range of individuals and organisations which include: Dr S Barlow, International Fishmeal and Oil Manufacturers Association; Mr Schougrrd, Association of Fish Meal and Oil Manufacturers, Denmark; Dr R Bailey and Dr H Sparholt, International Council for the Exploration of the Sea; Mr H Berg Madsen and staff, 999 Factory, Esjberg; Mr E Kirkegard and Dr H Gislason, Danish Institute for Fishery and Marine Research; Mr N Vickman and Mr H Krog, Danish Fishermens Association; Scottish Fishermens Federation; Mr A Smith, Scottish Whitefish Producers Association; Scottish Office Agriculture Environment and Fisheries Department; Ministry of Agriculture Fisheries and Food; Mr C Rose and Dr M MacGarvin, Greenpeace; Mr T Hay, Fishermens Association Ltd; Scottish Pelagic Fishermens Association; V Mackinlay, S Wilson and staff at Univation; Ian Cargill, the Design Consultancy Unit at Grays School of Art, Aberdeen; and to many others we extend thanks. The analysis was aided by ongoing research under EU programmes on the market chain and ecological impacts of by-catch of mammals and other organisms in certain North Sea fisheries including scientists from EU Member States and ICES scientists.

We are also most grateful for critical review and suggestions on a earlier version of the chapter by Dr V Christensen, Dr D Raffaelli, Dr D Pauly, Mr R McColl and Prof. J Rowan-Robinson.

References

Anon. (1991). *Report* of the Multispecies Assessment Working Group. ICES CM 1991/Assess:7.
Anon. (1992). *Report* of the Multispecies Assessment Working Group. ICES CM 1992/Assess:16.
Anon. (1994). *Report* of the study group on seabird/fish interactions. ICES CM 1994/L:3
Anon. (1994). *Report* of the Multispecies Assessment Working Group. ICES CM 1994/Assess:9.
Anon. (1995). *ICES Cooperative Research Report* No. 214.
Anon. (1995). *Report* of the ICES Advisory Committee on Fishery Management, 1995. Part 1.
Anon. (1996). *Report* of the Multispecies Assessment Working Group. ICES CM 1996/Assess:3.
Anon. (1996). *Report* of the Working Group on the Assessment of Demersal Stocks in the North Sea and Skagerrak. ICES CM 1996/Assess:6.
Anon. (1996). *Report* of the ICES Advisory Committee on Fishery Management, 1996. ICES Cooperative Research Report No 214.
Basimi, R.A., and Grove, D.J. (1985). Estimates of daily food intake by an inshore population of Pleuronectes platessa L. off Anglesey, North Wales. *Journal of Fish Biology* 27:505-520.
Bromley, P.J. (1991). Gastric evacuation in cod. *ICES Mar., Sci., Symp.* 193:93-98.
Christensen, V. (1995). A model of trophic interactions in the North Sea in 1981, the Year of the Stomach. *Dana* 11 (1): 1-28.

Christensen, V. and Pauly, D. (1992). ECOPATH II - a software for balancing steady-state ecosystem models and calculating network characteristics. *Ecological Modelling*, Vol. 61, pp 169-185.

Coull, K.A et al. (1989). Length/weight relationships for 88 species of fish encountered in the North East Atlantic. *Scottish Fisheries Research Report* 43.

Commission of the European Communities. (1992). Assessment on the biological impact of industrial fisheries in the North Sea and in the Skagerrak and Kattegat. SEC(92) 2406.

Cressie, N.A.C. (1990). The origins of kriging. *Mathematical Geology*, Vol. 22, p. 239-252.

Cushing, D.H. (1980). The decline of the herring stocks and the gadoid outburst. *J. Cons. int. Explor. Mer*, 39: 70-81.

Daan, N. (1973). A quantitative analysis of the food intake of North Sea cod. *Netherlands Journal of Sea Research* 6:479 - 517.

Daan, N. (1989). Data base report of the stomach sampling project 1981. ICES Cooperative *Research Report* 164.

Daan, N. et al. (1990). Ecology of North Sea fish. *Netherlands Journal of Sea Research* 26:343-386.

Elliott, J., and Persson, L. (1978). The estimation of daily rates of food consumption for fish. *Journal of Animal Ecology* 47:977-991.

Fogarty, M.J., Cohen, E.B., Michaels, W.L., and Morse, W.W. (1991). Predation and the regulation of sand lance populations: an exploratory analysis. *ICES Mar. Sci. Symp.*, 193: 120-124.

Food and Agriculture Organisation of the United Nations. (1994). Review of the state of the world marine fishery resources. FAO Fisheries *Technical Paper* No 335.

Food and Agriculture Organisation of the United Nations. (1994). A global assessment of fisheries by-catch and discards. FAO Fisheries *Technical Paper* No 339.

Food and Agriculture Organisation of the United Nations. (1995). *Code of Conduct for Responsible Fisheries*.

Food and Agriculture Organisation of the United Nations. (1995). Precautionary approach to fisheries. FAO Fisheries *Technical Paper* No 350/ pts 1 & 2.

Gislason, H and Kirkegard, E. (1996). The industrial fishery and the North Sea sandeel stock. In: *Seminar Report*: The precautionary approach to North Sea fisheries management. Oslo 9-10 Sept 1996. Fisken Og Havet NR 1-1997.

Greenstreet, S.P.R. (1996). Estimation of the daily consumption of food by fish in the North Sea in each quarter of the year. Scottish Fisheries *Research Report* Number 55.

Greenstreet, S.P.R., Bryant, A.D., Broekhuizen, N., Hall, S.J., and Heath, M.R. (1997). Seasonal variation in the consumption of food by fish in the North Sea and implications for food web dynamics. *ICES Journal of Marine Science*, 54: 243-266. 1997.

Hannon, B., and Joiris, C. (1989). A seasonal analysis of the southern North Sea ecosystem. *Ecology* 70:1916-1934.

Hansson, S. et al. (1996). Predation rates by North Sea cod - prediction from models on gastric evacuation and bioenergetics. *ICES J. mar. Sci.* 53: 107-114

Hewett, S. W. and Johnson, B.L. (1992). Fish bioenergetics model. University of Wisconsin Sea Grant *Technical Report* No. WIS-SG-92-250.

Hildén, M. (1992). Nordic workshop on bioenergetics of fish. *Nordiske Seminar-* og Arbejdsrapporter 517.

Hislop, J.R.G. et al. (1983). A preliminary report on the analysis of the whiting stomachs collected during the 1981 North Sea stomach sampling project. *ICES CM 1983/G:59*.

Jobling, M. (1986). Mythical models of gastric emptying and implications for food consumption studies. *Environmental Biology of Fishes* 16:35-50.

Jones, R. (1974). The rate of elimination of food from the stomachs of haddock, cod and whiting. *Journal du Conseil* 35:225-243.

Jones, R. (1984). Some observations on energy transfer through the North Sea and Georges Bank food webs. *Rapp. P.-v. Reun. Cons. int Explor. Mer* 183:204-217.

Journel, A.G. (1989). *Fundamentals of geostatistics in five lessons.* American Geophysical Union, Washington D.C.

Knijn, R.J., Boon, T.W., Heesen, H.J.L., and Hislop, J.R.G. (1993). Atlas of North Sea Fishes. *ICES Cooperative Research Report* No 194.

Last, J.M. (1989). The food of herring in the North Sea. 1983-1986. *Journal of Fish Biology* 34: 489-501.

Mace, P.M. (1994). Relationships between common biological reference points used as thresholds and targets of fisheries management strategies. Canadian *Journal of Fisheries and Aquatic Sciences* 51:110-122.

McGlade, J. and Shepherd, J. (1992). Techniques for biological assessment in fisheries management. *Berichte aus der Oekologischen Forschung* Band 9. Forschungszentrum Juelich Gmbh.

Mehl, S. and Westergard, L. (1983). Gastric evacuation rate in mackerel in the North Sea. *ICES CM 1983/H:34*.

Meadon, G.J. and Chi, T.Do. (1996). Geographic information systems - Applications to marine fisheries. FAO Fisheries *Technical Paper* No 356.

Myers R.A. et al. (1994). In search of thresholds for recruitment overfishing. *ICES J. mar. Sci.* 51: 191-205.

Myers, R.A., Bridon, J. and Bamman, N.J. (1994). Summary of worldwide stock and recruitment data. *Can. Tech. Rep. Aquat. Sci.*

Persson, L. (1986). Patterns of food evacuation in fishes: a critical review. *Environmental Biology of Fishes* 16: 51-58.

Robertson, J.H.B. (1993). Design and construction of square mesh windows in whitefish and prawn trawls and seine nets. Scottish Fisheries Information *Pamphlet*. No 20 1993.

Rumohr, H. et al. (1987). A compilation of biometric conversion factors for benthic invertebrates of the Baltic Sea. *The Baltic Marine Biologists Publication* 9.

dos Santos, A.J. (1990). Aspects of the ecophysiology of predation in Atlantic cod. *Dissertation* from Tromso University.

Glossary of Terms

ACFM	Advisory Committee on Fishery Management (ICES).
Benthos	Organisms living on or in the sea bottom.
Biomass	Fish stocks expressed as a weight.
Blackfish	Illegally landed over quota fish.
By-catch	Species caught incidental to the main target species.
CFP	Common Fisheries Policy (of the European Union).
Discards	Small or over quota fish thrown thrown back into the sea (usually dead).
DG XIV	European Directorate for Fisheries Management.
ECOPATH	Food web computer model.
FAO	Food and Agriculture Organisation of the United Nations.
ICES	International Council for the Exploration of the Sea.
MAGP	European Union Multi-Annual Guidance Program.
MBAL	Minimum Biologically Acceptable Spawning Stock Level.
MSVPA	Multi Species Virtual Population Analysis.
Plankton	Small organisms carried in the water (Phyto/Zooplankton)
Pout Box	Area of the North Sea where fishing for Norway Pout is illegal.
Pre-recruit	Juvenile before joining the parent spawning stock.
Recruitment	When fish reach maturity and enter the parent spawning stock.
SSB	Spawning Stock Biomass (of mature fish).
S-R curve	Stock recruitment curve.
TAC	Total Allowable Catch.
Trophic trickle	Top-down effect between trophic levels.
TSB	Total Stock Biomass (mature fish plus pre-recruits).

Appendix 1

ICES Area IV a b c.

Data used in analysis for periods 1977-86 and 1987-94.

Sources: ICES CM 1996/Assess:6, ICES CM 1996/Assess:3, ICES CM 1992/Assess: 16 and ICES CM 1994/Assess:9, ICES CM 1991/Assess:7, ICES Cooperative Research Report (1995) 214.

R = directed roundfish; D = discards; FR = flatfish with roundfish bycatch ( );
MnI = minor industrial; MjI = major industrial with bycatch ( ).
O = overexploited; C = close to safe; S = at safe limit; U = unknown.

Evaluation of Ecological Effects of the North Sea

Species	Fishery	Categ	Lndgs	HC	IB	Disc	Effort	Period
Cod	RD	O	•		•*		•	63-94
Haddock	RD	S	•	•	•	•	•	63-94
Whiting	RD	U	•	•	•	•	•	63-94
Saithe	R	C/S	•	•	•		•	70-94
Plaice	FR (cod)	O	•		•*		•	63-94
Sole	FR (cod)	S	•				•	63-94
Herring	MnI	O	•				•	70-94
Sprat	MjI(herr)	U/S?	•				•	70-94
Mackerel	MnI	O	•		•*		•	63-94
H mackerel	MnI	U	•				•	70-94
Norway pout	MjI	S	•				•	63-94
Sandeel	MjI	S	•				•	63-94

Data available for the estimation of the daily consumption of food by fish in the North Sea in each quarter of the year. (source: Greenstreet (1996). Scottish Fisheries Research Report No 55).

Species	Diet(%)A/LQ	P(t/d)AQ
Cod	• (A)	•
Haddock	• (A)	•
Whiting	• (A)	•
Saithe	• (L/A)	•
Plaice	• (A)	•
Sole	• (-)	•
Herring	• (L/A)	•
Sprat		•
Mackerel	• (A)	•
H mackerel	• (L)	•
Norway pout	• (A)	•
Sandeel		•

A age; L length; Q quarter; P prey consumed; t/d tonnes per day

This chapter presents an analysis of the ecological impacts of the industrial fisheries of the North Sea.

    Published data from the International Council for the Exploration of the Seas and the literature, indicate that the demersal, pelagic and industrial fisheries of the North Sea fisheries have undergone significant changes during the last two decades. To examine the potential effect that this might have had on the productivity and ecological dynamics of the key fish species for human consumption, a multispecies analysis of diet composition, consumption rates, trophic efficiencies and impacts was undertaken. The analysis was broken into three historical periods, 1983-85 which was used to consider the specific diet requirements of human consumption species and the two periods 1977-1986 and 1987-1994, to reflect changes in the observed trends in diet and landings from the different fisheries and establish possible changes in food availability in the two major periods.

Results indicate that: up to 60% of the fish component of the diet of human consumption fish species consists of sandeels and Norway pout in approximately equal amounts; and that there was a 30% decline in overall biomass consumption in the second period 1987-1994, coincident with changes in the transfer efficiencies between trophic levels and the loss of several trophic interactions. The observed changes were largely associated with the dynamics of the top predators, suggesting that a trophic trickle had appeared in the ecosystem, and that prey switching and shifts in the suitabilities of the predatory fish species had occurred. One potential reason may be that because sandeel and Norway pout are the two species which are consumed in the greatest quantities, there may be direct competition between the industrial fishery and the human consumption fishery.

The potential for recruitment and growth overfishing was also analysed in terms of biomass thresholds; it was concluded that whilst the major human consumption stocks in the North Sea do not show any significant signs of recruitment failure, most of them are certainly at risk.

The interpretation of these results is compounded by the level of over quota landings, by-catches and discards of human consumption species within the industrial and human consumption fishery. Results from interviews with the industry indicate that levels are as high as 60-70% in some areas and times of year, with an overall level of not less that 30%. To fully assess the impacts of this and the industrial fisheries on human consumption species three scenarios were examined using the MSVPA and ECOPATH models: i) a 30% increase in the historical landings of human consumption species for the 1987-1994 period; ii) a forward MSVPA run using these revised data and with a 50% decrease in the industrial fisheries (i.e., Norway pout, herring and sandeel) and based on the revised 1991 stomach survey data; and iii) another run with 50% decrease in the industrial fisheries, using 1977-1986 ecosystem structure as a starting condition.

However, given the large uncertainty in estimating the potential reaction of fleets to any large changes in regulations, forward projections of the impacts of reducing the industrial fishery were not undertaken.

Results show that the introduction of the 30% over quota landings and discards into the models have a far larger and more widespread effect on the transfer efficiencies between trophic levels compared to a reduction in the industrial fisheries under reported 1994 conditions. The removal of 50% of the industrial fishery has the effect of reducing the transfer efficiency of the trophic level where these species exist, suggesting that many of the gains would be dissipated rather than passed on up through the ecosystem. However, any changes that have been observed in the ecosystem over the past two decades appear to be attributed not only to shifts in predator-prey interactions but also to the higher levels of extraction of top predators.

The overall conclusion of this analysis is that the high levels of over quota landings and discards from the roundfish fishery will mitigate against the full effect of any reduction in the reported industrial fishery being observed. Unless there is a full accounting of the losses within both the industrial and human consumption fisheries, it will be extremely difficult to predict the long-term functioning of the ecosystem. Using increased estimates of unreported landings and catches, it is clear that some improvement would occur with a reduction of the industrial fisheries in the North Sea. This would ensure that the human consumption species would encounter levels of food similar to those historically consumed when the cod, haddock, whiting and other stocks were at higher levels. The overall spatial and temporal coherence of the North Sea predator-prey interactions should improve as a result.

It is recommended that i) improved surveys of the North Sea sandeel stock be undertaken immediately and that the surveys cover the four quarters of the year, ii) a detailed analysis of the ecological changes in the North Sea be undertaken with full participation from the industry and scientific community, so that consensus on the state of the system can be obtained, iii) new analysis on the impact of over quota landings on the economic and ecological aspects of sustainable management and the adoption of a precautionary approach should be undertaken with full EU Member State support, iv) interim technical measures should be examined including the use of an east-west division of the North Sea along the 4 degree east longitude line, westwards of which reductions in the sandeel extraction should be allowed particularly in sensitive areas supporting large populations of birds, mammals and fish dependent on sandeel, v) the United Nations FAO Code of Conduct for Responsible Fishing be adopted for the North Sea, vi) an integrated GIS system for the North Sea fish resources be established to hold all relevant data and contain analysis programmes for the calculation of future resource strength.

CHAPTER 24

Digital Elevation Models by Laserscanning

U. Lohr

ABSTRACT: The TopoSys laserscanner system is designed to produce digital elevation models (DEMs) at a maximum accuracy of 0.5m in x and y and 0.1m in z. The regular scan pattern and the measurement frequency of 80,000 measurements per second (on average 4 measurements per m^2) allow to generate high quality DEMs. The onboard sensor package is completed by a SVHS video camera which records the survey area during the flight. The mainly automated data processing allows to generate DEMs of large areas at a short production time. The produced DEMs come into common use as basic data for different applications, some of which are water resources management, shoreline control, planning of utility lines and urban planning (simulation of noise and pollution distributions). The performance of the system is illustrated with help of DEM sections produced with the TopoSys system.

Introduction

Lasers are well known for their ability to measure distances with high accuracy. Laserscanners in aircraft can be used to scan the Earth's surface in order to produce digital elevation models. Due to their measurement principle, laserscanners support an entirely digital processing. Therefore, laserscanning is the appropriate choice for a largely automatic DEM generation: Even without post-processing the calculated raster DEMs can form a basic data set for a GIS used to perform monitoring and / or simulation tasks.

Description of the System

The TopoSys system can be divided into a flight and ground segment. The main elements on-board the aircraft are the subsystems: navigation & positioning, laserscanner, video camera and data handling and recording (see also Fig. 1).

The system design of the navigation and positioning system is based on the integration of dGPS with an Laser Inertial Navigation System (LINS) for position and attitude determination. GPS, LINS and laserscanner data are stored digitally on hard discs, while the video images are stored with help of a standard SVHS recorder. A time synchronisation facility ensures the precise synchronisation of these different data streams during the post-processing.

GPS data are registered with help of Novatel L1/L2 receivers at a data rate of 1 Hz, while the high precision LINS collects position and attitude data at a data rate of 64 Hz. So, after flight path restitution and merging of dGPS and LINS measurements, position and attitude data are available at 64 Hz. This means that (at a cruising speed of about 70 m/sec) the systems orientation is known at each meter.

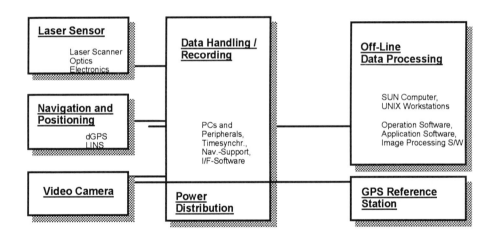

Fig. 1. TopoSys System Components

The laserscanner is the key element of the flight segment. The eye-safe sensor operates in the near infrared region at 1.54 µm at a pulse rate of about 80 kHz. This type of fibre-optical linescanner has been developed for more than 10 years and proved its quality and reliability in various non-topographic applications. For DEM generation the basic scanner has been slightly modified in order to meet the special requirement. So, for example, the scan angle was fixed to +/- 7° in order to minimise shading effects

Digital Elevation Models by Laserscanning 351

at the borders of the scan. The small scan angle in combination with the high measurement rate forms the basis for reliable distance measurements and the high quality DEMs.

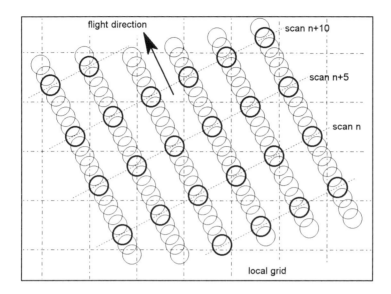

Fig. 2. Scan pattern on the ground

The off-line processing system includes several UNIX workstations. The key software packages are the software package for flight path restitution and the TopoSys package for DEM generation and elimination of vegetation cover. Further tools allow the fusion of digital data from various sensors and an image processing software package supports DEM visualisation and quality control.

The primary output of the DEM production process is a 16 bit raster DEM, typically at a raster width of 1 m and a height accuracy of about 0.1 m. The distribution of measurements points on ground is shown in Fig. 2. As the distance between neighbouring scan is less than 0.15 m, the TopoSys scanner provides on average four measurements per m^2. In other words: about four elevation measurements are available to perform statistical and plausibility analyses when calculating a single DEM value for 1 m grid. Depending on the customer, either this basic product or value-added products are delivered - in various co-ordinate systems and data output formats.

As the final accuracy of the DEMs is also determined by the precision of the position and attitude measurements during the survey flights, these components have been selected carefully. The TopoSys LINS allows a pointing accuracy of 0.2m on ground (from a survey altitude of 1000m), while dGPS provides a maximum accuracy of 0.1m for the position measurement.

A summary of performance parameters is given in Table 1.

TABLE 1.
Performance Parameters

sensor type	pulse modulated laser radar
Range	☐ 1,000 m
scanning principle	fibre optic line scanner
Transmitter	solid state at 1.5 μm
measurement principle	run-time measurement
laser pulse rate	80,000 Hz
scan frequency	600 Hz
field of view	+/- 7°
number of pixels per scan	127
swath width (at 1'000 m flight height)	250 m
accuracy of a single distance measurement	< 0.3 m
resolution of a distance measurement	< 0.06 m
position accuracy of TopoSys DEMs	< 1.0 m
height accuracy of TopoSys DEMs	< 0.15 m
laserscanner classification	class 1 by EN 60825 (eye-safe)

DEMs for Various Applications

TopoSys DEMs, geocoded and delivered at a raster width of 1m to 3 m, are intended to be used as one layer in a GIS. Presently, these kinds of high resolution DEMs are used as basic data set for:

- **water resources management and monitoring of coastal areas**
- **3-D city models**
- **obstacle detection (e.g. in approach tracks of airports)**
- **DEM generation in forestry areas**

Further applications are developed in close contact with customers.

DEM Example of the Coastal Zone

Fig. 3 shows a section of the DEM of the northern part of the island Langeoog, North Sea. The image is a relief presentation of the raster 1m DEM.

As the survey flight was performed during low tide, the DEM includes not only parts of the village Langeoog, the dune area and beach, but also the mud bottom. Due to the different texture, even the land / water boundary can be identified in the digital data (the upper left corner in Fig. 4).

Digital Elevation Models by Laserscanning

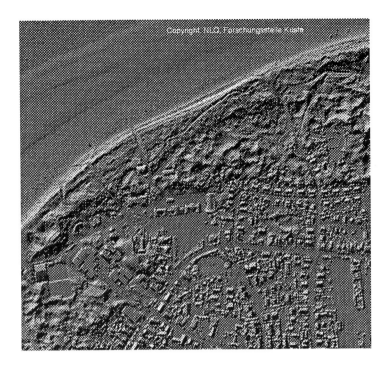

Fig. 3. DEM of The Northern Part of The Island Langeoog bottom area and the beach and a distance of 1 m in the dunes.

On the basis of the DEM, contour lines had to be generated for the dune area, the beach and the mud bottom. In order to allow the calculation of contour lines, vegetation and buildings were eliminated from the DEM. Fig. 4 shows the superposition of the postprocessed DEM with contour lines which have a distance of 0.25m in the mud The processing of the Langeoog data was finished in January 1997. The quality control proofed that the requested accuracy of $z \leq 0.1$ m was met. Up to now DEMs of that quality have been delivered for coastal areas which total up to 400 km^2.; a further 1000 km^2 of DEM for a land application will be delivered until end of summer 1997.

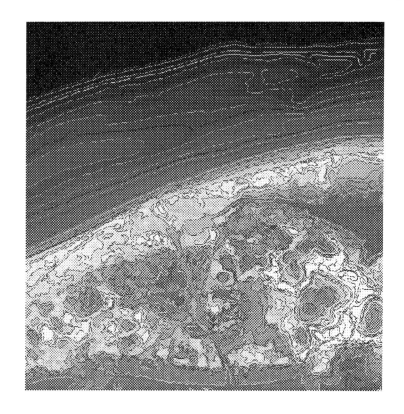

Fig. 4. Postprocessed DEM Overlaid with Contour Lines
(Copyright: NLÖ, Forschungsstelle Küste)

CHAPTER 25

Error Modeling and Management for Data in Geospatial Information Systems

M.A. Chapman, A. Alesheikh, and H. Karimi

ABSTRACT: This chapter critically examines some current methods for modeling spatial data errors with particular emphasis on line and areal objects. An error management strategy for handling errors in Geospatial Information Systems (GISs) is presented. This strategy serves to identify all the interrelated components of error modeling and uncertainty reduction. In applying such a management strategy, it is important to consider, not only the causes and effects of uncertainty in spatial data, but also the procedures used to measure and model these errors. The ability to successfully manage and reduce a particular form of error in GIS output is dependent on the nature of the error and the availability of a proper error indicator.

Introduction

Geospatial Information Systems (GIS) technology is becoming of vital importance in all disciplines that employ spatial data. Information derived from GIS is frequently used in decision-making processes and as such can have far-reaching effects on the land, its people, and the environment. Uncertainty in GIS creates a complex problem that is directly concerned with the quality and suitability of this information. Therefore, the future utility of GIS as a decision-making tool depends on the development of error models for GIS data and its related operations.

This chapter proposes a strategy for error management in GIS. Furthermore, it is argued that while error modeling and communication are very important for the sustainability of the GIS technology (Veregin, 1989), they constitute only a part of the strategy for error handling. The chapter is then critically reviews some existing error models for geographical data (particularly, lines, and polygons). The use of such models in solving point-in-polygon problems is subsequently addressed.

A Decision Support Strategy

Similar to all problem solution approaches, the proposed strategy is comprised of the key elements of problem identification, evaluation of alternative methods, and implementation of the adopted solution to simplify a decision making process. It is clear that the way in which a GIS is used varies between users with different skills and responsibilities, similar to the effect of different types of decisions and application areas. However, it is argued that to reach subsequent steps of the strategy, the lower steps should be addressed first, and if these needs are not met, the attainment of the higher steps may not be achievable. The proposed decision support strategy is presented in Fig. 1.

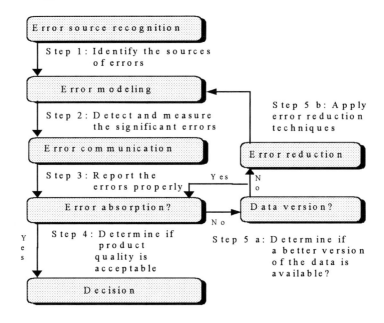

Fig. 1. The Proposed Decision Support Strategy

Error Source Recognition

The first level of the proposed strategy is concerned with the basic problem of identifying the causes and effects of errors with respect to spatial data. In this regard, the data, system, model and user are important causes of errors in GIS products. Depending on the applications at hand, one or a combinations of the error sources may be assumed to be the prominent source of errors. Sources of errors may manifest themselves in a variety of error forms (e.g., geometric, attribute and so on) which has been extensively studied by many researchers (Hunter and Beard, 1992; Goodchild et al., 1992; Chrisman, 1991). One of the most basic forms of error classification distinguishes between "geometric" error, or error in positional features such as points and lines, and "attribute" error, or errors in the values of thematic attributes. Chrisman (1989) refers to these two forms of errors as "positional" and "attribute" errors respectively while Bedard (1987) calls them "locational" and "descriptive" errors. However, it should be remembered that spatial data (such as coordinates of points) and aspatial data (such as landuse cover) are both attributes of a spatial entity.

ERROR MODELING

The second stage of the strategy is concerned with methods of assessing accuracy levels and error modeling. For aspatial error, the measurement of error is frequently done by building confusion matrices. This method is common when evaluating the performance of a classification algorithm for digital images. While confusion matrices are appropriate for categorical data, a variety of methods have been proposed for assessing spatial error (Alesheikh, 1997). In addition, different applications will place different priorities on the various uncertainty forms. While positional errors are the focus of this chapter and more generally the focus of spatial modelers, attribute errors are a major concern in other applications and probably the combination of the two may be significant. In addition, other forms of uncertainty such as logical consistency may need more attention in applications where topological information is needed.

This step is also concerned with evaluating and modeling the propagation of error as a consequence of applying GIS operations that involve spatial data. The magnitude of errors may be drastically changed due to different operations that the data are subjected to. Examples of error propagation may be found when determining the uncertainty of a straight line object in a GIS and solving the point-in-polygon problem. These issues will be elaborated upon in the following sections.

ERROR COMMUNICATIONS

The concept of reporting the quality of a product is not new, with some of the earliest forms being positional accuracy statements as found on hardcopy topographic maps. To realize the fitness of data for a specific application the accuracy of the data (i.e. metadata) should be provided. It should be the producer's responsibility to provide the necessary information to allow users to determine whether the product is suitable for their use in the first place - which brings up the notion of "truth in labeling".

Though many data transfer standards have been in place for reporting errors, the US Spatial Data Transfer Standard appears to be the most complete (Fegeas et al., 1992). Regarding the data accuracy and quality, the standard has a specific provision for information about the following components: data lineage, positional accuracy, attribute / temporal accuracy, logical consistency; and completeness. The rationale is that this detail will constitute a complete data quality report that gives sufficient information to make users able to assess the data's fitness for use.

Different aspects of the source data and their respective applications are important for communicating error information. The data lineage describes source and update material (with dates), methods of derivation, transformations, and other elements relating to the processing history. Positional accuracy deals with how closely the locational data (encoded coordinate values) represent true locations. Attribute accuracy is similarly concerned with non-locational descriptive data. Logical consistency refers to the fidelity of encoded relationships in the structure of spatial data (e.g. where appropriate, the degree to which topological relationships have been verified should be reported). The completeness part of the quality report includes information about the geographic area and subject matter coverage (e.g., selection criteria and other mapping rules that are relevant completeness indicators).

Indeed, uncertainty can be conveyed in different formats such as statistical indicators and visual aids. Some useful mechanisms for displaying this information are given in subsequent sections.

ERROR ABSORPTION

The next stage of the decision strategy deals with deciding whether product quality satisfies the requirements of the current task and recognizing what can and cannot be modified. Indeed, if the quality is found to be acceptable, then the issue of uncertainty has effectively been dealt with. However if the quality is not acceptable, then further action is required. Regardless of the efforts put on the error reduction techniques, it should be realized that there will always remain some residual uncertainty which users must decide to either absorb (i.e. accept) if they wish to use the data, or else reject it. This stems from the fact that all measurements are of limited accuracy and no model will ever perfectly reflect the real world. The amount of uncertainty absorbed can be considered to be the risk associated with using the data or product.

DATA VERSIONS AND ERROR REDUCTION

The purpose of this stage is to decrease the magnitude of uncertainty to an acceptable level. If the uncertainty of the product does not meet the application requirements, versions of data that may satisfy the objectives should be tested. Different versions of data may already be created using techniques such as image pyramids or multi-resolution methods, and they are stored in the database. Since a top-to-bottom approach is the usual theme in GIS - users look at the global picture first then zoom to view finer detail - to find the specific data needed, accurate data can be found in the lower level of the pyramid. However if the object shows fuzzy characteristics, upper level data is needed.

If no version of data meets the users need (i.e. the user reaches the lowest level of the pyramid) then error reduction techniques should be applied. For example, in the case of parcel-based systems, Bedard (1987, p. 181) recognized that actions such as field checking of observations, strengthening geodetic control networks, defining and standardizing technical procedures, mandatory registration of all rights in land, and improved professional training, all contribute to the confirmation of "... the precision and crispness in the description and location in space and time of a spatial entity". Other provisions such as the use of better data processing methods, collecting additional data, sampling at a higher frequency, improving the spatial/temporal resolution, using better models, and improving procedures for model calibration may be required to improve the accuracy of the data/product. However, it should be pointed out that more accurate data needs more effort, time and expense.

Uncertainty Modeling of Geospatial Objects

The implications of the proposed strategy may be illustrated with reference to the modeling of geospatial objects (particularly, lines and polygons). Point uncertainty modeling has thoroughly examined in disciplines such as photogrammetry, surveying and geodesy (Mikhail and Gracie, 1981) and, as such, this chapter will focus on modeling the uncertainty of line and polygon objects in a GIS database. The following is a brief evaluation of existing line uncertainty models namely the epsilon band and error band models.

THE EPSILON BAND MODEL

The fundamental concept of the epsilon band model is based upon the principle that a cartographic line is surrounded on each side by an area of uncertainty of constant width, epsilon (ε), similar in appearance to a buffer zone. The concept may be visualized as the orthogonal projection of a ball with a radius ε that is rolling along the line as shown in Fig. 2. The model was designed to provide users with a measure of the error associated with digitizing cartographic lines. This concept has been developed and enhanced by researchers such as Perkal (1966), Blakemore (1984), and Chrisman (1989) and many other authors have since applied the idea in a variety of ways.

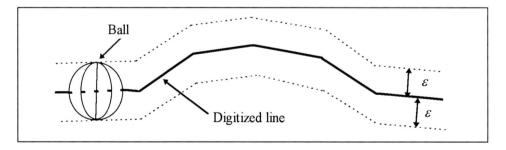

Fig. 2. The Concept of the Epsilon Band

Though many interpretations of the epsilon band exist, they can be categorized in two groups: deterministic and probabilistic. In the deterministic case, the true position of the line is considered to lie somewhere in the buffer zone. Deterministic interpretation of the epsilon band sounds counter intuitive, because:

1) **it provides no model of error distribution inside the band**

and

2) **it proposes that the true line is definitely located within this region**

In the probabilistic interpretation of the epsilon band, the width of the zone is assumed to be a function of different variables, and that their uncertainties accumulate into the final stage. For instance Alai (1993) assumed scale, digitization, slope and attribute of the polygons adjacent to the lines are the related variables while Blakemore (1984) simply related the band width to the digitizing error, round off error and generalization error. The probabilistic interpretation of the epsilon band has been inconsistent with what analytical procedures suggested (Shi, 1994).

However, in spite of its weaknesses, the epsilon band model has the following advantages (Carver, 1991):

- **it involves minimal extra processing time**
- **it uses existing spatial operations for implementation (e.g. buffer zone operation)**
- **different feature categories can have different epsilon values assigned to them as attributes**

and

- **the concept is easily understood and can be applied during the execution of many spatial operations**

THE ERROR BAND MODEL

In an effort to further develop the epsilon band model Dutton (1992) proposed the error band model. In the error band model, the digitized end points of a straight line are drawn from a random sample of possible positions, having a circular normal distribution that forms a population of connected line segments. Unlike the epsilon band that assumes constant width along a line segment, Dutton's model (Fig. 3) suggests a narrower band in the middle of the line segment. Shi (1994) computed the error distribution along the line segments and through analytical derivations demonstrated the shape of Dutton's model. Dutton's simulated model and Shi's analytically-derived model are based on the assumption of circular errors at the endpoints of a line segment. A circular error assumption neglects the correlation

Error Modeling and Management for Data

between the end points and assumes the magnitudes of uncertainty in perpendicular directions (e.g., X, Y) are similar (Alesheikh, 1997). However, statistical dependence does not apply if the endpoints are determined using a similar method or instrument. Another drawback of the error band model is that it does not accommodate any modeling errors. In another words, assuming a straight line connecting end points may not be justifiable.

THE RIGOROUS UNCERTAINTY MODEL OF A LINE SEGMENT

A straight line segment may be defined as a combination of points conforming to a linear function. Hence, one may determine the uncertainty of a line segment as the aggregate of the uncertainty of points constructing the line. Point uncertainties are usually represented by error ellipses that are computed from the variance-covariance matrix of the point coordinates. So, the problem of line uncertainty modeling turns out to be the determination of the uncertainty of any arbitrary point, Σ_u, located along the assumed straight line segment, based on the statistical information Σ_{AB} of the endpoints coordinates. However,

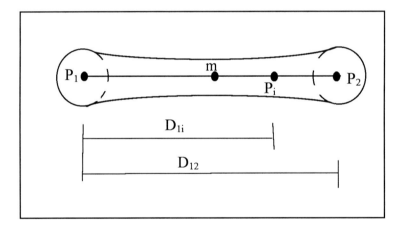

Fig. 3. The Concept of Error Band

modeling error is another aspect that should be considered in the determination of the line uncertainty model. If U is an arbitrary point located along a line segment AB (Fig. 4), its coordinates may be determined by :

$$\begin{bmatrix} X_U \\ Y_U \end{bmatrix} = (1-r) \begin{bmatrix} X_A \\ Y_A \end{bmatrix} + r \begin{bmatrix} X_B \\ Y_B \end{bmatrix}$$

(1)

where, $r = \dfrac{D_{AU}}{D_{AB}} = \dfrac{\sqrt{(X_U - X_A)^2 + (Y_U - Y_A)^2}}{\sqrt{(X_B - X_A)^2 + (Y_B - Y_A)^2}}$ and

$0 \le r \le 1$.
(2)

The covariance matrix of U may be derived by applying the law of error propagation as follows;

$\Sigma_U = J * \Sigma_{AB} * J^T$
(3)

where J is Jacobian matrix and Σ_{AB} is the variance-covariance matrix of the coordinates of endpoints.

Once the error matrix is defined, the error ellipse of the arbitrary points may be represented by:

$\left(\dfrac{X - X_U}{\sigma_{X_U}}\right)^2 - 2*\rho*\left(\dfrac{X - X_U}{\sigma_{X_U}}\right)*\left(\dfrac{Y - Y_U}{\sigma_{Y_U}}\right) + \left(\dfrac{Y - Y_U}{\sigma_{Y_U}}\right)^2 = (1 - \rho^2) * C^2$
(4)

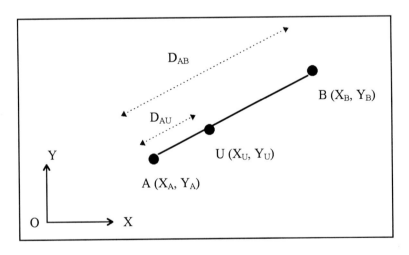

Fig. 4. A Line Segment

where:

σ_{X_U}, and σ_{Y_U} are the diagonal elements of the covariance matrix of U,

$0 \leq \rho = \dfrac{\sigma_{X_U} * \sigma_{Y_U}}{\sigma_{x_U y_U}} \leq 1$ is the correlation factor,

and

C is a constant that determines the probability levels of the error ellipses

Once the error ellipses for arbitrary points along the straight line segment are defined, the area that these ellipses encompass creates the confidence region around the line. This region is used to represent the uncertainty of a line segment. Fig. 5 illustrates this concept for an arbitrary line segment.

A very interesting property of this modeling procedure is that it can be easily extended to determine the uncertainty of a curve. If the curve function, y = f(x), is known, for example in the case of a road design, the only modification that occurs in the method is that the distance from two points on the curve should be computed by the following equation:

$$D_{AU} = \int_{X_A}^{X_U} \sqrt{1 + (\dfrac{dy}{dx})^2} dx$$

(5)

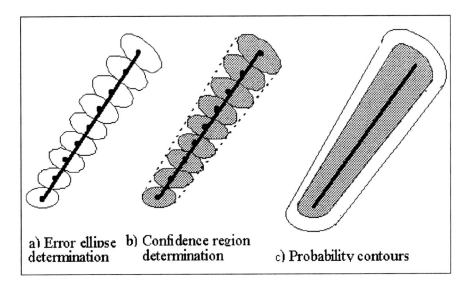

a) Error ellipse determination   b) Confidence region determination   c) Probability contours

Fig. 5. Line Uncertainty Model Generation

EVALUATION OF THE PROPOSED UNCERTAINTY MODEL

Under the assumption that the uncertainty of a line segment is affected only by the errors of its two endpoints, the above uncertainty model has been derived. The proposed model contributes to the improvement of existing boundary uncertainty models in the following ways:

1) **it provides the error distribution of the uncertainty (unlike the deterministic interpretation of the epsilon band model)**
2) **the model is analytically derived (contrary to Dutton's simulation-based model)**
3) **the model respects the correlation among the contributing parameters, and finally**
4) **the model is a superset of the currently used ones. All other models can be assumed as special cases of this proposed model**

Moreover the model has undergone several tests and proved its superiority. In practice, the initial assumption of the model can be realized when the function (linear in this case) is assumed perfect or the line segment is long, such as the railroad or township street, where modeling error is minimized. Because of this assumption, the uncertainty of the points located along the line should not exceed the uncertainty of the endpoints.

However, if the modeling error is under question or the modeling error is of a considerable magnitude, then the line uncertainty model should be improved. To accompany modeling error, Equation 2 is generalized to;

$$U = \alpha P_A + \beta P_B$$
(6)

and the covariance matrix of point U will be:

$$C_U = \{\alpha^2 C_{P_A} + B^2 C_{P_B} + 2\alpha\beta C_{P_A P_B}\} + \{P_A^2 C_\alpha + P_B^2 C_\beta + 2P_A P_B C_{\alpha\beta}\}$$
(7)

where Equation 6 is what we considered in the above uncertainty model and, Equation 7 is the contribution of the modeling error. Indeed, if there is a correlation between modeling error and endpoints error, the summation will add a new term indicating the correlation. The modeling error in this study is nothing but the error attributed to the slope and intercept of line segments. It can easily be found that based on the magnitude of the modeling error the shape of the line uncertainty model may well be represented by an ellipse rather than a hyperbola. Fig. 6 demonstrate the differences between line uncertainty models. The modeling error in this case is estimated by a Gauss-Markov process.

Error Modeling and Management for Data

Polygon Uncertainty Model

Uncertainty of the boundary of a polygon object in GIS may be defined as the combination of the uncertainties of the line segments circumscribing the polygon. In other words, the rigorous boundary uncertainty model of a polygon object is simply the combination of all the line uncertainty indicators enclosing the polygon. It is important to notice that the uncertainty of a polygon is different than the uncertainty of the boundary of a polygon, since the interior of polygons can also have uncertainty. However, the interiors of polygons are more homogeneous than the boundary of polygons, causing the magnitude of uncertainty to be greater on the boundary than the interior. Consequently, emphasis will be put on modeling the uncertainty of the boundary of the polygon. Fig. 7 shows one realization of the polygon uncertainty model.

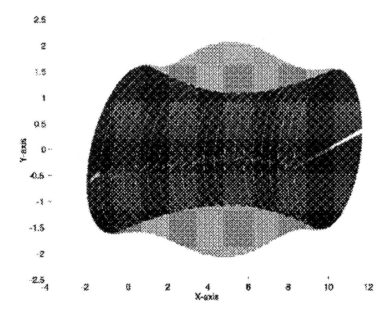

Fig. 6. Line Uncertainty Models (C = 1).

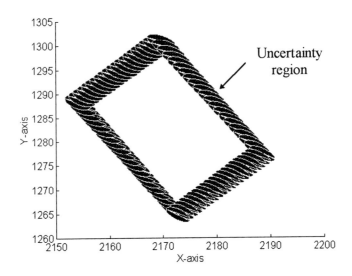

Fig. 7. Uncertainty Visualization of a Polygon.

The Point-in-Polygon Problem

In order to determine the position of a point with respect to a polygon, the polygon may first be decomposed into several triangles (2-D primitive objects), and then, the position of the point is examined with respect to the triangles. A triangle can be represented in a normalized homogeneous coordinate system by its vertex coordinates: (0, 0, 1), (0, 1, 0), and (1, 0, 0), as shown in Fig. 8. Assuming the coordinates of the points to be examined are given as $p(\alpha,\beta,\gamma)$, then if any coordinate of the point is equal to 1, the point is located on the boundary of the triangle. If any coordinate of the test point is greater than 1 or less than zero, it indicates that the point is located outside the triangle. Finally, if all of the coordinates of the desired point in the homogeneous coordinate system are less than one, and greater than zero the point is located inside the polygon. The above procedure has been used to address a point-in-polygon problem presented in Table 1.

Error Modeling and Management for Data

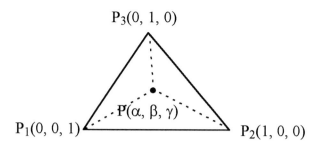

Fig. 8. Representation of a Triangle in a Homogeneous Coordinate System

However, the quality of the solution to the point-in-polygon problem has yet to be addressed properly. A simple computation of the quality of the point-in-polygon solution could avoid placing buoys on dry land or rivers outside their floodplain (Chrisman, 1989).

The position of a point inside, on or outside the triangle $P_1 P_2 P_3$ can be determined by:

$$P = P_1 + \lambda(P_2 - P_1) + \mu(P_3 - P_1) = \alpha P_1 + \beta P_2 + \gamma P_3,$$
(8)

with $\alpha + \beta + \gamma = 1$, $P_1 = \begin{bmatrix} x_1 \\ y_1 \end{bmatrix}$, $P_2 = \begin{bmatrix} x_2 \\ y_2 \end{bmatrix}$, $P_3 = \begin{bmatrix} x_3 \\ y_3 \end{bmatrix}$, where

$\lambda = \beta = det\,(P - P_1 \;\; P_3 - P_1) \,/\, det\,(P_2 - P_1 \;\; P_3 - P_1),$
$\mu = \gamma = det\,(P_2 - P_1 \;\; P - P_1) \,/\, det\,(P_2 - P_1 \;\; P_3 - P_1),$
$\alpha = 1 - \beta - \gamma.$

Given $P_1, P_2,$ and P_3 are non-collinear, and *det* stands for determinant (Blais, 1996).

Equation 8 represents the position of point P as a linear function of triangle points. By applying the error propagation law, the uncertainty of point P with regard to other points can now be readily computed, hence the solution to the point-in-polygon problem can been improved by providing accuracy information.

To decrease the computation time and apply the polygon uncertainty model, instead of computing the covariance matrix of each test point, it is proposed to determine whether the point falls inside the specific probability region of line segments. That is the minimum distance e between the test point and the line segments is computed to see if it falls inside the probability region. The test involves the evaluation of each of the coordinates of the point in question to see if any of the coordinates are less than or equal to |e|. Therefore, the proposed solution has two steps:

1) determine whether a point is inside, outside, or on the boundary of a polygon, and
2) determine in which probability region the point falls.

Table 1 presents the results of a few tests performed on the polygon.
Indeed, if the results do not fit for use, more accurate data should be employed. If accurate data are available in the database they will be used, otherwise error reduction procedures should be followed.

TABLE 1
Results of Point-in-Polygon Tests

Pt	X Coordinate	Y Coordinate	In *	Out *	Bound. *	C = 1	C = 2	C = 3
1	2172.76	1264.86			*	*	*	*
2	2172.70	1264.9	*					*
3	2172.74	1264.89		*			*	*
4	2160	1280.926		*		*	*	*
5	2160.5	1281	*				*	*
6	2178	1290	*			*	*	*
7	2177.5	1290	*					
8	2179.095	1289.045			*	*	*	*

Conclusions

This chapter has attempted to clarify, within the proposed error management strategy, how to model and manage the uncertainty of linear objects in GIS databases. The proposed error models are rigorous and determined by applying the laws of error propagation. The chapter has elucidated that the strategy for error management cannot be separated from the methods employed for modeling the uncertainty of spatial objects, which in turn depends on the detecting of the significant forms of error deemed in the objects.

References

Alai, J. (1993). Spatial Uncertainty in a GIS. *M.Sc.E. Thesis*, Department of Geomatics Engineering, The University of Calgary, Calgary, Alberta, Canada.
Alesheikh, A.A. (1997). "Uncertainty modeling of line and polygon objects in GIS." *Fourth International Conference* on Civil Engineering, Tehran, Iran.
Bedard, Y. (1987). "Uncertainties in land information systems databases." *Proceedings of the ACSM-ASPRS Auto-Carto 8 Conference*, Baltimore, Maryland, pp. 175-184.

Blakemore, M. (1984). "Generalization and Error in Spatial Databases." *Cartographica*, Vol. 21, No. 2, pp. 131-139.

Blais, J.A.R. (1996). "Design and implementation of land information systems." *Lecture* materials, Department of Geomatics Engineering, The University of Calgary, Calgary, Alberta, Canada.

Carver, S. (1991). "Adding error handling functionality to the GIS toolkit." *Proceedings of the Second European Conference on GIS (EGIS '91)*, Brussels, Belgium, pp. 187-196.

Chrisman, N. (1991). "The error component in spatial data." Eds. D.J. Maguire, M.F. Goodchild and D.W. Rhind, *Geographic Information Systems: Principles and Applications* (Longman: London) Vol. 1, pp. 165-174.

Chrisman, N.R. (1989). "Error in categorical maps: Testing versus simulation." *Proceedings of the Ninth International Symposium on Computer-Assisted Cartography (Auto-Carto 9)*, Baltimore, Maryland, pp. 521-529.

Dutton, G. (1992). "Handling positional uncertainty in spatial databases." *Proceedings of the 5th International Symposium on Spatial Data Handling*, Charleston, South Carolina, Vol. 2 pp. 460-469.

Fegeas, R.G., Cascio, J.L., and Lazar, R.A. (1992). "An Overview of FIPS 173, the spatial data transfer standard." *Cartography and Geographic Information Systems*, Vol. 19, No. 5. pp. 278-293.

Goodchild, M.F., Guoqing, S., and Shiren, Y. (1992). "Development and test of an error model for categorical data." *International Journal of Geographical Information Systems*. Vol. 6, No. 2, pp. 87-104.

Hunter, G.J., and Beard, K. (1992). "Understanding error in spatial databases." *The Australian Surveyor*, Vol. 37, No. 2, pp. 108 - 119.

Mikhail, E.M., and Gracie, G. (1981). *Analysis and Adjustment of Survey Measurements.* Van Nostrand Reinhold Company, New York.

Perkal, J. (1966). On the Length of Empirical Curves. *Discussion Paper* Number 10. Michigan Inter-University community of Mathematical Geography.

Shi, W. (1994). Modeling Positional and Thematic Uncertainties in Integration of Remote Sensing and Geographic Information Systems. *Ph.D. Thesis*, ITC, The Netherlands.

Veregin, H. (1989). "Error modeling for the map overlay operation." Eds. M. Goodchild and S. Gopal, *Accuracy of Spatial Databases*, Taylor and Francis, London, pp. 3-18.

CHAPTER 26

The Use of Dynamic Segmentation in the Coastal Information System: Adjacency Relationships from Southeastern Newfoundland, Canada

K. A. Jenner, A. G. Sherin, and T. Horsman

ABSTRACT: The Coastal Information System (CIS) has been developed at the Geological Survey of Canada (Atlantic) (GSC Atlantic) to store data on shore-zone geomorphologic form and material from the Atlantic Provinces of Canada. Data are interpreted from coastal aerial video imagery, stored as lines and points and spatially referenced using the dynamic segmentation feature of the geographic information system (GIS) ArcInfo. An application of dynamic segmentation is explored in the along-shore and across-shore identification of coastal form relationships. Examples of binary and tertiary adjacency relationships are presented from the southeastern coast of Newfoundland, Canada. These data quantify predominant form relationships, and when tabulated independently for specific sections of coast, document similarities and differences in coastal form sequences.

Introduction

The application of geographic information systems (GIS) to coastal zone research and coastal zone management is well established. Rickman and Miller (1995) published a comprehensive bibliography demonstrating the use and impact of GIS to the understanding and management of the coastal zone. Jones (1995) concludes that GIS is being utilized or developed specifically for a variety of roles in coastal management. The advantages of dynamic segmentation for managing linear spatial objects and associated attributes are presented by Dueker et al. (1992) with specific application to transportation systems. The use of dynamic segmentation for representing the coastal zone as a linear spatial object was suggested by Bartlett (1993). McCall (1995) discussed the application of dynamic segmentation to the coastal zone and McCall (1995) and Sherin and Edwardson (1997) demonstrated the application of dynamic segmentation for County Cork, Eire and the Atlantic Provinces of Canada respectively.

Dynamic segmentation was implemented for coastal zone data storage at the GSC Atlantic, in the summer of 1994, with the development of the ArcInfo -based CIS (Sherin and Edwardson, 1994; Sherin and Edwardson, 1997). Using dynamic segmentation, the coast is modelled as a single, linear spatial object with a distance referencing system. The purpose of this chapter is to further explore the use of dynamic segmentation in the query of binary and tertiary adjacency relationships between coastal geomorphologic features from the southeastern coast of Newfoundland, Canada. Three queries are presented. The first explores binary and tertiary sequencing between geomorphologically dissimilar sections of coast. The second explores binary relationships within three bays with a similar coastal geomorphology including coarse clastic bayhead barrier beaches. In addition, the OVERLAYEVENTS command is used to search for coastal form and material within two specified linear extents from the coastal form 'barrier.'

Mapping

The coastal classification developed at the GSC Atlantic builds largely upon the existing classifications of Howes et al. (1994) and Owens (1994). Coastal units are categorized by 'form' and 'material' and are subdivided into 'solid' and 'unconsolidated' supertypes. These supertypes are further subdivided into types and each type may contain one or more subtypes and one or more features, in addition to a height class designation.

Line and point data for the nearshore, foreshore and backshore are interpreted from oblique aerial video surveys of the coastline. Where available, vertical aerial photographs, 35 mm photographs and groundtruth data are used to supplement video survey interpretation. These data are then entered into the database using a 'heads-up' digitizing method which was developed at the GSC Atlantic (Sherin and Edwardson, 1994).

Study Area

The study area is located along the southeast corner of the Avalon Peninsula, Newfoundland, Canada (Fig. 1). It encompasses the coastline of nine, 1:50 000 National Topographic System (NTS) map sheets and spans 913 km along the mainland. The study area is characterized by two geomorphologically distinct regions. The east coast is dominated by steep, high (> 10 m) bedrock cliffs and irregular bedrock outcrops with

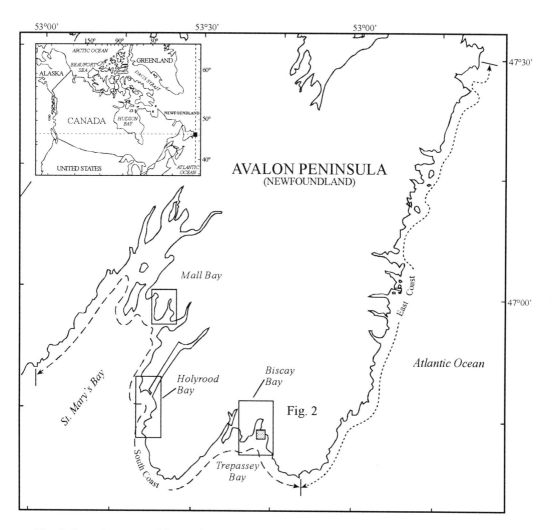

Fig. 1. Location map of the study area showing the boundary between the east coast and south coast data sets, the areal extent represented in Fig. 2 and the location of the index maps presented in Fig. 5

elevations between 5 and 10 m. The bedrock defines a relatively straight section of coastline interrupted by several narrow bays. Unconsolidated high cliffs, beaches and bedrock platforms are less extensive. Southward, the coastal geomorphology changes. A distinctive feature of this section of coastline is a series of northeast-southwest trending

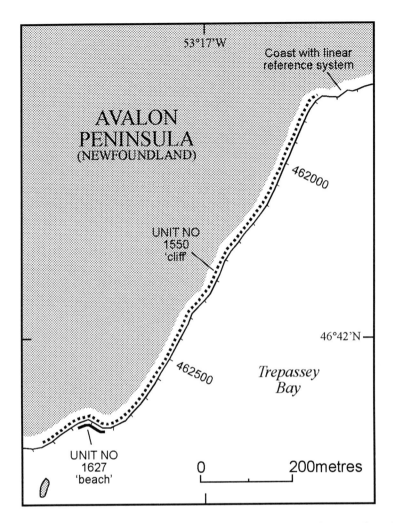

Fig. 2. The linear reference system, used in dynamic segmentation, for a section of coast along Trepassey Bay, Newfoundland (refer to Figure 1 for location). Tick marks are placed every 50m and the distance along-shore is indicated every 500m. The along-shore extents of the two coastal units described in Tables 1, 2, and 3 are also shown

fjords and bays, particularly along the perimeter of the larger Trepassey and St. Mary's Bays, many of which are closed by coarse clastic bayhead barrier beaches. Headlands comprise steep to vertical, high bedrock cliffs and steep, high, eroding unconsolidated cliffs; both often fronted by irregular bedrock outcrops up to 10 m in height. Between the headland and the barrier, steep high (> 10 m), eroding or partially stabilized unconsolidated cliffs, often fronted by low (< 5 m), irregular bedrock outcrops, predominate.

Dynamic Segmentation

Dynamic segmentation links attributes, such as geomorphologic form and material, to a linear spatial object, such as the coast, by use of a distance referencing system. In the example shown in Fig. 2, the coastal form type 'cliff,' represented on the landward side of the coast, extends from a distance of 461839.406 m to 462877.188 m. Similarly, the coastal form type 'beach,' on the seaward side of the coast, extends from a distance of 462719.000 m to 462789.000 m. These distances are stored in a database table as 'from' and 'to' items. Tables 1, 2 and 3 show the definition of items in the database and attribute data associated with the two linear coastal form types presented in Fig. 2. Coastal forms represented as points can also be managed this way by storing a single value for the distance along the coast in the database table. The linear extent or position of these forms can be dynamically created by the CIS as distinct linear or point objects. Accordingly, adjacency relationships between coastal geomorphologic forms and materials are inherent in the employment of dynamic segmentation.

In this study three types of adjacency relationships are used. Adjoining relationships are end-to-end relationships between coastal units. Across-shore relationships are those between coastal units in the foreshore and backshore. Overlapping relationships are those where coastal units overlap along the coast. We have not distinguished between these three types of adjacency relationships for the analysis presented in this chapter.

Methods

A program was written in Arc Macro Language to analyze the along-shore and across-shore sequences of coastal geomorphologic forms. The program was designed to include only linear coastal forms sorted in ascending order of distance along the coast (database item 'to' in Table 1). The program steps along the coast examining the binary and tertiary sequences as each linear form is encountered. It then stores the number of occurrences of each unique relationship in one of the two database sequence tables. Sequence data were collected for coastal form 'type' and material 'supertype' on several selected data sets outlined in the discussion below. In this program, a binary adjacency relationship is the relationship between a coastal unit and an adjacent coastal unit with the next highest 'to' value regardless of whether the relationship is across shore or along shore. A tertiary relationship occurs between a coastal unit and the following two coastal units irrespective of whether the relationship is across shore or along shore.

TABLE 1.
Database contents for selected records from the Coastal Information System

Database Item Name	Description	Record Entry for UNIT_NO = 1550	Record Entry for UNIT_NO = 1627
ROUTE#	indicates which shoreline the coastal unit is on 1 = mainland; 2 = island	1	1
UNIT_NO	unique system-supplied number for the coastal unit	1550	1627
FROM	the beginning distance measure for the coastal unit (m)	461839.406	462719.000
TO	the end distance measure for the coastal unit (m)	462877.188	462789.000
NTS_SHEET	National Topographic System map sheet number	1 K/11	1 K/11
ZONE	the across-shore zone	backshore	foreshore
MAPPER	coastal scientist interpreting the source documents	Tracy Horsman	Tracy Horsman
OPERATOR	person entering the data into the CIS	Tracy Horsman	Tracy Horsman
QUALITY	coastal scientist responsible for quality control	Kimberley Jenner	Kimberley Jenner
SOURCE	source documents interpreted	video	video
SOURCE_DATE	date the source document was acquired	07/03/81	07/03/81

TABLE 2.
Database contents for selected records in the Coastal Information System from the *Form* table

Form Database Item Name	Record Entry for UNIT_NO = 1550	Record Entry for UNIT_NO = 1627
SUPERTYPE	unconsolidated	unconsolidated
TYPE	cliff	beach
SUBTYPE	steep	pocket
HEIGHT (height class)	2 (5-10 m)	
FEATURES	not stabilized	
COMMENTS		

TABLE 3.
Database contents for selected records in the Coastal Information System from the *Material* table

Material Database Item Name	Record Entry for UNIT_NO = 1550	Record Entry for UNIT_NO = 1627
SUPERTYPE	unconsolidated	unconsolidated
TYPE	clastic	clastic
SUBTYPE		mixed sand, pebble, cobble
COMMENTS		sand component in lower intertidal

The ArcInfo OVERLAYEVENTS command was also used to examine adjacency relationships. This command compares the along-shore extent of two or more database tables (referred to as event tables in ArcInfo terminology) which have 'to' and 'from' distances. Using the INTERSECT subcommand of the OVERLAYEVENTS command, a resultant database table is created containing records of coastal units which are common to the event tables chosen for the query. A record is created at every end point of an event, or

TABLE 4.
Binary and tertiary relationship abbreviations for form 'type' and material 'supertype'

Abbreviation Form	Coastal form 'type'
a	anthropogenic
b	beach
c	cliff
f	flat
o	outcrop
p	platform
l	slope
wb	waterbody
wc	watercourse
Material	**Material 'supertype'**
s	solid
u	unconsolidated

in our case a coastal unit, for all of the input event tables. If any events, or coastal units, overlap in extent, the total number of events or coastal units will increase in the resultant database table. For example, a single unit representing a backshore cliff could be divided into several units, all with the same cliff attributes, if this coastal unit is overlain with several smaller units in the foreshore. The consequences of this to our analysis will be discussed below.

Discussion

Adjacency relationship queries were completed separately for the east and the south coasts of the study area to quantify the predominant form relationships and to use these relationships to compare and contrast the coastal geomorphology. Binary and tertiary along-shore and across-shore sequence relationships were tabulated for form 'type' and material 'supertype' on all linear coastal units from the foreshore and backshore. Data for the most frequently occurring adjacency relationships for both regions are presented in Fig. 3. On the east coast, the ten most frequently occurring relationships accounted for 70 percent of the total population whereas on the south coast the ten most frequently occurring relationships represented 72 percent of the total population. Eight of these binary relationships occur in both the east and south coast populations. The remaining two binary relationships for the south coast are represented by unconsolidated material whereas the

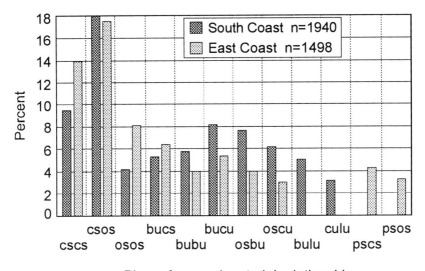

Fig. 3. Histogram comparing the most frequently occurring binary relationships of form 'type'/material 'supertype' from the east and south coasts of the study area. Refer to Table 4 for an explanation of abbreviations

two unique relationships for the east coast are represented by solid material. These data also show that of the eight binary relationships common to both areas, the percentage of adjacency relationships which contain unconsolidated material appears to be higher in the south coast data set with the exception of unconsolidated beach/solid cliff (bucs). This tends to support the visual analysis of coastal geomorphologic foreshore and backshore data from the CIS which suggests a higher proportion of adjacent unconsolidated forms along the south coast.

Along-shore and across-shore tertiary sequence relationship data were tabulated separately for the south and east coasts and are presented in Fig. 4. Because of the large number of unique relationships, the data were combined into four categories: solid/solid/solid relationships (s/s/s), solid/solid/unconsolidated relationships (s/s/u), unconsolidated/unconsolidated/solid relationships (u/u/s) and unconsolidated/unconsolidated/unconsolidated relationships (u/u/u). Tertiary s/s/s relationships appear to predominate along the east coast. Meanwhile, s/s/u relationships tend to occur at a similar frequency on both the east and south coasts. Conversely, u/u/s

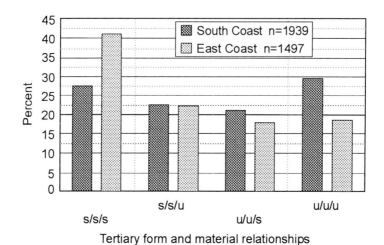

Fig. 4. Histogram comparing tertiary form 'type'/material 'supertype' adjacency sequences from the east and south coasts of the study area. Data are presented in four material 'supertype' tertiary relationship categories.

and u/u/u relationships seem to occur more frequently along the south coast. These data tend to support the observations for binary relationships presented above.

Binary sequence relationships were tabulated for three bays containing coarse clastic barriers to determine whether specific relationships of coastal form 'type' and material 'supertype' characterized the bays. These data are presented in Fig. 5. Within the three bays no specific coastal form 'type' and material 'supertype' relationship predominates; however, two binary relationship trends are prevalent. The solid cliff/solid outcrop (csos) relationship identified in Fig. 3 also occurs most frequently in Holyrood and Biscay Bays. In Mall Bay this relationship represents the third most frequently occurring relationship; unconsolidated beach/solid cliff (bucs) is the most frequently occurring relationship and reflects the higher number of pocket beaches along this section of coast. The second most frequently occurring relationships in all three areas are those containing unconsolidated cliffs. In Holyrood Bay, unconsolidated beach/unconsolidated cliff (bucu) and unconsolidated cliff/unconsolidated slope (culu) account for about 17 and 14 percent of the binary relationships respectively, whereas in Biscay Bay unconsolidated cliff/solid outcrop (cuos) accounts for 14 percent of the binary relationships. In Mall Bay unconsolidated beach/unconsolidated cliff (bucu) represents roughly 15 percent of all binary relationships. The higher percentages of binary relationships containing unconsolidated cliffs suggest that these cliffs may be potential sources for the clastic barriers.

A further example of using adjacency relationships was explored with the OVERLAYEVENTS command. This command was used to identify coastal units and selected attributes within a 10 km and 4 km along-shore extent from coastal units with the form 'barrier.' Two new database or event tables were created by selecting only the units which represented barrier beaches. The 'from' and 'to' distances of these units were modified to increase the extent of these units by 4 km and 10 km in each direction respectively, for the two queries. These newly created event tables were then used with the complete coastal unit event database in the OVERLAYEVENTS command using the INTERSECT subcommand. The resulting tables contained records of all coastal units found that were within the boundaries of the extended barrier limits. The ArcInfo FREQUENCY command was then run on the resulting tables to analyze qualitatively for dominant coastal forms.

In this application of the OVERLAYEVENTS command, a coastal unit may exist within the expanded extents of more than one barrier and, therefore, more than one record will be created for that unit in the resultant table. Because a particular unit may be counted more than once, a higher frequency of that type of unit is produced precluding frequency data from being used quantitatively.

A total of 141 records with the form 'barrier' occur in the database for the study area. The two most predominant forms found within 10 km of barriers are steep, unconsolidated cliffs in the backshore and fringing beaches in the foreshore. In descending order of occurrence these forms are followed by: steep, solid cliffs in the backshore; solid outcrops in the foreshore; steep, solid cliffs in the foreshore; and solid platforms in the foreshore. The two most predominant forms found within 4 km of barriers are lagoons and fringing beaches in the foreshore. These are followed, in descending order of occurrence, by: steep, unconsolidated cliffs in the backshore; smooth, unconsolidated slopes in the backshore; flats in the foreshore; and wetlands.

In general terms, query on the 10 km extent demonstrates the dominant solid forms that characterize the coast and includes unconsolidated coastal forms which may nourish the barriers. The results from the 4 km extent query provide more coastal form information specific to the barrier itself, as reflected in the predominance of lagoons, foreshore flats and wetlands.

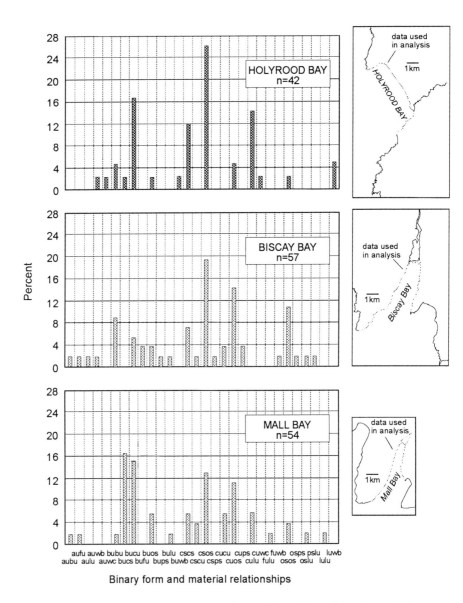

Fig. 5. Histograms showing the binary relationships of form 'type'/material 'supertype' from Holyrood Bay, Biscay Bay and Mall Bay

Conclusions

We have presented two applications of the use of adjacency relationships in coastal zone studies. Our application of binary and tertiary sequence relationships, for comparison between the more regional east and south coast data, demonstrates the ability to characterize predominant coastal forms and materials within the study area. Analysis of tertiary sequence relationships tends to support the conclusions identified from the binary sequence relationship data. The use of the OVERLAYEVENTS command to examine in more detail coastal form and material within 4 km and 10 km of each end of all barriers, provided a qualitative picture of the predominant adjacency relationships between barriers and other coastal form types. Quantitative analysis of these data was not possible because of the over representation of coastal units when two or more database tables were overlain and duplication of coastal form types existing within the expanded coastal extents of more than one barrier. Examination of binary sequence relationships in three specific extents of the coast containing a single barrier beach, refines the coastal characterization to those units which are part of the larger barrier system. Consequently, relationships containing unconsolidated material, which may be the source material for the barrier, were some of the more frequently occurring binary relationships.

The use of dynamic segmentation for querying form or material adjacency relationships enables the user to establish, characterize and quantify alongshore patterns. Our use of dynamic segmentation within the CIS has made possible the exploration of binary and tertiary adjacency relationships. We have used these relationships to compare and quantify, where applicable, the similarities and dissimilarities of the coastal zone at both a regional and local scale. This same technique may have a wider application to studies specific to coastal zone management including the identification of potential hazards and sensitive areas.

Acknowledgments

The authors would also like to acknowledge reviews of the chapter by P. Moir and D.L. Forbes.

Geological Survey of Canada Contribution No. 1997083.

References

Bartlett, D. (1993). Space, time, chaos and coastal GIS. In: *Proceedings, 16th International Cartographic Conference and 42nd Deutscher Kartographentag,* May 3-9, Köln, Germany, pp. 539 551.
Deuker, K.J., and Vrana, R. (1992). Dynamic segmentation revisited: a milepoint linear data model. *Journal of the Urban Rural Information System Association,* Vol. 4, pp. 94-105.
Howes, D.E., Harper, J., and Owens, E. (1994). *British Columbia physical shore-zone mapping system.* Victoria, British Columbia Resources Inventory Committee, Province of British Columbia, 72 pp.

Jones, A.R. (1995). GIS in coastal management: a progress review. In: *Proceedings, CoastGIS '95,* edited by R. Furness, February 3-5, University College, Cork, Ireland, pp. 165-178.

McCall, S. (1995). The application of dynamic segmentation in the development of a coastal geographic information system. In, *Proceedings, CoastGIS '95,* edited by R. Furness, February 3-3, University College, Cork, Ireland, pp. 305-312.

Owens, E.H. (1994). Coastal zone classification system for the National Sensitivity Mapping Program. *Report* prepared for Environment Canada, Dartmouth, Nova Scotia, 14 p. and appendices.

Rickman, T.L. and A.H. Miller (1995). *A Catergorized Bibliography of Coastal Applications of Geographic Information Systems.* Madison, Wisconsin: Sea Grant Institute, University of Wisconsin.

Sherin, A.G., and Edwardson, K.A. (1997). A coastal information system for the Atlantic Provinces of Canada. *Marine Technology Society Journal,* Vol. 30, pp. 20-27.

Sherin, A.G., and Edwardson K.A. (1994). Using GIS and dynamic segmentation to build a digital coastal information database. In: *Proceedings (abstract), Coastal Zone Canada '94,* edited by P.G. Wells and P.J. Ricketts, September 20-23, Halifax, Nova Scotia, Vol. 5, p. 2378.

CHAPTER 27

Consideration on Satellite Data Correction by Bidirectional Reflectance Measurement of Coastal Sand with a Remote Sensing Simulator

H. Okayama and J. Sun

ABSTRACT: In the field of remote sensing, there is a need to correct the intensities of bidirectional reflectance because the satellite data collected in nadir and off nadir directions show different values in spite of the measurement of the same point on the Earth's surface. We measured the characteristics of bidirectional reflectance from coastal sand using a remote sensing simulator. From the results of the measurements, we obtained indicatrices, hence Minnaert constants (k's) were calculated. We corrected the reflected intensity by assuming that the Minnaert constant (k) of the same area is equal.

Introduction

In papers on reflection and scattering, Lambertian surfaces are usually assumed, but natural surfaces are generally non-Lambertian and their spectral properties obtained from remote sensing data from aircraft or satellite are dependent on view and sun angles (Jackson et al., 1990; Royer et al., 1985; Slater and Jackson, 1982; Nicodemus, 1970). It is, therefore, necessary to take into account the bidirectional reflectance distribution function (BRDF) of the target to compare quantitatively the measurements acquired under different illumination and observation conditions. Knowledge of the bidirectional properties of natural surfaces are especially important.

In the field of remote sensing there is a difference between the data in the nadir and off-nadir directions for the observation of the same point on the Earth's surface by side-angle imagery sensors such as the Advanced Very High Resolution

Radiometer (AVHRR) on board the NOAA Satellite, and the High Resolution Visible (HRV) instrument on board the SPOT satellite.

In this chapter, as a fundamental experiment of a bidirectional reflectance, we obtain the bidirectional reflectance characteristics for the sand and we discuss how to correct the reflected intensities observed in different nadir directions. Our correction method has merit in that it can be applied to a non-Lambertian surface.

Experiments and Results

The experiment was done by using a remote sensing simulator shown in Fig. 1 (Genda and Okayama, 1978). This simulator has a halogen lamp lit by a stabilized power supply. The detector of the light is a photomultiplier. The optical source and the detector can move along the guides, and required incident and reflected angles are obtained. The detector can also move in the azimuth direction, enabling the bidirectional reflectance of an object to be measured. The size of the sample stage is 60 cm in diameter. The data are processed by a computer.

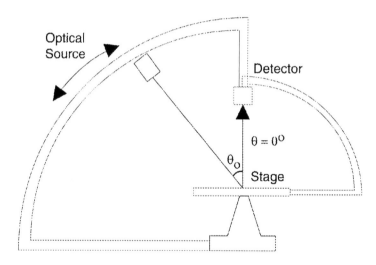

Fig. 1. Schematic diagram of the simulator for remote sensing

The sand collected at the Kujukuri coast in Japan was used for the experiment. To determine the wavelength of experiment, a Wratten No.25 filter with the same wavelength band as Landsat MSS (5) was used. In the measurement of the reflectance, a white barium sulphate board was used. The reflectance of a sample target is calculated by the following equation:

$$\rho(\lambda) = \frac{I_{\text{target}}}{I_{\text{white}}} \times \rho_{\text{white}}(\lambda)$$

(1)

where $\rho(\lambda)$ is the reflectance of target. I_{target} is the reflected intensity of the measured target. I_{white} is the reflected intensity of the white board. $\rho_{white}(\lambda)$ is the reflectance of the white board. The incident angle, reflected angle and azimuth angle are shown in Fig. 2.

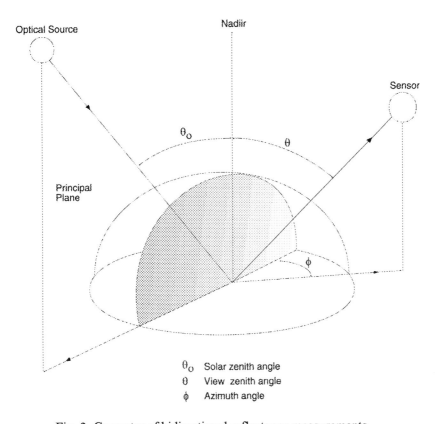

Fig. 2. Geometry of bidirectional reflectance measurements

First, we measured the reflectance in nadir ($\theta = 0°$) direction and at $0°$ azimuth angle. The indicatrices of the sand are obtained by changing incident angles. The result in the nadir direction is shown in Fig. 3, and the Minnaert constant obtained from the indicatrix is k = 1.26.

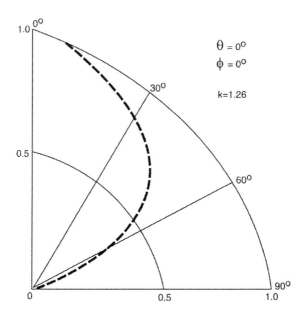

Fig. 3. Indicatrix of sand when the detection angle is 0° and the azimuth angle is 0°

Next the experiments were made at detection angles (θ = 10°, 20°, 30°, 40° and 50°). The experimental results at the detection angles 20° and 50° are shown in Figs. 4, and 5, respectively. As is shown in Figs. 3 to 5, when the detection angle is constant, the indicatrices hardly change by azimuth angle.

The measurement results of the reflectance are shown in Figs. 6 to 8. As is shown in Figs. 6 and 7, where the detection angles are 0° and 20°, respectively, the reflectance varies between 0.2 and 0.5 with change of the incident angles. When the detection angle is 50°, as is shown in Fig. 8, the reflectance varies between 0.3 and 0.9 and the distribution depends on the azimuth angle.

Next, the change of Minnaert constants by the azimuth angle is shown in Fig. 9. Minnaert constant (k) is higher when the azimuth angle is larger. In

Consideration on Satellite Data Correction

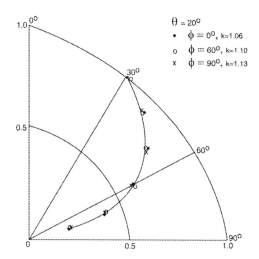

Fig. 4. Indicatrices of sand when the detection angle is 20° and the azimuth angle angles are from 0° to 90°

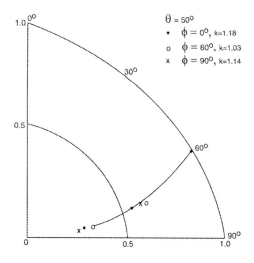

Fig. 5. Indicatrices of sand when the detection angle is 20° and the azimuth angle angles are from 0° to 90°

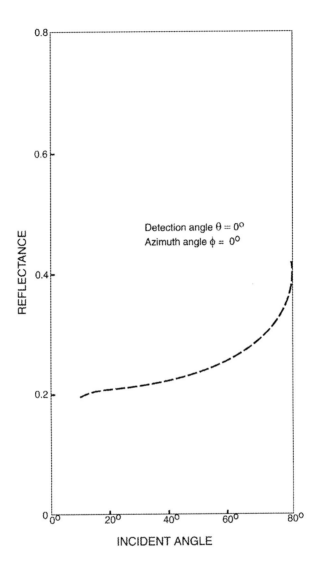

Fig. 6. Reflectance of sand when the detection angle is 0°

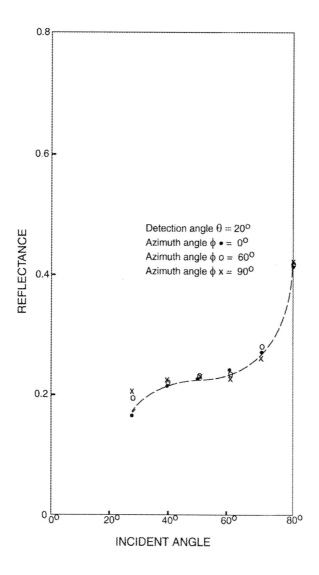

Fig. 7. Reflectance of sand when the detection angle is 20° and the azimuth angles are from 0° to 90°

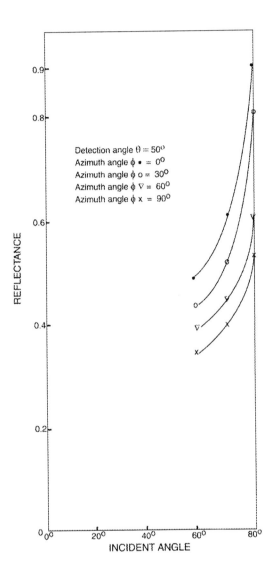

Fig. 8. Reflectance of sand when the detection angle is 50° and the zenith angles are from 0° to 90°

the case of the detection angles 40° and 50°, when the azimuth angle is 0 degrees, the Minnaert constant shows a large value and when the azimuth angle is 30°, the value becomes the minimum and with an increase in the azimuth angle the value becomes larger.

Correction of Reflected Intensity

The satellite data collected in nadir and off-nadir directions differ in spite of observation at the same point on the earth's surface. Then the correction must be done for these data. To correct them, we measured bidirectional reflectances by a remote sensing simulator. The equation of the reflected intensity is represented by use of Minnaert constant (Minnaert, 1941) such as the following:

$$I = I_o \cos^k \theta_o \cos^{k-1} \theta$$
(2)

where k is a Minnaert constant, and θ_0 and θ are incident and detection angles, respectively. Since two different Minnaert constants k_1 and k_2 are obtained from the indicatrices at the same point measured in nadir and off-nadir directions, the two reflected intensities I_1 and I_2 of nadir and off-nadir direction, respectively, are:

$$I_1 = I_o \cos^{k_1} \theta_o$$
(3)

$$I_2 = I_o \cos^{k_2} \theta_o \cos^{k-1} \theta$$
(4)

As these reflected intensities are assumed to be equal, so

$$\cos^{k_1} \theta_o = \cos^{k_2} \theta_o \cos^{k_2 - 1} \theta$$
(5)

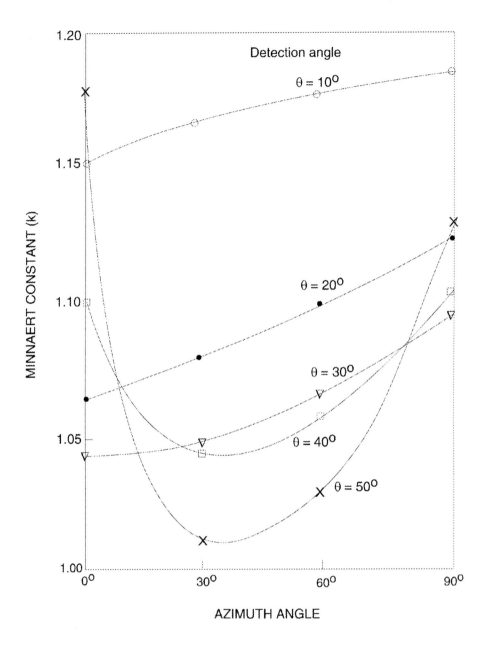

Fig. 9. Minnaert constants of sand when the detection angles change from 10^0 to 50^0

Minnaert constants are also assumed to be equal, so

$$k_2 \cdot x = k_1 \quad (6)$$

From the Equation (5).

$$\cos^{k_2 \cdot x} \theta_o = \cos^{k_2} \theta_o \cdot \cos^{k_2 - 1} \theta \quad (7)$$

Hence

$$k_2 \cdot x \log \cos \theta_o = k_2 \log \cos \theta_o + (k_2 - 1) \log \cos \theta \quad (8)$$

Therefore

$$x = \frac{k_2 \log \cos \theta_o + (k_2 - 1) \log \cos \theta}{k_2 \log \cos \theta_o} \quad (9)$$

From Equation (9) x is obtained and k_1 in Equation (6) is calculated, but due to experimental errors x must be multiplied by a correction coefficient c.

Then,

$$k_2 \cdot x \cdot c = k_1 \quad (10)$$

These c values are shown in Tables 1 to 4.

The correction of the reflected intensity is possible by substituting k_1 into Equation (3). The correction coefficients, c, obtained from experimental results are shown in Tables 1 to 4. These coefficients obtained based on Minnaert constant k = 1.26 given by the indicatrix for nadir direction ($\theta=0°$) and azimuth angle ($\phi=0°$) for sand. They are not much varied for the same reflected angle.

Conclusions

The experimental results of the bidirectional reflectance have shown that the reflection characteristics differ by the reflected and azimuth angles as represented by use of Minnaert constants. The Minnaert constants are larger when the reflected angles are smaller and when the azimuth angles are larger.

In this chapter, we have introduced a correction method, which enables the correction of data from different satellites. Even though observed at the same point, there is a disagreement in the data collected at two different reflected angles, nadir and off-nadir. In these corrections the Minnaert constant obtained from an indicatrix must be used.

References

Genda, H., and Okayama, H. (1978). Estimation of Soil Moisture and Components by Measuring the Degree of Spectral Polarisation with a Remote Sensing Simulator, *Applied Optic.s.* Vol. 17(21): 3439-3443.

Jackson, R.D., Teillet, P.M., Slater, P.N., Fedosejevs, G., Jasinski, M.F., Aase, K., and Moran, M.S. (1990). Bidirectional Measurements of Surface Reflectance for view Angle Corrections of Oblique Imagery, *Remote Sensing of Environment.* Vol. 32:189-202.

Royer, A., Vincent, P., and Bonn, F. (1985). Evaluation and Correction of Viewing Angles Effects on Satellite Measurements of Bidirectional Reflectance, *Photogrammetric Engineering and Remote Sensing.* Vol. 51: 1899-1914.

Slater, P.N., and Jackson, R.D. (1982). Atmospheric Effects on Radiation Reflected from Soil and Vegetation as Measured by Orbital Sensors using Various Scanning Directions", *Applied Optics.* Vol. 21: 3923-3931.

Nicodemus, F.E. (1970). Reflectance Nomenclature and Directional Reflectance and Emissivity, *Applied Optics.* Vol. 9(6):1474-1475.

M. Minnaert, M. (1941). The Reciprocity Principle in Lunar Photometry, *Astrophysics.* 3.93:403-410.

TABLE 1.
Correction coefficients when azimuth angle is 0°

θ_o	$\theta=10°$	$\theta=20°$	$\theta=30°$	$\theta=40°$	$\theta=50°$
30	1.09	1.17	1.18	0.98	0.73
45	1.09	1.18	1.19	1.08	0.89
60	1.10	1.18	1.20	1.11	0.97

TABLE 2.
Correction coefficients when azimuth angle is 30°

θ_o	$\theta=10°$	$\theta=20°$	$\theta=30°$	$\theta=40°$	$\theta=50°$
30	1.07	1.14	1.17	1.14	1.21
45	1.09	1.16	1.19	1.18	1.24
60	1.08	1.16	1.20	1.20	1.24

TABLE 3.
Correction coefficients when azimuth angle is 60°

θ_o	$\theta=10°$	$\theta=20°$	$\theta=30°$	$\theta=40°$	$\theta=50°$
30	1.05	1.10	1.17	1.08	1.13
45	1.06	1.12	1.18	1.15	1.18
60	1.07	1.13	1.18	1.17	1.20

TABLE 4.
Correction coefficients when azimuth angle is 90°

θ_o	$\theta=10°$	$\theta=20°$	$\theta=30°$	$\theta=40°$	$\theta=50°$
30	1.04	1.06	1.05	0.96	0.80
45	1.05	1.10	1.10	1.05	0.95
60	1.06	1.11	1.12	1.09	1.02

CHAPTER 28

Constructing a Geomorphological Database of Coastal Change Using GIS

J. Raper, D. Livingstone, C. Bristow, and T. McCarthy

ABSTRACT: Coastal systems are amongst the most dynamic in geomorphology and the most troublesome for environmental managers and civil engineers. This is because their landform and process regimes are characterised by rapid change over the short term (days and weeks) - making them difficult to predict and manage over the medium term (periods of months and years). Although there has been significant progress in the understanding of nearshore processes and landforms on low depositional coasts over the last 20 years (Carter 1988), physical models have proved difficult to extend from the timescales of a single tidal cycle to the multi-year prediction periods required by environmental managers and engineers. By monitoring actual morphological changes and correlating them with real energy inputs from waves and tides it has been possible to study the nature of medium term change directly by exploring the correlations of energy input and morphological change. Birkbeck College and Babtie Group are currently employing both strategies in parallel in a study of the morphodynamics of spits and nesses on the East Anglian coast of England as part of the UK Ministry of Agriculture, Fisheries and Food (MAFF) sponsored Coastal Area Modelling for the Long Term (CAMELOT) programme. This chapter aims to show how data on morphodynamics can be collected at the coast and how such data can be integrated for a holistic analysis. Examples of preliminary work in this area will be presented from spit morphodynamic data at Scolt Head Island on the North Norfolk coast

Introduction

Coastal systems are amongst the most dynamic in geomorphology and the most troublesome for environmental managers and civil engineers. This is because their landform and process regimes are characterised by rapid change over the short term (days and weeks) - making them difficult to predict and manage over the medium term (periods of months and years). This dynamic behaviour is determined by the inherent variability of coastal process forcing: the combined action of reversing tidal flows and directionally varying wind waves on coastal sediments means that the whole shape of the coastline can be rapidly changed under storm conditions. This is especially true on low depositional coasts and in estuaries which are characterised by beach barriers, dunes and salt marshes as in both cases the unconsolidated materials are particularly mobile. Around the low depositional coasts of the North Sea large tidal ranges and frequent storms mean coastlines exhibit complex spatial and temporal behaviour.

Although there has been significant progress in the understanding of nearshore processes and landforms on low depositional coasts over the last 20 years (Carter, 1988), physical models have proved difficult to extend from the timescales of a single tidal cycle to the multi-year prediction periods required by environmental managers and engineers. One research strategy has been to attempt to scale up coastal modelling from the short term spatial and temporal scale (where the full mechanics of all the processes must be represented) to medium term models (where the mechanics of the processes must be aggregated in some way) (Martinez and Harbaugh, 1993). A second research strategy has focussed on the investigation of the medium term morphodynamics of specific coastal sites. By monitoring actual morphological changes and correlating them with real energy inputs from waves and tides it has been possible to study the nature of medium term change directly by exploring the correlations of energy input and morphological change.

Birkbeck College and Babtie Group are currently employing both strategies in parallel in a study of the morphodynamics of spits and nesses on the East Anglian coast of England as part of the UK Ministry of Agriculture, Fisheries and Food (MAFF) sponsored Coastal Area Modelling for the Long Term (CAMELOT) programme. The aims are firstly to develop a better understanding of the interactions between coastal processes and form on low depositional coasts, and secondly, to study the spatial and temporal patterns of landform response (spits and ness shape and elevation) to process forcing (wave and tide energy inputs). This knowledge will allow more informed management intervention (by hard or soft engineering) on low depositional coasts by ensuring that developments are appropriate for the spatial and temporal scales of the change occurring.

Examples of such benefits could be the design of sea defences so that they have a lifetime equal to the time that a natural feature remains in a supporting position. An example would be the case of a sedimentary barrier such as a spit (e.g. the Spey spit, Scotland) which may progressively extend across a vulnerable urban frontage providing natural protection from wave energy (Riddell, 1995). If and when the spit is breached and is starved of sediment supply it may shorten and decline in height so as to give no protection to the frontage. This is the time at which the defences should be renewed, as the defences will age while a new spit forms and extends across the frontage. In this fashion the overdesign and inappropriate timing of construction can be avoided.

However, studies of actual morphodynamic change must be founded upon a consistent database of geomorphological information. This chapter aims to show how data on morphodynamics can be collected at the coast and how such data can be integrated for a holistic analysis. The key datasets for the morphodynamic analysis of the landforms of the nearshore intertidal zone are firstly, dense measurements of terrain elevation (especially at breaks of slope), and secondly, descriptions of sedimentary materials of which the landform is made, and thirdly, data on wave energy and tidal heights in the vicinity of the landform. Such data must be georeferenced, integrated into a GIS and referenced to time so that queries concerning the change of terrain elevation and sedimentary materials can be undertaken. Examples of preliminary work in this area will be presented from spit morphodynamic data at Scolt Head Island on the North Norfolk coast (for location see Fig. 1).

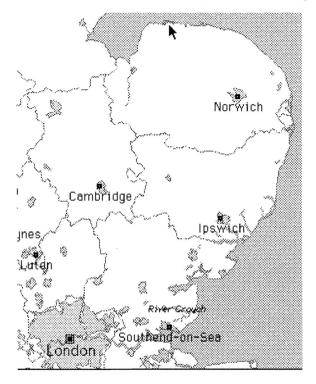

Fig. 1. Location of Scolt Head Island on the North Norfolk coast
(denoted by the arrow)

Data Sources for Spatio-Temporal Analysis of Changing Coastal Terrain

MAPS AND CHARTS

Maps and charts contain much valuable information for the study of geomorphological change. However, the update cycles of mapping/hydrographic agencies are not usually appropriate to the timescales of geomorphological investigations and the precise nature/time of the update may not be documented in full for the end user. Furthermore the mapping of coastal landforms or mean tide lines and the sampling of sedimentary materials is carried out for navigation, asset recording or leisure information and not for geomorphological purposes. Many of the mapped features are interpreted from air photography and are not formally defined for the end user. As a consequence maps and charts are probably most valuable in the coastal zone as a source of control information and the positions of fixed assets/buildings. The researcher must in most cases collect information on geomorphology as and when needed.

COASTAL FORM AND COMPOSITION

Data on the shape/composition of landforms at the coast can be obtained directly and indirectly. Indirect methods include satellite remote sensing, aerial photography, aerial videography and laser surface profiling. *Satellite remote sensing* can currently provide multispectral imagery at up to 10m pixel resolution (in panchromatic form), although several new sources at 1m resolution will become available within the next couple of years. Satellite remote sensing is expensive to purchase and the user generally has no choice about the time of capture: Fig. 2 shows 10m panchromatic SPOT data for Far Point on Scolt Head Island which was taken at high tide, therefore concealing the whole intertidal area. Elevation data can be derived from satellite interferometry with 1-2m accuracy but must be averaged over the area of the horizontal pixel. Hence where the surface slope is high, interferometry will have a lower accuracy due to the areal averaging effect (Zebker et al., 1994). Satellite remote sensing also provides access to multispectral reflectance data that can be used to distinguish between sediment composition and vegetated areas.

By comparison, *aerial photography* is cheaper to purchase from archival collections and is available at a variety of scales down to a pixel size of centimetres, although almost all the data is from the optical band. The main drawback in its use is that it is usually collected according to logistical priorities rather than research priorities, hence the state of the tide seen in photography of the coast is often less than ideal. It is also rarely collected at geomorphologically significant intervals such as immediately after storms as it is expensive to commission. Experiments with the use of model aeroplanes are an exception to this rule (Green and Morton, 1994)

Constructing a Database of Coastal Change Using GIS   403

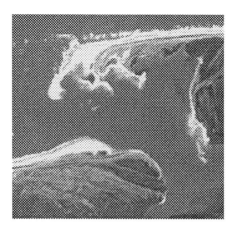

Fig. 2. A 1990 10m resolution panchromatic SPOT image for Far Point on Scolt Head Island taken at high tide and concealing the whole intertidal area (copyright SPOT Image).

and are a potentially valuable coastal imaging method. Fig. 3 shows the Far Point of Scolt Head Island in a 1990 vertical photograph at high tide showing wave refraction around the spit but concealing the intertidal area.

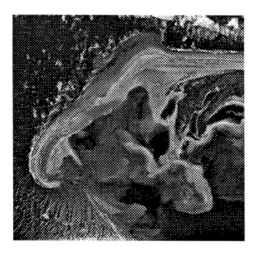

Fig. 3. Far Point of Scolt Head Island in a 1990 vertical air photograph at high tide showing wave refraction around the spit but concealing the intertidal area (photography courtesy of Cambridge University Committee for Aerial Photography).

Until very recently, aerial photography has been an analogue capture method using large format cameras, with the imagery being recorded on film and reproduced as photographic paper prints. Such prints need to be scanned for use in digital image processing systems. Digital photography has recently been tested as a capture method (Koh and Edwards, 1996) achieving 1524 by 1012 pixel 8 bit images at a rate of one frame every 2.75 seconds which required 1.54Mb of storage space. Historically, aerial photographs are also not georeferenced precisely, photograph positions being plotted on maps from flightlines and by approximation from photograph content. Some modern aerial photography (such as the digital photography by Koh and Edwards (1996)) is now referenced by the Global Positioning Systems (GPS) coordinates for the plane at the time the photograph was taken. However, in general historical aerial photographs can only be mosaicked together to make an imagemap if there is sufficient overlap between the photographs, and if there is ground control visible in the overlap between the photographs.

Without ground control (or georeferencing of the photographs by GPS) pattern matching techniques to overlap photographs can be used, although these techniques are difficult to use when there are no regular geometric shapes such as buildings visible in the imagery. If there is sparse ground control that is only on some photographs then aerotriangulation techniques can be used to transfer control from one photograph to another. However, this is not very accurate when the control is always on one edge of the photographs, as is often the case in coastal photography. Aerial photography can be stereocorrelated to generate a surface elevation model accurate to centimetres if there is heighted ground control visible in the overlap between photographs. In general the lack of photograph georeferencing or ground control in air photography at the coast can be a significant barrier to its use in mapping or stereocorrelation.

Aerial videography is an alternative low cost, lower resolution form of (generally optical band) machine readable imagery (recorded on magnetic tape) which is more cost-effective for the end user to obtain than aerial photography (King, 1996; Um, 1997). Imagery can be shot either vertically or obliquely depending on the aerial platform available. For the cost of a set of 150 aerial photographs taken with a large format camera the user can commission multi-hour aerial videography runs along the coast at the appropriate state of the tide or on geomorphologically appropriate occasions. Fig. 4 shows a 24 bit videography image of Far Point from April 1997, Scolt Head Island at low tide with a 1m ground resolution and requiring 1.2Mb of storage space. This allows ground control to be placed at key sites of interest before the flight and allows flight planning to ensure that control appears in the actual imagery obtained. Since aerial surveyors can review the captured imagery whilst in the air, sites can be reflown to obtain the necessary coverage and ground control visibility.

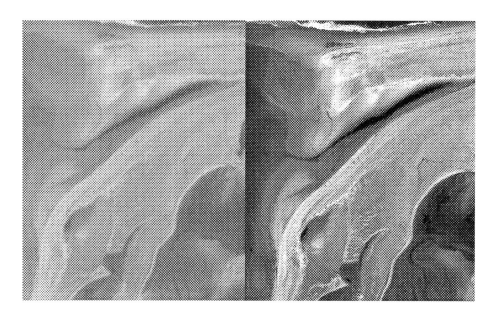

Fig. 4. A 24 bit videography image of Far Point, Scolt Head Island taken in April 1997 at low tide with a 1m ground resolution.

Most videography rigs are equipped with GPS position logging of the survey plane's flight. This record of GPS positions along the plane's track at 1 second intervals can be cross-correlated with the individual frames of the video either in real-time (for example using the Navtech Systems 'Telenav' unit - Cooper et al. (1996)) or by post-processing (for example using a Horita time code generator - Brunner et al. (1995)). The position fix and video data can be recorded for subsequent playback or can be fed directly into a real-time plotting system such as RTGIS from AVL (www.avxs.demon.co.uk). This allows the survey plane's track to be plotted on a map: since 25 video frames are captured for each GPS position (PAL video is 25 frames per second), each plotted point will correspond to a specific video frame.

Semi-professional video cameras have a resolution of 500 lines or more vertically, which, when 'framegrabbed' (converted from analogue television signal to computer images) in the aspect ratio of a standard computer monitor can give images of 576 by 768. Using slightly wide angle lenses, video cameras can obtain images roughly as wide as flying height. Hence, flying at 500m will generate a swath width of 500m that will give a pixel size of around 70cm for typical imagery framegrabbed using normal aspect ratios. The imagery can also be recorded as analogue video onto VHS tape or digitised into a digital video format such as MPEG, although digital movies require around 10Mb per minute of storage. Aerial videography can also be stereocorrelated if heighted ground control is available since the considerable along-track redundancy of the frames ensures adequate overlap. Livingstone et al. (1999)

give examples of a optical band videography-draped surface model for vegetated sand dunes on Scolt Head Island and show the obtained accuracies for this purpose.

Laser surface profiling is a new and promising method of elevation survey which can be integrated with other remotely sensed imagery captured at the same time or previously. Using a laser distance measuring device that sweeps back and forwards over a scene at 80-200Hz relative heights can be determined instantaneously (Lohr, 1997). Used aboard an aerial platform the laser device scans a swath defined by the forward movement of the aircraft and can be corrected by on-board attitude sensors. Currently a research technology it is likely to grow in importance rapidly.

Ground survey is the key method of direct landform survey and is usually based on the observation of points relative to local or national control frameworks. Typical ground surveys employ a total station with electronic distance measurement and the computer logging of data points. The horizontal and vertical angles subtended to the backsight line (a line between 2 known points) and the distance to the surveyed point are recorded in the data logger. These observations are converted into plane coordinates by a survey reduction program. Over 20 ground surveys in the form of point measurements of position and elevation have been collected at Far Point on Scolt Head Island since 1992, each survey involving the collection of 700-1000 data points over an area 1km square. The data points collected in these surveys have been located on the breaks of slopes and crests of the spits in the study area by the survey pole carrier and assigned a code for the mean grain size of the sediment by visual inspection. The coordinate frame used is a local coordinate system (the 'Scoltframe') defined by reference to surveyed monuments on the island, but rotations to National Grid are available. Accuracy is dependent firstly on the fidelity of the ground monumentation network and secondly on the precision of the instruments: both are checked frequently in the Scolt Head surveys.

Surface models for the surveys carried out at Scolt Head Island have been generated using triangulation methods. Triangulation is an exact local surface modelling method in which the surface is made of inclined triangular elements constrained by the data points. This method of surface modelling has been used here and is recommended for assessment of morphological change (as opposed to visualisation) because the surface fits exactly to the data points with no extrapolation (Gold, 1989). The data points were selected in the field in order to mimic the surface morphology as closely as possible and with the exact fit principles of triangulation representation explicitly in mind.

COASTAL PROCESS DATASETS

Data on wave height and period is sometimes available from direct measurement at sea and a number of wave platforms and wave buoys are permanently stationed offshore (see for example http://www.nws.fsu.edu/). This data is rarely if ever gathered in the nearshore zone close to landforms under study due to the operating constraints in this area, hence wave shoaling models are required to estimate wave heights at the shoreline. Wave heights at user-defined points can now be hindcast from wind records: the UK Meteorological Office has developed a wave hindcasting system which has been validated against a number of physical wave measurement stations (Reeve et al., 1996). Tidal stage in the form of an 'average' tidal curve is available for a number of standard ports around the coast of Britain; however, these are dominated by estuarial rather than open coast sites where the tidal behaviour is often quite different. Tidal flow rates are available for a number of points around the coastline, usually in shipping lanes and not in the nearshore zone.

Integration of Coastal Spatio-Temporal Data Sources

While capturing data on coastal landforms and processes is problematic given the difficulties of ground control for indirect methods and the difficulties of access and instrumentation for direct methods, the integration of the collected data is also challenging. Specific concerns include the handling of coordinates and datums, the regionalisation of data collected at points and the structuring of spatio-temporal datasets.

COORDINATE SYSTEMS AND DATUMS

Coastlines are difficult environments in which to integrate different data sources. In general terrestrial and marine coordinate systems are different and require transformations between projections and datum spheroids. There are different mapped coastlines that differ by date of survey (problematic in a highly dynamic environment like the coast) and also by the definition of sea level used. In elevation terms the terrestrial map datum is usually related to mean high tide while the marine chart datum is expressed as positive value below lowest astronomical tide (LAT). The difference between the terrestrial datum and the marine datum varies according to tidal range - which is spatially variable and often only measured at certain points e.g. ports. This makes it difficult to create integrated terrestrial and marine elevation models.

    The ideal solution to this integration problem is to add the actual tidal range at each point to the measured depths: this would then allow the terrestrial and marine elevation models to be connected along the mean high tide line. Since tidal range data is only available at a limited number of points on the coastline a method of interpolation is required. One method that has been developed experimentally in Norfolk is to linearly interpolate tidal range between known points along the coastline. Where there are barrier islands, only the coastline with a long fetch to the open sea is used and inlets are ignored. Using this method, where interpolated steps of 0.25m change in tidal range are reached a line is drawn orthogonal to the coast until it exits the area, reaches another coastline or until it intersects another such line.

    This semi-automated procedure defines zones within which the tidal range can be considered to be within 0.25m of each other. By adding this tidal range to all measured depths below LAT within each zone, and changing the sign on the aggregate value to negative, a depth below the mean high tide terrestrial datum is derived.

Where there are 'negative depths' on the chart i.e. height values on drying intertidal bars which are above LAT these values are subtracted from the tidal range to give a 'negative depth below the terrestrial datum' which is then smaller than the tidal range.

THE REGIONALISATION OF DATA COLLECTED AT POINTS

Many coastal datasets are collected at points but need to be regionalised i.e. the values interpolated between data points, and sometimes, extrapolated into areas with no data. However, the regionalisation needs to be appropriate to the phenomenon being observed, for example, depending on whether the data is discrete or continuous. The available procedures range from interpolation for continuous data to region building techniques (for example, Thiessen polygon creation) that can be used to create discrete zones with sharp boundaries.

When discrete phenomena that have well-defined boundaries in some places (such as sediment type) are nonetheless sampled at points, perhaps for logistical reasons, regionalisation cannot be achieved using interpolation techniques. This type of data can be regionalised using Thiessen techniques by assigning the value of a sample point to the Thiessen polygon produced around it. The result is a set of discrete polygonal tiles within which sediment type is constant - see Fig. 5 for an example of sediment type Thiessen polygons for Far Point, Scolt Head Island. By dissolving the boundaries between the polygons with the same attribute, discrete zones can be created from (originally) point data. These zones can then be converted from vector to raster form to permit grid-based modelling of the attributes or comparison with grid-collected data such as remotely sensed imagery.

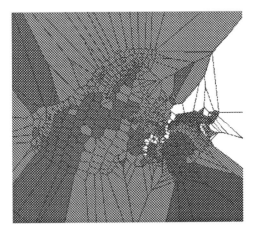

Fig. 5. Thiessen polygons constructed around surveyed point measurements of position and elevation for Far Point, Scolt Head Island collected from a survey in April 1992 and shaded by sediment type (white = dune, dark grey = sand, light grey = gravel).

When a phenomenon sampled at points is continuous (such as elevation) then interpolation techniques can be used. There are a wide range of 'local' interpolation techniques available (those that produce new values on the basis of close-by sampled points), however, not all of them are appropriate. In the case of elevation surveyed at points the value obtained from measurement and reduction to plane coordinates is likely to be within a few centimetres if the local benchmarks are correct and the instrument is being used correctly. If there is a good coverage of sample points an exact fit approach to interpolation should be preferred, i.e. triangulation or spline fit approaches, as the points are known with a high degree of accuracy. Where the points are collected at terrain-sensitive positions during the survey e.g. at breaks-of-slope, then triangulation techniques produce conservative surface models with nothing added through extrapolation as is sometimes the case with splines. Breakline constraints can be added to triangulation models to force the triangles to honour topographic form.

Approximate interpolation techniques should be reserved for visualisation or circumstances in which the points are not known with any degree of certainty, or when they are badly distributed. Figs. 6a and 6b show a surface model interpolated from the surface data collected in September 1993 using distance weighting and triangulation techniques respectively. Dummy points were added at the four corners of the study area to force the interpolation to fit to a set of edge constraints; by using these same points to interpolate other survey data for the same site at different times it was possible to compare the different surveys.

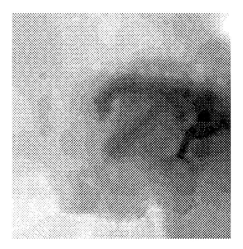

Fig. 6a. Contour map of a surface model for Far Point, Scolt Head Island made using distance weighting techniques from data surveyed in March 1995.

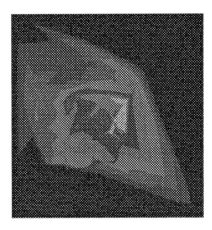

Fig. 6b. Contour map of a surface model for Far Point, Scolt Head Island made using triangulation procedures from data surveyed in March 1995.

STRUCTURING OF SPATIO-TEMPORAL DATASETS FOR MORPHODYNAMIC STUDIES

The study of actual coastal morphodynamics requires that morphological change and process inputs be correlated and explored for process signatures in the morphology and lags between the timing of process energy input and form response. Such studies could potentially show that such coastal geomorphology is chaotic: this may mean that patterns of energy input to coastal morphology are more important that the sheer magnitude at any given time. However, such studies depend critically on temporally structured spatial databases so that spatio-temporal behaviour can be explored.

In the research at Scolt Head Island the methodology of Raper and Livingstone (1995) has been put into practice. This approach called for the creation of firstly, a low level distributed database of morphology, composition and process which was integrated in terms of coordinate system and spatial/temporal datum, and secondly, a high level data model allowing the user to specify spatio-temporal queries on this data. Data collected across space in short periods of time has been stored in the ArcView GIS as geometric layers: in figure 7a surveyed data points from September 1995 are shown together with points from January 1996. ArcView has been used to measure distances and height differences between points at a similar position in absolute space for the two different times: in figure 7b the two cross sections show the profiles of the spit shoreface for the pairs of points identified by the white arrow in figure 7b.

This high level data model allows the user to compare spatial locations at different times (for example, by the construction of time-difference maps) or to track the movement of spatial configurations through time (for example, by analysing slope angle changes through space and time). The comparison of surfaces through time and the tracking of specific features has been examined, revealing the potential of virtual

Constructing a Database of Coastal Change Using GIS 411

reality tools to visualise change, as well as the morphodynamic changes that have been observed on Scolt Head Island.

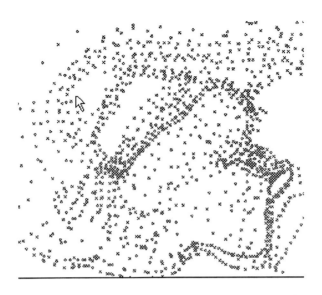

Fig. 7a. Surveyed data points at Far Point on Scolt Head Island from September 1995 shown together with points from January 1996

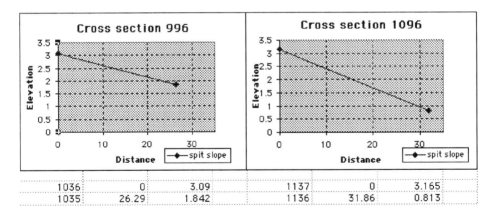

Fig. 7b. Cross sections showing the profiles of the spit shoreface for the pairs of points identified by the arrow in figure 7a

Conclusions

There is no doubt that GIS offer the coastal manager, engineer or modeller useful tools for storage, display and analysis. However, in practice the construction of a coastal database requires a considerable data gathering effort as many of the standard national mapping or remotely sensed imagery sources are unsuitable for the special needs of the dynamic coastal zone. The primary data collected or received form other sources usually requires a secondary restructuring to convert from one data model to another or to regionalise the data from points to areas. GIS may also need to be interfaced to modelling systems such as those used to shoal waves in from deep water monitoring stations.

However, while capture and integration issues are problematic, practical experience has shown the way to handle these problems- as demonstrated by the Scolt Head case study in this chapter. The real theoretical problems that must surely define the key priorities for future research into coastal GIS are concerned with morphodynamic change. In this area there are all too few methods available to study the precise issues (such as surface evolution) that may provide some of the answers to key questions about coastal geomorphological change.

References

Bristow, C., Raper, J., and Livingstone, D. (in prep) Morphodynamic change in a dynamic spit complex at Scolt Head Island, Norfolk, England. To be submitted to *Geomorphology*.

Brunner, J., Dalsted, K. ,and Arimi, A. (1995). The use of aerial video for land us/land cover characterisation and natural resource management project impact assessment in Niger. *USAID/Niger Research Report*, October 9, 1995.

Carter, R.W.G. (1988). *Coastal Environments: An Introduction to the Physical Ecological and Cultural Systems of Coastlines*. London, Academic Press.

Cooper, R., McCarthy, T., and Raper, J.F. (1995). Airborne Videography and GPS. *Earth Observation* 4 (11), 53-55.

King, D.J. (1996). Airborne Multispectral Digital Camera and Video Sensors: A Critical Review of System Designs and Applications. *Canadian Journal of Remote Sensing*. September 1996, 100-119.

Green, D., and Morton, D.C. (1994). Acquiring environmental remotely sensed data from model aircraft for input to GIS, *Proceedings of AGI '94*, 15.3.1-27.

Gold, C. (1989). Surface interpolation, spatial adjacency and GIS. In, Raper, J.F. (Ed.) *Three Dimensional Applications of GIS*, 21-36.

Koh, A., and Edwards, E. (1996). Integrating GPS data with fly-on-demand digital imagery for coastal zone management, *Proceedings of AGI '96*, 6.1.1-5

Livingstone, D.L., Raper, J.F, and McCarthy, T.M. (1999). Integrating aerial videography and digital photography with terrain modelling. *Geomorphology*, Vol. 29(1-2), 77-92.

Lohr, U. (1997). Digital elevation models by laser scanning: principles and applications. *Proceedings of the Third International Conference on Airborne Remote Sensing*, Copenhagen, Denmark, 7-10 July 1997, I-174-80.

Martinez, P.A., and Harbaugh, J.W. (1993). *Simulating Nearshore Sedimentation*. Oxford, Pergamon.

Raper, J.F. and Livingstone, D (1995). Development of a Geomorphological Data Model Using Object-Oriented Design. *International Journal of Geographical Information Systems*, Vol. 9 (4), 359-83.

Reeve, D., Li, B. and Fleming. C. (1996). Validation of Storm Wave Forecasting. *Proceedings of the 31st MAFF Conference of River and Coastal Engineers*, Keele, 3-5th July, 4.3.1-12.

Riddell, K.J., and Fuller T.W. (1995). The Spey Bay Geomorphological Study. *Earth Surface Processes and Landforms*, Vol. 20, 671-686.

Um, J-S. (1997). *Evaluating Operational Potential of Video Strip Mapping in Monitoring Reinstatement of a Pipeline Route*. Univ. of Aberdeen, Ph.D Thesis.

Zebker, H.A., Werner, C.L., Rosen, P.A., and Hensley, S. (1994). Acuracy of topographic maps derived from ERS-1 inteferometric radar. *IEEE Transactions on Geoscience and Remote Sensing*. Vol. 32(4), 823-36.

CHAPTER 29

Development of a DSS for the Integrated Development of Thassos Island

H. Coccossis and K. Dimitriou

ABTSRACT: The development of a decision support system for the integrated development of Thassos Island (Greece) is the focus of a research program undertaken by the University of Aegean. The system will be based on the GIS ArcInfo. The evaluation of economic and social activities and ecological processes affecting the island will be based on the use of a database of Sustainability Indicators linked to the GIS. Since Thassos represents a typical Mediterranean ecosystem with intensive human pressure the principles of Integrated Coastal Zone Management will be the base for the DSS. The proposed system will serve strategic planning purposes.

DSS in ICAM

Coastal areas are extremely valuable as they concentrate a rich diversity of natural habitat areas and a large variety of natural resources (Carter, 1988). As the interface area between land and sea, coastal areas are extremely important and fragile from an ecological perspective and should be carefully managed. Among the main limiting factors in the sustainable development of coastal areas are the conflict of interests and the complexity of jurisdiction between the various national, regional and local authorities dealing with the regulation of different activities and uses of coastal resources (UNEP/MAP/PAP, 1995).

The main objective of Integrated Coastal Area Management (ICAM) is to bring these concerns together within a unified management plan. Although the co-ordination of the different decision making forums and economic interests of the coast for the multisectoral planning of sustainable development is assumed, in practice, ICAM involves often the preparation of a strategic plan for the coastal area stating the general objectives and policies to achieve sustainable development, and then their elaboration into area-specific management actions for land and sea uses, mobilizing various policy instruments, procedures and controls, legal institutional and financial requirements etc. (OECD, 1993).

During the implementation phase the long-term management strategy is transformed into specific actions and projects which are supplemented by monitoring and a continuous information feedback. Several tools can be used for the sustainable multi-sectoral management of coastal activities. Regulatory instruments include land-use planning, building regulations, licensing activities based on environmental impact statements, construction guidelines for the coastline, conservation regulations, etc.

At this phase the use of DSS could be proved essential in Integrated Coastal Area Management. Although a vital aspect of any information system for effective management is clarity and simplicity, the complex patterns of interactions between natural and human ecosystems occurring in coastal areas, require complex and integrated approaches (Coccossis, 1985).

The differentiation and diversification in the scales of the complex patterns of interactions in both time and space make the use of GIS ideal for coastal management purposes. The ability of GIS to store, handle and analyze spatial data (geographical and attribute) together with their real time performance boosts the decision making process. The linkage, combination, intersection etc of various layers for the extraction of the objective in parallel with the built-in capability of algebraic operations makes GISs a necessary tool for the direct evaluation of the management process.

It is generally regarded that the idea of Decision Support Systems started with the work of Gorry and Scott-Morton (1971). Nowadays it is accepted that these systems focus on decisions and on supporting rather than replacing the decision-maker. Decision theory focuses on finding the best solution to any problem. Orthodox DSS use the so-called constrained optimization and are firmly established on a single criterion approach. The ICAM process implies conflicting interests and presents multidimensional characteristics where a single criterion is not sufficient and multiple criteria are needed. Multicriteria evaluation methods are powerful tools in conflict management and could be embodied to DSSs. The aim is the development of a DSS capable of handling multiple criteria.

As there is a multitude of interests and management objectives, a coastal GIS should reflect to the extent possible such concerns in order to link cross-impacts. Although some interactions are measurable and therefore quantifiable a number of linkages and phenomena can only be expressed in a qualitative manner. This fact generates the necessity to embody a multiobjective DSS to the GIS (Despotakis, 1991). The variety of decision making situations especially where spatial information is of crucial importance generate the so called Spatial Decision Support Systems (SDSSs) which are an important subset of DSS.

A crucial aspect of sustainable coastal area development is the type of information (data) that is going to be used in the SDSS for the evaluation and supervision of the procedure. The use of Indicators of Sustainable Development (ISDs) primary focus and purpose is to facilitate decision making and should be considered as ideal tool for integrated coastal area management. ISDs are classified in three main categories (Gouzee, 1996):

- **The driving force indicators (also mentioned as pressure indicators) refer to human activity processes and patterns that impact on the state of sustainable development. The use of these indicators can raise awareness and support for coastal management action by improving communication and understanding on coastal issues, resulting in eliminating the complexities involved**

- **The state indicators illustrate the existing socioeconomic, environmental and institutional condition of the site. State of the system information can concentrate on key coastal characteristics**

- **The response indicators reflect policy options and other responses to changes in the state of sustainable development. It is recommended that these indicators should be restricted to measures of policy, technical institutional or management action, and avoid indicating changes or trends in driving forces or state measurements which represent only feedback from response actions**

The main feature of the approach described above is the use of ISDs for the development and operation of a SDSS. The use of all three sets of indicators would generate a spatial information system capable of supporting decision-making process in three environmental management levels: pressure, state, and response. Since in most cases there will be no solution optimising all the development criteria at the same time the system would have the ability to handle multidimensional data in order the environmental decision maker to find compromise or/and alternative solutions (Paruccini et al., 1997).

Description of the Case Study

Thassos is one of the biggest Greek islands with an area that covers 382 km^2, it is located seven miles south from the costs of East Macedonia region. Its maximum width, east to west, is 22 km while its maximum height, north to south, is 26 km. The total perimeter of the island is 100 km.

The capital of the island is the city of Thassos. Administratively Thassos belongs to the prefecture of Kavala and consists of one municipality and nine communities with a total of 38 settlements. The linking of the island with the mainland is accomplished through regular services of car ferries.

The island is mountainous and can be divided into three geomorphologic areas. The first includes the coasts and occupies little space, the second hilly area is extended almost parallel to the first and is also relatively small. The third area consists of massif rock and occupies most of the island.

The hydrogeologic conditions on the island depend on the composition and the type of rock. The northeast sector of the island that is structured from marble formations presents a satisfactory underground water horizon. On the contrary the rest of the island presents low underground deposits of water.

The climate of the island can be characterized as meso-Mediteranean with an annual average of 600 mm height of rain. The mean temperature is 14.25°C while the mean relative humidity is 64.5%. There are no strong winds affecting the island (58.5% apnea).

Marine ecosystems of Thassos are characterized by their near natural condition and their relatively large diversity of fish and shells.

Regarding the composition of land-uses, 25% of the area of the island is covered by variable agricultural uses, 57% by various types of forests, 11% by grasslands and only 5% could be characterized as urban land.

Forests are mainly composed of coniferous trees (*Pinus nigra and Pinus brutia*) and occupy the middle of the island. Other phytosocial formations that can be found in the island are: the formation of broad-hardleafed with typical species such as *Pistacia lentiscus*, *Olea europea* and *Querqus ilex*, the *Erica arborea and Erica verticilata* formation, the formation of deciduous broad-leafed and riparian formations with typical species such as *Platanus orientalis*, *Polulus sp.*, *Salix sp.* etc. It is important to notice that in spite of its relevant small size the island presents 12 endemic plant species.

Thassos has a remarkable fauna that consists of mammals (such as *Lepus europeus*, *Martes foina*, *Erinaceus europeus*, etc), reptiles, amphibians, and birds. Mountain Ipsarion has been characterized as a Corine biotope because of the significant biological and natural functions that it supports.

Thassos supports a total population of 13,527 that appears almost stable for the last two decades. The distribution of manpower to each sector of economy is 27.8% to the primary sector, 7.8% to secondary sector, 12% to construction, 44% to tertiary sector (tourism, transportation, services, commerce) and 8.4% of people appears to be unemployed.

Thassos, regarding its insular nature, supports a wide range of economic activities. The major economic activities affecting the island are:

- **Marble extraction. This economic activity occupies 2000 ha, an area that represents 5% of the total surface of the island. There are 12 fixed areas for extraction mainly located to mountainous and forest areas. The white marble that Thassos produces is one of the best marble qualities worldwide. The rational marble extraction constitutes for the national economy a major source of foreign exchange since the entire volume is exported. A total of 55,000 m^3 of marble are produced every year from 25 quarries and most of this quantity is treated in 8 factories that are located in the island. A total of 500 people work in the extractions while 200 people are occupied in the factories**

- **Oil extraction. This economic activity is located to the sea area between Mainland and Thassos. This relatively new activity (first established 1981), is** of extreme importance for both Thassos economy and national economy. The oil deposits are considered to be satisfactory for the continuation of this activity for the next few decades while there are certain plains for future expansion. This activity includes risks for both the coastal and marine environment. There is an immediate probability of leakage of oil or hydrogen sulfide. Environmental impacts are possible at all stages of oil exploitation. The method of exploitation is from steel or concrete production platforms. The impacts of these structures stem partly from operational releases and partly from accidents. In addition to these environmental effects, oil exploitation has other impacts. The presence of rigs and pipelines creates exclusion zones for fishing vessels and other shipping, while the debris associated with offshore oil operations can damage fishing gear or entangle ships' propellers. A further problem, only now emerging, is the question of decommissioning and disposal of oil installations. Finally it must be mentioned that this activity counteracts other activities such as tourism and recreation

- Agriculture. There are 8063.4 ha of agricultural land in Thassos with an average lot of 2.1 ha. The main agricultural product is olive oil. The area that is covered by olive trees represents 91.4% of all cultivated areas while the total production of oil and salted olives is 60000 tons and 20000 tons. There are 20 oil-presses scattered across the island. The disposal of the remains of that process into streams is responsible for the pollution of groundwater deposits and affects nearshore marine ecosystems through runoff. It should be mentioned that there is no evidence of the overutilization of soil or excessive use of fertilizers and pesticides

- Stock breeding. Although stock-breeding is an economic activity that supports the internal market to a great extent, it represents only 3% of the total gross income of the island and is last to the economic valuation. There are a total of 49,000 breeding animals scattered across the island in various nomadic formations. The main impact of this activity is over-grazing. This could be prove rather harmful to the terrestrial environment due to the desertification of the landscape and the loss of soil, especially in areas with steep slopes. This activity shows a strong competition on the use of land with both agriculture and forestry. A traditional economic activity, which was very profitable in past decades, was that of bee breeding. There are a lot of apiaries scattered over the island, but after a large fire (in the Summer of 1985) that burnt a significant proportion of forest areas, this activity declined

- Fishery. Fishing boats of various sizes form a relative large fleet that supports not only the market of the island but also a part of national needs for fish. Fishing is an activity that is tied to coastal locations. Although nearshore fishing is an activity that has significant impacts on coastal marine resource mainly due to overfishing or following certain practices, in most cases illegal;

(small sized nets, dynamite use, etc), it still has a certain positive significant on island's economy. Another aspect of the fishing sector with potential negative impacts on the environment is aquaculture which could compete for the use of sea space with other uses, mainly recreation, but also with traditional fishing activities, industry, etc. The impacts from aquaculture installations are not restricted to the use of sea space but also include possible pollution from excessive use of fishfood, a problem which can be particularly important in enclosed bays and estuaries

- Secondary sector. There is a concentration of small industries (timber treatment, dockyards, marble treatment, storehouses) at the town of Thassos in an industrial park within an area of 20 ha, the only one on the island. There are possible impacts of industry on the marine and underground water resources pollution through effluents. Indirect impacts of industry such as overconcentration, land-use conflicts and urban development on the coasts have to be also considered

- Tertiary sector. The island's economy is supported to a great degree by this sector and more precisely by tourism. There is an average increase of 5% to the arrivals each year. A total of 60,000 tourists visited the island last year. The main tourist attraction can be considered the nature and the landscapes of the island with the unique combination of sea and forest. Relevant research revealed that 41% of all tourists visit the island because of the combination described above. There are a wide variety of other tourist attractions such as ancient Greek, Roman, Byzantine temples, fortification constructions, monuments etc. Tourism in Thassos appears strongly seasonal and becomes increasingly more intensive. This results in a reduction of natural sites and open space, substantial alterations of coastal landscapes and conflicts between the use of land, water and other resources. These adverse effects could be further exacerbated by the associated indirect effects of tourism related urban development for trade, transport facilities, vacation houses, infrastructures, residences for those working in the tourism sector etc. Recreation and leisure activities would also have adverse impacts on coastal ecosystems due to physical effects or disturbances from human presence. Pressures on the coastal zones of Thassos are likely to increase in the future

The impacts of urban development on Thassos coastal resources are very limited. They are restricted to phenomena such as: loss of natural habitat areas due to reduction of vital space or pollution and waste disposal, which are localized to few urbanized sites.

The impacts of various facilities for sea transport requiring coastal locations have not yet been studied. Theses facilities can affect the marine environment through pollution or as a result of the coastal engineering works required. Coastal habitats and beaches of Thassos can be severely affected due to loss of vital space and changes in coastal processes.

Environmental deterioration can have a significant impact on development prospects of Thassos. Although for the moment there are no major conflicts of the use of coastal resources there is speculation for the future development of the island. The long-term future of Thassos depends on the rational management of both economic and natural resources.

Development of DSS for Thassos Island

The main objective of the DSS is to contribute to the Sustainable Development Process of the island. The scope is to register most (if not all) of the components that affect the development process. Considering the insular nature of Thassos the Integrated Coastal Area Management principles will be fully embodied in the DSS. These principles can be summarized as follows:

- **Coastal areas are unique and require special management approach**
- **Coastal management should concentrate on the coastline and the critical area of interface between land, water and air**
- **Coastal management should include an area on the seaside and an area on the landward side**
- **The geographical boundaries of the area for coastal management should respect the function of coastal ecosystems and could reflect administrative boundaries as well**
- **Special emphasis should be taken into consideration and combined with the protection of coastal resources**
- **The approach to the development of the coastal areas should reflect and respect the functioning of natural ecosystems**
- **A multiple use of coastal resources could be a good basis for coastal management**
- **Traditional means to manage coastal resources should be incorporated to the extent possible in the coastal management scheme**

In that context the island is divided in four zones and each zone is characterized in the GIS database with a different identification. This way coverage with four Ids is created. The result can be considered as the first thematic entity of the proposed methodology.

The critical zone is a narrow band of land and sea 100 meters wide, adjacent to the shoreline, characterized by high ecological value and intense development pressures. Except for absolutely essential construction such as marinas any other kind of structures should be prohibited at this zone.

The dynamic zone extends inland, 2 kilometers wide, where there is strong dependence and/or influence of human activities and natural processes on coastal features and resources.

Special regulations for land-use planning, building regulations, licensing activities based on environmental impact statements, conservation regulations, resource-use quotas, emission standards, transport regulations, etc should be adopted for that zone.

The wider zone of influence, extends up to the municipal borders, which influences in part directly or indirectly the other two zones and serves as buffer zone between the inland and the coast.

The inland mountainous area where the development pressure is not so intense.

During the next stage all the critical factors affecting the development of the island are charted. The charting methodology is composed of three additional individuals thematic entities.

The second thematic entity demands that all the economic activities will be charted so as every one is to be represented by a different coverage for the entire surface of the island. The database for each activity will include quantitative information about the past and present state of each polygon regarding the particular activity. Some fields of the database will be devoted to the synergy, antagonism and conflict that the particular activity shows against all the other economic activities. The activities that will be charted are agricultural, stock breeding, various form of industry, tourism development, marble extraction, oil extraction etc. The use of ArcInfo union command (Overlay module) will form one final coverage including all partial information described above. The result will be the disecting of the island's surface into polygons with identical economic characteristics.

In the third entity terrain and landscape natural or man-made characteristics will be charted. Each characteristic will be charted resulting in the formation of a unique coverage. The database will include spatial quantitative information about each characteristic together with information about the compatibility of each category of the particular characteristic for all possible forms of economic development (e.g. tourism). At this point it must be mentioned that linear and point terrain characteristics (e.g.road network, springs) will be transformed to polygons using buffer of influence. The items that will be included to this unity are road network, hydrographical network, water springs, slopes, underground water deposit, actual and potential natural vegetation, sites of high cultural, historical or ecological significance, zone(s) under special legislative or protection status, landfills, etc. The use of the ArcInfo union command (Overlay module) will form one final coverage including all partial information described above. The result will be the chopping of island's surface to polygons with identical terrain and landscape natural or man-made characteristics to their entirety.

The fourth entity will include spatial quantitative and statistical information (including urban data) for each administrative division. For each administrative division (or for each polygon that is formed by the boundaries of the settlements if there are available and sufficient data) SDIs will be included in the database. The database will include pressure, state and response indicators in the fields of society, economy, environment and institution. The indicators will be fully compatible with the equivalent chapters of Agenda 21 (for example: The equivalent chapter for the category environment, subcategory water, is the chapter 17 (Protection of the oceans, all kinds of seas and coastal areas) of Agenda 21).

The GIS software that is used for the process described above is ArcInfo while the processing of aerial photographs, the digitization of primary maps (scale: 1/25000-1/50000), the performance of spot surveys, the utilization of questionnaires and the collection of statistical data from various sources are used for data acquisition.

At the end of the process a database structure of the SDSS will be ready. The SDSS will support multiple scenario evaluation and analysis capabilities. This can be achieved by the development of a modelbase that will contain interaction relations for the four thematic unites. The basic principles that will guide the configuration and formation of the modelbase are described below:

- **The coverage of the first thematic entity (Coastal perimetric zones) represents the proximity to the shore. This will be used as a criterion for:**
  - the state and future development pressure and
  - the focusing of the management process
- **The second thematic entity (Economic activities) expresses the state economic situation and will be mainly used for the finding of:**
  - the economic conflicts and
  - the economic and development trends.
- **The third entity (Terrain and landscape natural or man-made characteristics) reflects the carrying capacity of the development niche. This will be used for the tracking of**
  - the limited or favourable factors for any potential development process and
  - the areas that require special protection or conservation.
- **The fourth entity (SDIs) will serve as an objective criterion for the evaluation and supervision of the state and future sustainable development process.**

Under consideration is the way that the weighting of alternative scenarios will be succeeded. The examined solutions are the embodiment of a multicriteria analysis technique (Jansen, 1991) or the development of an evaluation procedure using the macro language of Excel.

The ability of ArcView to interact with other applications and to support spatial queries makes it ideal as an SDSS generator. It is proposed that the decision maker will basically use ArcView package from ESRI for the utilization of SDSS features. Finally, it is hoped that in it's mature state the SDSS will support extensive Artificial Intelligence features through the use of fuzzy logic.

It is estimated that the proposed SDSS will prove essential for identifying problems (state of the environment), targeting actions, monitoring progress towards desired goals and periodic reviewing of policies.

References

Agenda 21. (1992). Programme of Action for Sustainable Development. Rio Declaration on Environment and Development. *The final text of agreements negotiated by governments at UNCED.* 3-14 June, Rio de Janeiro, Brazil.

Carter, R.W.G. (1988). *Coastal Environments: An Introduction to Physical, Ecological and Cultural Systems of Coastlines.* Academic Press, London.

Coccossis, H. (1985). Management of coastal regions: the European experience in nature and resources *(UNESCO Journal on Environmental Affairs)* Vol.XXI, No. 1 (Jan.-Mar. 1985), pp. 20-28.

Couzee, N. (1996). Indicators of sustainable development: An institutional approach. 1st International Conference of Applied Econometrics Association, Lisbon, Portugal, April 10-12, 1996.

Despotakis K. (1991). Sustainable Development Planning Using GIS *Ph.D. Dissertation*, Amsterdam.

Gorry, A., and Scott-Morton, M. (1971). *A Framework for Information Systems.* Sloan Management Review, Vol. 13, Fall 1971, pp. 56-79.

Jansen R. (1991). Multiobjective decision support for environmental problems. *Ph.D. Dissertation*, Amsterdam.

OECD (Organization for Economic Cooperation and Development) (1993). *Coastal Zone Management Integrated Policies.* OECD: Paris.

Paruccini, M., Haastrup, P., and Bain, D. (1997). Decision support systems in the service of policy makers. *IPTS report*, Vol. 14, May 1997.

UNEP/MAP/PAP (United Nations Environment Programme/ Mediterranean Action Plan/ Priority Actions Programme) (1995). *Guidelines for integrated management of coastal and marine areas, (with special reference to the Mediterranean basin).* PAP: Split, Croatia.

CHAPTER 30

Development of a Spatial Decision Support System for the Biological Influences on Inter-Tidal Areas (Biota) Project within the Land Ocean Interaction Study

N. J. Brown, R. Cox, A. G. Thomson, R. A. Wadsworth, and M. Yates

ABSTRACT: The Land Ocean Interaction Study (LOIS) is a major NERC initiative seeking to understand the exchange, transformation and storage of energy and materials between the land and the ocean. In LOIS the BIOTA programme studies the fluxes of the inter-tidal zone and especially the interactions between biotic and abiotic components of the system. To facilitate communication and co-operation the BIOTA data centre has developed a spatial Decision Support System (DSS) based on a hybrid distributed database and a Geographical Information System (GIS). Special attention has been paid to ensuring: known data quality, the existence of appropriate meta-data, the integration of process based models and spatial analysis and visualisation tools.

Introduction

The coastal zone is an area of potential conflict between human activities and the natural environment. Many major cities have coastal locations. It provides resources for, recreation, transport, waste disposal, industrial and agricultural activities. However, many coastal environments are suffering from degradation due to human activities. These effects may be exacerbated by predicted climate change and consequent sea level changes. In the past, sea levels have fluctuated considerably. For instance, during the Pleistocene period sea levels were more than 120m below current levels and it has been suggested that the sea level may rise by a metre over the next century (Warrick and Oerlemans, 1990). The Natural Environment Research Council (NERC) of the United Kingdom has recognised the importance of this environment and has funded The Land Ocean Interaction Study (LOIS). LOIS was the largest project within NERC's extensive community research programme. In all, over 300 scientists are involved in the core programme and associated special topics over the six year (1994 - 1999) life of the study.

ITE scientists and, in particular, their GIS specialists, have extensive experience with collaborative projects. Developments such as the DOE Core Model (Parr and Eatherall, 1994) have paved the way to encouraging collaboration through sharing of data and construction of GIS based analysis techniques. LOIS is the most ambitious collaboration attempted in the UK between terrestrial, freshwater, marine, atmospheric and geological scientists (NERC, 1994).

Linking disciplinary studies, examination of fluxes through distinct geographic regions and the coupling of data and models has been encouraged by development of easily used interfaces to GIS. These developments are described in this chapter.

LOIS

LOIS is a six-year programme to quantify the exchange, transformation and storage of materials between the land and the deep ocean and determine how these parameters change in time and space. Scientific research within the BIOTA programme concentrates on the inter-tidal zone but must also address influences from both the land and sea.

Fig. 1 shows the relationships of the environmental zones that are of interest to LOIS. The flux of materials will have implications for the variability of ecological, geochemical and geomorphological processes that are being studied within LOIS. Spatial scales vary from the effect of tube worms on sediment stability to the topography of the Continental shelf and associated land masses. The processes that are investigated range from individual storm events of a few hours to the evolution of the coastline over the last ten thousand years

The primary objectives of LOIS are (NERC, 1994):

- **to estimate the contemporary flux of momentum and materials (sediments, nutrients, contaminants) into and out of the coastal zone, including transfers via rivers, coasts, groundwater, the atmosphere and the shelf-ocean boundary;**
- **to characterize the key physical and biochemical processes which govern coastal morphodynamics and the functioning of coastal ecosystems, with particular reference to the effects of variations in sediment supply and inputs of pollutants;**
- to describe the evolution of coastal systems from the Holocene to now in response to changes in climatic conditions, changes in relative sea-level and the impact of human activities;
- to develop coupled land-ocean models to simulate the transport, transformation and fate of materials in the coastal zone, and provide the basis for predicting hydrological, geomorphological and ecological conditions under different environmental scenarios for the next 50 - 100 years.

The main products of LOIS research are to be:

Development of a Spatial Decision Support System

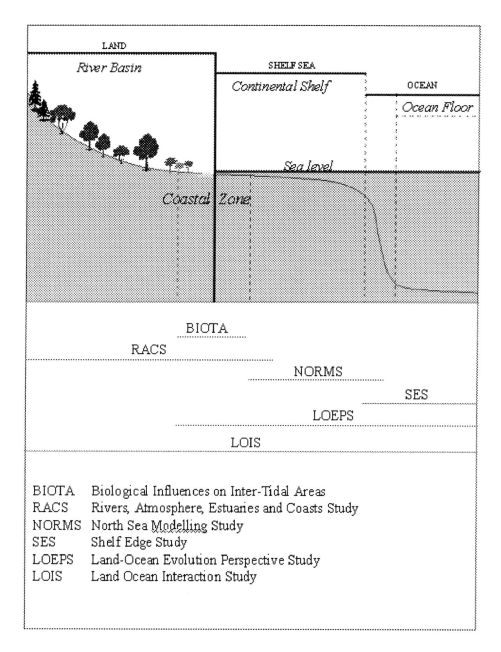

Fig. 1. The relationships of the environmental zones which are of interest to LOIS

- improved understanding of multi-disciplinary processes in catchments, estuaries and coastal oceans
- integrated models to simulate the flux, transformations and effects of materials from the land, through rivers, estuaries, the coastal zone and the atmosphere to the shelf edge
- integrated databases and models within GIS, and other systems, to make the understanding and information accessible for the purpose of coastal zone management
- improved technologies for monitoring variability and change in river and coastal environments, particularly with respect to the transport and fate of suspended sediment, nutrients and organic pollutants
- integrated databases for the LOIS study areas that will be made available on CD-ROM

BIOTA

The BIOTA programme studies the fluxes within the inter-tidal zone. It concentrates on the interactions between biotic and abiotic components of this system. The main study sites of the BIOTA programme are found on the east coast of England. Fig. 2 shows the location of these sites. The study sites cover a range of coastal environments including river estuaries, bays and exposed coasts. They include the North Norfolk coast, the Humber estuary, the Wash and Northumberland. Within the study areas are sites containing mature and developing saltmarsh and a variety of bare sediments and shellfish and invertebrate beds.

To facilitate communication and co-operation, the BIOTA data centre has developed a spatial Decision Support System (DSS) based on a hybrid distributed database and a GIS. Example analyses from this DSS are described in this chapter.

The main theme of research within BIOTA is to examine the relationships between plants and animals and the physical environment within their habitat, the intertidal area. There are a number of main research areas but here we have described a small selection of example analyses.

The Decision Support System - Design and Construction

Because of the requirements within LOIS for the integration of scientific research, a simple spatial Decision Support System (DSS) is being developed by the BIOTA data centre staff (the authors) to provide access to data, models and decision support functionality. The key reason for a DSS is that it provides assistance in conceptualising the problem, supporting intelligence, design and choice in decision making, and production of memory and control aids. In the particular case of the BIOTA programme, a DSS has been preferred to an "expert system" because, within environmental problems, there is rarely a tractable algorithmic solution, there is usually a multiplicity of objectives, there is a need to generate and evaluate alternatives and to justify choices made.

Data for the BIOTA project come from a wide variety of sources. Some of these, such as remote sensing and ground surveys carried out within LOIS, are freely available

Development of a Spatial Decision Support System

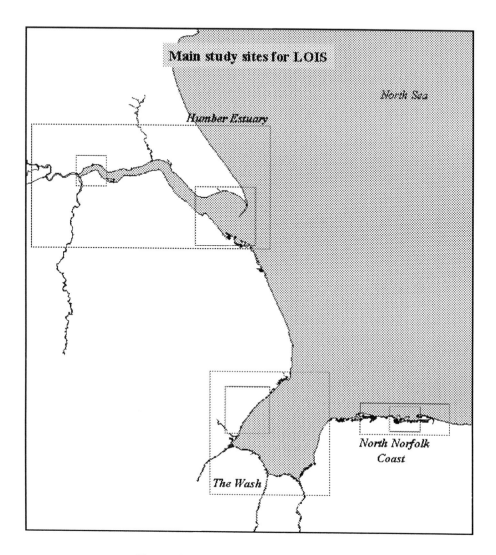

Fig. 2. The location of the LOIS study sites

while other datasets have to be leased from their owners and are available to the LOIS community.

Within LOIS, data centres act to coordinate and assist with requests for data to external bodies and the flow of data between research scientists. Emphasis during the first stage of the DSS development has been on the integration of data and models. Currently the user has access to 300 Mb of data spread across the network, and this total is likely to increase to several gigabytes as data sets are extended to cover the entire coastline of interest. There is also the potential for using the system to distribute and access data across the wide area network (WAN). Interrogating external databases (such as Oracle) across the LAN does degrade response times compared to transforming the data into some Arc specific format: however, the increase in response time is out-weighed by the benefit of having only one image of the data. The system has now reached a point where external decision makers could be productively involved.

ITE has created similar products for the analysis of various environmental subjects at a range of scales; these have included:

- **the DOE funded climate change Core model (Parr and Eatherall, 1994)**
- **the MAFF funded ITE Wetlands GIS (Brown et al., 1997)**
- **the EC funded LANDECONET demonstrator (Brown and Firbank, 1997)**

In one important respect the DSS has adopted an alternative approach to that used in these projects. These three products have used various software including ArcInfo and ArcView GIS, but in each case the majority of datasets used have been converted into the format preferred by the host software. In the DSS, data created for LOIS/BIOTA are maintained in their original form.

While the system is designed primarily to stimulate scientific cooperation and information exchange between researchers, it also provides secure and convenient access to a wide variety of data and models and will provide an aid for effective coastal management.

The conceptual model of the flow and storage of information within the system is shown in Fig. 3.

The DSS is being developed in the Arc Macro Language (AML) of the ArcInfo GIS system on UNIX workstations. The decision to implement the system in AMLs was mainly prompted by the speed and ease with which it can be developed and modified in response to the needs of its users, the BIOTA research community.

Within the DSS a series of menu interfaces have been developed, which incorporate some of the analysis methods appropriate for the areas of study. Some of these GIS facilities are standard operations such as raster overlay, whereas more complex facilities have also been implemented.

Some of the main facilities offered by the GUI are LOIS specific maps, physical data (such as profiles), biological data, national maps (such as Ordnance Survey), and GIS and modelling functions.

A central feature of the design is that data are stored in their "native" format. Within every directory a "readme" file is kept which provides a brief description (source, date, accuracy, resolution etc.) of all the data, files, programmes and sub-directories which are present. UNIX shell programmes are started at regular intervals by the "cron" facility to ensure that "readme" files are kept up to date. Users of the system only have "read" access to the data. Use of "permissible" combinations or analysis of data is controlled by the menu system. Because of this design approach the system can be distributed across different machines on the local area network (LAN).

More complex analysis includes the comparison of the estimated distribution of a plant species derived from: remotely sensed data (spectral response), quadrat samples (botanical survey) and modelled from exposure and inundation rates.

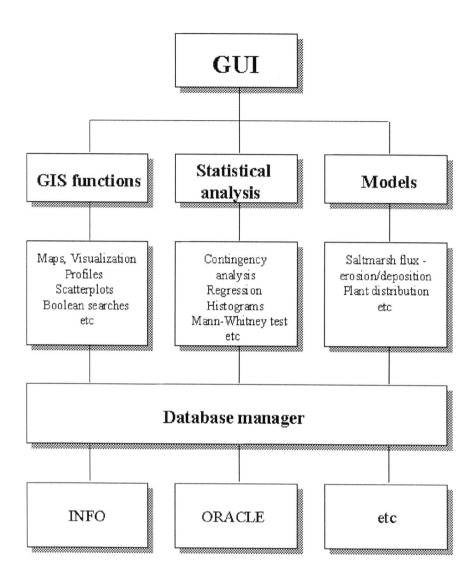

Fig. 3. Conceptual diagram of the flow of information within the system

The Decision Support System - Example Use

At its simplest a GIS can be used to visualize spatial data. This could, for example, the distribution of flocks of dunlin observed over some time period with additional information showing how these relate to the elevation of the shore (see Fig. 4). A simple quantitative exploration of this type of data can include the use of statistics (such as Mann-Whitney) to demonstrate the significance of perceived differences in preference between species.

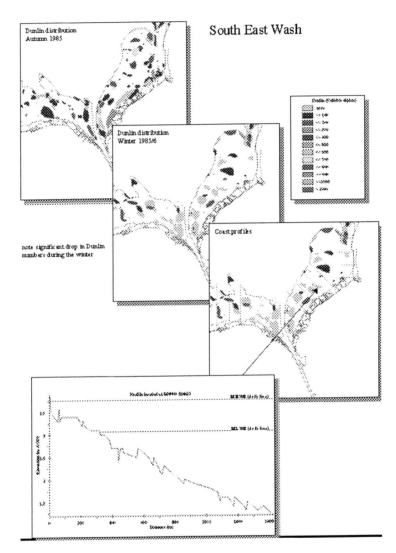

Fig. 4. The distribution of flocks of dunlin observed over some time period with additional information showing how these relate to the elevation of the shore

The DSS also provides the ability to perform Boolean searches. The Boolean search has been designed to illustrate how an area suitable for managed retreat could be found. The user specifies the parameters and values they wish to include in the analysis. These might include distance from high water mark, elevation, current land cover etc. The system searches through the available data to identify those regions that meet the user requirements. Fig. 5 shows a screen dump of the Boolean search panel and the resultant map.

One of the more complex models included in the system describes the accretion of sediments on a saltmarsh (Wadsworth and Pakeman, 1996). This hybrid model consists of AML code and an external C programme and makes use of remotely sensed data from the Compact Airborne Spectral Instrument (CASI), a digital terrain model (DTM), published parameters for predicting tide levels and field data on sediment trapping in vegetation.

CASI data have been classified into 18 types, ranging from well established mature saltmarsh, through developing and pioneering saltmarsh to algae-covered muds, bare sediments and water (Thomson et al., 1994).

The saltmarsh model outputs an accretion rate that can be used to estimate new beach profiles and generate a new DTM. As plant species exhibit different abilities to tolerate inundation and exposure, a new beach form can be used to predict the movement of different vegetation zones. The revised location of the zones will in turn affect the accretion and erosion rate. In this way the interdependent interactions between biota and the physical environment can be simulated. Preliminary results from the North Norfolk site indicate average deposition rates three to four times lower than the mean prediction for sea level rise over the next century. A more detailed description of this type analysis is found in Wadsworth and Pakeman (1997). An example is shown in Fig. 6.

Discussion

The LOIS Decision Support System has adopted a significantly different approach to data access that has proved effective during the lifetime of the project. The policy of retaining data in its native form has not presented us with insurmountable problems or lead to significant decrease in efficiency in data processing, analysis or display.

Distributed hybrid data bases are feasible across the local area network (LAN). Such data bases allow a single image of the data to be maintained and curated, simplifying data audits and quality control, while the increase in access time is still manageable. Access across the LAN shows the potential for practical applications across the Internet and the dissemination of data and models both within the research community and to a wider audience.

The complexity of commercial GIS can be adequately managed by the development of simple graphical user interface (GUI). An adequate GUI can be constructed quickly using facilities within ArcInfo. Any loss in robustness, flexibility and speed are not, for this application, significant deterrents.

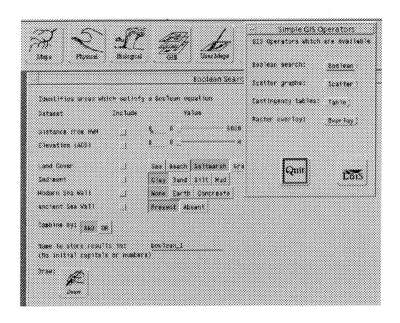

Fig. 5. A screen dump of the Boolean search panel

Conclusions

GIS systems can provide a focus and a stimulant to cooperation and collaboration across academic disciplines. The success of the system can be measured by the willingness of scientists to contribute data and models and to participate in new collaborative research.

It is possible to identify potential areas of cooperation and gaps in coverage only if the system framework is started early. An early start is more likely to lead to cohesive, comprehensive and coherent systems.

LOIS is an ambitious project that will, when completed, demonstrate and describe the processes that link the land and the ocean. While science is being advanced in each component of LOIS, a significant novelty is the explicit use of GIS in fostering integrated research and models across academic disciplines.

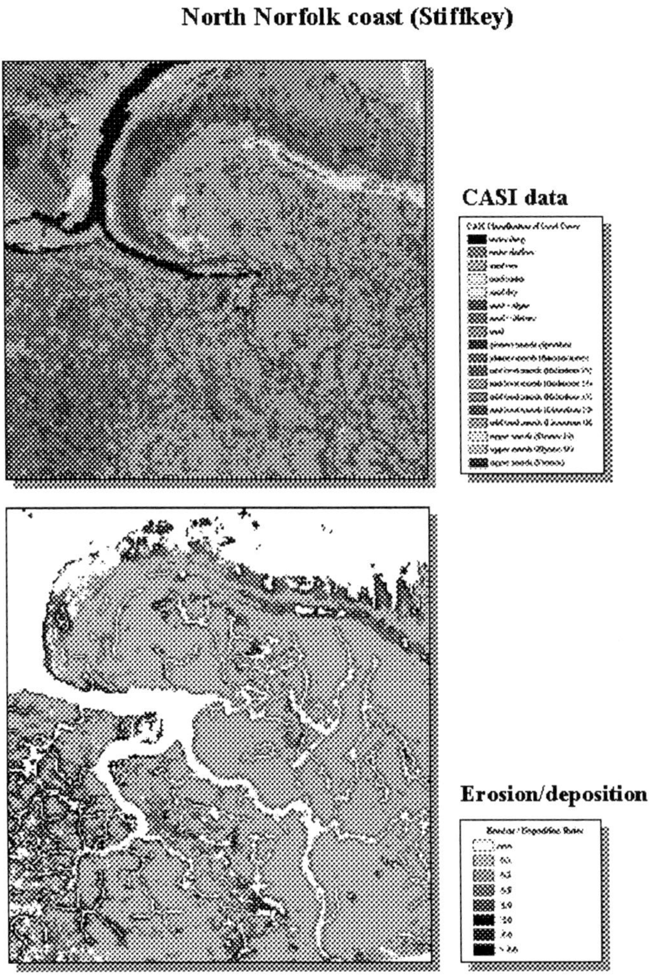

Fig. 6. An example output from the saltmarsh model.

Acknowledgments

This work was carried out as part of the Land Ocean Interaction Study funded by the Natural Environment Research Council. We wish to thank the Environment Agency for access the their transect information. We would also like to thank Laurie Boorman, Jim Eastwood, Robin Fuller, Selwyn McGrorty and Tim Sparks for help with data collection.

References

Brown N.J., Swetnam R.D., Treweek J.R., Mountford J.O., Caldow R.W.G., Manchester S.J., Stamp T.R., Gowing, D.J.G., Soloman, D.R. and Armstrong A.C. Issues in GIS development: adapting to research and policy-needs for management of wet grasslands in an Environmentally Sensitive Area. *International Jounal of Geographical Information Science.* Volume 12(5):465-478

Brown N.J, and Firbank, L.G. (1997). Farm Landscapes for Biodiversity, the LANDECONET demonstrator. A GIS demonstrator constructed in conjunction with the LANDECONET manual of farm landscapes for biodiversity. This product has brought together GIS (and other) data from the component countries of the LANDECONET group. The demonstrator supports the manual entitled Farm Landscapes for Biodiversity. A guide to using landscape ecology to assess and improve the quality of northern European farmed landscapes for biodiversity. A Manual produced by the LANDECONET group, April 1997.

NERC (Natural Environment Research Council) (1994). Land Ocean Interaction Study (LOIS) Implementation plan for a community research project. NERC, Swindon, Wiltshire UK

Parr, T., and Eatherall, A. (1994). Demonstrating Climate Change Impacts in the UK. The DOE Core Model Programme. *Final Report* to the DOE.

Thomson A.G., Eastwood J.A., and Fuller R.M. (1994). Further developments of airborne remote sensing techniques. *First Progress Report.* Environmental Information Centre, Institute of Terrestrial Ecology, Monks Wood. Cambs. PE17 2LS UK

Wadsworth, R.A., and Pakeman, R.J. (1996). Integration of a process based model of saltmarsh development with remotely sensed data in an environmental GIS. *Paper presented at ISPRS Conference: New Developments in GIS* . March 6-8, Milan, Italy.

Wadsworth, R.A., and Pakeman, R.J. (1997). Assessing the significance of changes in the relative abundance of plant species characteristic of mature salt marshes. *Proceedings of CoastGIS'97.* Second International Symposium on GIS and Computer Mapping for Coastal Zone Management. University of Aberdeen, Scotland, August 1997.

Warrick, R.A., and Oerlemans H. (1990). Sea level rise. In, Houghton J.T., Jenkins G.J. and Ephraums J.J. (Eds.) *Climate Change: The IPCC Scientific Assessment.* Cambridge University Press, UK pp. 257-282.

CHAPTER 31

User Assessment of Coastal Spatial Decision Support Systems

R. Canessa and C. P. Keller

ABSTRACT: Given the current trend in the development of Spatial Decision Support Systems (SDSS), a study was undertaken to investigate the development and implementation of Coastal SDSS for multiple-use coastal management. Imperative to this topic are the expectations and requirements of coastal managers for Coastal SDSS. This paper reports on results of such a user assessment gleaned from an international questionnaire survey of coastal managers. Assessments were made for decision making, analytical, and data components of Coastal SDSS, and implementation factors. General characteristics of coastal management relating to these components are first described. The discussion then relates each of the components to GIS and SDSS covering both strengths and weaknesses.

Introduction

Almost from their initial development in the 1960s GIS have been heralded as decision support technology (Eastman et al., 1993; Cowen, 1988), even as an "essential" technology for decision-makers (Mumby et al., 1995). Following from these accolades, GIS have evolved from inventory-based information systems to decision support tools within a broader decision-making environment. The decision support capabilities of GIS were formalised in Spatial Decision Support Systems (SDSS). SDSS are computer systems which combine a database management system, spatial and non-spatial analytical models, and graphical and tabular reporting. The components are integrated with human-expert knowledge within a decision-making framework and accessed by a user-friendly interface to provide alternative solutions to a decision problem (Densham, 1991). SDSS have been developed for various applications such as locational planning, land consolidation and environmental management. However, Fedra (1995) notes that "Success stories of actual [SDSS] use in the debate and policy-making process, such as water resource management, are somewhat more rare, in particular at the societal rather than commercial end of the spectrum of possible applications". Several explanations can be proposed.

SDSS primarily have been designed for single users usually experienced in GIS. It is now generally accepted that public planning and resource management, including coastal management, should be undertaken by multi-party task forces involving non-scientist decision makers and in some cases the general public. The decision-making process involving multiple parties tends to be less structured involving more oral exchange. In many cases decision-makers do not have access or cannot operate GIS due to inadequate experience or interface complexity. To address some of these issues, a new generation of GIS-based DSS are being developed, namely collaborative spatial decision making systems (CSDMS) aimed at converging the divergent views of decision makers towards an agreeable solution to a specific problem in an often iterative and interactive process (Densham et al., 1995).

While GIS technology has been advancing, this has not been matched by an ability to integrate non-GIS specialists and their expectations. The need to consult and involve decision-makers in the development of GIS decision support technology has been emphasised by several researchers (Green, 1995; Raal et al., 1995; Basta, 1990). As part of a study to develop Coastal SDSS (Canessa, 1997), it was thus recognised that one of the key factors would be to consult with intended Coastal SDSS users. A questionnaire survey was undertaken to answer this call and ascertain from likely users of Coastal SDSS an assessment of technical and implementation factors. This included decision making, analytical and inventory components, and the attitudes and expectations of respondents to using a GIS-based Coastal SDSS. Results of this survey formed the basis for the development, implementation and evaluation of a pilot Coastal SDSS in an actual community-based coastal management initiative (Canessa and Keller, 1997).

Questionnaire Design

A two-phased mailed questionnaire was undertaken to elicit information from potential users of Coastal SDSS. The objective of the first phase questionnaire was to gain general information on the coastal management process and the manner in which decisions and strategies are derived. In addition, respondents were also questioned about their knowledge, experience and expectations of GIS. The purpose of the second questionnaire was to follow up on some of the themes emerging from Phase I and to focus on tools and techniques used in coastal management, and to explore the implementation of coastal decision support systems. Because of the exploratory nature of the questionnaire, open-ended questions were primarily employed. Responses were evaluated using content analysis to narrow and consolidate them into categories that could be reported and summarised using descriptive statistics.

Participants were selected from the literature, conference proceedings, coastal management programs and personal contacts. The emphasis was to seek coastal

User Assessment of Coastal DSS

decision-makers rather than GIS experts *per se*. One hundred and fifty questionnaires were mailed to coastal decision-makers in April 1994. Forty-four questionnaires were returned completed (29% response rate). Phase II was administered by mail in March 1995 to 83 coastal decision makers comprising all 44 respondents from Phase I and an additional 39 subjects. Twenty-four participants returned completed questionnaires (29% response rate) of which 14 had completed Phase I.

Respondents in both phases exhibited a similar professional profile in terms of geographical range, organisational representation, responsibilities and professional disciplines. Almost half of the respondents came from British Columbia, with the remainder from eastern Canada, the United States, Europe and South Africa. Organisations represented were primarily provincial/state, and federal governments. Other organisations represented comprised regional governments, independent commissions, universities and consulting firms. Most respondents were at that time engaged in planning and regulatory work or research primarily focused on providing scientific information. Respondents worked in a wide range of fields related to coastal management with primary emphasis on ecology/environment, and comprehensive resource and land use planning. In addition, sectors such as aquaculture, conservation, tourism, navigation, geology and geomatics were represented. GIS experience of Phase I participants ranged from none (20%) to post-secondary education (14%).

The following sections report on the results of the questionnaires. They are presented according to the components of Coastal SDSS, namely decision making, analysis and data. General characteristics of coastal management relating to these components are first described followed by their relevance to GIS and SDSS including strengths and weaknesses. The last section deals with method of system implementation.

Decision Making

Prior to exploring the potential application of SDSS to coastal decision making, it was necessary to characterise the decision-making context particularly the goals of coastal management, the general process followed and the participants.

The predominant impetus for coastal management revealed by the respondents is the occurrence of environmental impact interactions. These are addressed through the establishment of protected areas, use strategies and activity siting. Coastal management initiatives described involved a multitude of sector interests dominated by commercial fisheries, and recreation and tourism. Coastal management is generally approached by generating and evaluating alternative strategies by a group of decision-makers. Within the general process, there were various tasks identified. Responses were distilled into ten tasks comprising:

1) **Define problem**
2) **Establish process**
3) **Collect data**
4) **Analyse/assess data**

5) **Synthesise results**
6) **Review draft**
7) **Revise strategy**
8) **Adopt strategy**
9) **Implement strategy**
10) **Monitor implementation**

Respondents made it clear that these tasks do not represent a prescriptive process of ordered and discrete steps for coastal management. Some of these tasks may be addressed simultaneously, backtracked, or omitted altogether depending on the issues addressed and the process established. This has implications for the development of Coastal SDSS. Unlike business applications for which DSS were originally designed, the coastal management decision process has been described as ill-defined, variable and even *ad hoc*. This can lead to two avenues. Firstly, the development of Coastal SDSS would need to focus on flexibility in order to accommodate the characteristics of coastal decision making. Secondly, the development and implementation of Coastal SDSS themselves could prompt the coastal management process to become more structured and defined.

Initiatives described in the questionnaire were dominated by government agencies primarily at the provincial/state level. Nevertheless, comments supported the growing trend towards involving community stakeholders in the decision-making process in various capacities ranging from public presentations to participatory involvement. Direct participation of community stakeholders within a decision-making group has implications for the implementation of Coastal SDSS. A large and varied decision-making group requires that a wider range of expertise needs to be accommodated, and that more participant input needs to be 'funneled' through a Coastal SDSS.

Respondents felt that GIS could play a role in supporting decisions about multiple use on the coast. The most common roles identified by Phase I respondents for GIS in coastal management decision making were for the storage of an integrated inventory, and for communication and display (Fig. 1). In the former role, it was advocated that GIS accommodate the compilation into a common database of the wide range of data needs typical of multi-disciplinary initiatives such as coastal management. In addition, several respondents noted that GIS facilitate data sharing, accessibility, standardisation and co-ordination. The emphasis on communication and display was placed on the production of easily understood and customised output particularly for a wide ranging audience from "professional scientists to environmental planners and politicians to various public interest groups" (respondent's quote). Other roles identified included 'what-if' exploration and evaluation of alternative planning scenarios, and spatial identification of conflicting or competing interests. Of those who identified roles (32), 28% expected GIS to improve the efficiency of decision making through flexibility and speed in providing analysis and results to the extent that GIS allow for a "dynamic decision-making process". In addition, several respondents noted that GIS lend a greater degree of objectivity and accuracy to the decision-making process. Decisions become based on data widely "perceived as objective" and precise. For example, " 'It's too close' becomes '2.7 miles away' " (respondents quote).

Despite these expectations of GIS, several respondents cautioned against the role of GIS in coastal decision making with such comments as:

- "GIS will be *only* one tool"
- "Many factors that go into the final decision are non-spatial"
- "GIS could be used to *influence* decision making, but the decision would still be political"

and

- "GIS provides the background information but the actual decision making about multiple use is still made by the agencies after technical overview and public discussion".

This last comment illustrates one perceived role of GIS as inventory systems rather than technical or analytical systems. On the other hand, GIS were also perceived to contribute to technologically or quantitatively driven processes to the exclusion of qualitative and human dimensions; "The 'data right' answer is not always the 'people right' answer" (respondent's quote). Additional impediments identified to using GIS were the resources required in terms of cost and time. In addition, GIS were seen to reside in the domain of technical experts, which distances the managers and decision makers from its use. GIS were also noted as being a seductive tool with unrealistic expectations as a panacea for coastal management, and encouraging a false sense of security in its users, leading to misguided use. The lack of data in terms of accessibility and quality was also found to limit its use.

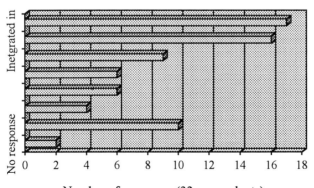

Fig. 1. Role of GIS in multiple-use coastal decision making (Phase I)

Analysis

Despite the proliferation of coastal management initiatives, there are surprisingly few examples published on specific methods and techniques used to develop the strategies that are produced (see for example Clark (1990)). Whereas the general decision-making process may be described in reports and papers, methodological details are passed over with phrases such as, "what eventually emerged from the year-long process was a zoning plan" (respondent's quote) or "a plan was delivered by a consultant" (respondent's quote). Therefore, in Phase II of the questionnaire specific information on analytical techniques for Coastal SDSS was directly sought from decision-makers. Based on the general decision-making approach that emerged from Phase I, emphasis was placed on techniques to generate and evaluate alternatives.

Despite the analytical emphasis of the question, there were few responses which specifically referred to methods of generating alternatives as opposed to information or process elements that would contribute to an understanding of alternatives. Although the responses were characterised by great variability and in many cases generality, the most common groups of techniques to generate alternatives comprised ecological and environmental assessment such as environmental impact assessment, carrying capacity assessment, and principles of ecology. Almost as many responses were received on spatial analysis with particular emphasis on mapping techniques and GIS, however no details were given on specific methodologies.

Recipients were also asked to identify methodologies, techniques or procedures for evaluating alternatives. As with the responses for generating alternatives, these responses were diverse and in many cases very general such as "compare scenarios", or use "evaluative criteria". However, a dominant method identified for evaluating alternatives was impact assessment, namely such as economic, environmental and/or social. Traditional cost-benefit analysis was suggested in terms of natural resources, environmental quality and economic viability. Ranking and weighting methodologies were suggested for specific criteria such as economic, social impacts, regional strategies, jurisdiction, political will and research data as well as overall ranking of alternatives. Group processes also mentioned include negotiation, consensus building, round table procedures and simulations.

The general responses to these questions did not provide the specificity desired in order to incorporate appropriate analytical techniques. Further direction was provided when questioned on the analytical strengths of GIS to contribute to coastal management. The most common analytical strength identified was their general efficiency and flexibility in manipulating and analysing data particularly towards exploring data, and testing new ides and plans. The overlay procedure was the most often-mentioned GIS analytical procedure. Overlaying several themes could allow users to compare maps thereby reveal relationships, cross-correlations and spatial conflicts. The overlay could be used to identify target areas such as sensitive areas, priority areas, suitability areas and 'no-go' areas based on specified criteria, and to explore 'what-if' scenarios with pre-defined parameters and interactively. The general expectation of GIS to develop models such as predictive models and to perform complex analysis was recognised, although no specific applications were suggested.

By far the most often-mentioned weakness of GIS with respect to manipulating and analysing coastal data by Phase I questionnaire respondents was the expertise required and difficulty in learning software. GIS analysis also was seen as having the potential to mislead decision-makers either unintentionally through seductive "smoke and mirrors", or intentionally through the bias of those controlling the analysis and data manipulation. Several respondents noted the inadequacy of GIS to easily support advanced modelling, particularly with respect to GIS coastal models. The derivation of GIS coastal models was noted to be complicated by the complexity and dynamic nature of coastal data. Again, GIS were regarded as promoting spatial technology as a "technical fix", while neglecting "subjective" or "fuzzy logic" analyses of multivariate problems and data that cannot be easily represented spatially, such as spiritual and cultural values.

Data

The importance of data for coastal management was emphasised by a third of Phase I respondents who identified data availability as a critical success factor for coastal management decision making. This was second only to co-operation and commitment to the process. Data availability factors included:

- **sharing information among participants**
- **all participants having the same information**
- **obtaining accurate data**
- **obtaining up-to-date data**

and

- **providing technical and objective information to substantiate discussions, rather than basing on opinion**

Furthermore, as described earlier, in Phase II participants were asked to describe techniques and methods for exploring alternatives. The majority of responses focused on information support topics rather than analytical topics. Information support topics included:

- **the compilation of a comprehensive database which covers the broad range of natural and human elements**
- **data acquisition techniques such as identifying information needs, identifying information sources, data availability and sampling techniques**

and

- **data management including the evaluation of database accuracy**

The importance of data was re-inforced by twenty-five percent of respondents who identified effective communication of information and results as a critical success factor for coastal management. 'Effective' was characterised as "understandable", "easy to read" and "consistent". In most cases maps were identified as the most effective and efficient format by which to convey information and results, such as defining areas of conflict, delineating proposed alternatives and presenting information to the general public.

As was discussed earlier, the primary role for GIS in coastal management that was most often mentioned in Phase I was inventory. Inventory functions of GIS mentioned included storing and displaying an integrated inventory, and to a lesser degree facilitating information sharing. The role of GIS in these capacities was raised repeatedly during other parts of the questionnaire. For example, inventory characteristics of GIS were identified as strengths of GIS related to analysis and decision making.

The main strength of GIS with respect to managing and maintaining a coastal inventory identified by Phase I respondents is the ability to integrate a variety of data sets into common information base (Fig. 2). The geographical context was seen to provide a common denominator for disparate data sets. A single source and multi-layered organisation was regarded as important to facilitate the accessibility and efficiency of storing and retrieving large volumes of data.

It has already been established that the compilation of a comprehensive database which accounts for the broad range of resources and users in the coastal environment is an important requirement for coastal management. The contents and taxonomy of such an inventory are well covered in the literature particularly with respect to biological and physical resources (see for example, Harper et al., 1993; Hale, 1991; Ricketts, 1991). Of interest to note is the importance placed on law, institution and policy by questionnaire respondents. This supports the commonly held view that the governance framework comprising numerous agencies and dispersed legislation presents a major impediment to integrated coastal management. Nevertheless, these elements are rarely incorporated in a coastal management inventory.

By far the most often-mentioned weakness of GIS with respect to managing and maintaining an inventory of coastal data was the cost involved in acquiring and inputting the data. As with weaknesses of its analytical use, the expertise and training required also applied to its inventory use. On the one hand, some respondents noted that data stored in GIS were easier to update than data stored on paper maps. On the other hand, it was also noted that the requirement to continually update the data was a weakness of GIS in terms of the time, cost and effort involved in maintaining a current database. In addition, incompatibility among GIS systems, data formats and data exchange procedures hindered the acquisition of an inventory.

Implementation

The preceding three sections give directions for technical specifications of Coastal SDSS with respect to decision making, analytical and data components. Equally, if not more important, to technical specifications is the means by which Coastal SDSS are incorporated in the process. For example, it was recognised that Coastal SDSS could impact the decision making 'playing field'. It was thought that Coastal SDSS could be used to exert influence on the process by directing its use according to the goals of those most familiar with the technology. Considerations of system implementation were pursued in Phase II by presenting three options of system incorporation based on formats supported in Group Decision Support Systems (GDSS) (Nunamaker et al., 1991).

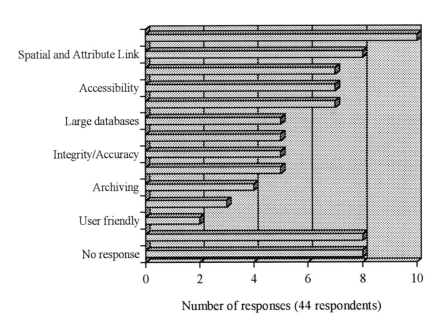

Fig. 2. Strengths of GIS for coastal management inventory (Phase I)

Interactive Used directly and interactively by the decision-makers at the decision-making table for display, query, and analysis and as a tool to seek consensus.

Chauffeured Used at the decision-making table by technical staff for display, query, and analysis and as a tool to seek consensus upon the direction of decision-makers.

Background Used as a background tool by decision-makers and/or technical staff, the results of which are brought to the decision-making table.

As an *Interactive* system, each decision-maker has access to a terminal for display and input which is networked to a central processing unit. One could argue whether the *Background* system is a true GDSS since the decision makers themselves do not interact or even see the system, but rather see the results brought to the table.

Respondents were asked to choose the scenario they would prefer to be involved in as decision-makers noting advantages and suggestions for the option chosen. The preferred option of Phase II respondents was *Background* in which there is least involvement by decision-makers with the Coastal SDSS (Fig. 3). This format was preferred because information would be provided without clutter and seduction of GIS at the table. It would place GIS as a *support* tool and not as a decision-*making* system and was seen to be a more time efficient and effective use of the technology than the other two options given the prospective technical expertise of decision makers. However, a drawback of this approach is that "technical committees can only be populated by those participants for whom the technology is accessible, such as industry, thereby putting other groups at a disadvantage" (respondent's quote).

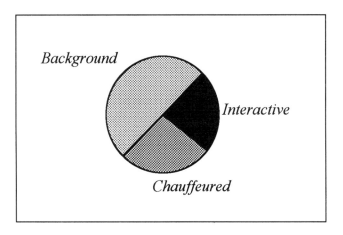

Fig. 3. Method of Coastal SDSS incorporation (Phase II)

Twenty-seven percent of respondents preferred a *Chauffeured*-style implementation. They felt that with the availability of a system operator it would be more efficient than the *Interactive* format and the technical staff would be on hand for explanation of constraints and limitations. It would also encourage collaboration between decision-makers and technicians, and provide decision-makers with a better understanding of the data than would either of the other two options. There would be less tendency of seduction, the output would be readily available for decision-makers and it would encourage involvement for group participation. According to the respondents, implementing this option would require easily understood output or screen display, and application specific training.

An *Interactive* implementation, which represents the most direct and dynamic involvement of decision-makers, was preferred by 23% of respondents. It was noted that this option would provide a transparent box for decision-makers, and the flexibility and immediacy of display. This option was determined to be the most inclusive and participatory of the implementation scenarios, and would provide equal access and a level playing field. On the other hand, one respondent noted that an interactive system could in fact create an unbalanced playing field giving an advantage to those 'techno-philic' decision-makers and a disadvantage to the 'techno-phobic'. Suggested requirements for implementing an interactive system include training at various levels, the flexibility and methodology to update, and a mediator to manage the flow of information.

Conclusion

The results of the user assessment questionnaire show that potential users of Coastal SDSS have both a guarded and optimistic view of their use. The perceived expertise required to operate GIS, their potential seductive nature and their technocratic emphasis led some respondents to be cautious about using a GIS-based decision support system in coastal management. The expertise required specifically represented a barrier to the implementation of Coastal SDSS directly by decision-makers. The barrier was exemplified by respondents' overwhelming choice of Coastal SDSS format, namely as a background tool handled by technical staff away from the decision-making table. The cautions can be summarised by the view that GIS should be considered one decision *support tool* among many, rather than an exclusive decision *making system*.

The over-riding strength of GIS identified, and the one most familiar to decision-makers, was the ability to efficiently and flexibly stored, display and query information. It is clear from the findings that inventory should be an important function of Coastal SDSS both in terms of the manner in which data are maintained and, more importantly, the manner of communication. In fulfilling this role, the ease with which GIS output, whether hardcopy or softcopy (screen), could overwhelm decision makers to the detriment of its communication goal should be considered. With respect to inventory content, the findings indicate that human elements of coastal management such as public perception, local knowledge, and the administrative framework of coastal management, somehow should be included in the inventory. Finally, the potential to use Coastal SDSS for generating and evaluating alternatives both analytically and interactively through the exploration of 'what-if' scenarios especially by taking advantage of overlay capabilities was clearly and repeatedly noted.

In conclusion, technical requirements of developing Coastal SDSS are only half the challenge towards getting a system to the decision-making table and used effectively by decision makers. More important is the method of incorporating the system within the decision-making process including the commitment and knowledge of an effective facilitator to direct the system's use, the education of decision makers on the functionality of the system, the encouragement of realistic expectations, and an awareness of short comings.

References

Basta, D. (1990). Information technology for multiple-use decision making. In S.D. Halsey and R.E. Abel (Eds.), *Coastal Ocean Space Utilization*. Elsevier Science Publishing Co., Inc, New York. 309-313.

Canessa, R. (1997). Towards a Coastal Spatial Decision Support System for Multiple-use Coastal Management. *Doctoral dissertation*. Department of Geography, University of Victoria, British Columbia, Canada.

Canessa, R. and Keller, C.P. (1997). Implementing GIS as a decision support tool for community-driven multiple-use coastal management. *Pacific Coasts and Ports Conference: Proceedings*. September, 1997. Christchurch, New Zealand.

Clark, M. (1990). North Carolina's Estuaries: A Pilot Study for Managing Multiple Use in the State's Public Trust Waters. Albermarle-Pamlico *Study Report* 90-10. UNC Sea Grant College Program, Raleigh, North Carolina.

Cowen, D.J. (1988). GIS versus CAD versus DBMS: What are the differences? *Photogrammetric Engineering and Remote Sensing* 54(11):1551-1555.

Densham, P.J. (1991). Spatial decision support systems. In D.J. Maguire, M.F. Goodchild and D.W. Rhind (eds.), *Geographical Information Systems: Principles and Applications*. Longman Scientific, London. Vol. 1, pp. 403-412.

Densham, P.J., Armstrong, M.P., and Kemp, K.K. (1995). Collaborative Spatial Decision Making - *Scientific Report* for the Initiative 17 Specialist Meeting. NCGIA Technical Report 95-14. NCGIA, Santa Barbara, California.

Eastman, J.R., Kyem, P.A.K., Toledano, J., and Jin, W. (1993). Explorations in Geographic Information System Technology Vol. 4 *GIS and Decision Making*. UNITAR, Geneva.

Fedra, K. (1995). Decision support for natural resources management: models, GIS and expert systems. *AI Application* 9(3):3-19.

Green, D.R. (1995). User access to information: a priority for estuary information systems. In R.A. Furness (ed.), *CoastGIS'95: Proceedings of the International Symposium on GIS and Computer Mapping for Coastal Zone Management*. International Geographical Union Commission on Coastal Systems, Sydney. pp. 35-59.

Hale, P. (1991). *Information Requirements for Marine Resource Management*. Prepared for the Sub-committee on User Needs and Applications, Inter-agency Committee on Geomatics, Mineral Policy Sector, Department of Energy, Mines and Resources, Ottawa, Canada.

Harper, J.R., Peters, S., Booth, J., Dickins, D.F., and Morris, M. (1993). Coastal Information Resource Inventory - Draft Report. *Report* prepared for the

Aquaculture and Commercial Fisheries Branch, British Columbia Ministry of Agriculture, Fisheries and Food, Victoria.

Mumby, P.J., Raines, P.S., Gray, D., and Gibson, J.P. (1995). Geographic information systems: A tool for integrated coastal zone management in Belize. *Coastal Management* 23:111-121.

Nunamaker, J.F., Dennis, A.R., Valacich, J.S., Vogel, D.R., and George, J.F. (1991). Electronic meeting systems to support group work. *Communications of the ACM* 34(7):40-61.

Raal, P.A., Burns, M.E.R., and Davids, H. (1995). Beyond GIS: Decision support for coastal development, a South African example. In R.A. Furness (ed.), *CoastGIS'95: Proceedings of the International Symposium on GIS and Computer Mapping for Coastal Zone Management*. International Geographical Union Commission on Coastal Systems, Sydney. pp. 273-282.

Ricketts, P.J. (1991). Managing information for regional coastal management in the Gulf of Maine. In O.T. Magoon (ed.), *Coastal Zone '91: Proceedings of the Seventh Symposium on Coastal and Ocean Management*. American Society of Civil Engineers, Long Beach, California. pp. 1946-1957.

CHAPTER 32

Internet-Based Information Systems: The Forth Estuary Forum (FEF) System

D.R. Green and S.D. King

ABSTRACT: In recent years much has been written about the use of the WWW (World Wide Web) or the Internet as a global information resource. Initially a primarily static textual and graphical information resource, developments in both the hardware and software technology have facilitated the delivery of information in a multimedia environment, including the use of animation, video and, most recently, interactive spatial information in the form of maps. In 1996, a Forth Estuary Forum Pilot Information System was developed. Provision of a simple navigational interface offers access to a wide range of information presented in the form of text and graphics. With the recent developments in HTML, Javascript and JAVA applets, considerable potential now exists to extend this type of information system into a low-level Geographical Information System (GIS). In 1997 through 1998, the FEF Info System went 'live'. Further developments, whilst the system is being updated, expanded, and enhanced, will ultimately aim to offer users a more dynamic form of information delivery, using the latest developments in Internet map servers. This chapter considers the use of the Internet and the accompanying information browsers as the basis for constructing an environmental information system using the FEF project at the University of Aberdeen as an example.

Introduction

The advantages of networking information are not new and they have been widely discussed in the context of coastal zone and marine management (see e.g. Amber, 1993). With developments in networking and communications technology, together with the recent developments in Internet technology both hardware and software, the benefits of traditional networking are now possible using computers.

This chapter discusses the potential of developing computer-based spatial information systems in the context of coastal zone management. More specifically the paper examines the concept of networking in the framework of a pilot information system, the Forth Estuary Forum Information System. Initially developed as a pilot information system, the FEF pilot provides an on-line information resource for the wide range of users of the Forth Estuary in Scotland. Initially planned as a means to provide an electronic resource for commerce, government and education, the FEF in conjunction with Edinburgh City Council is now developing the pilot into a fully functional geospatial information system that will make a wide range of geographical information (largely in the form of maps) available. The system will take advantage of the most recent developments in Internet software that now permit WWW site developers with map server tools, e.g. Autodesk, GRASS, ESRI, and GeoConcept, to deliver maps interactively on the fly to the user.

Access to Information

Without a detailed knowledge and understanding of our marine and coastal environment it is very difficult to make full use of the resources available to us; but this requires the availability of both data and information.

Information, much of it spatial, is essential in the workplace as the basis for planning and decision-making and access to it is vital. Together Geographical Information Systems (GIS) and the related technologies of remote sensing, Global Positioning Systems (GPS), digital mapping, and cartography offer one means by which it is possible to improve our capability in this area; and information systems the means to deliver such information to the appropriate people responsible for decision-making and planning.

Growing interest has been shown in the application of GIS to coastal environments. Of the many aspects of the marine environment where GIS is thought to be of potential benefit, estuarine environments (e.g. estuaries and firths) are of particular interest. During the last few years, in the UK especially, research has been undertaken by English Nature (EN) and Scottish Natural Heritage (SNH), and a variety of other organisations (e.g. English Nature, 1993a, b, c; SNH's Firths Initiative).

But, whilst GIS offers a new, innovative and potentially useful way to help manage our marine and coastal environmental resources, the complexity and sophistication of much of the commercially available GIS software really only makes it suitable for the GIS specialist involved in data analysis and research projects.

Information delivery, e.g. to Marine Resource Managers, however, is often more important in the workplace. In most cases to date, the information delivery mechanisms using GIS have not been entirely suitable for the increasingly broad user base who need access to spatial information in their work environment. A number of developments to provide 'information systems interfaces' to existing Geographical Information Systems, e.g. ArcView (ArcInfo), SmallworldView (Smallworld) and

Internet-Based Information Systems 453

others have gone some way toward overcoming the need for practical and easy information delivery by offering a much simpler user-interface for the individual who wishes solely to use the information stored in the GIS, but who does not necessarily want or need to be familiar with a GIS to the same extent and level that a specialist would need to be.

Such software developments have provided the decision-maker with a simple desktop tool whereby they can retrieve and display information, and subsequently incorporate it into other documents, e.g. a planning report.

Decisions require information, and in the modern workplace information must be readily available to those responsible for supplying information to the decision-makers, or to the decision-makers directly. Computers have revolutionised our workplace and have had a major impact upon work practices. With the rapid developments in desktop computers, networking and communications technology access to information has become increasingly easy. Desktop GIS running on PCs have been developed and as demand has grown more and more of these systems have focused on emphasising ease of use, not so much as GIS, but as information retrieval and display systems, which in some cases have been designed to deliver information in a decision-making environment.

Recent developments in Information Technology (IT) (hardware, processors, software), Human Computer Interfaces (HCIs), and multimedia have all enabled greatly improved access to information stored on desktop computers, and have enabled access for a much wider range of users than ever before. Whilst most of the above examples offer single users access to information provided on a disk or CD-ROM, the LCGISN (Louisiana Coastal Geographical Information System Network), however, has shown the benefit of a networked approach for access to information.

Networking, Communications and Information Technology

The concept and benefit of network provision for access to marine and coastal data and information is not new (see for example, Amber (1993)). It is widely recognised as an ideal way to provide for data exchange, and can operate at any level or scale, whether it is local, national or global. With the development of the appropriate Information Technology (IT), the potential to develop digital equivalents of the networking concept has rapidly become a reality.

With recent developments in Information Technology (IT), both software and hardware, it is now possible with the aid of a Client-Server system (Fig. 1) in conjunction with Internet Browsers, e.g. Netscape, to deliver multimedia information over a global network. Although the technology is not that old, the adoption of MS Windows (3.1, 95, 98, 2000 and NT) for example, in conjunction with Internet browsers (Netscape and Internet Explorer) offers the potential to deliver information in a simple yet effective manner. The advantage of a Windows environment is that it provides a user with a simple, intuitive interface, which together with the Internet Browser, offers tools designed to simplify the search for, retrieval, display, capture and output of a information - all the basics needed. Such information can easily be 'cut and pasted' into other software, e.g. data into a spreadsheet, text into a word processor, graphics into image processing software.

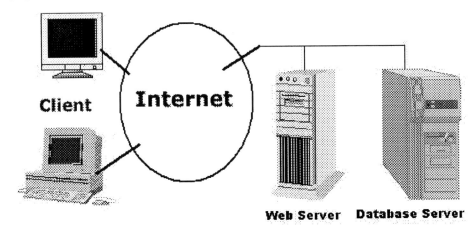

Fig.1. Client-Server Architecture

In effect the Internet, the accompanying tools, and also the browsers provide an ideal environment in which to develop firstly a simple information system, and secondly an information system that can provide an on-line decision support system, in this case for coastal zone management and marine environments.

Decision Support Systems

Over the last five or so years much emphasis has been placed on the development and use of GIS as Decision-Support Systems (DSS). Very simply Decision Support Systems represent a GIS with (a) a user friendly interface, and (b) customisation of a GIS for a particular application e.g. estuary management.

In the past many criticisms have been levelled at GIS for being too specialised, too complicated, and too difficult to use by the average person in the workplace; this is despite recognition that GIS is an invaluable tool in the workplace.

To counter such criticisms, but more specifically to make GI more usable in the workplace, a variety of different applications have focused on the development of more user-friendly systems, using more user-friendly interfaces, and customised for specific applications. Emphasis has been placed on developing interfaces in such a way that they either minimise the difficulty of using a system, limit the functionality of the GIS, or simply provide point and click environments that offer the user simple enquiry, information retrieval and display systems.

Green (1994, 1995), and Raal et al. (1995), amongst many others, have recognised the need to provide users with even more customised desktop applications which tailor the information system interface to the day to day or operational needs of specific user groups e.g. the Marine Resource Manager. These systems offer the user a hypercard/hypertext type of interface to information (e.g. Gardener and Paul, 1993).

The COMPAS and Louisiana Coastal Geographical Information System (LCGISN) systems are excellent examples of such operational systems. Another good example of a computer-based marine and coastal information system is the UK Digital Marine Atlas (BODC, 1992).

With the development of the Internet Hypertext Markup Language (HTML) it has now become possible to develop Internet-based systems that offer the developer a number of distinct advantages over previous software and systems.

To begin with the widespread adoption of the Windows-based operating systems has provided a common work environment, under which most software now runs. Netscape/Netscape Navigator (or other browsers) is one such piece of software. Running under a common operating system environment means that the user soon becomes familiar with most software, finds it easy to use and to navigate, and whether a casual or frequent user is able to make use of the software without continual recourse to manuals or tutorials. Whilst there are some differences in the software for, e.g. MS-Windows, most software operates in the same way and uses the same Human-Computer Interface (HCI).

Netscape likewise conforms to the MS-Windows metaphor, but in addition offers a secondary interface which again, like other software, offers the user a specific set of tools or functionality; in this case aimed at information query, retrieval, display, capture and output.

The Internet

In recent years there has been very rapid growth in the use of communications technology and particularly the Internet and its associated browsers e.g. Netscape. As noted by Manger (1995, p.1) it is the World's largest computer network and information resource and it is predicted by Akass (1994) that some 20 million people will have access to this information resource in the near future.

The WWW, commonly referred to as W3 or the Web, is a distributed multimedia hypertext system (Kelly, 1993). It was developed using a 'Client-Server' architecture, where a client sends a request to a World-Wide Server, which typically runs on a powerful computer system. Files retrieved will be displayed on the local machine, or are spawned to an external viewer (Kelly, 1993). Access can be open or restricted. Many applications currently exist, ranging from use as an information resource, for publicity, virtual libraries, commercial uses, Government information, to use for on-line teaching and training.

WWW documents are written in HTML (Hypertext Markup Language) using a range of HTML 'authoring' tools. Both text and graphics can be included, as well as animated files. Forms can be created, and can include email addresses. CGI scripts offer possibilities to add, e.g. search tools, and questionnaires.

Information 'posted' to the Internet is accessed through one of the many Internet Browsers, standardisation of which now appears to be settling on either Netscape e.g. Netscape Navigator Gold 3.0, or Microsoft Internet Explorer. A benefit of this software is that firstly it runs under the standard MS-Windows interface, and secondly it provides a standard secondary interface in itself offering users a customised set of tools to access, retrieve, display, and output information. Within the Netscape or other browser interface, individual Internet pages offer the user a third level of interface using point and click buttons/text which is simple, intuitive, and ideally suited to the casual or infrequent system user.

Internet and GIS

The Internet, or the World Wide Web (WWW) has also been widely discussed in the context of a Geographical Information System (e.g. Murnion and Munroe, 1994a; 1994b). New hardware, software, and communications technologies have facilitated the rapid transfer and movement of all kinds of data and information including large map and image data files, e.g. satellite imagery.

There are a number of ways the Internet can be used in the context of GIS (Table 1).

TABLE 1
Some of the ways in which the Internet can be used in the context of GIS

Online educational and training resources
Access to information and spatial information (text, graphics, video)
Access to remote databases (text, numerical, image)
Interactive GIS

With developments in HTML and JAVA, even greater opportunities for information delivery over the Internet are now becoming possible. See, for example, the WWW sites of Autodesk (Mapguide), ESRI (Map Objects) and GRASS, demonstrating their interactive map-based technology for the Internet.

Beyond this, however, the HTML language, CGI (Common Gateway Interface) scripts, and VRML (Virtual Reality Modelling Language) offer an Internet-based system developer a set of construction tools with which it is possible to design, customise in the context of data and information delivery, a 'tertiary interface' in the form of an information system. Other developments include the products of Macromedia e.g. Authorware, Flash, Director, and Dreamweaver.

In a similar fashion to tool boxes such as Hypercard/Supercard, and Toolbook, HTML offers a developer all the tools with which to develop an information system and to link this information in a flexible and yet structured way - one which also offers full functionality associated with information delivery and navigation. The advantage of being able to provide access to information in such a way is clearly more advantageous than e.g. information systems delivered on disk, or CD-ROM.

The FEF Pilot Information System

In 1993, the Forth Estuary Forum produced a folder containing detailed information about the users of the Firth of Forth in the File on Forth. Although a very useful reference document, the File on Forth had a number of distinct disadvantages. The most obvious of these was the paper format, which by the time it reached the Forth Estuary Forum user base, was already becoming out of date.

Acquiring new information and reprinting a paper-based product can be quite costly. Updates are time consuming and the format means that such updates could only realistically be produced every so often rather than at the times when they are most needed and would be most useful. Another disadvantage is that such information may not necessarily reach the widest community who would be interested in the Firth of Forth.

A logical development of this initiative was to investigate the production of a computer-based equivalent to the File on Forth. A number of options were open. One was to utilise a standard software package such as a database. However, although this offers users a digital equivalent of the paper-based information that can be queried for selective retrieval this does not really overcome the need to reach as wide a range of users as possible. Furthermore, it does not make use of the full range of computer-based information delivery possibilities that now exist, e.g. delivering multimedia information in the form of text, images, graphics, and video; nor does it permit the use of, e.g. animation in the presentation of information. The Internet and the WWW browsers, however, offer an ideal solution providing users with a simple interface, tools, multimedia presentation, and also the benefits of networking, thereby maximising the delivery of information to as wide a user-base as possible.

Some Basic Considerations

STRUCTURE

For an information system to 'work' on the WWW there is a need to provide a clear structure for the material that will be posted. One of the current drawbacks of the Internet is the difficulty of locating information.

The initial structure defined by the FEF is shown in Fig. 2. This provides a starting point for the separation or break down of the proposed information into HTML WWW pages.

SYSTEM DESIGN

Within the Windows framework, there is an obvious need to ensure that the information system, as a working, customised system, is easy to use for a wide range of different users from different backgrounds; 'usability' being the appropriate term to use in this context. This is above and beyond the need to have a familiar and intuitive

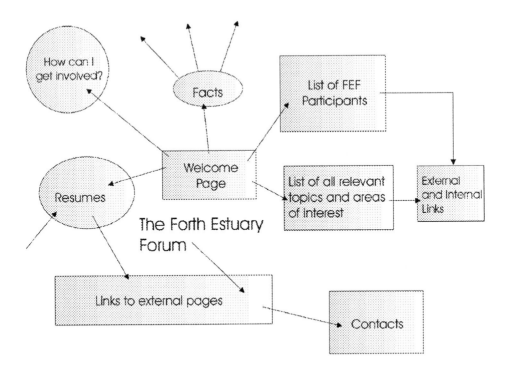

Fig. 2. The Forth Estuary Forum Pilot Information System structure

environment such as provided firstly by MS-Windows and secondly by Netscape or any other browser selected for access to the Internet, the aim being to:

❑	provide a familiar working environment	❑	maximise system navigation
❑	maintain continuity between screens (or pages)	❑	retain simplicity
❑	maximise flexibility	❑	to provide structure
❑	to make it aesthetically attractive	❑	logical
❑	consistent layout	❑	minimum amounts of information either graphics or text so that (a) it can be read easily and (b) the page does not take a disproportionate amount of time to load.
❑	Background	❑	fonts
❑	colour		

There is a need to adhere to

- **Basic Principles of Style**
- **Consistency**
- **Regulations and Legality**

In this sense, the design of not only the 'home page', which is the starting point for the navigational journey, and the focal point for the user, but also the additional pages must be designed in such a way that they fulfil these criteria.

Adherence to such criteria will undoubtedly lead to a much more interesting page for the system user, one that is pleasant to look at, easy to understand, has continuity between pages, is easy on the eye, is selective in what it displays and conveys to the eye, and does not take too long to display; which may introduce a feeling of impatience and therefore limit the desire of a person to access the page, which in the case of an information system would be a severe disadvantage and would therefore defeat the purpose.

The benefit of the HTML Language is that it allows the system designer (the WebMaster) to create a customised structure and interface, not only in terms of the screen layout, but also the amount of information displayed at any one time, the highlights, the balance between text and graphics, as well as the degree of interaction the user will have with the system. The system designer can easily customise the interface to suit the application and also to meet the requirements of the potential user base.

NAVIGATIONAL TOOLS AND THE INTERFACE

HTML permits the user to establish links between additional pages containing information, other Home Pages, email, hotwords and so on, in a semi-structured fashion; that is it has been structured by the designer but there is freedom for the user to move between information pages at random.

Access to information necessitates the provision of a suitable HCI (Human Computer Interface). The design of the HCI (see e.g. Green and Rix (1995)) is very important because it facilitates access to a system and ultimately the information contained in it. As an information system, the tasks people will wish to perform might be e.g. browsing, retrieval, display, downloading, and output. Interface design in terms of e.g. screen layout, provision of key tools, colouring, menu structure, syntax, and terminology, and error messages are all important considerations to help the user. It is essential to make the work environment as pleasing and as simple to use as possible, and to help overcome the potential to bore the user, to distract from the task being carried out, and so on.

One of the major benefits of the Internet Browsers e.g. Microsoft Internet Explorer and Netscape is that they run under MS-Windows, for example, and provide a very user-friendly, task-specific interface with a range of simple, yet powerful, tools. These tools allow the user to view, display, retrieve, send and capture information over the Internet. Not only do they allow for simple navigation of the Internet, but they also allow for on-screen customisation e.g. colours and fonts, and can be used easily by nearly anyone with little background in computing or information systems.

Relatively standard menus, terminology, screen layout, colours, and functions provide users with a basic shell that is well suited to the casual or the frequent user. Adherence to the Windows style by software developers means that most software appears familiar, in the most general sense, to the user. Internet browsers e.g. Netscape, Mosaic, Internet Explorer, developed to run under Windows retain the 'feel' of any MS-Windows software, but in addition provide a 'customised' and very specific interface which is application-oriented; that is the interface is designed to offer the user a set of tools specifically for the task of information retrieval. In this sense the browser provides an additional interface for the user that is task specific. The interface is simple, easy to grasp, and easy to use. Individuals can of course customise the interface with the aid of the tools provided.

But, beyond this, there is also another level of interface, that of the actual Internet pages. The appearance and control of the interface - the information delivery interface - is controlled by the HTML, Javascript, or JAVA applets which make up the Internet pages. Within the Internet browsers, navigational tools are provided via HTML (Hypertext Markup Language) links established between the WWW pages. HTML also provides the means to structure data and information, and to implement the design layout of each page. Through the use of CGI scripts it is also possible to add in links e.g. email, forms, and multiple choice questions. A more recent development has been that of JAVASCRIPT (an extension of HTML), and the JAVA programming language. JAVA applets offer the information system developer more tools to enhance overall information delivery through, e.g. animation, interaction etc. The HCI (Human Computer Interface) is an important component of any information delivery system.

Such information can be a static delivery, taking the form of text, images, or graphics. In the case of e.g. the background, colour or enhanced wallpaper can be used to provide aesthetic appeal, identity and continuity between the many links. Text can be sized, coloured, aligned, and the fonts altered to maximise the visual impact of the screen. Highlights and breaks in structure can be provided through the addition of spacing lines and graphics. Small graphics and thumbnails can be added to make the screens more aesthetically appealing, eye-catching, and informative. Beyond this, text and graphics can also be linked (hot links) to other text and images. But information can also be delivered in a much more dynamic way e.g. as a video clip, or through the use of animated graphics. These may be simple animations that enhance the visual appeal of the screen, or animations that deliver information in a way that could not be so effectively delivered in any other way. Inclusion of Java applets can be used to extend the delivery of information in an interactive fashion.

The Interface provided by the browser provides a multiple-level entry interface for the user. At the first level is the MS-Windows interface. At the second level is the Browser interface, and at the third, and in this case perhaps the most important level is the Internet Information System interface which focuses the user on the information contained within via e.g. links, hotwords, graphics, text and so on. Figs. 3-5 illustrate some aspects of the Forth Estuary Forum Information System.

Potential Problem Areas

Naturally there are likely to be associated problems with the provision of information over the Internet which must be considered. Kelly (1993), for example, mentions a number of legal and ethical issues. These include questions about the legality of the WWW service, liability, the Computer Misuse Act, Copyright, the Designs and Patents Acts, the Data Protection Act, equality of access to information, and advertising. Not all are relevant in the case of this example, but nevertheless where data or information owned by an individual or an organisation is being distributed, potentially with free on-line access, care must be taken to ensure that access is provided to only that information and not other information.

On a practical level questions must also be asked about the costs, the co-ordination, the role of Government, and the Private Sector in the Internet. There are also problems with being able to move large volumes of data around e.g. the bandwidth, from the service provider to the user and from the users back to the servers (Pillar, 1994) which need to be considered if the Internet is to be seriously considered as a major point of access for information from a wide range of users.

Some concern has been expressed in the past by people about provision of access to an organisation's spatial databases, which is possible over the Internet. Whilst the Internet can be used in a variety of different ways as the basis for a GIS, providing access to remote spatial databases via a network, the scenario outlined above offers an alternative and simpler route for the provision of spatial information, and one that is

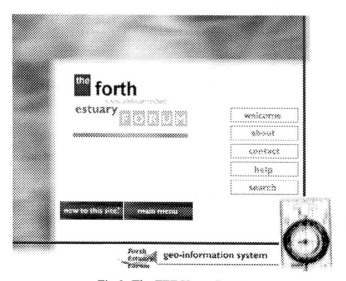

Fig.3. The FEF Home Page

under the full control of the information provider. This can be achieved by allowing the individual organisations to decide exactly what data and information they make available over the Internet. The posting of information, and the way in which this information is presented can also be controlled by the individual organisations. However, this does not mean providing access to databases that contain sensitive or confidential information.

Summary and Conclusions

The pilot information system briefly outlined above successfully illustrates one of a number of different ways in which the Internet and its associated information browsers can be used in a very simple, yet highly effective, way in the workplace to store, structure, and deliver information to a wide range of users from many different backgrounds. The networking technology and the browser interfaces are clearly ideal as the basis to permit the development of an information system. The Internet browser tools also provide one way to customise the WWW page interface for specific applications, allowing users not only to freely access information in a very flexible, but structured way, but also to focus on a specific task or set of tasks related to their area of work. Much of the information required by such users is spatial and this can also be included as part of the information resource. Careful design of the information interface within the browser interface offers navigational tools which can be made as flexible as the designer permits.

Information can be structured in both a flexible, but also semi-rigid way for a particular application. In the case of the FEF Pilot Information System, the objective has been to provide a detailed information resource for many different users, with wide-ranging backgrounds who may wish to acquire information about the Firth of Forth. To begin with the Internet offers a means to make such information available electronically. The benefits this offers over the presentation of such information in a traditional bound text form are: flexibility, delivery of images (colour), regular updating, additional interaction, and so on. But the Internet also offers access to information from a much wider range of sources. It also offers users possibilities to communicate electronically with the FEF, either commenting on or In the future, extensions to the HTML language, Javascript and the JAVA language will provide the information system developer with an even more powerful construction toolbox to design a customised navigational interface for the delivery of information. Such developments will also permit the delivery of information in even more interesting and dynamic ways providing the tools to, for example, deliver animated graphics, and interactive maps. adding to the information resource. In this sense the FEF can gather information for its own use far more easily.

Internet-Based Information Systems

Fig. 4. An example webpage from the FEF website (courtesy of Scottish Natural Heritage (SNH))

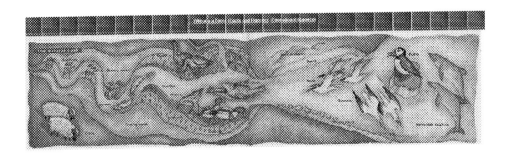

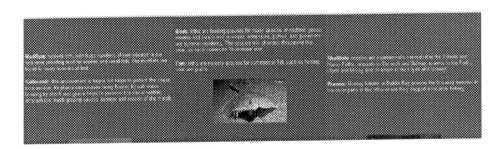

Fig. 5. An example webpage from the FEF website
(courtesy of Scottish Natural Heritage (SNH))

References

Akass, C. (1994). The Whole World in His Hands. *Personal Computer World.* December 1994. pp. 380-383.
Amber, (1993). Economic Development and Environmental Protection in Coastal Areas: A Guide to Good Practice - *Draft Report.* Commission of the European Communities. 63p. Appendices and Bibliography.
BODC (1992). United Kingdom Digital Marine Atlas *Software (CD-ROM and disks).*
COMPAS (1992). NOAA's Coastal Ocean Management, Planning and Assessment System. *User's Guide* for the Texas Product. U.S. Department of Commerce. NOAA. 6-3 pages.
English Nature (1993). *Estuaries: Information: English Nature's Estuaries Initiative.* 4p.
English Nature (1993). *Strategy for the Sustainable Use of England's Estuaries.* Campaign for a Living Coast. English Nature. 43p.
English Nature (1993). *Estuary Management Plans - A Co-ordinator's Guide.* Campaign for a Living Coast. English Nature. 88p.

Gardener, L. A., and Paul, R. J. (1993). Developing a Hypertext Geographic Information System for the Norfolk and Suffolk Broads Authority. *Hypermedia.* Vol. 5(2):119-143.

Green, D.R. (1994). Using GIS to Construct and Update an Estuary Information System. *Proceedings* of the Conference on Management Techniques in the Coastal Zone. Centre for Coastal Zone Management, University of Portsmouth, October 24-25, 1994. pp. 129-162.

Green, D.R. (1995). Internet, the WWW and Browsers: The Basis for a Network-Based Geographic Information System (GIS) for Coastal Zone Management. *Proceedings* AGI Conference and Exhibition, 21-23 November 1995, pp. 5.1.1-5.1.12.

Green, D.R., and Rix, D. (1995). Access to Information: The Human Computer Interface and GIS. *AGI Source Book* 1996. pp. 78-95.

Joint Nature Conservation Committee (1994). *Coastal Conservation.* September 1994. 2p. JNCC: Coastal Conservation Branch.

Kelly, B. (1995). Running a World-Wide Web Service. *SIMA Report Series.* Number 6. January 1995. Edition 1.2. Support Initiative for Multimedia Applications (SIMA). Advisory Group on Computer Graphics. 76p.

Kleiner, K. (1994). What a Tangled Web they Wove. *New Scientist.* 30th July 1994. No. 1936. pp. 35-39.

LCGISN (1993). Louisiana Coastal GIS Network *Newsletter.* December 1994. Vol. 4(2).

LCGISN (1994). Louisiana Coastal GIS Network *Newsletter.* September 1993. Vol. 3(2).

Manger, J. (1995). *The Essential Internet Information Guide.* McGraw-Hill Book Company: London. 515p.

Murnion, S., and Munroe, G. (1994a). Surfing on the 'Net': Information Resources for GIS Users. *Proceedings* of the Eighth Annual Symposium on Geographic Information Systems. February 21-24, 1994. Vancouver, British Columbia, Canada. pp. 32-35/54.

Murnion, S., and Munroe, G. (1994b). GIS and the World-Wide Web: Information Resources for GIS Users. *GIS Europe.* Vol. 3(2):23-26. March, 1994.

Raal, P.A., Burns, M.E.R., and Davids, H. (1995). Beyond GIS: Decision support for coastal development, a South African example. In R.A. Furness (ed.), *CoastGIS'95: Proceedings of the International Symposium on GIS and Computer Mapping for Coastal Zone Management.* International Geographical Union Commission on Coastal Systems, Sydney. pp. 273-282.

CHAPTER 33

Mike Info *Coast* - A GIS-Based Tool for Coastal Zone Management

R. Andersen

ABSTRACT: MIKE INFO *Coast* is a product from the Danish Hydraulic Institute (DHI) specifically designed for the handling of coastal data. It is constructed as an extension to ArcView 3.0, and it thus has all the benefits of a GIS. Apart from geographical maps, MIKE INFO *Coast* handles data such as bathymetric surveys, coastal profiles, images and various types of hydrographical measurements (e.g. wind and wave data). With MIKE INFO *Coast* it is possible, in a user-friendly manner, to manipulate (heterogeneous) coastal data in various ways (e.g. calculation of volume changes from survey to survey). Reports and standard presentations for decision support can also be easily generated. Naturally, MIKE INFO *Coast* exchanges data with DHI's other software products, such as LITPACK and MIKE 21.

Background

Historically, the Danish Hydraulic Institute (DHI) has focused its computing efforts in the computer simulation area. Thus, since the end of the 1960s, DHI has developed models of various kinds. Some of the most well known are MIKE 21 (a modelling system of 2D free surface flows), MOUSE (a modelling system for pipe networks), MIKE 11 (a system for the 1D modelling of rivers, channels and irrigation systems) and LITPACK (an integrated modelling system for littoral processes and coastline kinetics).

With the spread of computer technology in recent years, however, it has become apparent that the task of evaluating model result data is no longer solely the task of the expert. An explicit need for user-friendly access to complex models and their results has been expressed. In order to accommodate this DHI has developed various interfaces to the models. As all the models of DHI ultimately express geographic information, a GIS (*ArcView*) has been chosen for the presentation of the model-related data. Thus, several modules have been developed as add-ons to the DHI model, e.g. MIKE 11 GIS for MIKE 11, MOUSE GIS for MOUSE. Common to the above-mentioned GI-modules is that they require a model to supply data.

Inspired by the recent EAGLE-project where a GIS-based system was developed for the presentation of many different kinds of data (such as hydrographical and biological data, model result data, images etc.) DHI has set out to create a suite of information systems for the presentation of environmental data. This suite of products has been called MIKE INFO, and it is the intention to develop MIKE INFO products for various areas of data management. One of these areas is the management of coastal data, and the corresponding product is MIKE INFO *Coast*.

What is MIKE INFO *Coast*?

MIKE INFO *Coast* is a tool for managing coastal data. This covers specifically:

- **Data Management:** All coastal data are stored and maintained within one system
- **Data Manipulation:** The coastal data can be edited and modified
- **Data Presentation:** The data can easily be presented in various ways
- **Report Generation:** Reports can be generated and saved for later use

So far MIKE INFO *Coast* has been used in-house in two projects. One in the United Arab Emirates and the other in Malaysia. In both projects a stretch of around 30 km shoreline is to be maintained. Hydrographical data, images, maps and surveyed points were covered in both projects. The MIKE 21 modelling system was also applied in both cases. Having developed and tested MIKE INFO *Coast* for about two years in-house, DHI has now released the first version in the public domain.

MIKE INFO *Coast* has been constructed as an extension to the GIS ArcView 3.0 from ESRI (Environmental Systems Research Institute). This means MIKE INFO *Coast* and ArcView are completely integrated, and that all the built-in functionalities in ArcView are available in MIKE INFO *Coast*. As an example you can use ArcView's standard facility for digitising maps, and then later on have them presented within MIKE INFO *Coast*.

In ArcView/MIKE INFO *Coast* the user interface depends on the current active document, i.e. the user interface changes as the active document changes. ArcView contains several types of documents, which all have their individual user interface (Fig. 1). When loading the MIKE INFO *Coast* extension, you actually load a new type of document into ArcView. This type of document is called "coast", and has its own tailor-made user interface especially suited to work with coastal zone data. For instance the user interface covers options such as "calculate differences between surveys", "import sediment size data" and other functions specific to coastal zone data.

Mike Info *Coast* - A GIS Tool For CZM

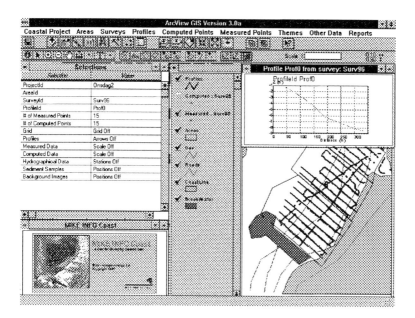

Fig. 1. ArcView with the MIKE INFO Coast-extension loaded.

A Situation of Typical Use

The standard scenario would be a coast, where bathymetric surveys record the depths in a not necessarily uniform way. However, there may be cases where some or all of the depths measured are structured in profiles. Typically several surveys are carried out for a longer period of time (many years), though it is not expected that data will be measured at the same locations from survey to survey. Often there will be some correlation when surveys consist of profiles. After each survey, data will be imported into MIKE INFO *Coast* and processed. At a later stage data might be exported for use in LITPACK or MIKE 21.

As MIKE INFO *Coast* handles coastal data it is necessary for the users to have a professional background in this area. The users must have some education and academic knowledge, e.g. as an engineer. The user will be expected to enter data into the system, manipulate the data using the functionalities provided and to make presentations. This requires that the user is familiar with computerised data handling.

Data Handled by MIKE INFO *Coast*

MIKE INFO *Coast* handles data in the following categories:

- **Maps.** This covers the various kinds of maps needed for presentation, e.g. maps of coastlines, buildings, groynes. Maps are stored in shape file format, which is ArcView's native format.

- **Other geographically related data.** This covers measurements of depths/heights (with geographic reference) and profiles. Also included is any data derived from the depth measurements, such as interpolated surfaces, results from difference calculations etc. Survey related data with a geographical reference are also stored as shape files.
- **Administrative data.** This covers data with no geographical reference (e.g. general survey information). These data are stored in dBase IV-format, which is a well-known database format and immediately supported by ArcView. However, these data could also be stored in an external database and reached through ODBC.
- **Regularly measured hydrographic data.** This covers measurements at fixed positions of variables such as wind speed/direction, wave height etc. Hydrographic data are stored in ASCII-format which makes on-line data acquisition easier, should the need arise.
- **Sediment data.** In the case were sediment samples have been collected and analysed it is possible to keep this information within MIKE INFO Coast as well.
- **Background images.** These images are related to the maps, and can be satellite photos and/or aerial images.
- **Directional images.** This covers photographs taken at fixed ground locations in a certain direction. If several photo surveys have been performed it is possible to compare photographs from the same location but taken at different times.
- **Model result data.** This covers data to and from LITPACK and MIKE 21. Model data are imported/exported in the ASCII-format exported/imported from the relevant model.

Functionalities in MIKE INFO *Coast*

MIKE INFO *Coast* covers a range of functionalities that can be used for managing and presenting coastal zone data. Some of these functionalities are listed below:

- **Import of measured or surveyed point.** These points are imported from a simple ASCII-file containing XYZ-format data.
- **Editing of measured points**. It is possible to edit the imported surveyed points in various ways. You can delete erroneous points, add new ones, shift the datum of all or some of the points etc.
- **Organising the measured points in profiles**. You can define a set of profiles in various ways (interactively by pointing, automatically generated or imported). Then you can organise your measured points relative to these profiles, by projecting the points onto the profiles.
- **Calculation of differences between profiles.** Having organised your measured points you can compare data from various surveys. Thus, you can calculate the difference between two profiles and get an estimate of the erosion/deposit for that particular profile.

Mike Info *Coast* - A GIS Tool For CZM   471

- **Generation of reports**. Note that generated reports can always be edited using the built-in ArcView functions. An example of a generated report page is seen in Fig. 2.

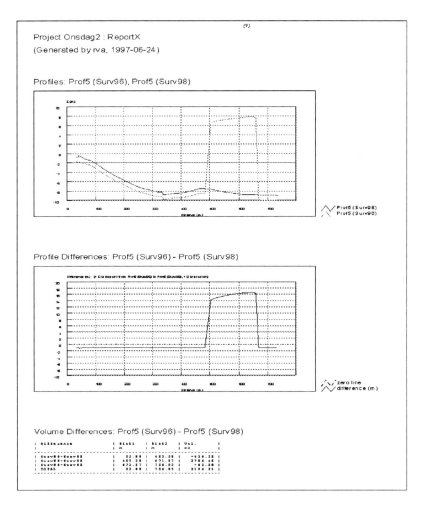

Fig. 2. Sample report page generated by MIKE INFO Coast showing a difference calculation between two profiles. Note that the data are purely artificial.

The functions mentioned so far primarily concern data organised in profiles. However, MIKE INFO *Coast* does not require this kind of organisation. Other functions are listed below:

- **Generation of contours.** It is possible to generate a set of contours by interpolating the measured or projected point data.

- **Difference calculation covering an area.** You can calculate differences between two surveys by generating difference maps from interpolated maps.
- **Edit and view hydrographical data.** You can specify stations were hydrographical data are measured. A dedicated tool for managing the often very large time-series is supplied with MIKE INFO Coast.
- **Image support.** Viewing and report generation of background images and directional photos is supported.
- **Handling of sediment sample data.** Apart from storage and editing, various presentations of sediment sample data are included, such as a standard particle size distribution curve.

Concluding Remarks

This chapter has briefly described the product MIKE INFO *Coast* version 1.0 as it has been developed by the Danish Hydraulic Institute. It is the first member of the MIKE INFO family, which is a suite of product that attempts to present data in an easy, informative and user-friendly manner. It has been the aim to construct a system, where the user can move quickly from import of data to reports presenting these data appropriately. It is easy to predict that more systems like this will appear in the future, and DHI has committed itself to make an effort toward this development.

References

Andersen, R. (1996). EAGLE - A GIS for Environmental Impact Assessment. *Geographical Information. From Research to Application through Co-operation,* 1, 310-317.

Foster, T.M., and Skou, A.J. (1992). LITPACK, An Integrated Modelling System for Littoral Processes and Coastline Kinetics. 3rd International Software Exhibition for Environmental Science & Engineering. Como, Italy.

Skou, A., Hedegaard, I.B., Fredsøe, J., and Deigaard, R. (1991). Applications of Mathematical Models for Coastal Sediment Transport and Coastline Development. Proceedings of COPEDEC III. Mombassa Kenya.

Warren, I. R., and Bach, H. K. (1992). MIKE 21: A modelling system for estuaries, coastal waters and seas. *Environmental Software 7.* Elsevier Science Publishers Ltd.

CHAPTER 34

Matching Issue to Utility: An Hierarchical Store of Remotely Sensed Imagery for Coastal Zone Management

S.D. King and D.R. Green

ABSTRACT: The ultimate goal of any environmental scientist, sociologist or manager is to understand how the environment works. As knowledge is the key to understanding, data must be collected about the environment and analysed in the context of the specific issues being investigated. This will yield information that can be gathered and stored, and will further contribute to the body of knowledge surrounding the said issue. A simplified way of regarding the environment is as a collection of objects that relate to each other, and interact with each other over space and through time. This being the case, much of the data collected is geo-spatial in nature; that is, it has a geographical variability. Remotely sensed imagery of the Earth is one source of geo-spatial data that can be used to help explain how the environment works, and can therefore be used for environmental management. Over the past twenty years, the number of different sensors used to collect remotely sensed data has grown considerably, and this proliferation has led to a vast store of multi-spectral, multi-spatial and multi-temporal data of different scales which is theoretically available for use in environmental management. Such a large database is a unique resource for the environmental scientist because real world processes operate at a range of different scales, and hence all the differently specified data will have their own particular uses. Despite this, the vast array of data available has arguably limited the practical use of remote sensing in the management arena, largely because the non-specialist user does not know what data source will aid in the problems that they wish to solve. The coastal zone is an example of a unique and dynamic environment, the effective management of which requires a thorough understanding of how it works. For the coastal zone manager, geo-spatial data and information is therefore a necessity. Although remote sensing has the potential to provide much of this information, the coastal manager will not necessarily know which sensor will provide the best imagery (in terms of spatial, spectral and temporal resolution) for each particular problem that has to be resolved. There is a need, therefore, for a generic set of guidelines that will help the coastal manager decide what set of imagery, from which sensor, will best provide the information required to solve the specific problem. In essence, the issue to be resolved must be matched with the utility (or characteristics) of the sensor. Initially, such a set of guidelines may be constructed using a paper-based matrix. However, the need for rapid access to the data sets and the correct information suggests that an electronic equivalent would be more far efficient and useful. Such a system would allow the user to define their problem attributes by answering a set of basic questions, and would then be guided to the imagery that would best help solve the problem. In effect, the user would be able to simply and quickly interrogate a geospatial, multi-scale hierarchical database of satellite imagery, providing quick and efficient access to the data and information required.

Introduction

The sociological and physical constraints of modern-day living dictate that human beings will attempt to manage the natural environment. Most environmental management today is operated on the basis of sustainable development, with the objective of providing for the needs of humans without detriment to the wellbeing of the environment. Understanding how the environment works, and understanding and predicting future environmental processes is essential to successful management. This, in itself, is heavily reliant upon the provision of data, information extraction from that data, knowledge acquisition and understanding.

Most data collected about the environment is geographical in nature because it relates to objects and features that are visible to humans on the Earth's surface. Prior to the 1960s, much of the data collected was arguably very static in nature, being highly descriptive and paper-based. Since the 1960s, the advent of computers, Geographical Information Systems (GIS), and new data collection devices, such as satellites, have given us the ability to collect, store and manipulate data in many different ways. Data has become more interactive and 'spatialised' because we are now more readily able to simultaneously visualise the geographical relationships between earth surface objects at varying scales.

Although aerial photography has been used for environmental data acquisition for most of this century, the more recent development of satellite technology, as well as new airborne sensors has left the environmental management community with an extremely large resource from which data can be drawn. This data is held in various spatial, spectral and temporal resolutions, all of which have potential for use. However, in many environmental applications, remote sensing in not fully utilised because those who could benefit from it do not know which source of data is best suited to the issue that they are attempting to resolve. Using coastal zone management as an example, this paper will demonstrate how remotely sensed data may be organised into a hierarchy to which coastal issues may be matched. Finally, a paper-based matrix that matches coastal issues to remotely sensed data is presented, and the future development of an electronic version is suggested.

The Importance of Geo-Spatial Information in Environmental Management

From the earliest times, maps (a source of geographic information) have been used to essentially record data about the world in which we live. Couclelis (1998) argues that geographic data is used by people because the geographic world is a generally accessible and comprehensive realm of experience, offering "a complete array of forms, colors, textures and patterns" as well as "a wide range of basic sensorimotor, cognitive and affective experiences" (p.209) and because things that have the look and feel of geographic space make it easier for people to understand.

The use of geo-spatial data appears to be a natural human instinct and we all make use of geographic data and information on a daily basis, be it maps for directions, plans for building, and ultimately, our own mental maps of our surroundings (Green and King, 1998). It is, therefore, not surprising that geography and geographical tools have become essential in our day-to-day living, decision-making, and management of the environment.

Using Remote Sensing to Collect Environmental Data

Remote sensing has witnessed a rapid development since the late 1970s, and is now frequently used as a source of environmental data. A much vaunted advantage of remote sensing, particularly satellite remote sensing, is that data from parts of the electromagnetic spectrum other than solely the visible, such as thermal, microwave and near-infrared (which cannot be seen by the human eye) can be utilised. This allows information to be gathered about the surface features that may otherwise go unnoticed (King, 1998).

Despite this, it is still perhaps the spatial resolution of different remotely sensed imagery that is of greatest importance in the context of geo-spatial data. This is because the spatial resolution of an image ultimately controls the amount of detail that is shown, and the information that may be extracted from it.

Scale and Spatial Resolution in Remote Sensing

Before defining scale and resolution, it is important to understand the distinction between data and information, so that the controls scale and resolution have on information content may be fully comprehended. A typical definition offered by Heywood et al., (1998) and Olivieri et al., (1995) is that data is a set of measurements describing an object or process, whereas information is data that has been analysed and given context. On the other hand, Couclelis (1998) suggests that data is something that can be "automatically manipulated and processed by a machine, whereas information presupposes the involvement of a cognitive agent" (p.211), or in other words, someone recognising information held in the data.

Scale and resolution, like data and information, are closely related terms (although they have different meanings). When we think of scale in relation to geo-spatial data, maps frequently come to mind. Maps take up less physical area than they represent. The reduction in areal coverage is reflected by the scale of the map, which indicates the number of metric units on the ground that are represented by a unit on the map (Joao, 1998). Spatial resolution on the other hand refers to the smallest object that is discernable in the data. For example, in remote sensing, a resolution of 30m means that features and objects less than 30m across would not be seen as separate to their surroundings. Scale and resolution are therefore related because scale sets the lower limit to the size of an object that may be shown on a map (Joao, 1998). In terms of information content, a large-scale map of 1:10,000 will usually contain more detailed information than a 1:50,000 map covering the same geographical area. Likewise, an aerial photograph with a spatial resolution of 1m will potentially contain more detailed information than a subset of a satellite image covering the same area, with a spatial resolution of 30m. In some cases, however, a small-scale map (with greater generalisation) will arguably provide more information, especially when considering large-scale processes (Goodchild and Quattrochi, 1997). This would also be true of a larger resolution image, where some patterns become more apparent when a larger area of the Earth's surface can be seen.

Scale and resolution consequently play a crucial role in determining which data is used to provide the information for a given situation. For instance, most people, when trying to find out a road route between two locations will turn to a road atlas, perhaps with a scale of 1:125,000. However, if a specific street in a city had to be found, then a street map, with a scale of perhaps 1:15,000 would be more appropriate. The need for greater information content requires the user to move from a small-scale to a larger scale map. In essence, the user has matched the issue (or information requirement) to the utility of the map, the process being made more simple because maps are organised hierarchically by scale, with most users having the knowledge that large-scale maps offer greater detail (information) than small-scale maps.

The same principle may conceivably be applied to remotely sensed imagery for environmental management, particularly where the information required from the imagery has a spatial component. In theory, remotely sensed imagery could be arranged using a hierarchy based on spatial resolution, rather than scale, with detailed information being available on high-resolution imagery, and more generalised information available on low-resolution imagery. Such a scheme does rely heavily upon the principle that most of information required from remotely sensed imagery will be the information that is visually obvious. While this by no means disregards the importance of spectral information (outside the visible), the work of King (1998) has argued that most non-specialist users of the data will find it easier to visually extract information from the imagery, either by manual or on-screen, computer-based, interpretation. Again, this is largely because remotely sensed imagery provides us with an instantaneous recognition (through shapes, patterns and colours) of geographical features and relationships within the natural environment (relating to the geographic metaphor suggested by Couclelis, 1998). Even where spectral information is important, the spatial resolution will still inevitably govern the information that can be extracted, as spectral patterns will have a multi-spatial and therefore multi-scale component.

Matching Issue to Utility – Using Coastal Zone Management as an Example

The coastal environment is perhaps one of the most unique in the world. The multiplicity of biological and landscape types which the interaction between land, water and air produces, together with human use of the marine and land coastal resource (for infrastructure development, leisure and recreation, fishing, mineral extraction and power generation), make it one of the most difficult areas of the world to manage and care for (King, 1998). Management difficulties essentially arise from the high physical diversity of the coastal system, with different coastlines varying in width, length, processes at work and natural habitat. To coordinate a successful and holistic management of the coastline, there is a need for data to be gathered within the natural coastal framework which is up-to-date and relevant, and can help coastal managers formulate the procedures that will balance human actions with the needs of the environment (King, 1998).

Despite the complicated nature of the coastal zone, its natural dynamics and processes do in theory lend themselves quite well to scientific enquiry using geographical principles. The coast is very spatially and temporally dynamic, with human and natural objects, features and landforms all affecting each other in some way across space and through time. Data collected about these will inevitably have a geo-spatial component, and this data can be stored, manipulated and analysed within a GIS to produce models of reality which aid in our understanding of the environment.

Both GIS and remote sensing have been quite widely used in coastal research. However, although GIS is good for the integration, management and analysis of data, it is not a source of raw data or method of primary data collection. Remote sensing, on the other hand is. Its many advantages include up-to-date data provision, or data on demand; data acquisition without field visits; early identification of environmental problems; the ability of satellites to image large tracts of the Earth (allowing managers to view whole coastal systems); near real-time and real-time imagery, and more detailed monitoring of coastal landscapes, habitats and processes using higher resolution imagery such as aerial photography and videography (King, 1998). However, perhaps its greatest advantage is that the data delivered is usually in a form that most people will instantaneously recognise as a 'snap-shot' of the Earth. It is this instinctive recognition that makes remotely sensed imagery one of the best sources of geo-spatial data, because it can be interpreted to a certain extent by most people, and this level of interpretation yields valuable information.

Such an instantaneous, and possibly instinctive recognition and comprehension of features shown on a remotely sensed image is what, in many cases, makes this source of data far more valuable than a map, because maps are a more abstract representation of reality, and not everyone can read maps. Yet, lack of exposure to remote sensing beyond perhaps the occasional aerial photograph (be it oblique or vertical), limits the use and application of remote sensing by the non-specialist, to environmental management. Indeed, it is the relative familiarity of true colour or black and white aerial photography above satellite imagery and other airborne sensors that prevents a more inventive use of remote sensing, because people automatically think of aerial photography as the best imagery for the job.

There are a number of reasons why this might be the case. Firstly, both true colour and black and white aerial photography is more likely to be easily understood by most people. Secondly, many people prefer to work, or perceive that it is easier to work with hard copy photographs, rather than a digital copy on a PC or workstation; and thirdly, the greater spatial resolution of aerial photography, which provides better detail than most other current sensors, is perceived as being more useful. Finally, there is a widely held belief that because there are relatively few sensors developed specifically for coastal zone monitoring, then the imagery available from other sensors will not provide useful information (see for example Doody and Pamplin, 1998).

All of these problems can be relatively easily overcome with better education, greater use and practice. For example, most remotely sensed imagery can be displayed using colour combinations that more closely represent what the human eye sees, and most imagery can be output as a hard copy. Now and in the future, more simple techniques for delivering and manipulating digital data will become available that make digital imagery easier to use. In terms of resolving the last two problems, it is necessary to return to the amount of information required from an image, and this relates to the issue to which the information held in the image is being applied. If, for instance, the position and general quality of a section of sea wall was required, then aerial photography would most likely provide the answer, as only the high spatial resolution of aerial photography can really show such detail. However, if the total area of an estuary or firth was needed, then a satellite image would be more appropriate to use, because far more aerial photographs would be needed to cover such a large area, with no benefit in information content. What this shows us is that people need to be made aware that all remotely sensed imagery can have some use, as long as the correct imagery is applied in the correct situation. Nevertheless, this requires a thorough understanding of the nature of the problem being investigated, an understanding of the information required, and a reference guide as to which imagery will provide that information. In essence, the issue must be matched to the utility or characteristics of the sensor.

There are a whole range of issues, both natural and anthropogenic, which need to be resolved at the coast. To match these issues to the imagery, one must understand the information required to solve the issue and then the spatial, spectral and temporal characteristics needed to provide the information. These characteristics can then be compared with and matched to the spatial, spectral and temporal characteristics of the different sensors. For example, a coastal issue might be the perceived increase in coastal flooding due to saltmarsh erosion. The information required to investigate this issue would be, where are the saltmarshes, are they eroding, which properties and land are vulnerable? The objective would be to find imagery that could be used to identify saltmarsh areas, analyse coastal change, and identify vulnerable property or land. To identify the general pattern and location of saltmarsh areas in an estuary would usually require the imagery to have a spatial resolution of at least 30m (Fig. 1) (this would provide enough detail as well as area coverage). The spectral characteristics would not be too significant in this case, as areas of saltmarsh can often be identified by pattern and apparent texture. However, different band combinations can be used to highlight certain vegetation and the vegetation/water boundary. A series of images over a period of time would be needed to identify saltmarsh erosion. Having identified any areas of possible erosion (Fig. 2), it would then be possible to look at these areas in more detail using higher resolution imagery which could then be used to locate any land or property in danger (Fig. 3). The information provided by such an analysis would then be available for the coastal manager to formulate a management plan for this issue.

Using Remotely Sensed Imagery for CZM 479

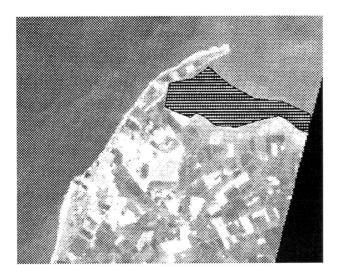

Fig. 1. A 30m Resolution Landsat TM image with saltmarsh areas identified by horizontal hatching (LANDSAT data © NOAA. Distributed by CHEST under license from Infoterra International)

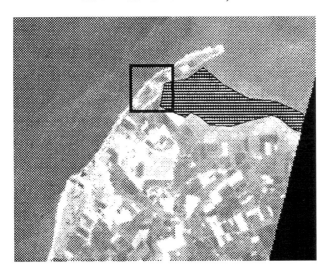

Fig. 2. Identifying an area of interest (AOI) on the Landsat TM image (LANDSAT data © NOAA. Distributed by CHEST under license from Infoterra International)

Fig. 3. Higher resolution (1-2m) aerial photography of the AOI, used to identify areas of saltmarsh (vertical hatch), areas of settlement potentially vulnerable to flooding (horizontal hatch), and flood defences (thick black line)
(© Environment Agency (EA) and English Nature 1997)

Using this method of enquiry, it is possible to build up a table of the issues that a coastal manager faces, the information required to solve these issues, and the objectives needed to extract the information from remotely sensed imagery. These can then be matched up to the sensor to provide the imagery necessary to help solve the issue. Table 1 below is an example of a matrix, developed for the Solway Firth. What becomes apparent from the matrix, and the saltmarsh example, is that different levels of imagery will be applicable to each part of the enquiry. It is therefore feasible to develop an image hierarchy on the basis of spatial, spectral and temporal characteristics from which the imagery may be drawn. This can be envisaged as a database containing layers of information, from the general to the more detailed, each layer having its own content and use.

Moving Towards a Geo-Spatial Coastal Information System

Although the paper-based matrix is a guide to the imagery that could successfully be used to provide the information required to help solve specific issues, it does have a number of problems associated with it. Firstly, the problem criteria are very specific to one coastal area, and therefore may not be applicable in other situations. Secondly, there is no direct link to the database of imagery, which means that the user would still have to search for the image source, and retrieve it. A more efficient and effective approach would provide the means to interrogate a database of imagery using generalised search criteria via a computer interface which would automatically present the user with the correct image source and information.

A possible way of envisaging such an approach is to think of an Internet search tool, such as Altavista (http://www.altavista.com) (Fig. 4), or perhaps the online BIDS (Bath Information and Data Services) service (Fig. 5). In the case of Altavista, the user is required to enter a search term, such as remote sensing, and the search engine will then return a set of results relating to all Web sites that cover remote sensing. The user is also provided with a direct link to each web site. BIDS is slightly different as it actually presents the user with a set of search criteria, such as whether to search by author, publication or title. There is also the choice to search by topic (such as remote sensing) and year of publication. A similar system for retrieving remotely sensed imagery for coastal applications could be designed, operating on the basis of a spatial hierarchy. Table 2 shows examples of a number of basic questions that would allow the user to search the system.

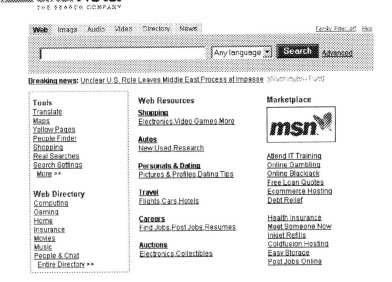

Fig. 4. The Altavista WWW search engine (courtesy of Altavista - http://www.altavista.com)

TABLE 1

A paper-based matrix matching coastal issues to remote sensors

Theme	Issue	Information Required	Objective	Aerial photos	Aerial Videography	CASI	SPOT	Landsat TM	Resurs-01
			Resolution Spectral (Bands)	3	3	15	4	7	4
			Spatial	1-2m	1-2m	4m	10m	30m	160m
			Temporal (Days)	Variable	Variable	Var.	26 (Var.)	16	21 (Var.)
Coastal Development	How to encourage and accommodate economic development	Where is economic development currently situated	Identify areas of economic development	Y	Y	Y	Y	?	N
		Where might new development be situated	Identify areas where new development may be situated	Y	Y	Y	Y	?	N
	The inadequacy of transport links	What are the current transport links	Identify current transport links	Y	Y	Y	Y	?	N
		Where might new transport routes be developed	Identify areas where new transport routes could be developed	Y	Y	Y	Y	?	N
	Social economic and environmental impacts of oil and gas exploration	The proximity of oil and gas exploration to human habitation and environmentally sensitive areas (ESAs)	Identify areas of oil and gas exploration, proximity to human habitation and ESAs	Y	Y	Y	Y	?	N
	Impacts of waste effluent on water quality and public health	Where does waste effluent enter the sea	Identify waste discharge outfalls	Y	Y	Y	?	?	N
									N
	The effects of afforestation on landscape, nature conservation and agriculture	Where are the current forest plantations	Identify forest plantations	Y	Y	Y	Y	Y	?
		Where are the planned plantation areas	Identify areas of planned plantation	Y	Y	Y	Y	?	N
		Proximity to areas of concern	Measure proximity to areas of concern	Y	Y	Y	Y	?	N
	Conflicts between development and conservation	Where are the conservation areas	Identify conservation areas	Y	Y	Y	Y	?	N
		Proximity to development areas	Measure proximity to development	Y	Y	Y	Y	?	N
	Preventing development which is unsympathetic to the countryside	Which areas should not be developed	Identify areas which are sensitive to development	Y	Y	Y	Y	?	N
	The impact of sea-level rise on property	Which areas are vulnerable to sea-level rise	Identify low-lying areas	Y	Y	Y	Y	?	N
	Increased flooding due to saltmarsh erosion	Where are the salt marshes	Analyse coastal change	Y	Y	Y	?	Y	?
		Which properties/land are vulnerable	Identify vulnerable property/land	Y	Y	Y	?	?	N
	The extent to which coastal structures are exacerbating coastal erosion	Where are the coastal structures	Identify coastal structures	Y	Y	Y	?	N	N
		Which areas are eroding	Identify erosion	Y	Y	Y	?	N	N
		Are coastal processes affected	Monitor coastal processes	Y	Y	Y	?	Y	?
	The impact of sand and gravel extraction	Where does sand and gravel extraction occur	Identify areas of extraction	Y	Y	Y	?	N	N
		Does extraction affect natural processes	Monitor natural processes	Y	Y	?	?	N	N
	The need to understand physical processes of erosion and accretion	Where does erosion/accretion occur	Monitor coastal change and sediment movement	Y	Y	Y	?	Y	?
		What are the patterns of sediment movement		Y	Y	Y	Y	Y	?
Pollution and Water Quality	High levels of unsightly and potentially dangerous beach litter, domestic waste and sewage outfalls	Where are the problem areas	Identify pollution	N	N	N	N	N	N
		Where are the outfalls	Identify outfalls	Y	Y	Y	?	?	N
	Improving the quality of bathing waters	Where are the bathing water areas	Identify bathing waters	Y	Y	Y	?	?	Y
		What is the water quality like	Assess water quality	?	?	Y	?	?	N
	Problems of radioactive discharges	Where is radioactive waste discharged	Identify outfalls	Y	Y	Y	?	?	N
		Which areas does it affect	Identify radioactive waste	?	?	Y	?	?	?
	The potential for oil pollution from tanker traffic in the Irish Sea	Are there areas affected by oil pollution	Identify oil pollution	Y	Y	Y	?	Y	?
	Agricultural pollution (nitrates, silage effluent, slurry)	Where are the farms	Identify farms	Y	Y	Y	Y	?	N
		Is pollution identifiable	Identify forms of pollution	Y	Y	Y	?	?	?

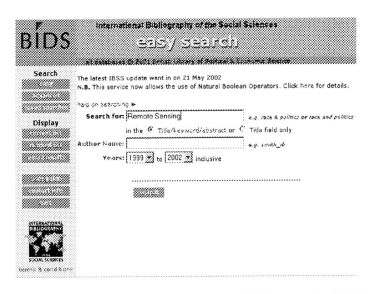

Fig. 5. The BIDS document search engine (courtesy of BIDS - http://www.bids.ac.uk/)

TABLE 2

Examples of questions used to interrogate a database of remotely sensed imagery for coastal zone management

Spatial Questions	Options/Answer
For what geographic area do you require imagery?	a map showing the areas that imagery is available for, and numbered for choice
For what size of area do you require imagery?	Offering a choice of area in km^2
What is the minimum size of the feature(s) you wish to resolve?	Offering a choice of image spatial resolution
Temporal Questions	Options/Answer
From which year do you require imagery?	Offering a choice
From which month do you require imagery?	Offering a choice
Spectral Questions	Options/Answer
How many bands do you require and at what intervals?	Offering a choice

After answering the questions, the system would provide a set of search results with direct links to the imagery for viewing. Once the image had been retrieved, more advanced search options would perhaps allow the user to identify the information required from the imagery, with options for information layers, such as coastal defence positions, to be added. In essence, the system would act rather like a 'Juke Box', where the user enters a request, and the record, or data layer is retrieved on demand, thus providing a Layered Information Search and Retrieval System (LISaRS).

Arguably, many systems for retrieving remotely sensed data do already exist (see for example NOAA's image catalogue on the Internet (Fig. 6)). However, the human computer interface (HCI) is frequently too complex for the non-specialist user to understand and use. Moreover, the data supplied is not geared toward a specialist

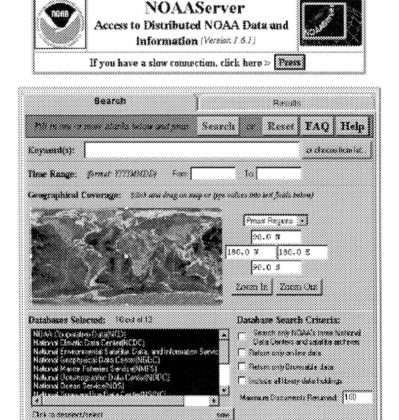

Fig. 6. NOAA's image retrieval system (courtesy of NOAA - http://www.epic.goc/cgi-bin/NOAAServer)

application, and it is not often arranged in a hierarchical fashion. Such systems are limited because the potential user still needs to know what data would be necessary for their application (requiring a certain knowledge about the sensor and its characteristics) and would also have to know how to get the information out of the data. The LISaRS, on the other hand, would be specially developed for coastal zone management, and would provide a simple HCI with the prompts that are needed to deliver the data and information to solve a specific issue. The data types, having already being matched to an issue, would be exactly what the user required.

Discussion and Conclusions

Human beings appear to have an in-built and instinctive ability to gather and use geo-spatial data. Indeed, the use of such data is an essential and invaluable part of day-to-day living. Maps are perhaps one of the most widespread methods of displaying and using geo-spatial information. However, the advent of new technologies, such as remote sensing and GIS has dramatically increased our ability to understand the world around us. In the modern world, the effective management of the environment has become a necessity. Commensurate with this is the need for geo-spatial data about the environment that will eventually become the information and knowledge that is required to carry out environmental management. Remote sensing is an excellent source of data about the environment. One of its key advantages is that because it is a 'snap shot' of the real world, people will instinctively relate to and recognise what is shown in the image. Despite this, remotely sensed data is not as widely used as it could be, because those who need to use it (environmental managers) are not familiar with its utility, or automatically assume that aerial photography is the best source of data for everything. In reality, understanding that each sensor provides imagery with its own set of benefits and limitations is the key to its use. The spatial resolution of an image (rather like the scale of a map) is the basic limit to the amount of information that an image contains when considering spatial relationships, even between spectral data. On this basis, imagery can be organised with a hierarchical structure, from a high spatial resolution to a low spatial resolution. By determining the spatial, temporal and spectral requirements of an environmental issue, it is possible to understand the information requirements for solving that issue, and to match these up with the corresponding characteristics or utility of an image. In this way, criteria can be developed to guide the user to the correct image for a particular issue. A further development of this is to construct a computer-based Layered Information Search and Retrieval System that will automatically retrieve an image using the users own search criteria. Such a system would provide greater utility than a standard image catalogue, GIS or Decision Support System (DSS). Its advantage would be in the simple interface to visual data at a range of scales that people can use in a management situation, like coastal zone management.

Many problems still exist when trying to utilise remote sensing technology in the environmental management arena. These include the lack of knowledge surrounding the types of imagery and what they can be used for, image processing techniques to extract useful information, and a general 'slowness' in accepting information technology (IT) in the management workplace. However, the system proposed would effectively help to 'operationalise' remote sensing in this area, an objective that it is highly important to meet.

References

Couclelis, H. (1998) 'Worlds of Information: The Geographic Metaphor in the Visualisation of Complex Information', *Cartography and Geographic Information Systems*, 25 (4), pp.209-220

Doody, J. P., and Pamplin, C. (1998) 'Information and ICZM - Lessons from the European Union's Demonstration Programme on Integrated Management of Coastal Zones', In: *Proceedings of the 27th International Conference on Remote Sensing, 8-12th June 1998*, Tromso, Norway

Goodchild, M. F., and Quattrochi, D. A. (1997) 'Introduction: Scale, Multiscaling, Remote Sensing and GIS', In: Quattrochi, D. A., and Goodchild, M. F. (eds.) *Scale in Remote Sensing and GIS*, Boca Raton, CRC Lewis Publishers, pp.1-11

Green, D. R., and King, S. D. (1998) 'Public Access to Information - Developing a National GI-based Resource', *Government IT*, 2 (1), pp.124-126

Heywood, I., Cornelius, S., and Carver, S. (1998) *An Introduction to Geographical Information Systems*, Harlow, Addison Wesley Longman Limited

Joao, E. M. (1998) *Causes and Consequences of Map Generalisation*, London, Taylor and Francis

King, S. D. (1998) *Remote Sensing as an Information Source for Better Coastal Zone Management*, University of Aberdeen, Unpublished Masters Thesis

Olivieri, S. T., Harrison, J., and Busby, J. R. (1995) 'Data and Information Management Communication', In: Heywood, V. H., and Watson, R. T. (eds.) *Global Biodiversity Assessment*, Cambridge, Cambridge University Press, pp. 607-670.

CHAPTER 35

Predicting the Distribution of Marine Benthic Biotopes in Scottish Conservation Areas Using Acoustic Remote Sensing, Underwater Video and GIS

C. Johnston and A. Davison

ABSTRACT: To implement the European Directive on conservation of natural habitats (92/43/EEC), government conservation agencies, in this case Scottish Natural Heritage (SNH), must identify areas as possible Special Areas of Conservation (pSACs). To inform the future management and monitoring of these sites, SNH need to know the distribution of habitats and species in Scottish marine SACs. Distribution of marine benthic biotopes (habitats and their associated species) was predicted for two physiographically different pSACs in Scotland: Papa Stour in Shetland, and Lochs Duich, Alsh and Long, NW Scotland. The methodology involved acoustic remote sensing, ground validated by remote video and grab sampling, with analysis and display of results using a PC based GIS. Field surveys were carried out in 1996. Acoustic track point data obtained were interpolated to obtain a continuous polygon coverage of the acoustic properties of the sea bed within each pSAC area. An acoustic signature for each biotope type, or biotope complex was determined by buffering the ground validation sites, intersecting buffers with acoustic track data, and deriving descriptive statistics for each acoustic variable. Spatial analysis of acoustic and ground validation data within the GIS was then used to produce a continuous coverage polygon map of predicted biotope types for each pSAC.

Background to the Work

The UK government is committed to the implementation of the European Council Directive (92/43/EEC) on conservation of natural habitats and of wild fauna and flora, commonly known as the EC Habitats Directive. This Directive was a major contribution from the European Community to the Biodiversity Convention at the Rio Earth summit of 1992. The requirements of the Habitats Directive were adopted into UK law under the Conservation (Natural Habitats &c) Regulations 1994 (1995 in Northern Ireland) - referred to here as the Regulations.

The Directive requires that member states select and present to the EC a list of sites for consideration as Special Areas of Conservation. These sites are selected to represent certain habitats and species that are listed in Annex I and II of the Directive. Seven of the 168 habitat types and nine of the 623 species listed in the annexes of the Directive are found in the marine environment of the UK.

In compliance with the Directive, the UK government has charged its conservation agencies (Scottish Natural Heritage (SNH) in Scotland) with the task of identifying and selecting a UK wide suite of marine *possible* Special Areas of Conservation (pSACs). These were identified in 1995 and extensive public consultations were undertaken on their initial selection. In October 1996, on completion of the consultation process, fourteen Scottish marine SACs were presented to the EC as *candidate* Special Areas of Conservation (cSACs).

To inform the future development of management schemes for marine SACs and to assist in the formulation of conservation objectives (required under the Regulations), it is necessary to obtain broad scale survey information of the extent, distribution and quality of the habitats and associated communities (biotopes) within the selected sites. This process will allow the identification of areas of particular sensitivity and also allow a more targeted approach to site management. There is a requirement within the Directive to monitor and report upon the condition of the sites once every six years. To facilitate this monitoring, baseline information had to be obtained. This baseline includes information from detailed Phase II surveys (Hiscock, 1996) of sample sites within the intertidal and subtidal areas, and more broad scale mapping of the intertidal and subtidal areas to obtain an overview of the extent and distribution of the habitats and associated biota.

Description of pSAC Areas Surveyed

Distribution of subtidal marine benthic biotopes (habitats and their associated species) was predicted for two physiographically different SACs: Papa Stour in Shetland (a *candidate* SAC); and Lochs Duich, Alsh and Long, on mainland western Scotland near Skye (a *possible* SAC), using acoustic remote sensing, ground validated by video of the seabed and grab sampling of sediments.

Papa Stour is a small, remote island in western Shetland, separated from Mainland Shetland by Papa Sound, through which strong tidal currents flow. The cSAC area (see Fig. 1) encompasses the island and part of the adjacent Mainland coast. Both the island and Mainland coasts are very exposed to winds and sea swell from the west and south, and subject to strong tidal currents. The island itself is also very exposed to northerly winds and sea swell. Deep water occurs close inshore, with maximum depth within the cSAC area to the north and west approximately 60m below chart datum (bcd). The shores are rugged and predominantly rocky, with high cliffs to the west, lower cliffs to the east, numerous stacks, arches and caves, and several islands and offshore skerries. In the shallow sublittoral (down to around 20 m bcd) the substratum is largely rugged bedrock with coarse mobile sand. Offshore the substratum is predominantly coarse sands and gravels with low outcrops of bedrock. The main island of Papa Stour also has four shallow voes or inlets, which provide shelter from prevailing wind and waves.

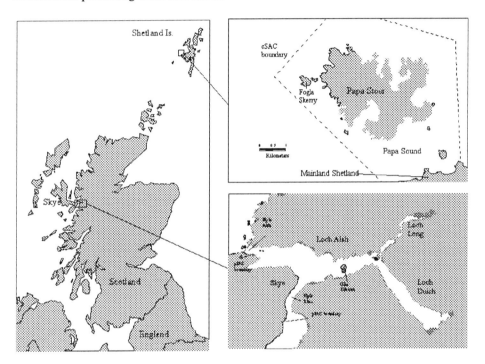

Fig. 1. Locations of Papa Stour cSAC and Lochs Duich, Alsh and Long pSAC.

The Lochs Duich, Alsh and Long pSAC (see Fig. 1), in contrast, is very sheltered from wind and sea swells by the island of Skye, and the hills and mountains which surround the lochs. The shallow sublittoral areas of all three of the lochs are predominantly steep and rocky, and all have deep sediment basins separated by shallower sills. Lochs Duich and Alsh are amongst the deepest of the Scottish sealochs, with maximum depths of over 100m bcd in outer Loch Alsh and the centre of Loch Duich. Lochs Duich and Long in particular have very steep, vertical or overhanging rock shores extending into the sublittoral, with submarine cliffs of around 80 m depth in upper Loch Duich. There are relatively shallow sills in the middle narrows and at the entrance of Loch Long; between Lochs Duich and Alsh; in mid Loch Alsh between the mainland and Glas Eilean; and between Skye and the mainland at the two entrances to Loch Alsh (Kyle Akin to the north west and Kyle Rhea to the south west). In these narrows very strong tidal currents occur, with currents at spring tides up to 7 knots (kn) in Kyle Rhea, and 3 kn in Kyle Akin (Admiralty Chart 2540). Freshwater inputs to the system can be high, and due to the sheltered nature of the lochs, incoming freshwater is often not mixed within the water column. This may result in up to around 6m depth of fresh water lying on top of salt water within Loch Long and upper Loch Duich at times of high freshwater inputs.

The biological interest of the two SACs relates to their differing environmental conditions. Papa Stour cSAC was selected for its reefs and caves, with biological communities characteristic of northern wave exposed conditions. Lochs Duich, Alsh and Long pSAC was selected also for its reefs, but particularly those characteristic of very sheltered conditions and subject to considerable freshwater influence.

Surveys

Due to time and financial constraints, and the large areas of seabed needing to be surveyed, Scottish Natural Heritage (SNH) required a broad scale survey strategy to be adopted, and preferably one which could incorporate previous data. Remote sensing was therefore required, however such methods using visual or other spectra are not suitable to map seabed types in deep and turbid waters (Davies, 1997). Acoustic remote sensing with analysis of data within a Geographic Information System (GIS) appeared to be the best available option. The methodology selected was developed by the *BioMar* team at the University of Newcastle-upon-Tyne, England (now *BMap* team), and involved acoustic remote sensing, ground validated by remote video, grab sampling and diver recording, with analysis of data and display of results within a GIS (Davies et al., 1997). This combination of methods was also capable of incorporating previous survey data where accurate site positions were available. Such biological data were available from previous surveys by various methods for seven sites in Papa Stour, and sixty eight sites in Lochs Duich, Alsh and Long.

The acoustic and ground validation surveys were carried out by Entec personnel and subcontractors from 10th to 19th August 1996 for Papa Stour cSAC, and from 15th to 22nd September 1996 for Lochs Duich, Alsh and Long pSAC.

An acoustic ground discrimination system (AGDS), combined with a Global Positioning System (GPS) for position fixing, linked to a personal computer with navigation and data collection software recorded data on acoustic properties of the seabed at 5 second intervals. Whilst the boat traveled along a set path at a speed (over ground) of approximately 4 kn., continuous sets of measurements (or tracks) of the physical nature of the seabed were recorded and displayed on the computer using navigation software.

For Papa Stour, the seabed over most of the survey area was at less than 30 m bcd, with a maximum depth at the boundary of the cSAC of approximately 60 m bcd. A *RoxAnn Groundmaster*™ AGDS was used, which analyses the primary and multiple return echoes from a high frequency (200 kHz) echo sounder to quantify the depth, roughness (echo 1 or E1) and hardness (echo 2 or E2) of the seabed (Chivers et al.

1990). The maximum operating depth is limited by the power output of the echo sounder used. The power output of high frequency echo sounders is generally, and was in this case, too low to give sufficiently good second echo (E2) returns in water deeper than approximately 45 m, depending on the nature of the seabed. The *RoxAnn*™ system requires both the 1st and 2nd echo returns to function properly, and this system was therefore operating at the limit of its capability in the deep areas at the boundary of the Papa Stour cSAC.

For Lochs Duich, Alsh and Long, acoustic data were required from shallow waters down to over 100 m depth. A *RoxAnn*™ system with high frequency sounder would not give ground discrimination in such deep waters. A *RoxAnn*™ with low frequency sounder could discriminate in deeper waters, but would not provide sufficient ground discrimination in shallow water. The system employed was a *QTC*™ acoustic ground discrimination system, also with a 200 kHz echo sounder, which analyses the shape of only the primary echo return (Collins et al. 1996), and can therefore operate successfully in both shallow waters and down to over 100 m depth. The *QTC*™ system outputs depth and three parameters Q1, Q2 and Q3, to describe the shape of the echo.

AGDS are designed to map in real time the physical characteristics of the seabed, and do not take account of depth in their classifications of different seabed types. This poses problems in the use of such systems for mapping of biological communities, as the same biological communities may occur on very different seabed types, and water depth is a fundamental factor in determining the type of biological community present. It is therefore necessary to post-process the acoustic data and calibrate them with ground validation data, for successful mapping of biological communities.

Ground Validation

The acoustic data were ground validated by high quality drop-down video for epifaunal information and/or grab sampling areas of seabed for infaunal information. Sites were located where preliminary examination of the acoustic data indicated areas with acoustically different properties; and in areas with apparently the same acoustic properties, but with a wide depth range. In the latter case, validation sites were located in both deep and shallow water within these areas. In sedimentary areas, grab samples for infaunal and particle size analysis were taken in addition to the video sample.

Video recordings obtained were analysed to describe the physical and biological characteristics of the seabed to compile an inventory of biotopes present within the survey. Sites were assigned to biotope types according to the Marine Nature Conservation Review (MNCR) biotope manual, Version 96.7 where possible (Connor et al., 1996), or provisional new biotopes were described if a record did not fit into any of the existing biotope types. In kelp forests and parks where the substratum type was not visible, and where the variety and detail of understorey flora and fauna was not clearly distinguishable from the video, some difficulty was encountered in allocating sites to biotope types. This was because the biotope classification (Version 96.7) did not have enough suitable broad categories in which to fit observations where Phase II level information (Hiscock, 1996) on a site was not available. Difficulties were also encountered in relating infaunal data from grab samples to epifaunal data from videos to fit biotope descriptions in the biotope manual.

Interpolation of Acoustic Data

The acoustic track data points obtained as described above (Fig.s 2 and 4) were converted to a continuous polygon coverage of the acoustic properties of the sea bed by interpolating between adjacent data points to calculate values for intermediate areas within *Surfer for Windows*™. Standard geo-statistical procedures were employed for the interpolations; a review of geo-statistics by Rossi et al. (1992) suggested that the procedure *kriging* was most suited to random data points (Davies et al. 1997). *Surfer for Windows*™ provides a kriging algorithm to reduce the track data to a rectangular grid of data points for the survey area. Display of the interpolated results as a grid, rather than smoothing the data to provide hard boundaries, which in these areas do not usually exist, subliminally reflects the resolution of the predicted data. The grid size, in this case of 50 m by 50 m, should be selected as appropriate for the acoustic track spacing achieved during the survey (which was between 100 m and 500 m in these cases). A larger grid should be selected where track spacing is greater, to indicate the lesser resolution of the track data. The resulting map is interpreted as the topographic and physical habitat map.

Matching of Acoustic and Ground Validation Data

The map of acoustic properties of the seabed was analysed with biotope data from ground validation sampling within *MapInfo*™ GIS, to produce a predictive map of biotope distribution. Part of this analysis was to derive an acoustic signature for each recorded biotope or group of biotopes. The *MapInfo*™ buffer capability was used to create a 50 metre buffer around each ground validation site or line. Fifty metres was selected as an appropriate size buffer, to reflect the accuracy of position fixing during the ground truthing. Each buffer was then coded according to the biotope recorded at that ground validation site. Within *MapInfo*™, acoustic data points falling within each buffer were then captured, and assimilated into a table. In the few cases where more than one biotope was distinguished from one video tow, the buffer around the video tow line was split logically into sections representing each biotope. Minimum, maximum, mean and variance values of the acoustic variables E1, E2 and depth (for *RoxAnn*™ data) or Q1, Q2, Q3 and depth (for *QTC*™ data) were generated within *MapInfo*™ from the captured acoustic data for each buffer, to form an acoustic signature for each biotope or group of biotopes.

Using the acoustic signatures for each biotope group to construct Boolean queries, areas which matched each signature were selected in turn and shaded according to the corresponding biotope. An example of a query would be "select all areas where E1 is between 0 and 0.5, and E2 is between 0.15 and 0.75, and depth is less than 5 m, code these areas with biotope type 1". These codes were used to prepare the maps (Figs. 3, 5 and 6) showing the distribution of the biotopes or biotope groups, by assigning a shade or fill to each individual code.

The boat used for the survey of Loch Long was different to that used for the rest of the survey due to a low bridge at the entrance of the Loch preventing access by the main survey vessel. It is generally accepted that acoustic data obtained from different boats under different environmental conditions may not be comparable, therefore acoustic signatures were derived separately for Loch Long using only biotopes found in that Loch, and acoustic signatures for Lochs Duich and Alsh using only acoustic and ground validation data obtained in those lochs.

The range of data points within different signatures may overlap with each other. When data are mapped, this may result in some areas with narrow data ranges being obliterated by others with a very wide spread of data points within a signature. Therefore manipulation of the signatures for each biotope was required, in particular for those biotopes with only a relatively small number of associated data points, but a very wide data range. This was an iterative process, of reviewing the data range for each biotope, checking the ground validation data to see how well particular records fitted with the relevant biotope description, possible re-allocation of ground validation sites to another biotope, then re-calculating the statistics for the particular signature, and reviewing again.

Some difficulty was encountered in this process, as we found that at this level of survey, the acoustic data ranges for some biotopes were not distinguishable separately. This was found where different biotopes have similar life forms (Bunker and Foster-Smith, 1996), and occur on the same substratum types. An example of this occurred on steep bedrock and boulders in the Lochs Duich, Alsh and Long pSAC. Ground validation recorded at least two biotopes on steep rock below the kelp zones: dead men's fingers (*Alcyonium digitatum*) where the rock or boulders were wave exposed or tideswept, and brachiopods (*Neocrania anomala*) and solitary sea squirts (ascidians) where steep bedrock or boulders occurred in sheltered, still water conditions. These two biotopes are very distinct, yet both consist of animals with similar form and size, and occur on the same substratum type, albeit in differing environmental conditions. Unfortunately, the range of acoustic data associated with the buffers for ground validation sites allocated to these two biotope types was wide, and overlapping, and we could not distinguish acoustically between the two on the basis of the acoustic data. This problem was overcome by allocating one acoustic signature (with a consequently wide data range) to several biotopes with similar life form, in this case termed 'deep rock and mixed substrata with faunal turf'.

Tables 1, 2 and 3 relate to the legend of the maps in Fig.s 3, 5 and 6, and show the MNCR biotope codes (Connor et al., 1996) for each lifeform category used on the maps. Biotope codes prefixed with DAL or PS are provisional new biotopes identified during the Duich, Alsh and Long survey, or the Papa Stour survey respectively. Phase II MNCR forms were completed for these biotopes during our surveys, and the forms submitted to MNCR for verification.

Where there were known to be sublittoral cliffs along the eastern edges of Loch Duich, with deep soft sediment on the seabed below, the interpolated acoustic data close to the shoreline were misleading. The acoustic system, because it was directed vertically downwards from the sea surface, could not detect the cliffs, and showed only the soft sediment adjacent to them. This was felt to be misleading when represented on the maps, therefore a blanked-out buffer zone of 50m width was applied to the maps within the GIS along the whole coastline of Lochs Duich and Alsh. This method was thought valid as acoustic data were not obtained consistently very close to the shoreline during the survey. This was not, however, implemented for the Loch Long data, because acoustic data were obtained close to the shore in this case, and as the loch is so narrow, this procedure would remove all data for some narrow stretches of the Loch.

TABLE 1 (see accompanying Fig. 3)

Papa Stour lifeform descriptions (Fig. 3)	Constituent biotope codes (Connor et al., 1996)
Kelp forests on rock	MIR.LhypGz.Ft, MIR.Lhyp.Ft
Kelp parks on rock	MIR.Lhyp.TPk, MIR.LhypGz.Pk
Kelp on deep scoured mixed substrata	MIR.LsacScrR
Kelps on shallow mixed substrata	MIR.XK, SIR.Lsac.X
Faunal and algal crusts on rock	ECR.AlcC
Dense faunal turf on mixed substrata	ECR.AlcTub, MCR.FaAlc, MCR.Oph
Coarse sands	PS.CGS, PS.IGS
Shallow fine sands	PS.IMS
Shallow fine sand and mud	PS.IMS.Are

TABLE 2 (see accompanying Fig. 4)

Lochs Duich and Alsh lifeform descriptions (Fig.s 5 and 6)	Constituent biotope codes (Connor et al., 1996)
Kelp on rock	MIR.Lhyp.Pk, Lhyp.TPk
Kelp on mixed substrata	MIR.Lhyp.Ft, SIR.Lsac.Ft, SIR.Lsac.X
Scoured/tideswept mixed kelps	MIR.XK
Deep rock and mixed substrata with faunal turf	ECR.AlcTub, MCR.Oph, SCR.NeoPro, DAL.SCR.ModOph
Deep muddy sand and boulders with faunal turf	SCR.ModHo, DAL.CMS.ModEch
Deep mixed substrata with sparse fauna	DAL.CMX.Fa
Maerl	IGS.Phy
Limaria beds	IMX.Lim
Shallow muddy sands	DAL.IMS.PAB
Deep muddy sand with polychaetes and cockles	DAL.CMS.Pcer
Deep muddy sand with echinoderms	DAL.CMS.PE
Deep soft mud with burrowing megafauna	CMU.SpNep, CMU.Beg, DAL.CMS.Bur, DAL.CMS.BvPol

Predicting the Distribution of Marine Benthic Biotopes

TABLE 3 (see accompanying Fig. 5)

Loch Long lifeform descriptions (Fig.s 5 and 6)	Constituent biotope codes (Connor et al., 1996)
Sublittoral fringe mussel beds	SLR.Myt.X
Kelp on mixed substrata	MIR.Lhyp.Ft, SIR.Lsac.Ft, SIR.Lsac.X
Tideswept faunal turf	ECR.AlcTub
Deep silted rock with faunal turf	SCR.SoAs
Mixed substrata with polychaetes and bivalves	DAL.CMX.PolBv
Shallow muddy sand	DAL.IGS.PolVS
Deep burrowed soft mud	CMU.SpNep, CMU.Beg, DAL.CMS.Bur, DAL.CMS.BvPol

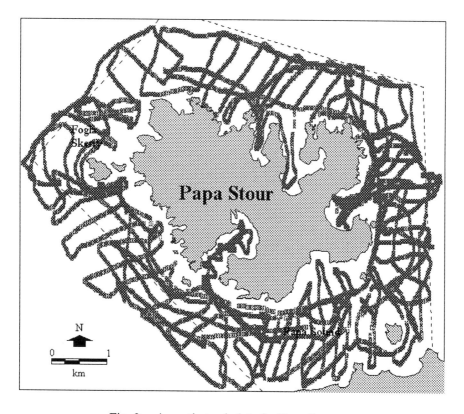

Fig. 2. Acoustic track data for Papa Stour

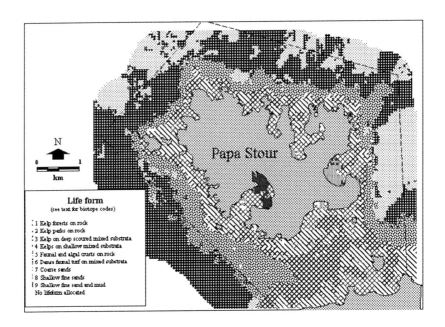

Fig. 3. Predicted distribution of life forms within Papa Stour cSAC

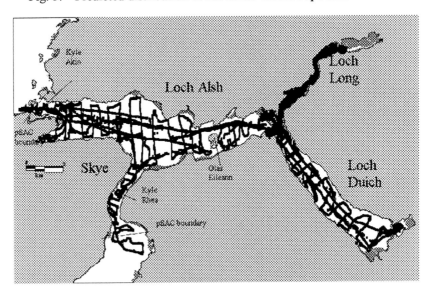

Fig. 4. Acoustic track data for Lochs Duich, Alsh and Long

Predicting the Distribution of Marine Benthic Biotopes

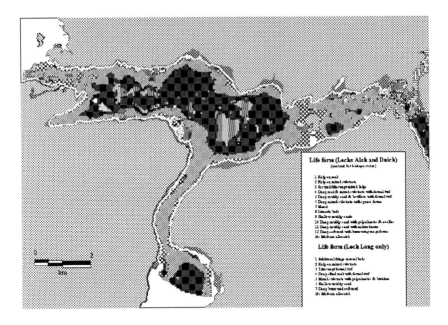

Fig. 5. Predicted distribution of life forms within Loch Alsh, Kyle Akin and Kyle Rhea

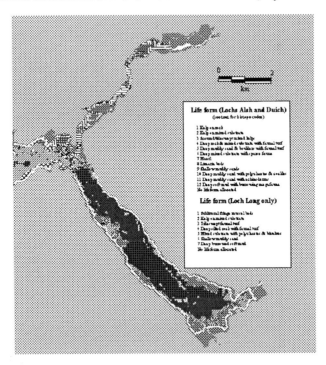

Fig. 6. Predicted distribution of life forms within Loch Duich and Loch Long

Use of the Information

The information gathered and presented from the two surveys will have a number of key applications. The surveys were commissioned primarily to gather information that would assist in the implementation of the Regulations (1995) which deliver the requirements of the Habitats Directive.

Under the Regulations the Relevant Authorities (local authorities and other regulating bodies) are expected to undertake their management of the site so that the conservation features of the site are maintained. The conservation agencies are required to provide the relevant authorities with a series of conservation objectives which will ensure the conservation features. The relevant authorities are encouraged to develop a more integrated approach to the management of the site and with advice assistance from the conservation agencies develop a management scheme to deliver the conservation objectives. This whole process is expected to involve detailed and wide consultation and discussion.

One of the most important uses of the data provided by the surveys is to provide a starting point for these discussions. Clearly it is difficult to initiate discussions with other interested bodies and individuals on the future management of a site without a comprehension of the distribution and extent of the habitats and communities found in the site.

SNH are using graphical data in a GIS environment to consider the approach to the development of targeted conservation objectives for both SACs. Both SACs have been proposed for their rocky reef habitats and communities. Having graphically mapped data showing the general distribution and extent of the rocky areas of the site will allow for conservation objectives and the future management scheme to focus on these areas.

The full use of the data gathered and graphically mapped by these surveys have fairly recently been received but already the following uses for the data have been identified.

- **the development of conservation objectives under the Regulations can become more focused and area specific using the information on the extent and distribution of the conservation features, habitats and communities of the sites**
- **the development of a management scheme for each site may use the data gathered during these surveys to develop a zoned approach to the practical management of the sites. This will enable appropriate management measures to be targeted to the most sensitive and vulnerable features of the sites**
- **it will allow the extent and distribution and, when considered with detailed Phase II information, the quality of the biotopes to be monitored**
- **it will be used as part of the baseline data for monitoring the condition of the SACs, which must be reported to the EC every six years**
- **under the Regulations, any new proposed development plan or project would be subject to the appropriate assessment of its likely impact on the conservation features of the site; the data gathered during the surveys**

described here will assist SNH to advise on marine impacts within the two SACs
- assist in the management of intertidal and adjacent SSSIs
- inform SNH staff in case work issues that effect the site
- the graphical mapped data can be incorporated into future interpretive material for the sites

Acknowledgements

This work was commissioned by Scottish Natural Heritage as part of its programme of research into nature conservation. The authors wish to thank Dr Jon Davies for his considerable assistance in the preparation of this chapter.

References

Bunker, F.StP.D., and Foster-Smith, R.L. (1996). *Field Guide for Phase I seashore mapping*. English Nature, Scottish Natural Heritage, Countryside Council for Wales and Joint Nature Conservation Committee.

Chivers, R.C., Emerson, N., and Burns, D.R. (1990). New acoustic processing for underway surveying. *The Hydrographic Journal*. Vol. 56: 9-17.

Connor, D.W. (1989). *Survey of Loch Duich, Loch Long and Loch Alsh*. Marine Nature Conservation Review report MNCR/SR/010/89. Nature Conservancy Council (now Joint Nature Conservation Committee), Peterborough.

Connor, D.W, Brazier, D.P., Hill, T.O, Holt, R.H.F., Northen, K.O., and Sanderson, W.G. (1996). *Marine Nature Conservation Review. Marine biotopes: A working classification for the British Isles, Version 96.7* Joint Nature Conservation Committee, Peterborough.

Davies, J., Foster-Smith, R.L., and Sotheran, I., (1997). Remote sensing in the dark: biological resource mapping in turbid water using sonar. *Proceedings of the 4th International Conference on Remote Sensing for Marine and Coastal Environments, 17-19 March 1997, Orlando, Florida USA*. Vol. II: 355-364.

Hiscock, K. ed. (1996). *Marine Nature Conservation Review: rationale and methods*. Joint Nature Conservation Committee (Coasts and Seas of the United Kingdom. MNCR series), Peterborough.

Hiscock, S., and Covey, R. (1991). *Marine biological surveys around Skye*. Marine Nature Conservation Review report MNCR/SR/003. Nature Conservancy Council (now Joint Nature Conservation Committee), Peterborough.

Rossi, R.E. et al. (1992). Geostatistical tools for modelling and interpreting ecological spatial dependence. *Ecological monographs*. Vol. 62: 277-314.

CHAPTER 36

Submerged Kelp Biomass Assessment using CASI

É. L. Simms

ABSTRACT: Remote sensing is an effective data source for location of submerged kelp beds. However, biomass assessment must be supported by in situ data. This chapter presents a method to estimate the biomass from remote sensing images. The method takes advantage of the synoptic coverage offered by remote sensing and the local variability of biomass measurable through field surveys. An application is presented with an airborne image of the south coast of Newfoundland, Canada. It is shown that biomass estimates vary with the species present in dense kelp beds. The biomass estaimates based on the entire survey is 802 ± 517 metric tons (t). However, an estimate of 1501 ± 77 t takes into account the species dominating the surveyed quadrats.

Introduction

The ecology and biology of submerged marine kelp beds have historically been obtained through field surveys, scuba diving, and laboratory analyses. Remote sensing data provide means to extract information on the location and the biomass of submerged kelp beds (Lambert et al., 1984; Belsher et al., 1996). *In situ* data are usually considered the standard against which the accuracy of remote sensing biomass assessments are compared. The spatial resolution and the expression of the biomass obtained from the *in-situ* data is often incompatible with the remote sensing data. Field data collected in small quadrants, of 0.25 to 1 m^2, report on the wet or dry weight (John et al., 1980; Williams, 1987). On the other hand, the spatial resolution of remote sensing images can be of a few meters, achievable with airborne data, to 20 or 30 m as obtained with satellite images. The discrepancy becomes a problem if the spatial variability of the kelp beds is scale dependent.

The vertical structure of underwater kelp species such as *Laminaria* and *Fucus* is not visible on a vertical image. Remote sensing methods usually derive the weight biomass of submerged macrophytes from the multispectral reflectance or radiance values (Grenier, 1987; Lavoie et al., 1991; Armstrong, 1993). The canopy coverage (horizontal variation) determines the spectral reflectance of submerged kelp beds, not the vertical structure and the biomass it contains.

The approach presented in this chapter uses remote sensing and field data as a complement. The kelp biomass, in terms of canopy coverage, was derived from a remote sensing image and the field data relate the canopy coverage to the wet weight biomass. rather than using remote sensing as a replacement of field data collection. The biomass of submerged kelp beds was assessed for an area of Fortune Bay, Newfoundland. An airborne image was analyzed to extract the area occupied by the population of macrophytes, and the field data collected characterized the biomass found in dense kelp beds.

Data

A CASI (Compact Airborne Spectrographic Imager) image was acquired on August 14,1992 between Fortune and Grand Bank, Newfoundland. The image centered at 47°05' N latitude and 55°48' W longitude covers approximately 4 km of coastline. Light wind and smooth sea state occurred during the flight. Stratus clouds at 1850 m prescribed that the flight altitude be set to approximately 1500 m. As a result, the data were acquired at a resolution of 2x2 m, over a ground swath of 1024 m. The selection of six, 20 nm wide, spectral bands as specified in Table 1 accounted for low light level due to the overcast weather. Proper coverage of the curved coastline required two flight lines. The first image covered 3.1 km and the second image 1.4 km. These were georeferenced in a mosaic image that provided a continuous coverage of the study area.

A private contractor in marine services carried out the field work on September 15, 1992. The survey included fourteen samples located within the area classified as kelp beds on the image. The bottom comprised large boulders to small rocks from 10 m depth to the intertidal zone. There was no measurable quantity of seaweed in the 6 to 10 m depth zone.

The survey was conducted in water depth varying from 3 to 6 m in populations anchored on large rocks and boulders. All sites surveyed displayed a canopy coverage of 95 to 100 %. Gravel and small stones composed the substrate in the shallower water, from the intertidal zone to 3 m depth, between the boulders having little algal growth.

TABLE 1
Spectral band selected

Band	Centre wavelength (nm)	Band limits (nm)	Band	Centre wavelength (nm)	Band limits (nm)
1	475	465 - 485	4	610	600 - 620
2	525	515 - 535	5	670	660 - 680
3	550	540 - 560	6	750	730 - 770

The surveys included the proportion of the bottom occupied by stipes and short plants in 1 m^2 quadrants. The wet weight biomass in kg/m^2 was reported for each species occupying more than 5 % of the surveyed quadrants.

Methods

A natural logarithm transformation applied to all spectral bands produced a set of depth dependent variables. A discriminant analysis based on pixels representing kelp free substrate at various depths resulted with a water depth image (first discriminant function) and a bottom type image (second discriminant function (Lyzenga, 1978; Lambert, 1992). The analysis applied to a sub image defined the first discriminant function coefficients representing water depth variation of 0 to 11 m depth (6 fathoms). The transformed images which values were strongly correlated with the water depth have been used to extract a set of coefficients. The analysis included three transformed images of the green and red spectral bands that displayed correlation coefficients of -0.72 to -0.86 (Table 2). Water turbidity affects the blue spectral band. The transformed far red band did not show a strong correlation with water depth, but conveyed information only related to very shallow water. Finally, band 6 was not used due to the attenuation of the infrared wavelength by water.

Given that the variance of the first discriminant function is explained by the water depth variation, most of the information on the bottom type is enhanced by the coefficients of the second discriminant function. A density slicing classification of the second discriminant function image identified groups representing the bottom types. On this image, low values corresponded to bare substrate that on all visible bands would have a higher reflectance value, and the highest values represent the kelp beds. A normal distribution characterizes the groups of pixels assigned to each bottom type category.

TABLE 2
Correlation coefficient between values of the water depth dependent images and the water depth

Transformed band	1	2	3	4	5
Correlation coefficient	+0.41	-0.77	-0.86	-0.72	-0.59

Non-metric statistical analyses were applied to the field data collected throughout the area classified as dense kelp beds. The median, quartile, and notches were identified with the objective of defining the characteristics of a dense kelp bed where canopy coverage is 100 %. The median represents the centre value of the distribution, and the hinges delimit the range covered by the middle half of the data. The notches define the group of values that represents the distribution with 95 % confidence level (Velleman and Hoaglin, 1981). The biomass was estimated with Equation 1 as follows:

$$B = AW$$
(1)

Where B is the total biomass estimate in kg, A is the kelp covered area classified on the image, and W is the median biomass value derived from the quantitative survey in kg/m^2.

Results and Discussion

The bottom type image displayed three groups (Figure 1). The class representing the dense kelp canopy was classified in water depth of less than 5 m. A few isolated pixels of that class distributed in deeper water, suggested that the image radiometric noise caused a misclassification. A 3x3 mode filter eliminated the isolated pixels without modifying the proportion of pixels in each class. The first category occupies 80% of the area classified with 95.7 ha. The second and third categories, representing the kelp beds, are shown on Figure 2. They include respectively, 9% (10.7 ha) and 11% (13.7 ha) of the area analyzed.

The species inventoried in the study area were *Alaria esculenta, Desmarestia viridis, Fucus spp., Laminaria digtata,* and *Laminaria longicruris. Alaria esculenta* dominated in eleven of the fourteen quadrants surveyed, closely followed by *Laminaria sp.*, that was found in ten quadrats.

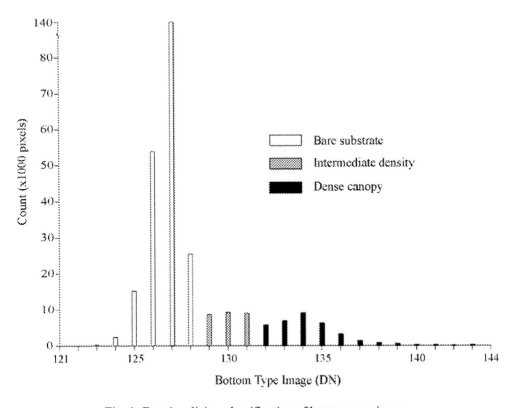

Fig. 1. Density slicing classification of bottom type image

The total weight recorded in sampled quadrants ranged from 1.8 to 14.5 kg/m^2. The total biomass in quadrant occupied only by *L. digitata* and *A. esculenta* was more important, up to twice the biomass occupied by other species composition. The presence ina quadrat of other species such as *D. viridis*, *L. longicruris*, *Fucus* sp. or *A. nodosum* resulted in biomass values of only 1.8 to 6.3 kg/m^2.

Equation 1 was applied to estimate the total biomass with the median value of sampled data. The minimum and maximum biomass were calculated with the lower and upper notches, respectively. The calculation is based on the area of 13.6 ha that corresponds to the dense kelp canopy on the classified image.

A median value of 5.9 kg/m^2 represents the sampled biomass, with a range of 2.1 to 9.7 kg/m^2 (α =0.05%). The total biomass value for the entire study area is therefore estimated to be 802 $\pm$ 517 metric tons (t).

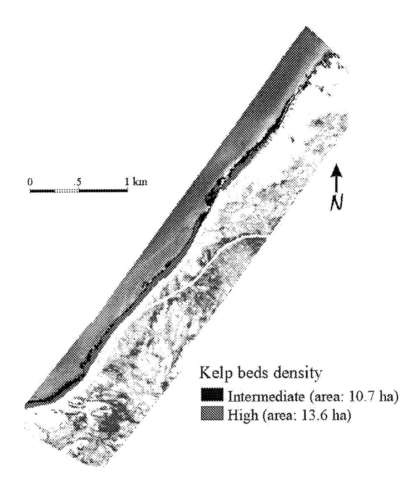

Fig. 2. Kelp beds distribution

A closer examination of the field data shows that two groups could be defined based on the association of species (Figure 3). The first group is characterized by the presence of one to three species that include *D. viridis*, *L. digitata*, *Fucus* spp. or *A. nodosum* which has a relatively low biomass. The median biomass of this group is 3.9 kg/m2 and it ranges from 3.5 to 4.4 kg/m^2 (α=0.05%). The second group displays the species *A. esculenta* and *L. digitata*, alone or in association (excluding all other species). The highest biomass values were present in that population, reporting a median biomass of 12.8kg/m^2 and a very limited variation given the lower and upper notches of 12.2 and 13.5 kg/m^2.

The two groups that were defined, based on the species association, resulted in the identification of populations that are significantly different from each other (α =0.05%). The analysis also showed that the biomass formed by *A. esculenta* and *L.*

digitata (Figure 3) is significantly different (α=0.05%) from the distribution representing the entire dataset. The information provided by a refined analysis of the data would allow increased accuracy of the biomass estimates. For example, if the dense kelp beds were occupied by the first group of species, the biomass estimate would be 530 ± 68 t. On the other hand, kelp beds entirely composed of *A. esculenta* and *L. digitata* could contain a biomass as high as 1741 ± 88 t. A proportional representation of these two groups based on the number of quadrat surveyed (8 for the first group and 6 for the second group) results in a total biomass of 1051 ± 77 t. Additional surveys conducted following a random distribution of the samples would allow a proper evaluation of the proportion occupied by each group of species.

Conclusion

The analysis of an airborne multispectral image provided the area occupied by dense kelp canopy. A series of field surveys was used to derive a median and range of biomass value representative of the submerged macrophyte population of the study area. The combination of the image and the field data resulted in an estimate of the biomass for the area covered with the image. It was found that the variability attached to the biomass estimate is very high, due to heterogeneity of the species composing the kelp beds. Examples of biomass estimates were derived for two groups of species. One group is composed of *A. esculenta* and *L. digitata*, and the other group includes *D. viridis, L. digitata, Fucus* spp. or *A. nodosum*. The biomass estimated for the first group is about twice the biomass of the second.

The results dictate the introduction of guidelines for the sampling strategy supporting the biomass evaluation of underwater kelp beds with remote sensing imagery. A quantitative survey conducted for a limited number of quadrats should be accompanied by a qualitative survey of larger observations. Both surveys would include the following observations and measurements:

- **Visual evaluation of the top of canopy coverage (%) at quadrant location**
- **Visual evaluation of the top of canopy coverage (%) surrounding the quadrant location**
- **Species abundance**
- **Relative and absolute location of the quadrant**

The quantitative survey would include two additional items of information as follows:

- **Wet or dry weight of species occupying the quadrant**
- **Total wet or dry weight in the quadrant**

The objective of a survey is to characterize the biomass and the composition of the macrophyte beds that are identified on a satellite or airborne image. Therefore, the location of dense canopies (100 % coverage) must first be sampled. Time allowing, populations displaying an intermediate density (near 50 %) should also be sampled, because they can be detected using remote sensing data (Lambert, 1992).

It is advantageous to add ancillary data such as water depth above the canopy, total water depth, and Secchi depth. These are easily collected and assist in the extraction of reliable information from the remote sensing data. Finally, data on substrate type and the presence of predators, provide an insight into the stability of the population being assessed.

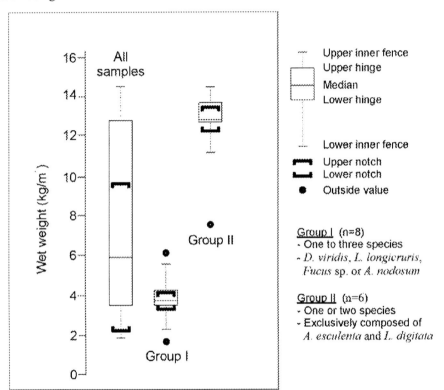

Fig. 3. Box-plot diagram of biomass data

Acknowledgments

This project was funded by the Greater Lamaline Area Development Association (GLADA) and by the Newfoundland and Labrador Department of Fisheries. The image was analyzed at the Geographical Information and Digital Analysis Laboratory (GEOIDAL), Department of Geography, Memorial University of Newfoundland.

References

Armstrong, R.A. (1993). Remote Sensing of Submerged Vegetation Canopies for biomass Estimation. *International Journal of Remote Sensing.* Vol.14(3):621-627.

Belsher, T.A., Lavoie, A., and Dubois, J.-M., M. (1996). La végétation marine: répartition, biomasse et gestion. In F. Bonn (Ed.) Précis de télédétection. Presses de l'Université du Québec/AUPELF, Québec, vol. 2, p. 427-474.

Grenier, M. (1987). *Cartographie Quantitative des Laminaires de la Baie des Chaleurs, Québec, avec le Capteur MEIS-II.* Mémoire de M.Sc., Université de Sherbrooke, Sherbrooke.

John, D.M., Lieberman, D., and Lieberman, M. (1980). Strategies of Data Collection and Analysis of Subtidal Vegetation in Langstone and Chichester Harbours, Southern England. In J.H. Price, D.E.G. Irvine, and W.F. Farnham (Eds.). *The Shore Environment,* Academic Press, London. Vol. 1:265-294.

Lambert, É., Lavoie, A., and Dubois, J.-M.M. (1984). Télédétection des alues marines des côtes du Québec: 1- Bibliographie mondiale annoté. Ministere de 'agriculture, des Pêcheries et de l'Alimentation du Québec. Cahier d'information no. 112, 44p.

Lambert, É. (1992). La localisation des peuplements de macrophytes immergés à l'aide de la télédétection et de données connexes. In F. Bonn (Ed.) Télédétection de l'environnement dans l'espace francophone, Presses de l'Université du Québec, 129-146.

Lavoie, A.F., Bénié, G.B., and Dubois, J.-M.M. (1991). Cartographie quantitative des macrophytes du lac Saint-Pierre avec le capteur MEIS-II. CARTEL, Université de Sherbrooke. Rapport: Centre Saint-Laurent, Environnement Canada, 41p.

Lyzenga, D.R. (1978). Passive remote sensing techniques for mapping water depth and bottom features. *Applied Optics.* Vol. 17:379-383.

Velleman, P.F., and Hoaglin, D.C., (1981). *Applications, Basics, and Computing of Exploratory Data Analysis.* Duxbury Press, Boston.

CHAPTER 37

Monitoring Coastal Morphological Changes Using Topographical Methods, Softcopy Photogrammetry and GIS, Huelva (Andalucia, Spain)

J. Ojeda Zújar, L. Borgniet, A.M. Pérez Romero, and J. Loder

ABSTRACT: This chapter deals with the use of different sources of data (field surveys with total station and GPS, aerial photographs and topographic maps) as well as their integrated digital treatment in a GIS context to quantify the morphological changes in a ridge of coastal dunes in the south-west of Spain. The results show very high and incrementing rates of foredune retreat, significant losses of foredune surface and a clear negative sedimentary balance (lowering and inland migration) in its recent evolution (1979-1996). Two processes can explain this evolution; (i) marine erosion and (ii) the reactivation of aeolian deflation. The combined use of GPS (code/phase) and softcopy photogrammetry seem to provide the best for monitoring future changes.

Introduction to the Study Area

The study area chosen to test the methodology is the stretch of coastal dune ridges between the settlements of Isla Cristina and La Antilla in the west of the province of Huelva (SW Spain) close to the Portuguese border (Figs. 1-3). The zone investigated in detail, approximately 700 m. long and to the west of La Antilla, was chosen for several reasons:

(i) It is **very dynamic**, being wide open to the hydrodynamic impact of the Atlantic Ocean. The average tidal range is 2.10 m (Borrego et al., 1992), a mesotidal coast exposed to wave forces of low to medium energy (76 % of waves are less than 50 cm. high) predominantly from the southwest. The combination of wave force and direction of approach with the alignment of the coast produces a strong dominant littoral drift which moves sediments eastward. The net eastward movement has recently been modelled mathematically at 260,000 m3/yr. (Medina, 1991).

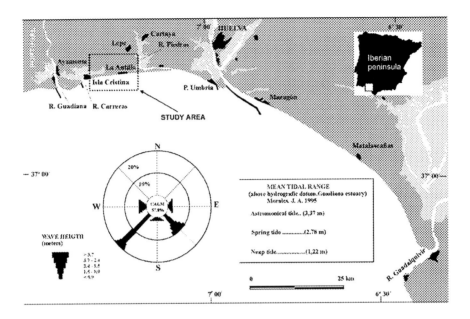

Fig. 1. The Study Area

(ii) The continuous belt of coastal dunes seen today is the result of the recent evolution from a **system of barrier islands** which was still active in the 19th century. Since the 19th century the tidal inlets separating the barrier islands have been closing to form **littoral spits** associated with the outlets of the principal rivers (Guadiana, Piedras, Carreras). Nonetheless between the river outlets we find a ridge of dunes which shelters old lagoons now infilled and converted into saltmarshes beyond which, inland, is the former cliffline marking the line of maximum postglacial transgression (c. 6000 BP). Over and above the ecological value of any coastal dune system and the important role they play in maintaining the morphodynamic equilibrium of beaches, in this case they also serve as a **natural dyke** protecting the areas behind from inundation, as they are below the mean level of high spring tides (tidal marshes) and in part are drained and dried out for various human activities.

(iii) These dune systems have been considered as **economically marginal areas**, the only activity being fishing centred at Isla Cristina with the rest unoccupied and in places, as in the study area, being zones of reforestation. However, since the 1960s, there has been a two way intensification of anthropic pressure.

a) increased fishing activity in Isla Cristina and the problems of maintaining a fixed frontier with Portugal in the late 1970s led to the construction of a system of jetties at the outlets of the Guadiana and Carreras which have had a major impact on longshore sediment transfer
b) development of tourist resort facilities has passed through two phases: a localized development in La Antilla in the 1970s and 1980s, followed in the 1990s by the creation of a major complex at Isla Antilla with associated golf courses and the proliferation of camping sites. For these sites the nearest beach is in the study area and the shortest way is over the dunes

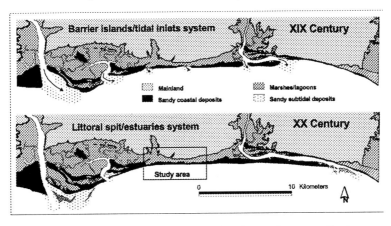

Fig. 2. Area of study: foredune evolution in southwestern Spain (western coast of Andalucia)

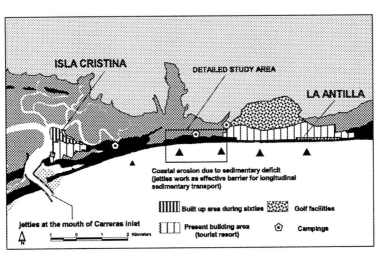

Fig. 3. Detailed area of study: Evolution of built-up areas and tourist facilities

Recent Foredune Evolution (1979-1996)

The location of the study site in relation to the aforementioned hydrodynamic of eastward littoral drift and the pressures of tourism, has meant that the processes which have controlled its recent evolution can be summarized into two morphodynamically distinct systems:

(i) **Marine erosion**: since their construction in the 1970s the jetties have been a total barrier to longshore sediment transport. The study zone being immediately downstream, has a definite sedimentary deficit and has been subjected to intense marine erosion in which the dunes serve as a sediment reserve. The results are:

a) the foredune is subjected to intense erosion processes with significant rates of foredune retreat and the duneface is effectively a cliff.

b) the marine erosion begins at the western end and moves downdrift

c) there is a marked loss of sediment from the foredune which reduces its role as a sediment reserve in storm events.

(ii) **aeolian deflation**: as can be seen in Fig. 4, the reactivation of deflation processes throughout the dune complex were practically non-existent until the late 1980s. Deflation increased throughout the 1990s linked to the increased tourist presence and, above all, to the uncontrolled access to the beach by campers and other temporary visitors from the nearby tourist facilities and villages. The destruction of sand-fixing vegetation along the different paths across the foredune has led to the development of blowouts. The morphological effects can be summarized as follows:

a) inland foredune migration because of blowouts

b) general lowering of foredune, increasing risk of inundation and the destruction of pinewood in the inner dune complex.

c) weakening of the dune by the loss of fixing vegetation increases vulnerability to marine erosion.

In general each morphodynamic system contributes to a typical landward migration of the foredune with significant losses of sediment from this sector of the coastal dynamic system, the destruction of vegetation and an increased risk of inundation.

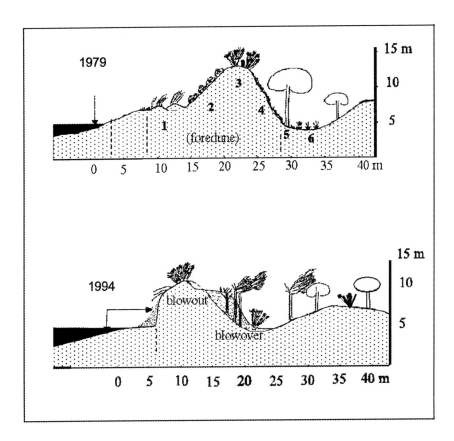

Fig. 4. Processes and effects in foredune evolution (1) marine erosion (sedimentary deficit): coastline retreat; loss of fordune sediments; (2) aeolian deflation: destruction of vegetation due to uncontrolled access to beach through foredune; migration of island due to blowout development; lowering of foredune (risk of flooding)

Objectives

To present the results and an evaluation of the sources, techniques and procedures are used to quantify the morphological impact of the aforementioned processes of marine erosion and aeolian deflation at a very detailed scale (1: 1000). The impact can be traced in the following ways:

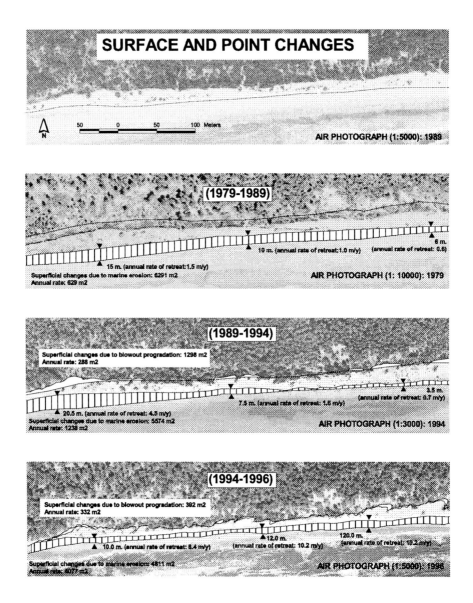

Fig. 5. A series of aerial photographs showing the changes

(i) **point measurement of changes** in the foredune/backshore line used as a coastline marker in particular locations

(ii) **measurement of surface changes** in the foredune either on the seaward face (marine erosion) or on the landward face (inland migration)

(iii) **temporal changes of foredune volume, DTM construction** and evaluation of **sediment budget**.

(iv) **spatial changes of volume**, relocation, etc. shown by **chart differencing analysis** i.e. overlay

Annual rates will be calculated for all these measurements and the result will be presented as **maps** to facilitate international comparison. Additionally the results will be used to produce **animations** to allow a fuller understanding of the patterns of change.

Sources and Methodology

The diversity of the sources of information used to meet these objectives and the need for integrated data processing makes the use of a GIS virtually essential even though some data require pre- or post-processing before integration into the GIS. For our study area the sources and techniques used were:

- **Cartography:** although cartography at 1:1000 is rare in non-urbanized coastal areas, we have found two reference documents with a 1m. contour interval which have provided a planimetric and geometric basis for the GIS[1].

    i) a topographic map at 1:1000 from the Ministry of Public Works used for the application of the Coastal Law of 1989. The map is based on field survey by classical topographic instrumentation and it was provided in analogue format. The map was digitzed and integrated into the GIS for subsequent analysis.

    ii) a topographic map 1:1000 from the new Andalusian Cartographic Institute produced by stereoplotting from air photographs and supplied in digital format.

---

[1]Projection is UTM, co-ordinates UTM zone 29

- **Detailed Air photographs:** we have available four photogrammetric flights, all in winter, a critical point because of the strong seasonality of the maritime climate in this zone.

February	1979	1:10000
April	1989	1:5000
December	1994	1:3000
February	1996	1:3000

The geometrically consistent integration of the air photos into the GIS required a preliminary digital processing by scanner and Desktop Mapping System 4.0 . This software eliminated tilt and radial distortion (ASPRS, 1996) and allowed the production of vector format output by stereoplotting (spot height, coastline/foredune planimetry etc.) or raster mosaics (orthophotos and DTM).

- **Field Survey:** this was carried out with a **Leica TC600 infrared Total Station** with a vertical or horizontal angular accuracy of 5" , a distance meter accurate to 3mm ±3ppm and an operating range between 1.1 and 1600m. This equipment was used to survey the ground control points for air photo rectification and then to assess the actual working accuracy of the Leica 200 GPS system. All GPS surveys were made by a mobile rover receiver (Leica 200 GPS) and postprocessed and differentially corrected (Cook et al., 1996) using data from a master receiver (Leica 200 GPS) fixed in a known location (bench mark). The operating range between the two receivers was 10-15 km.. The three available modes were used as follows:-

    i)**static mode: x,y,z accuracy = 5mm ± 1ppm.: measuring time 10-15 minutes: used for measuring ground control points for new air photos**

    ii) **stop-and-go mode: x,y,z accuracy = 3mm ±1ppm: measuring time 15 minutes to start and 25 seconds per new point: used to survey point heights on the foredune in order to improve DTM quality, the coastline using the foredune/backshore contact line, and the 1997 landward limit of foredune propagation**

    iii) **cinematic mode: x,y accuracy = 3cm ±1ppm and z accuracy = 6cm ±1ppm: measuring time is 15 minutes to start then automatic continuous measurement: used to survey coastline in 1997**

Results and Interpretation

Annual foredune retreat rate: each rectified or orthophoto was introduced into ArcInfo through a compatible format (TIF) and three measurements (west, centre, east) were made at each date with the following results:

	West	Centre	East
1979 - 89	15.0m (**1.5m/yr.**)	10.0m(**1.0m/yr.**)	6.0m(**0.6m/yr.**)
1989 - 94	20.5m (**4.5m/yr.**)	7.5m(**1.6m/yr.**)	3.5m(**0.7m/yr.**)
1994 - 96	10.0m (**5.0m/yr.**)[2]	12.0m(**6.0m/yr.**)	12.0m(**6.0m/yr.**)

Interpretation

We are clearly dealing with a regressive coast for which all dates show significant rates of retreat that can be explained by the construction of access breakwaters at the ports of Isla Cristina and Ayamonte. There is a progressive increment in the annual rates throughout the period that is consonant with the intensification of anthropic pressure, most notably after 1990. There is also differential net erosion from west to east with high rates in the west especially between 1989 and 1994 when this sector suffered the bulk of the erosion. However, these single annual figures are not necessarily representative and are certainly site dependent, especially given the complex dynamics of the erosion of these beaches over a very short time periods - presence of rip currents, giant cusps etc. Nonetheless, since they are used internationally these figures have been included to facilitate comparison with other sites.

Changes of surface area: the backshore/foredune line was interpreted and digitized from the rectified photographs or orthophotos for each date. An overlay process was carried out for each pair of dates and the results were overlaid on the orthophotos before printing.

Foredunes	Area lost by retreat	Area gained by inland propagation
1979-89	-6291m^2 (-629m^2/yr.)	nil
1989-94	-5574m^2 (-1238m^2/yr.)	+1298m^2 (+288m^2/yr.)
1994-96 [3]	-4811m^2 (-2405m^2/yr.)	+ 392m^2 (+196m^2/yr.)

as can be seen in Fig. 5 and in the table above, the spatial continuity of aerial photography produces results that are more detailed and interesting to interpret.

---

[2] The rates of retreat have been calculated by dividing by 2. In reality December 1994 - February 1996 requires division by 1.18 giving yearly rates of 8.4, 10.2, and 10.2 m/yr respectively

[3] As before the total was divided by 2. It should have been 1.18 which would give -4077 and +332 m^2/yr

i) during the first period (1979-89) all change was by marine erosion of the foredune's seaward side, as the foredune, and especially its landward side, was well controlled by vegetation. Anthropic pressure was very low in this period and concentrated in the built-up areas of La Antilla and Isla Cristina

ii) in the second period (1989-94), foredune retreat rate actually doubles but is spatially concentrated in the western sector. Yet, because of the higher incidence of tourist pressure and the opening of the first camping sites, the vegetation is beginning to be destroyed by overdune paths and landward dune migration begins although it is as yet only 1/6 of the seaward face's retreat rate

iii) during the third period (1994-94), rates of retreat on the seaward side are up to 6 times those of the first period, even though in this case the spread of marine erosion is more homogenous. Moreover, the rates of foredune landward migration continue at similar rates despite having a reduced width of dune to draw on for material

In summary the three phase evolution of coastal erosion in this zone is:

1979-89; a well developed foredune, with two ridges, and fixed by vegetation becomes a seaward eroded foredune with a landward slope still controlled by vegetation. This evolution is explained by sediment deficit caused by the combined impact of jetties and dam-regulated river discharges. Anthropic pressure is scant.

1989-94; expansion of built-up areas and the development of camp sites intensifies anthropic pressure, especially cross-dune paths to the beach. The sedimentary deficit continues to increase. At the same time the dunes are weakened by vegetation loss and lowered by deflation which together produce landward-moving blow-outs.

1994-96; sediment deficit continues to increase for the same reasons as before and anthropic pressure is yet stronger, because of more camp sites and increased occasional visits. Blow-outs increase and the heavy storms in the winter of 1995 meant that in some places the foredune disappeared leaving many of the pines to die subsequently from the combined effects of the wind, sand, and salt spray.

Foredune Volume Calculation and Sediment Budget

The possibility of obtaining altimetric data from whichever of the sources used has made it possible to characterize these changes in even greater detail. The first stage was to establish how much sediment the foredune had lost or gained, which required the construction of DTMs from the available data. Here we present just the DTMs for 1989 and 1994 (Fig. 6). For these two reference dates the measurements come from the

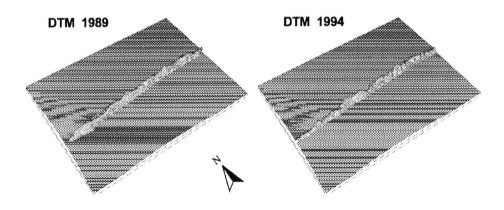

Fig. 6. 1989 and 1994 DTM

1989 1:1000 topographic map with 1m. contour interval, and from classical field survey and air photos for 1994. Both the contours and the spot heights were digitized and integrated into the GIS. The TIN module of ArcInfo was used to generate the DTM (Brandli, 1992) and the GRID module was used to convert to raster and to export as a geocoded image file for subsequent graphic use. For the other DTMs, and later when we return to the 1989 and 1994 ones, we are using a procedure to integrate the GPS data and an automatic stereocorrelator (Desktop Mapping System 4.0) with pairs of air photos. There are not yet any results from this process.

Results

DTM Volume 1989	2,686,094 m^3
DTM Volume 1994	2,675,713 m^3
Volume Change	**-10,381 m^3**
Annual rate	**- 2,077 m^3/yr.**
Annual rate per meter of coastline	**-2.6 m^3/yr.**

Interpretation

The absolute changes in volume and annual rates of loss reinforce the results calculated from surface area measurements, indicating forcibly the serious loss of material and thus the reduced effectiveness of the dune as a sediment reserve.

Volumetric Spatial Changes 1989-1994

The DTM allows not only the quantification of sediment loss but also its location using chart differencing in raster format (overlay). The graphics in Fig. 7 show clearly the four types of sediment redistribution.

> **i) areas of foredune lost since 1989 because of marine erosion on the seaward side of the dune. The losses have been calculated at pixel level in these areas and can be seen as dark grey.**
>
> **ii) areas added to the foredune since 1989 by inland migration are also quantified at pixel scale and are seen as light grey**
>
> **iii) areas of foredune which have lost height by aeolian deflation**
>
> **iv) areas of foredune which have gained height by aeolian action**

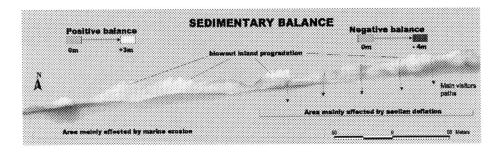

Fig. 7. Sedimentary Balance

Interpretation

Fig. 7 shows quite explicitly the clear predominance of areas with a negative balance, mostly by marine erosion in the western part and seen clearly in the 'cliffing' of the dune in 1994. In the east the loss is associated with deflation caused by tourist activity. In summary all the areas of positive balance are landward.

3D Graphic Presentation and Animation

Using the information from the 1989 and 1994 DTMs, a set of images has been assembled to help understanding of the dune evolution between these dates (Fig. 8). The DTM images were exported to VistaPro 3.0 to make four videos (*.avi) integrated into the GIS via hyperenlace in ArcView 3.0, but playable in any Windows environment. Two of the videos are used to show a flight over the dune, one for each date but with the view parameters held constant to aid comparison. The third video uses the 1994 DTM with the volumetric changes image overlaid. The fourth is used to model the change from one date to the other making use of the analysis tools from IDRISI to generate the intermediate stages.

Conclusions

The different techniques of analysis all show, very clearly, a foredune **undergoing high and increasing erosion**, causing the loss of a large amount of sediment from particularly significant areas of dune in a highly popular tourist area. Moreover, the increase in the rate of retreat is explicable in terms of human action, the building of jetties and tourist pressure, especially overdune paths which destroy vegetation and lead to blow-outs.

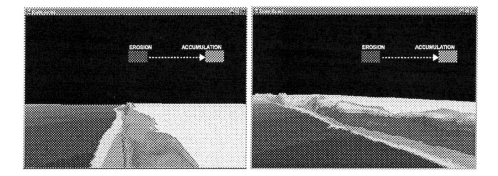

Fig. 8. Screenshot from VistaPro

These erosive processes and their consequential deposition lead to spreading of sediment and **loss of altitude** which eventually will lead to the dune failing to protect the landward zone against saltwater flooding. This has already occurred in some areas of the dunes. Among all the techniques used to trace the morphological changes in the foredune, the best results seem to come from a combination of high precision **GPS (code/phase)** and digitized high resolution air photos and **softcopy photogrammetric treatment**. The only really effective way to integrate and access the multi-source data is through a GIS but only after the full capability of other specialist software has been applied for specific functions.

References

American Society of Photogrammetry and Remote Sensing (1996). Special issue on softcopy photogrammetry. *Photogrammetric Engineering and Remote Sensing*. vol. 62.
Borrego, J , Morales J.A., and Pendon J.G (1992). Efectos derivados de las actuaciones antropicas sobre los ritmos de crecimiento de la flecha litoral de El Rompido. *Geogaceta*, 11: 89-92

Brandli, M. (1992). A triangulation-based method for geomorphological surface interpolation from contour lines. *EGIS'92 Proceedings*, 691-700.

Cook, A.E., and Pinder J.E. (1996). Relative accuracy of rectification using co-ordinates determined from maps and Global Positioning System. *Photogrammetric Engineering and Remote Sensing.* vol. 62 1:73-77

Dabrio, C.J. (1982). Sedimentary structures generated on the foreshore by migrating ridge and runnel systems on microtidal and mesotidal coast of Spain. *Sedimentary Geology* 32: 141-151

Medina, J.M. (1991). La Flecha de El Rompido en la dinamica litoral de la costa onubense. *Ingenieria Civil* 80:105-110.

CHAPTER 38

Characterization of Coastal Waters for the Monitoring of Pollution by Means of Remote Sensing; the Use of Satellite Imagery to Establish the Appropriate Pattern for Timing and Location of Sampling in Coastal Waters.

J. Ojeda Zújar, E. Sanchez Rodriguez, A. Fernández-Palacios, and J. Loder

ABSTRACT: This chapter presents a possible application of remote sensing to establish the pattern of sampling required to monitor the condition of coastal waters. The aim is to provide a dynamic characterization and delimitation of differing water masses in the estuary of the rivers Tinto and Odiel (Huelva, SW Spain) under specific synoptic conditions. This information will allow the planning of sampling routes which guarantee spatial and temporal representativeness. The method used for the dynamic characterization is based on a simple segmentation of one visible band. Standard Principal Components Analysis was used to zone the estuary. The results permit a series of recommendations as to how, when, and where to sample.

Introduction

In 1992 the Ministry for the Environment of the Junta de Andalucía (CMA), in collaboration with the Department of Physical Geography and Regional Geographic Analysis of the University of Seville began a line of investigation - Programme for Monitoring the Quality and Dynamics of Coastal Zone Marine Waters using remote sensing. The basic aims have been to:

1) **evaluate the possibilities of applying remote sensing to the study of various problems which affect the Andalucian coastal zone**

2) develop specific applications which allow the systematic exploitation of the ample collection of satellite images available to the regional ministry (Ojeda et al. 1994). Both the original images and the results of the thematic applications will be incorporated as information sources within the Andalucian System of Environmental Information (SinambA) developed by the Regional Environmental Ministry. Among the various thematic applications is one directed at the evaluation and monitoring of coastal water quality, to which there are two strands:

i) mapping water quality parameters such as turbidity, suspended solids, heavy metals etc. from satellite images (Fernández Palacios et al., 1994)

ii) assist the monitoring authorities to locate sampling sites which are spatially and temporally representative throughout the year

This chapter relates to this second aspect and presents the methods used, the conclusions reached concerning the dynamics of the estuary, and the spatial distribution of the various water masses under differing synoptic situations. This, in turn, will improve the sampling design on which the monitoring of water quality is based.

Study Area

The Tinto-Odiel estuary is formed by the confluence of the rivers Tinto and Odiel on the Atlantic Coast of SW Spain near Huelva (Fig. 1). This stretch of coast is semidiurnal mesotidal (average tidal range 2.10m.) with a slight daily irregularity as shown by the data taken at the harbour of Huelva (Borrego, 1992).

The presence of marshes, dunes, and beaches, associated with the estuarine dynamics, makes this area very valuable from an ecological point of view, and many of these areas have been declared Natural Protected Areas by the regional government. On the other hand, as in most European estuaries, many activities and intensive land uses are beside the estuary: harbours at Huelva and Puerto Umbría, a city at Huelva, tourist resorts at Puerto Umbría and Mazagón, petrochemical industries, settlements, and intensive agriculture. As usual, such an industrial and urban concentration causes high levels of pollution in the estuary. To remedy this situation it was necessary for the administration to take special measures through the Plan for the Management of Industrial Spillage, which is now concluded, having solved, in general, the environmental problems of the estuary. However, monitoring remains necessary because of the intensive industrial activity on the banks of the estuary.

Plan for the Monitoring of Andalucian Coastal Waters

The job of monitoring and controlling pollution in coastal waters in the Autonomous Community of Andalucía has been undertaken by the regional administration since

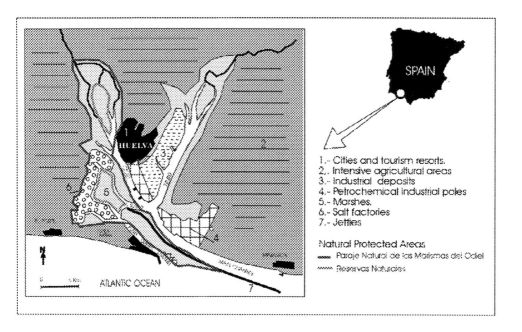

Fig. 1. The Study Area

1988 under the Plan for Monitoring Andalucian Coastal Waters. With the general aim of knowing the level of contamination of Andalucian coastal waters, samples of water and sediments were taken throughout the whole coast of Andalucía, including the tidal reaches of rivers.

For water, the aspect on which this chapter is concentrated, monitoring was based on analysis of samples from 133 sites, 73 in the sea and 56 in tidal reaches of rivers. The samples were analysed for copper, zinc, manganese, nickel, chromium, cadmium, lead, arsenic, mercury, pH, conductivity, suspended solids, dissolved oxygen, chemical oxygen demand, total coliforms, nitrite, nitrate, ammonia, phosphates, oils and greases, cyanide, fluoride, and phenols. The location of sample sites took into account the possible impact of human activity so that there was a greater concentration of sample sites in the more sensitive areas. Also depending on the sensitivity of the area and its exposure to contamination, the sampling was quarterly, bi-annual, or annual. Furthermore, in the zone affected by significant tides, the Atlantic coast, each sampling was made at both high and low water.

The Tinto-Odiel estuary is considered to be especially fragile and heavily affected by human activity, and so it has a more than average density of sampling sites, with one in the Tinto, two in the Odiel, six in the Padre Santo channel, and seven in the coastal zone facing the estuary (Fig. 2). As there is a significant tidal range, quarterly sampling is done at both high and low water. The collection and systematic analysis of samples since 1988 has provided an important and substantial database which, up to a point, shows the improvements made in water quality in this part of the Andalucian coast. Nonetheless the thorough use of this information is, on occasion, somewhat hampered by the nature of the sampling process.

1) **the littoral zone and the interior of the estuary are treated as independent areas, so the sampling days were different and, at times, far apart**

2) **sampling days were not chosen according to the hydrodynamic situation (e.g. tidal coefficient) but rather according to the availability of staff and boat**

3) **the sample points were based solely on visual reference points**

4) **the time of sampling was not recorded so that it is not possible to relate the sample to the exact state of the tide, only to a general statement of high or low tide. The time taken to sample the estuary sites meant that it was impossible for all to be taken at the top or bottom of the tide. Moreover knowing the time of sampling and thus the sequence would help to interpret the data in relation to water mass movement in response to the tide**

In a zone such as we are considering, in which the water dynamics are complex and the tidal currents reach a considerable speed (0.56m/sec - an average water mass displacement of 6km. in each half-tide cycle.) all the aforementioned characteristics of the sampling process mean that the available information leaves much to be desired, notably:-

i) the distribution of sample sites may not fully represent the estuary because of significant absences such as the El Burro and Punta Umbría channels, the upper sector of the Odiel and Tinto channels, the main channel mouth etc.

ii) taking the samples in different hydrodynamic circumstances means that comparability is lost

Clearly, a better awareness of the dynamics of the estuary and the disposition of the different water masses in various synoptic situations could help to improve the consistency and representativeness of the sampling process. Therefore the aim of this project is to evaluate the potential contribution of remote sensing to enhancing sampling representativeness both:

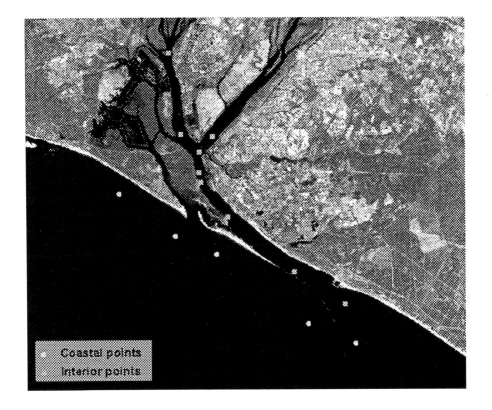

Fig. 2. Sample Points in the Tinto-Odiel Estuary

a) <u>spatially</u> by measuring water quality in the whole estuarine system at each date, and

b) <u>temporally</u> by obtaining site specific data that are comparable over the whole relevant time period. Satellite remote sensing has already been shown to be useful as a complementary source of information in this field (van Zuidam, 1993; Walker, 1996; Davies et al., 1997).

Methods and Results

ESTUARY DYNAMICS

The initial information was provided by 15 satellite images (Landsat TM path 202, row 34) covering the Tinto-Odiel estuary, and selected from the Image Library of the CMA to cover as many different hydrodynamic situations as possible.
Image processing was very simple. Firstly, a mask was applied to each image by choosing a threshold in Band 5 (1.55-1.75 μm), in order to separate land and water.

Then the values corresponding to the estuarine and coastal waters in a visible band (Band 2, 0.52-0.60 μm) were density sliced and colour coded. The intervals used in the density slicing are different for each image, because the atmospheric contribution to the water radiance varies and the minimum value is different for each image. The processed images show turbidity patterns which can easily be interpreted qualitatively from a dynamic perspective (horizontal movement of the surface waters), without the need for geometric and radiometric corrections that are required for quantitative studies, as the approach here concentrates on visual interpretation.

The variety of dynamic situations shown in the fifteen selected images and the cyclic character of some of the main factors that control turbidity patterns (e.g. tides), allow us to characterize the most typical situations in the dynamic behaviour of the estuary. The results have been examined fully in previous work (Ojeda et al., 1995). Only the more outstanding characteristics of ebb and flood tides are presented here.

FLOOD TIDE

Flooding starts in the main channel of the estuary some time after low water. This time lag can be over two hours, depending on the tidal coefficient (the greater the coefficient, the greater the time lag), and it is related to the inertia of the ebbing waters in the estuary. Flooding starts on the left side of the estuary mouth, close to the coastline, and, after a delay, most of the water volume is concentrated through the right side, next to the jetty. This means that there is a one-two hour period after low water during which a two-way circulation occurs in the estuary mouth (Fig. 3).

Once the flood tide currents are established in the whole of the main channel, coastal water passes through it homogenously for two-six hours, concentrating the turbid estuarine waters in the middle-upper sectors of the estuary. This homogenous propagation of the flood tide in the main channel contrasts with its different behaviour beyond the confluence of the two rivers:

1) **the tidal mass propagates more easily through the Odiel channel, because of its wider section and greater alignment with the main channel**

2) **the tidal mass is delayed and enters the Tinto channel less easily for the inverse reasons**

3) **part of the water that floods the Odiel channel enters the Burro channel and moves to the Punta Umbría channel, because of the different tide propagation model in the hypersynchronic Odiel channel and the delayed high water in the hyposynchronic Punta Umbría channel**

EBB TIDE

Flood tide waters continue for some time after theoretical high water but the time-lag during the establishment of the ebb currents is shorter than the delay in the establishment of the flood currents because of the estuary's configuration (less inertia, gravity etc.) (Fig.4).

The waters in the Tinto channel are partially blocked during the ebb tide. The water from the Odiel and El Burro channels flows to the main channel obstructing the Tinto waters. Only after a one-two hour delay does the Tinto channel manage to drain to the Padre Santo channel.

During ebb tide interesting patterns of turbidity plumes generated by the ebbing estuarine waters are found in the main channel mouth. The tidal phase and coefficient seem to be the main controlling factors as is demonstrated by the similarities between the turbidity patterns when both parameters have similar values.

1) **low coefficient: little turbid water reaches the end of the jetty and the turbidity maximum concentrates along the main channel**

2) **medium coefficient: the turbid water volume is higher and there is a sharp westward deflection of the plume**

3) **high coefficient: turbid water is pushed away from the jetty and the westward deflection becomes a clear spiral.**

The Punta Umbría channel turbidity plume spreads over a large area because the volume of water drained during ebbing water is higher than the volume of water that enters during flooding water. This means that this channel drains some of the Odiel water. This is important because, at low and medium coefficients, most of the estuarine water exchange takes place through this channel.

Zoning the Estuary

This is an attempt to establish the typical spatial distribution of water masses in the estuary and coastal zone for each stage of the tide. To characterize the different synoptic situations the first step was to group the available images according to the hydrodynamic situation which they represent. Having checked the state of the tide at each image, the images were grouped into three sets, high water (+/- 1hour), low water (+0-2 hours),and ebbing water (1-3 hours before low water) (Table 1). There are no images for flooding waters.

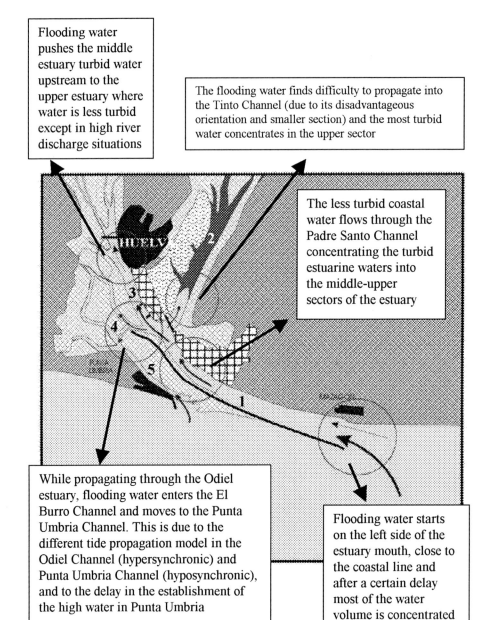

1. Padre Santo Channel, 2. Tinto Channel, 3. Odiel Channel
4. El. Burro Channel, 5. Punta Umbria Channel

Fig. 3. Hydrodynamic scheme during flooding water

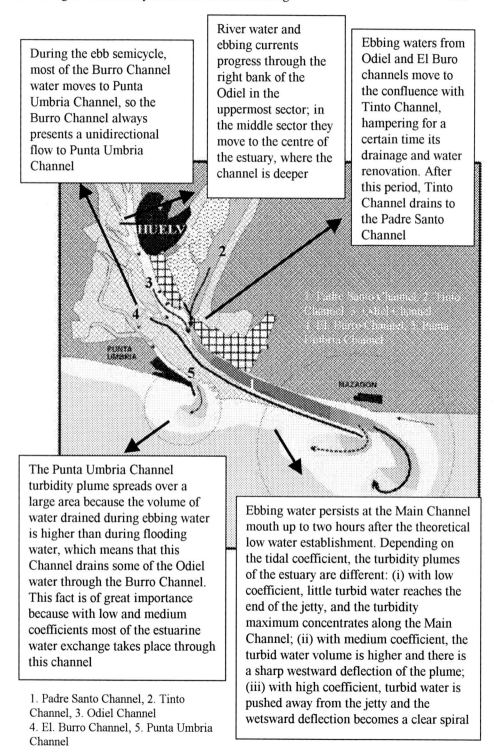

During the ebb semicycle, most of the Burro Channel water moves to Punta Umbria Channel, so the Burro Channel always presents a unidirectional flow to Punta Umbria Channel

River water and ebbing currents progress through the right bank of the Odiel in the uppermost sector; in the middle sector they move to the centre of the estuary, where the channel is deeper

Ebbing waters from Odiel and El Buro channels move to the confluence with Tinto Channel, hampering for a certain time its drainage and water renovation. After this period, Tinto Channel drains to the Padre Santo Channel

The Punta Umbria Channel turbidity plume spreads over a large area because the volume of water drained during ebbing water is higher than during flooding water, which means that this Channel drains some of the Odiel water through the Burro Channel. This fact is of great importance because with low and medium coefficients most of the estuarine water exchange takes place through this channel

Ebbing water persists at the Main Channel mouth up to two hours after the theoretical low water establishment. Depending on the tidal coefficient, the turbidity plumes of the estuary are different: (i) with low coefficient, little turbid water reaches the end of the jetty, and the turbidity maximum concentrates along the Main Channel; (ii) with medium coefficient, the turbid water volume is higher and there is a sharp westward deflection of the plume; (iii) with high coefficient, turbid water is pushed away from the jetty and the wetsward deflection becomes a clear spiral

1. Padre Santo Channel, 2. Tinto Channel, 3. Odiel Channel
4. El. Burro Channel, 5. Punta Umbria Channel

Fig. 4. Hydrodynamic scheme during ebbing water

TABLE 1
Images grouped by hydrodynamic situation

DATE	GROUP	HEIGHT (m)	COEFF.	TIME TO/FROM LOW WATER
28.06/92	HIGH WATER	2.47	0.81	+4h21m
15/06/93	HIGH WATER	2.57	0.57	+5h56m
10/10/89	HIGH WATER	2.69	0.69	-6h23m
17/11/91	HIGH WATER	2.61	0.66	-6h20m
09/05/91	HIGH WATER	2.59	0.65	-6h06m
09/07/96	HIGH WATER	2.68	0.49	-5h04m
09/09/95	EBBING WATER	-	1.06	-2h28m
25/11/94	EBBING WATER	1.44	0.49	-1h59m
07/06/96	EBBING WATER	-	0.64	-1h51m
26/08/90	EBBING WATER	1.08	0.45	-1h00m
15/02/90	EBBING WATER	0.97	0.51	-0h54m
22/05/96	LOW WATER	1.09	0.54	+0h17m
13/08/91	LOW WATER	0.41	0.83	+0h19m
09/07/90	LOW WATER	1.13	0.78	+1h39m
01/01/91	LOW WATER	0.89	0.97	+1h54m
15/10/85	LOW WATER	0.77	1.05	+1h56m

The images were processed in three phases:

i) **Preprocessing**: In a multitemporal study this phase is of supreme importance because pixel accuracy is needed between images and geometric and atmospheric correlation is essential. The first task was to correct all the images to a common geometric base, in this case the image from 15th. June 1993 previously corrected by first polynomial, to UTM co-ordinates zone 30 with radiometric interpolation by nearest neighbour in order to retain the original image values. Secondly all the images were subjected to a process of relative atmospheric correction or radiometric normalization from zones of constant reflectivity. The image used as the basis for correction was again that of 15th. June 1993, chosen for the particularly good atmospheric conditions at that time. The methods of atmospheric correction used is that developed in the University of Valencia (Lopez and Caselles, 1989)

ii) **Masking the land**: Band 2 was chosen for the analysis as it best revealed the turbidity structures which permit the identification of water masses. In order that the principal components analysis would extract the maximum information from the water area it was necessary to eliminate the land area from the analysis because its much greater variability impedes the extraction of information from the area that concerns us here. Because of the different stages of the tide the coastline does not match exactly even within the same group of images. To avoid a confusion of coastlines each group was given a mask generated from the image in that group with the lowest tide. The land-sea mask was generated by segmentation in Band 5

iii) **Standardized Principal Components Analysis:** The analysis of principal components is a technique widely used in digital image processing to synthesize the information from a group of bands into a reduced set of bands (Picchioltti et al., 1997). In the case of multitemporal applications such as this, the first principal component tends to show stable information that which is common to all images. The successive components show the information that is not held in common and thus are normally used as a change detection technique

The analysis of principal components was applied to each group on the previously normalized band 2 images with the land masked out. In this case we are concerned to detect precisely the elements which are stable within each group of images for which we are left with just the first principal component of each group. However, in all three cases, this contains almost 99% of the variation in the original images and shows the most common distribution of water masses in the estuary for each of the tidal circumstances considered. In all three cases the different images correlate positively with the first principal component in that high values in the principal component can be interpreted as zones of high turbidity in all images in the group and vice-versa. The high and low tide results, which are the most interesting for the sampling issue, are considered below.

HIGH TIDE

Four water masses are clearly discernable in order of increasing turbidity (Fig. 5)

- the upper Tinto channel has a very turbid water mass with a gradual transition to less turbid water at the confluence of the two rivers

- the upper Odiel channel has a medium-low turbidity water mass

- the lower reaches of the two rivers before their confluence have similar characteristics, with medium-high turbidity which in the case of the Odiel penetrates El Burro channel, demonstrating the link already detected with the Punta Umbría channel which, as the tide falls, will drain part of the Odiel's waters

- the channels of Padre Santo and Punta Umbría are also homogenous zones, in this case with medium-low turbidity because of the arrival of coastal waters which push the more turbid estuarine waters upstream

- the zone outside the estuary is occupied by clear coastal waters except for a narrow zone anchored to the coastline. The orange coloured zones in the mouth of the Punta Umbría channel relate to the sandy shallows of this zone

LOW TIDE

In the first principal component taken from all the low tide images (Fig. 6) three water masses can be identified in the estuary in order of increasing turbidity, apart from the cleaner sea waters outside the estuary.

- the external water mass of medium turbidity originated from mixing of the turbid estuarine waters with clear coastal water. This water mass occupies a wide zone parallel to the coastline and widens at the outlets of the principal and Punta Umbría channels, marking the maximum extent of the corresponding turbidity plumes

- a medium-high turbidity water mass located in a narrow zone anchored to the coastline, the final reaches of the Padre Santo channel, all of the Burro and Punta Umbría channels, and the middle reaches of the Odiel. This last drains more easily than the Tinto, displacing the turbidity maximum downstream towards the main channel. Clearly distinguishable is the turbidity plume generated by the Punta Umbría channel, much more so than in the case of the main channel

- the highest turbidity is found in the water mass which occupies the lower reaches of the Odiel channel, the whole of the Tinto channel, and much of the Padre Santo channel. Because the turbid Odiel waters drain early they arrive at the Padre Santo channel, concentrate the maximum turbidity in this zone, and link up with the waters from the Tinto channel which arrive later because of the Tinto channel's lesser transverse section and its non-alignment with the main channel

The distribution of the water masses mentioned so far can be considered representative of the situation at low tide because the ebb currents continue to flow for over two hours after the theoretical low tide point. However, if one considers the tidal phase of the images in this low-tide group, as seen in table 1, it is clear that there are two very distinct groups; two images only 17 and 19 minutes after theoretical low tide,

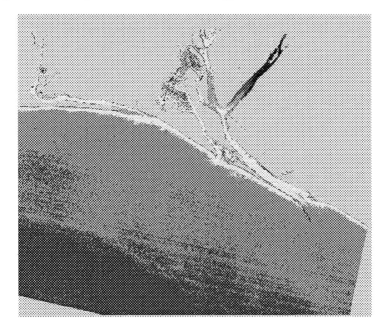

Fig. 5. First Principal Component extracted from High Water Images

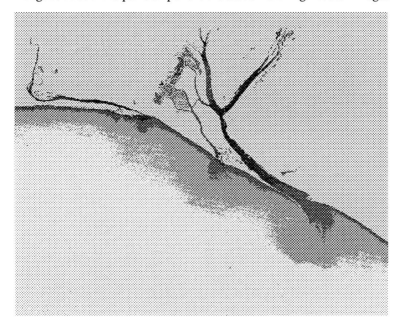

Fig. 6. First Principal Component extracted from Low Water Images

and three images acquired between 1.5 and 2 hours after low tide. The separate application of the same standardized principal components analysis to these two groups will provide an insight into the evolution of water masses and the process of estuary drainage during this period immediately after theoretical low tide.

- at the theoretical low tide, the Odiel, the Padre Santo channel and the Punta Umbría channel seem to be occupied by a homogenous mass of water of medium turbidity which also is beginning to form the corresponding turbidity plumes in the two outlets of the estuary

- the Tinto has not yet started to drain, so its turbidity is at maximum

- as time passes the Tinto begins to drain and the turbidity maximum extends through the upper and central parts of the Padre Santo channel

- the turbidity plumes in the two estuary outlets continue to extend and their contact with coastal waters becomes sharper

- the zone parallel to the coastline affected by turbid estuarine water grows wider

Conclusions : Proposals and Recommendations for Water Sample Collection

The information obtained from the images concerning the estuary dynamics and the spatial distribution of its water masses in different synoptic situations allows us to draw a series of conclusions which can be considered useful; recommendations as to how, where, and when to sample effectively and representatively over a long period. As to the collection of samples, slight modifications in the procedures used could improve the quality of the data and facilitate its interpretation.

- coastal waters in a broad belt parallel to the coastline are affected by the presence of the estuary which means that in no way should the samples within and outwith the estuary be taken on different dates. Therefore it is recommended that the data be taken as a group and that the whole area be considered a single zone for purposes of the monitoring plan

- the recording of timing and precise location of the sampling in each quarterly cycle is fundamental to the correct interpretation of the water quality data. Awareness if the exact tidal phase at sampling time allows the sample to be related to the distinct water masses within the estuary. Also knowing the time of collection allows the reconstruction of the route used in relation to general movement of water masses according to the state of the tide

As to the timing of sampling there are also some observations to be made.

- if the samples are taken at clearly identifiable and significant times such as high and low tide, the data can be more easily extrapolated to different water masses because the stability of the situations is greater than during the ebb or flood. Thus one can avoid taking several samples from the same water mass as it passes from one place to another. Given that sampling cannot be totally synchronous, it is recommended that sampling should start at the theoretical high or low tide and continue through the first, or even the second, hour afterwards, depending on the tidal coefficient, because the tidal currents continue longer when the coefficient is high

- although the information collected at high and low tide is different, both situations are interesting and sampling should be continued at both under the monitoring plan, in order to characterize different phenomena:

    i) at high tide: the quality of estuarine water and the interchange between the Odiel and the Punta Umbría channel

    ii) at low tide: the interchange process between estuarine and coastal waters is fundamental to any understanding of the estuary's capacity for renovation

Concerning the siting of sampling points, the results of the dynamic study reveal two problems (Fig. 7):

    i) there are zones of the estuary not covered by the current sampling procedure, notably the Punta Umbría and Burro channels

    ii) some zones have an excess of sample points giving redundant information e.g. the junction of the Odiel and Tinto rivers. To produce information that is basically different, it is proposed to have points located for high tide sampling and for low tide sampling (Fig. 8) even though in both cases the proposal would no longer use two of the current sampling sites

For high tide: five new sampling points are proposed.

- in the upper reaches of the Tinto where the most turbid waters are pushed by the rising tide

- in the middle reaches of the Odiel which, in effect, is a slight southward movement of an otherwise redundant site

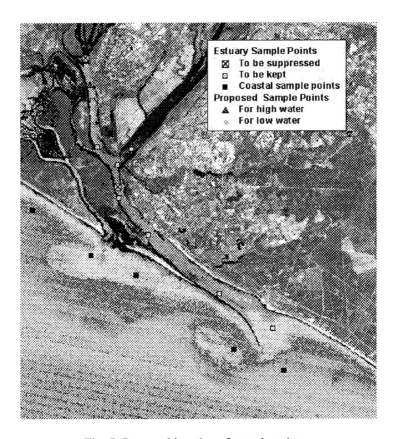

Fig. 7. Proposed location of sample points

- two sites in the upper reaches of the Odiel, to characterize the river water and marsh drainage channels which join at this point

- in the confluence of the Burro and Punta Umbría channels because of the significant interest of the interchange between the Odiel and the Punta Umbria channels

For low tide: the interchange between estuarine and coastal waters is the outstanding point of interest; so all new sampling points are all proposed for the exit of the estuary.

- one in the lowest reach of the Punta Umbría channel. Proposed because much of the interchange of estuarine and coastal waters passes through this point, especially the ebb of turbid estuarine waters. These are the ones most likely to be affected by high levels of contamination and the ones most in need of policing, not least because of their proximity to protected areas and an intensively used beach.

- two in the outfall of the main channel which will enable the monitoring of the distance reached by estuarine waters.

References

Borrego, J. (1992). *Sedimentologia del estuario del rio Odiel (Huelva, SO España)*. Ph.D. Thesis. University of Seville. Spain.

Davies, P.A., Mofor L.A., and Neves V. (1997). Comparisons of remotely sensed observations with modelling predictions for the behaviour of waste water plumes from coastal discharges. *International Journal of Remote Sensing*. 1987-2021.

Fernández Palacios, A. Moreira J.M., Ojeda, J., and Sanchez, E. (1994). Evaluation of different methodological approaches for monitoring water quality parameters in the coastal waters of Andalusia (Spain) using Landsat-TM data. *EARsel Workshop on Remote Sensing and GIS for Coastal Zone Management*. Rikswaterstaat. Survey Department. Delft. 114-123.

Lopez M.J., and Caselles, V. (1989). A multitemporal study of chlorophyll a concentration in the Albufera lagoon of Valencia., Spain, using Thematic Mapper data. *International Journal of Remote Sensing*, vol. 11:301-31.

Ojeda J., Fernández -Palacios, A., Moreira, J.M., and Sanchez, E. (1994). Programa de seguimiento de la calidad y dinamica del espacio marino a traves de imagenes de satelite. (Andalucía, de Medio Ambiente). *Revista de la Sociedad Española de Teledetección*, 3:9-15.

Ojeda, J., Sanchez, E., Fernández-Palacios, A., and Moreira J.M. (1995). Study of the dynamics of estuarine and coastal waters using remote sensing: the Tinto-Odiel estuary, SW Spain. *Journal of Coastal Conservation*, 1:109-118.

Picchiotti, R., Casacchia, R., and Salvatori, R. (1997). Multitemporal Principal Component Analysis of spectral and spatial features at Venice Lagoon. *International Journal of Remote Sensing*, vol. 18 1:183-196.

Van Zuidam, R.A. (1993). Review of remote sensing applications in coastal zone studies. *International Symposium "Operationalization of Remote Sensing"*.3-23.

Walker, N.D. (1996). Satellite assessment of Mississippi river plume variability: causes and predictability. *Remote Sensing of Environment*, vol. 58:21-35.

CHAPTER 39

European CZM and the Global Spatial Data Infrastructure Initiative (GSDI)

R. A. Longhorn

ABSTRACT: This chapter presents the latest information available about the marine data subject area in the Global Spatial Data Infrastructure (GSDI) initiative, which began at an international meeting in Bonn, Germany, 4 September 1996. The purpose of the GSDI initiative is to examine the degree to which a global information infrastructure for spatially referenced data can be implemented and which areas of use of spatial data are most relevant to international cooperation. Coastal zone management (especially in Europe, where national coastlines form the European coastline) and ECDIS (electronic charts) are two areas of special interest. A review of strengths and weaknesses in these areas, at global level, will be presented. The second global conference on GSDI took place in October, 1997 in North Carolina, USA.

Introduction

The concept of a Global Spatial Data Infrastructure (GSDI) has been discussed in one form or another for some years now, adopting the current title in 1994, when a proposal for a GSDI was published in the proceedings of a major European GIS/LIS conference. (Morrison 1995) The proposal made in Morrison's paper appeared only two months following the now famous US Presidential Executive Order to create the National Spatial Data Infrastructure, issued on 13 April 1994. Since that time, most major developed nations have either launched or are launching NSDIs in their respective countries and several initiatives exist at regional level, e.g. the EGII (European Geographic Information Infrastructure) (EUROGI, 1997) and the Permanent Committee on GIS Infrastructure for Asia and the Pacific, initiatives which began in 1995 and 1996, respectively (AUSLIG, 1996).

However, researchers into coastal zone management (CZM) and other marine research (MR) were already taking "global" views from the mid-1980's (at least) and major international conferences and conventions were being held or entered into from the early 1990s. Key globalisation issues revolve around climate change and other environmental aspects of ocean use - or misuse.

Several parallel activities led the way towards a more structured look at what a GSDI might entail, how it might be implemented, and how much effort would be involved, including:

- **rapid development of National Spatial Data Infrastructures, especially in the developed nations**

- **advances in GI standards work, at national and regional level, culminating in the work now underway at ISO at global level, especially regarding GI metadata**

- **rapid advances in information and communications technologies (ICT), including GIS and GPS, opening new data collection, processing, storage and transmission possibilities at much reduced cost**

- **increased focus on global environmental issues and the obvious need for GI and GIS to monitor global climate change, at sea, on land and in the atmosphere**

- **increased awareness by governments that no country can afford to ignore the global environment and that many regional (multiple cross-border) problems can be addressed more efficiently and more quickly if good quality GI exists (e.g. flood control throughout major river basins, coastal flooding, etc.)**

Thus, the time was considered right, in 1996, to convene a major international meeting to investigate the extent to which a GSDI could be implemented and/or should be implemented.

History of the GSDI Initiative

The first Emerging Global Spatial Data Infrastructure conference was held in September 1996, in Bonn, Germany. It was organized by the European Umbrella Organisation for Geographical Information (EUROGI), the German Umbrella Organisation for Geoinformation (DDGI), the Atlantic Institute (AI), the Institute for Land Information and its Land Information Assembly (ILI/LIA), the Open GIS Consortium (OGC), the Federal Geographic Data Committee (FGDC) and the Federation Internationale des Geometres - Commission 3 (FIG-COM3), under the patronage of Dr. Martin Bangemann - Member of the European Commission, responsible for Industrial affairs and Information and telecommunications technologies. The conference was attended by 63 invited representatives of

European CZM and the GSDI Initiative

organisations from around the world and from almost all the GI sectors, representing 20 countries

The main goals of the conference were:

- **to minimise duplication of national efforts and the cost of research and development on a global scale**
- **to identify the critical opportunities and threats inherent in creating a global spatial data infrastructure**
- **to create a standardised vocabulary with which the many issues could be unambiguously discussed across the globe**

and

- **to define the new concepts needed to facilitate an ongoing dialogue between diverse professions to design, implement and extend spatial data infrastructures needed in building and using geo-information products and service in a GSDI.**

Speakers from many nations presented their views on what GSDI might or might not be, how it could evolve in either a planned or ad hoc fashion, what benefits could accrue to global GI users if a GSDI followed some agreed development path, and even who had the right, let alone the legal mandate, to attempt such a venture as to define a GSDI implementation route. It was agreed that any GSDI would by default be built upon the scores of NSDIs already in place or being created around the globe. Therefore, it was important that all creators of NSDI policies should be made fully aware of the global ramifications of their evolving spatial infrastructures.

The meeting reached a consensus that it *is* the right time to start thinking about the major issues implied in the concept of a GSDI and a possible way forward, even if this can not yet be formalised as an 'implementation plan'. It was proposed that the participants to this first meeting comprised a GSDI Forum, which should be extended to all sectors in the Spatial Data/Geographical Information communities, world-wide. It was also accepted that there was no need to create new 'artificial' global projects in order to investigate the main issues. Rather, existing global projects, such as those of the G7 and other international projects (Earthmap, Global Mapping, other international environmental and geophysical projects, African projects like Africover, etc.), should be examined to see if common problems were being met and could be overcome.

Although this first GSDI Forum meeting arrived at no major conclusions in regard to GSDI, it became apparent that some body or Forum is needed to review and comment upon the practical problems being faced today by global GI projects. Exchange of data sharing and data use/re-use experiences from across disciplines was considered crucial, e.g. case studies prepared by participants in the Land-Ocean Interactions in the Coastal Zone (LOICZ) project could be most useful.

The will for good cooperation and of sharing ideas about infrastructure architectures and solutions between the nations was very evident at the meeting. However, because there was no formal mandate involved, and no fixed budget available for this initiative from national, regional or global institutions, advancement of the GSDI concepts, further communication and investigations into the issues, and practical work such as setting up Web sites to exchange experiences, must be effected on a purely voluntary basis. Fortunately, several participants were in the position to use their own project funds to help in this regard. A summary report of the first GSDI meeting can be found on the EUROGI Web site at URL http://www.frw.ruu.nl/eurogi/eurogi.html.

The second meeting of the GSDI Forum will be held in North Carolina, USA, 19-21 October 1996, again under the patronage of EU Commissioner Dr. Martin Bangemann, hosted by the Governor of North Carolina, chaired by Mr. Michael Brand (details on URL http://www.gov.state.nc.us/GSDI97).

Coastal Zone Management and Marine Research in the GSDI Agenda.

It goes without saying that much CZM and marine research and development is global in scale. Much of the work required in globalizing" marine GI relates to standards and to analysis of practical experiences in collecting, using (i.e. in various models) and disseminating marine GI, especially on a regional and/or global basis.

MARINE GI STANDARDS.

As to marine GI standards, a major player is the IHO. At the September 1996 GSDI Forum meeting, a presentation on the issues surrounding a Marine GSDI (Anderson and Evangelatos, 1996) focused primarily on the achievements of the International Hydrographic Organization (IHO), whose membership includes 60 countries of the 130 coastal states in the world. The IHO produced the first global series of bathymetric maps - the General Bathymetric Chart of the Oceans, the latest edition of which was produced by Canada. Several countries are now producing digital versions of this important body of marine GI.

The IHO's International Hydrographic Bureau, in Monaco has also achieved the IHO S-57 spatial data standard for use in Electronic Chart Display and Information Systems (ECDIS). Harmonisation work is also underway between IHO's S-57 standards and those of NATO's Digital Geographic Information Working Group (DGIWG) DIGEST exchange standard, via a joint IHO/DGWIG Technical Committee. The work of both these organisations has also been made available to ISO's TC 211 which is now working on GI metadata and interoperability standards.

The IHO also created a committee in 1992 to bring together national and regional Electronic Navigation Charts (ENC) infrastructure with the view to creating a global ENC infrastructure. The current concept is that Regional Electronic Navigation Chart Centres (RENC) will eventually be interconnected to form the global system. IHO's ultimate goal is to develop standards to provide the basis for all hydrographic data, which can also be extended to cover all other products and data related to hydrography and navigation.

Thus, international standards for marine geo-spatial data are progressing. Anderson and Evangelatos contend that "the hydrographic electronic chart infrastructure is the basis of the marine geo-spatial data infrastructure" and that "..international standards for marine geo-spatial data are progressing ... not yet mature, but far enough advanced to provide a clear direction for the development of the Global Marine Spatial Data Infrastructure". The question is, does the global CZM and marine research community agree with this summing up of the situation? If not, then they should make their thoughts known to the organisers of the forthcoming GSDI Forum meeting in North Carolina, perhaps by contacting the EUROGI President.

The CZM/Marine research community will probably have far less trouble concurring with the concluding statement in marine section of the GSDI forum report, namely:

"The requirement now is for an international program to conduct comprehensive mapping of proritized ocean and biomass resources, to create databases that can be made available internationally and to create the infrastructure to support a conservation and management regime".

PRACTICAL EXPERIENCE IN SHARING GLOBAL MARINE GI.

The Land-Ocean Interactions in the Coastal Zone (LOICZ) core project within the International Geosphere-Biosphere Programme (IGBP) of the International Council of Scientific Unions (ICSU) is but one example of a truly global research initiative which has much to contribute to the GSDI debate. This project has already faced the problems of acquiring, using and disseminating large volumes of multidisciplinary GI from across the globe. LOICZ began in 1993 and will run for 10 years. More than 400 scientists were involved in developing the Science Plan and more than 2000 scientists in over 130 countries are now involved in the project. The experiences of the participants in specific LOICZ research projects will be invaluable in regard to the GSDI discussions now underway. (Details of LOICZ can be found at URL http://www.nioz.l/loicz/).

EARTHMAP

One of the global earth observation programmes presented at the GSDI Forum meeting was Earthmap, a public-private consortium which proposes to advance the use of geospatial data and tools for decision makers. Earthmap is being promoted by the Global Environment and Technology Foundation in the USA, where further details can be found at URL http://www.gnet.org/earthmap. Earthmap is ultimately about better decision making for "sustainable development" which, on a global basis, requires the best use of huge amounts of multidisciplinary GI. The consortium felt that the time is right to exploit the rapid advances in satellite imagery and in computer and telecommunications technologies which now make it possible to systematically collect, organise, analyse, and share earth observation data with tremendous speed and from hundreds of locations, simultaneously.

Earthmap is organized around 17 major actions, three of which are of relevance to the CZM/marine research community. These are:

- **establish an Internet-based information sharing network of geospatial applications**
- **support and help coordinate international efforts to develop world-wide geospatial metadata and data standards, and promote improved management of large geospatial databases**
- **work with other national and international agencies to develop a comprehensive and continually updated inventory of digitised geospatial data**

Earthmap activities are divided into ten areas of sustainable development, of which CZM and marine research figure only slightly in Area A - Natural Resources Management (why not more CZM here, you ask?) and strongly in Area B - Environmental Monitoring (monitoring oceans, reefs, and coastal zones). In this latter area, mention is made of the Global Ocean Observing System (GOOS) and of the International Coral Reef Initiatives under the Coastal Zone Module of GOOS, under which initiative the World Conservation Monitoring Center is digitising existing maps of the world's reefs. The problems of data management and use are highlighted by the statement that "in just two minutes, a satellite can collect phytoplankton data for an area the size of Texas and New Mexico, which would have previously taken a ship an entire decade (to collect)".

International Initiatives/Programmes in Relation to the GSDI Discussion

Numerous global initiatives exist which include elements of CZM and marine research and which should be part of the GSDI discussions. The United Nations Environment Programme (UNEP) is involved in a series of major initiatives springing from the 1992 Rio Conference. The Baltic Sea programme underway at UNEP-GRID is a typical example of such work. Other UN organisations involved in marine or coastal research, training or education include UNESCO, the WMO - World Meteorological Office, FAO - Food and Agriculture Organisation, UNIDO - UN Industrial Development Organisation, IMO - International Maritime Organisation and more.

Unfortunately for the CZM and marine research community, there were no representatives of these major environmental or ocean-related organisations or initiatives in attendance at the GSDI meeting in Bonn in September 1996. The needs of this large and growing community were brought to the attention of those attending the second GSDI meeting in October 1997.

UNEP AND CORE ENVIRONMENTAL DATA.

Much of the work of UNEP is now focused on global climatic change and/or environmental monitoring at regional level. In 1994, a symposium was held in Thailand which focused on the core data sets which would be needed to intelligently analyse environmental and sustainable development issues. One of the five key topic areas where such data was considered essential was "fresh water and coastal zone management". One recommendation of the meeting was that national governments, donor agencies and international organizations should conduct surveys to document the status of core data sets. These surveys were concluded by 1997 and some results are already becoming available, although finding them can be difficult, as they are not listed in one convenient place (UNDP, 1994).

Another UNEP initiative is the Global Programme of Action for the Protection of the Marine Environment from Land-Based Activities. A major element of work in this programme has been the creating of a GPS information clearinghouse, which can be visited electronically at URL http://www.unep.org/unep/gpa/gpaich.htm. This major global initiative was created by an intergovernmental conference held in Washington, D.C., USA in December 1995. The goals, global plan of action, participants, and other relevant background documents can be found at the GPA home page at URL http://www.unep.org/unep/gpa/home.htm.

INTERGOVERNMENTAL OCEANOGRAPHIC COMMISSION (IOC) - UNESCO

IOC, founded in 1960 as a UNESCO initiative, comprises an Assembly, Executive Council and Secretariat (based in Paris) representing 125 member states and has established several subsidiary bodies. The IOC's activities are mainly global, with regional subsidiaries in the major ocean areas of the world. The IOC lends support to numerous national and regional programmes and initiatives, such as ICoD, ICCOPS, MEDCOAST and others in the Mediterranean region, and has a separate Regional Committee for the Black Sea. It also supports TEMA - Training, Education and Mutual Assistance in the marine sciences. IOC is also responsible for the Global Ocean Observing System (GOOS), including coastal zone activities.

The IOC is directly involved in management of marine data via various experts groups, sponsorship of conferences, workshops and numerous publications of "best practice" (see References section). The IOC's IODE Group of Experts on Marine Information Management (GEMIM), which first met in 1984, produces periodic reports setting out numerous aspects of global marine information, including standards, collection techniques, information analysis products, dissemination by new technology (CD-ROM and the World Wide Web) and much more. In discussing the role of global information for CZM and marine research in a Marine GSDI, the work of the IODE should not be ignored. Documents from IODE already address such information topics as: a directory of training opportunities in marine information management, WWW server and CD-ROM of marine information, global directory of marine institutions and scientists, ASFISIS - user-friendly package for the management of bibliographic information, print and electronic tools for the publication of marine science information, regional co-operation in scientific information exchange (RECOSCIX), document delivery over the Internet and reflections on IPR issues.

Because IOC/IODE operates in the global arena with a UN mandate, readers are advised to consider what its recommendations and achievements have been to date and where the focus is for the future. (UNESCO, 1996) IODE GEMIM reports can be ordered from the IOC Secretariat at e-mail: p.pissierssens@unesco.org and/or via the Web at URL http://www.unesco.org/ioc/.

INTERNATIONAL COUNCIL FOR THE EXPLORATION OF THE SEA (ICES)

Founded in 1902 and with current membership representing 19 countries from both sides of the Atlantic, including all European Coastal states (except the Mediterranean countries east of, and including, Italy), ICES is the oldest intergovernmental organisation in the world focusing on marine and fisheries science. Its multi-disciplinary work programme concentrates on hydrography, physical oceanography, population dynamics of fish stocks, standards of quality and comparability of ocean-related data. The ICES secretariat is located in Denmark, from which site three databanks are maintained for oceanographic, fisheries and environmental (pollution) data. More than 100 meetings are held each year by ICES working groups, study groups, workshops and committees, the latter of which advise national Member Country governments, international regulatory commissions and the European Commission. Many of the workshops deal with coastal or estuary problems, not only deep ocean research, and focus on regions such as the Baltic Sea and Mediterranean.

OTHER INITIATIVES/PROGRAMMES.

There are numerous other global and regional initiatives which should be mentioned in relation to formulating policies and practices for the GSDI. Some of these are:

- **GOOS - Global Ocean Observing System**
  (http://www.unesco.org/ioc/goos/gloss.html)
- **EuroGOOS - European regional GOOS**
  (http://www.minvenw.nl/projects/netcoast/eurogoos/euroconf.html)
- **GELOS - Global Environmental Information Locator Service, part of the G7 ENRM - Environment and Natural Resources Management project** (http://enrm.ceo.org/)
- **SEAWIFS - Sea Viewing Wide Field of View Sensor programme and the SeaStar remote sensing satellite for the colour of water due to changes in the amount of plankton, chlorophyll and sediment** (http://seawifs.gsfc.nasa.gov/SEAWIFS.html)

EU Initiatives in Relation to GSDI

The European Commission sent a Communication to the Council, to the European Parliament, to the Economic and Social Committee and to the Committee of Regions titled "GI2000: Towards a European Policy Framework for Geographic Information". This document set out Commission thinking in regard to "... a policy framework to set up and maintain a stable, European-wide set of agreed rules, standards, procedures, guidelines and incentives for creating, collecting, updating, exchanging, accessing and using geographic information. This policy framework must create a favourable business environment for a competitive, plentiful, rich and differentiated supply of European geographic information that is easily identifiable and easily accessible."(DG XIII/E, 1997)

The document refers to "global issues" in a very few places, as follows:

"The most important political actions needed are to achieve agreement between the Member States ... to ensure that European solutions are globally compatible."

"The G7 Ministerial Conference in Brussels on 25-26 February 1995 confirmed the opportunities the information society will offer and stressed the need for global cooperation. Several of the projects defined at this summit involve significant use of geographic information. This concerns especially the projects on Environmental and Natural Resources management, Global Emergency Management and Maritime Information Systems."

"The policy should take account of similar initiatives in other parts of the world and ensure European contribution to initiatives of global harmonisation of geographic information."

According to the proposed Communication, either the Commission or a public/private partnership coordinating group established by the Commission, will attempt coordination in regard to global geographic information policy and projects, such as those proposed via the G7 and discussions already initiated at global level for the Global Spatial Data Infrastructure (Chenez, 1996).

However, there is little evidence that the European Commission officials currently in charge of the GI2000 "initiative" are greatly concerned about globalisation issues, partly because these are at such an early stage of discussion and because there is still so much work to be done in the European arena on harmonisation of pan-European GI. To support this statement, it is worth noting that no European Commission official attended the first GSDI conference in Bonn, held in September 1996, even though the meeting was billed as being under the patronage of Commissioner Bangemann. European interests at this meeting were represented by EUROGI and senior academics from the European GI R&D community.

GSDI in the New Millennium?

What is the role of the CZM/marine research community in the current GSDI debate and how much can be achieved as we enter the next millennium? As with most "global" initiatives, there are hundreds of actors and scores of main actions to be involved in creating a GSDI. At this stage (mid-1997), we have not even defined what should be the major elements in a GSDI, let alone which of those elements will be of most interest to the CZM/marine research community. However, one thing is certain, CZM is multi-disciplinary and uses multiple types of GI, GIS tool sets and other modelling tools. The GI collected, analysed and disseminated by the CZM community varies widely in composition, scale, quality and content. The skills used in CZM span a wide range, from "simple" coastal plain cartography to hydrography, remote sensing and analysis of satellite imagery, biology, chemistry, hydrology, geography and oceanography, to name but a few. New tools, methodologies and techniques, theories and counter-theories are being proposed daily. A major task of the CZM researcher is simply to keep informed!

Where can the GSDI be most helpful to the CZM research community? Three areas spring to mind:

1) **Marine/CZM metadata - including adoption of relevant international standards for creating and disseminating marine information metadata**

2) **Global directory development - directories of information resources (with relevant and good quality metadata!), directories of researchers, directories of results, directories, directories, directories! And preferably on-line, please**

3) **Enabling the creation of truly global CZM user communities, via the Internet or Web, who share common goals, problems, experiences**

How is the CZM community represented today in regard to the GSDI? Widely throughout the many regional and global environmental, sustainable development and ocean observing programmes and initiatives which already exist. Yet nowhere in regard to the current GSDI Forum meeting(s). It is time that CZM and marine research staff took an interest in these issues and discussion fora, in order that all aspects of GI can be covered by or within the evolving Global Spatial Data Infrastructure. This means participation in the GSDI debates, which will become more numerous and (hopefully) more focused, from organisations at all levels, e.g. independent researchers, research institutions, national and pan-regional research associations, global organisations (UNEP, WMO, FAO, IHO, IMO and many others).

Based on your own research experiences, whether these be in purely local projects, or national, regional or global investigations, let the convenors of the next GSDI meeting know what it is that most concerns YOU regarding global marine GI.

Don't let the initiative for development of a Marine GSDI default into the hands of only those who take the time to attend.

Some of you are only now starting on your CZM research careers while others still have many years of productive work ahead. Technological advances continue apace. Digital data grows in volume almost exponentially. Collecting, managing, using and disseminating this data is already your direct concern and will continue to be more of a problem as time goes on. The job of a researcher becomes more multidisciplinary year by year. The data required to solve marine research problems becomes more interlinked year by year. New skills are required and "old" skills can be improved via new technology. Because of the tremendous importance of the oceans and the coastal zone to the needs of mankind, local, national and regional governments, research councils, international aid and development organisations and numerous global climate and environmental investigation programmes **all** need the best input from the CZM and marine research community. Ideally, an effective GSDI will assist that community in producing ever higher quality results, ever more quickly.

References

Anderson, N.M., and Evangelatos, T. (1996). "Marine Global Spatial Data Infrastructure", in *Proceedings of a Conference on the Emerging Global Spatial Data Infrastructure*, Bonn, Germany, 4-6 September 1996.

AUSLIG (1996). *Report of Proceedings from the second meeting of the Permanent Committee on GIS Infrastructure for Asia and Pacific, 29.9-4.10.1996, Sydney, Australia.* AUSLIG - Australian Surveying and Land Information Group, PO Box 2, Belconnen, ACT 2616, Australia.

Chenez, C.C. (1996). *Proceedings of a Conference on the Emerging Global Spatial Data Infrastructure,* held under the Patronage of Dr Martin Bangemann, European Commissioner for Industrial Affairs, Information and Telecommunications Technologies, Bonn, 4-6 September 1996.

EUROGI Wider Approach Task Force (1997). *The European Geographic Information Infrastructure (EGII) - Towards a wider approach: Raising awareness across Europe - Executive Summary - What is the EGII? - How do we realise the EGII?* EUROGI, PO Box 508, 3800 AM Amersfoort, the Netherlands.

European Commission, DG XIII/E (1997). *GI-2000: Towards a European Policy Framework for Geographic Information.* DG XIII/E, Luxembourg.

Langaas, S. (1993). *Global GIS Data Made Regional.* UNEP/GRID-Arendal, Stockholm, Sweden.

Morrison, J. L. (1994). 'The Global Spatial Data Infrastructure: A Proposal", *Proceedings of GIS/LIS '94 - Central Europe, Budapest, Hungary, June 12, 1994.* GIS World Books, Fort Collins, CO, USA, pp. 23-28.

Post, J. C., and Lundin, C.G. (1996). *Guidelines for Integrated Coastal Zone Management.* Environmentally Sustainable Development Studies and Monographs Series, No. 9. The World Bank, Washington, D.C., USA (ISBN 0-8213-3735-1).

UNDP (1994). *International Symposium on Core Data Needs for Environmental Assessment and Sustainable Development Strategies*, Bangkok, Thailand, 15-18 November 194. UNEP.

UNEP/GRID-Arendal (1996). *Annual Report 1996*, GRID-Arendal, Stockholm Office, Sweden.

UNESCO (1996). *IODE Group of Experts on Marine Information Management (GEMIM), Meeting report on Fifth Session, Athens, Greece, 17-19 January 1996*. IOC/IODE-MIM/V/3, 13.2.96, English only.

Wood, W. B., Freeman, P.H., and Miller, J.A. (1995). *Earthmap - Design Study and Implementation Plan*. Global Environment and Technology Foundation, Annandale, VA, USA.

CHAPTER 40

Access to Marine Data on the Internet for Coastal Zone Management: The New Millennium

D.R. Green and S.D. King

ABSTRACT: The Internet has evolved very rapidly from its early beginnings and relatively limited capabilities - delivering essentially static information - into a very dynamic information resource which now allows us to deliver images, text, video, and on-the-fly mapping at the local, national and global scale. Public awareness of the Internet has also grown rapidly, aided by the increased visibility of website addresses in adverts on television, school websites, and supermarkets, and a drive to encourage more and more people to have their own email address and even websites. The Internet is undeniably a tremendous breakthrough as far as access to data and information is concerned. In recent years, developments in Internet software have provided the capability to deliver interactive maps and images, Digital Image Processing (DIP), and GIS functionality over the web. The Internet has extended the possibility to provide online access to geospatial data and information for ICZM, offering many unique advantages. There are, however, a number of problem areas that require attention if the Internet is to be used for CZM in the UK. This includes, for example, the pathways for locating data and information resources on the Internet. One solution to this problem has been the development of 'portals' or 'gateways'. More difficult to overcome are the problems associated with data and Metadata standards, copyright and pricing, multiple data formats and filters, and in the UK a need to establish a government led initiative to co-ordinate the development and maintenance of a national geospatial data and information framework.

Introduction

Ten years ago the Internet, in its current form, simply did not exist. Today, however, many of us take the Internet for granted. A multimedia-based information resource, widely accessed at home and at work using a variety of Internet browsers (Netscape

Navigator/Communicator and Microsoft Internet Explorer), the Internet has evolved very rapidly from its early beginnings and relatively limited capabilities - delivering essentially static information - into a very dynamic information resource which now allows us to deliver images, text, video, remotely sensed imagery, geospatial data via online image processing and GIS, and on-the-fly mapping at the local, national and global scale.

Computer Technology

To a large extent the Internet has been enabled by the microcomputer and processor technology now available, lower prices, considerable foresight (on the part of a few people) and end-user demand. Newer and faster microprocessors, cheaper hardware (monitors, PCs, printers, scanners, digital cameras, mobile phones, digital assistants), and a wide range of software products - ranging from programming languages such as Java, scripting languages such as JavaScript, CGI programmes, HTML, DHTML, CSS, (Cascading Style Sheets) for fonts, and software packages such as Microsoft Frontpage, Visual Page, Hotmetal and a host of other graphics and web preparation software (e.g. Paintshop Pro, Photoshop, Dreamweaver) have provided a vast toolbox to use for the design and creation of websites and for 'posting' a wide variety of information from many disparate sources on the Internet.

Rapid Development of Technology

The speed with which the technology has developed and its adoption by many people has led to a proliferation of websites containing a vast array of information presented in an almost infinite different number of ways, including static images, dynamic and animated techniques, multiple pop-up windows and menus, and electronic shopping carts. This Internet 'toolbox' is continually being developed by companies such as Microsoft, Intel (Fig. 1), and Sun Microsystems, as well as organisations like the W3C (WWW Consortium) at the Rutherford Appleton Laboratories in the UK.

Greater Public Awareness

Public awareness of the Internet has also grown rapidly, aided by the appearance of website addresses (URLs) in adverts (e.g. Intel), on billboards, on television, on the sides of ocean going ships, courtesy cars (e.g. Adrian Smith Saab - Aberdeen and Carlisle), and instrument suppliers (e.g. Scotia Instrumentation - Aberdeen and Newcastle). Virtually every age group has now become aware of the Internet through the media (e.g. BBC Blue Peter, and the Newspapers), schools (e.g. Prudhoe Community High School - with its own website (Fig. 2)), packaging, and even via the supermarkets (e.g. Tesco).

Recently there has been a drive to encourage people to have their own email address and even website, with companies like Compuserve, AOL, Freeserve (www.freeserve.net), Lineone (www.lineone.net), and Tesco (Fig. 3), offering free webspace and free email addresses. Computer and special Internet magazines have also helped to promote greater awareness of the Internet, with articles on browser software,

Access to Marine Data on the Internet for CZM 557

featured Internet sites, and development software. The Saturday and Sunday newspapers also have special sections featuring websites.

Fig. 1. The Intel WebOutfitter website (courtesy of Intel Corp UK Limited)

Fig. 2. The Prudhoe Community High School website
(courtesy of Prudhoe Community High School)

Fig. 3. The Tesco website (courtesy of Tesco)

One must not forget also the considerable amount of information that is now available on the World Wide Web, which is a mix of academic, commercial, government, and educational material. Information on the web consists of text documents, tutorials (static and interactive), image and map databases, datasets (Figs 4a, 4b and 4c), mapping software, online digital image processing, panoramic views (using Quicktime and other viewers), online video cameras (Webcams), and even Geographical Information Systems (GIS). There is a vast amount of information 'out there' on the WWW, which is freely accessible, at least in most cases. Software, books, tickets, computers etc. can all be chosen and purchased, as can datasets including maps. Sites offering maps provide individuals with downloadable or printable directional maps for practical use, as well as interactive cartography. Some of the information available provides the end-user with help in creating and managing websites, as well as enhancing the presentation through the provision of page templates, JavaScripts, CGI scripts, animated GIFs, video and sound clips, buttons and graphics, fonts, images, maps, and in some cases dynamic graphics. This in itself has helped many individuals, with relatively little knowledge of the Internet, web design and creation, to make even greater use of the web technology in many different and often exciting and innovative ways.

Access to Information

Irrespective of one's likes or dislikes for the Internet, or indeed companies like Microsoft and Netscape, there is no doubt that the Internet constitutes a very significant and major revolution in the provision of access to, and presentation of, both data and information of all sorts. The impact and importance of the Internet is very considerable. Here we have an application of computer technology which is very wide reaching, one that has helped to move microcomputers into both the workplace and the home in a very short space of time, which has affected nearly everyone; the way in which they work, the way in which they source information, where they find it, accessibility to a wide range of different information that can be copied, cut and pasted into documents and reports, can be reworked, and transmitted to many others. However, there are still a number of problems, concerns and stumbling blocks, particularly for the widespread use of the Internet as the basis for information systems.

Concerns

There is the question of being able to find and locate both data and information. Helped by the various search engines e.g. Alta Vista, Yahoo, Lycos, Google (Fig. 5), Excite, and Hotbot, virtually anyone can find information based on the use of one or more keywords. Searches can also be made within a website which are also sometimes helpful. Unfortunately, searching the Internet can also lead to diverting attention away from the original quest for specific information. One is guided only by the short text description that is returned in the search listing. Whilst all this is very useful sometimes, at other times it is somewhat annoying to have to search hundreds of

Fig. 4a. The Ericsson Mobile Phone/Internet/Email website (courtesy of Ericsson)

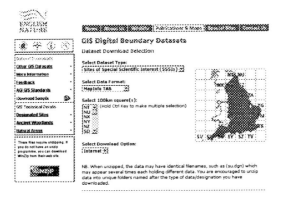

Fig. 4b. The English Nature website (copyright of English Nature)

Fig. 4c. The Getmapping Plc website
(courtesy of Getmapping plc)

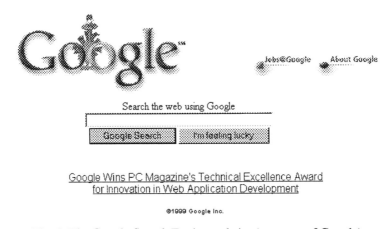

Fig. 5. The Google Search Engine website (courtesy of Google)

listed entries for useful information: indeed much of the information retrieved on the Internet could be considered to be of little real value.

Whilst the design, structure and layout of many sites is helpful when looking for information, many are unfortunately not, and even the most basic of information like a contact address can become very hard to find, often proving to be very frustrating for the end-user. Many sites are muddled and difficult to navigate, whilst others are simple, 'gimmick free' and include very clear labelling and navigation. Some sites comprise largely static material, which may be updated very infrequently; a solution to this can be through the provision of a noticeboard, or most recent update date on the main page that highlights recent changes to the site.

Often a novelty at the outset for many organisations, websites can unfortunately rapidly become outdated, unattended to, with broken internal and external links, and in some cases may even have moved to a new address or no address at all. Many sites fall into a state of disrepair when the originator moves on, or the company employed to do the work does not continue, or the client runs out of finances.

Although many people can see the obvious advantages that data and information system, GIS or Decision Support System (DSS) would bring to coastal management, as yet, GIS software is generally still too complex to be used successfully by the coastal applications specialist, or offers too many functions which are not necessarily needed for many applications.

In the context of maps, there are many concerns, especially in the UK, about provision of access to data. The main concern is one of giving unrestricted access to data and information, with the fear that (a) people may use the data for other purposes, which may include selling it on for commercial gain, (b) using it for a purpose other than that for which it was intended, and (c) that it will undermine attempts at cost recovery. To some extent such fears are perhaps a little unjustified, as one has to ask the question 'are end-users looking for the raw data or just information?' Unfortunately, judging by the number of images, scripts and graphics that are 'borrowed' (effectively stolen?) from websites, it would appear that many people do in fact not only seek information but also the data! There is genuine concern from organisations such as the Ordnance Survey and others about providing access to map data, especially if the original data was expensive to capture, and they are driven by the need to recover costs. Trusting people to be honest about getting access to data and paying later is not really a realistic option. Others are less concerned or are willing to provide data and information at lower costs. Unfortunately, and this may be a left over part of the UK culture from the days of government bodies like e.g. ADAS, many people expect to be able to get hold of data and information for free or for very little outlay; the same also goes for software. Where it becomes costly to purchase and there are recurring costs, then in all likelihood people will try to borrow datasets or copy them illegally. For the academic community some headway has been made through deals that provide the educational and research community with discounted datasets, although in the current climate of economic restraint and cutbacks, the ability to purchase even these is not always a possibility, and besides there may be restrictions placed on the data if the research work is part of a commercial consultancy. The question of copyright also always pops up; who owns the data/information? Copyright is limiting the amount of data and information that can be freely transmitted and shared for coastal management purposes. There is still no real data standard for geospatial data in the UK, data is still collected and stored in many different formats, held in inaccessible locations, and catalogued with no consistent Metadata standard. By and large, the 'real' geospatial datasets are few and far between on the Internet, only being available where the aim is to distribute datasets to a wider community. At present relatively few sites in the UK deliver maps that are in, for example, vector format, most offering instead small sections of raster maps.

Freedom to post information on the Internet is in many ways beneficial to many people, but also is unfortunately open to abuse, not only in terms of what people place on websites, but also what they get access to.

There is also concern about the misuse of data and information, whereby data might be used for an application for which it was not intended, and may produce false results, that may cause harm, even destruction, a wrong decision to be taken and ultimately lead to liability claims. This problem will need to be addressed through the inclusion of disclaimers, and Metadata cataloguing what a dataset was originally collected for, used for, its quality, and guidelines on how to use the data.

The current lack of a clear direction in the UK as to how to address the question of access to data and information (despite the recent Freedom of Information Act) poses a serious problem for getting the geospatial data and information resource via the Internet to become a reality. There are still a number of stumbling blocks. Many people and organisations are reluctant to embrace Internet technology to the extent that it has been adopted in countries such as the USA, Canada and Australia. No one has yet taken the lead on organising coastal zone management at a national level in the UK, and there is little evidence of a co-ordinated policy for marine and coastal data and information.

GIS and Geospatial Data and Information Delivery

In recent years, developments in Internet software have provided the capability to deliver interactive maps and images, Digital Image Processing (DIP) (Fig. 6), and GIS functionality over the web. Nearly all software developers now have either a raster or a vector product, or both. The means to provide end-users with interactive 'clickable' maps, or 'on-the-fly' layer-based maps has been a major development and has opened up considerable new avenues for use. Already the scope of use of such mapping technology is well developed and in widespread use. One only has to look at the examples provided by ESRI (http://www.esri.com) (Fig. 7).

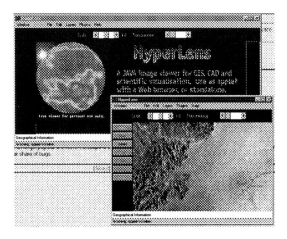

Fig. 6. Digital Image Processing on the World Wide Web - The Hyperlens website (courtesy of Hyperlens)

Access to Marine Data on the Internet for CZM   563

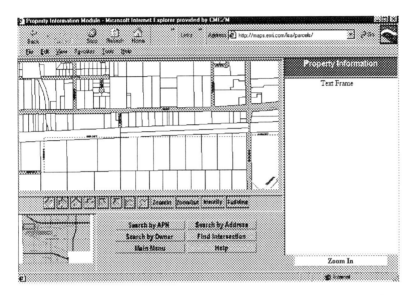

Fig. 7. The ESRI website (graphic image provided courtesy of ESRI)[i]

Internet technology has reached a stage where GIS, Digital Image Processing and Remotely Sensed imagery can now be found on the Web and utilised quite easily by applications specialists, rather than computer specialists. This means that everyone potentially has access to the data and information that they require, and has the ability to manipulate and use that data in the way that they need. The Internet can therefore now be used as more than an information system, it can also be used to provide decision support for coastal managers, by aiding in the formulation of answers to questions about coastal resource problems.

One of the best examples, in the context of coastal zone management is the Florida Marine Online Mapping website which utilises ESRI's Map Cafe (Fig. 8). This website epitomises in our opinion exactly what can be achieved using Internet-based technology, opening up the possibility to deliver geospatial data and information to a widespread end-user community. Some example screens reveal the potential. But this particular site is also a much more significant example in terms of what has been achieved in the context of integrated coastal zone management. Here we have (a) an identifiable point of access - a focus or portal to geospatial data and information, which is (b) catalogued, ordered and documented in a form that is (c) widely accessible, (d) in an electronic format (taking advantage of IT and the Internet), (e) includes Metadata, (f) is regularly updated, (g) is co-ordinated by a recognisable body, and (h) is user-interactive in the context of GIS and maps, offering information for planning and decision-making, whilst also providing the option for data should data be required in the context of a data analysis or modelling scenario.

Other examples include NOAA's Ocean GIS South East site (Fig. 9), which is in many respects very similar to the Florida Marine Online website, the Australian Coastal Atlas (Fig. 10), and Interwad (Fig. 11).

Integrated Coastal Zone Management (ICZM)

One of the primary requirements for Integrated Coastal Zone Management (ICZM) is data and information for decision-making and planning. Without access to either it is very difficult to undertake ICZM. Prior to the development of sophisticated communications and networking technology, and the desktop computer system, coastal managers have been reliant upon data and information in the form of paper files, documents, pictures and computer-based databases. Whilst adequate, the potential of desktop computers, GIS and the Internet has opened up many new possibilities to integrate many disparate data sources (archival and new), to store and manipulate huge quantities of data and information, to utilise remotely sensed imagery and digital maps in a single work environment. The Internet has extended the possibility to provide online access to geospatial data and information for ICZM, offering many unique advantages. This is how the Internet as a technology can help.

Through the provision of simple Internet browser interfaces, comprising task-specific functionality tools (Back, Forward, Home, Print, Save), together with the capability to design a sub-interface for the website with navigational buttons, indexes and pointers to structured information, it is possible to utilise the Internet to provide access to data and information.

CZM Information System Portals or Gateways

The concept of a portal or gateway is directed access to specific information - a sort of 'one-stop-shop' which acts as a collective signpost to similar or related data and information. A number have already been developed, with the aim of providing a directional pointer for the end-user seeking coastal data and information, thereby avoiding the need to search. In effect a portal or gateway provides the end-user with the much needed 'index' required.

An attempt to do this is that of SCOTCoast (http://www.abdn.ac.uk/scotcoast/) (Fig. 12), a single website with links to the Firths Initiative sites: Forth Estuary Forum (FEF); Solway Firth Partnership (SFP); and the Moray Firth Partnership (MFP). The idea behind this is to try and initiate a Scottish coastal zone management framework which will hopefully stimulate both a Scottish and a National interest in developing a national coastal data and information resource for the UK.

A National Coastal Data and Information Resource

In the last few years, we have at the University of Aberdeen been trying to promote the idea of the need for a nationally co-ordinated framework for geospatial data and information for the coastal zone. Why?

Access to Marine Data on the Internet for CZM 565

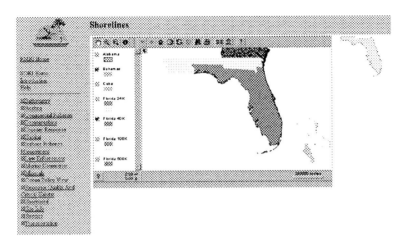

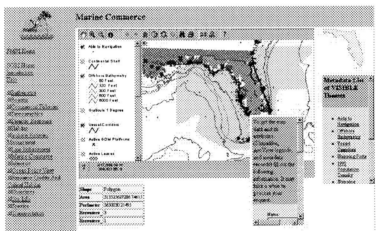

Fig. 8. The Florida Marine Online Mapping website (courtesy of Florida Fish and Wildlife Conservation Commission, Florida Marine Research Institute)

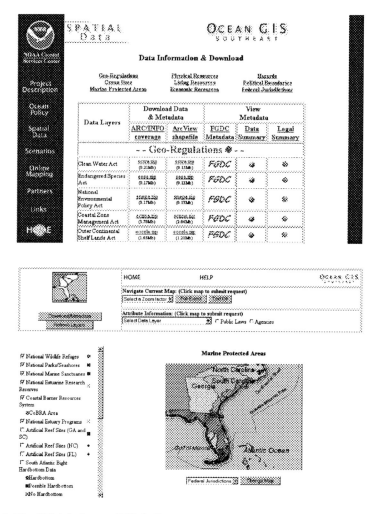

Fig. 9. The NOAA Ocean GIS Online Mapping website (courtesy of NOAA)

Fig. 10. The Australian Coastal Atlas (permission of Environment Australia)

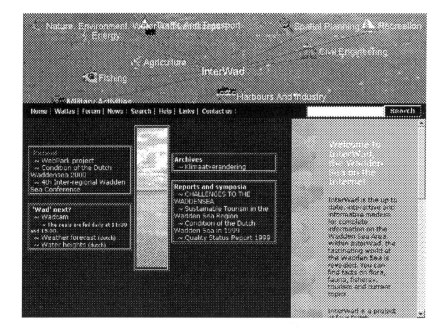
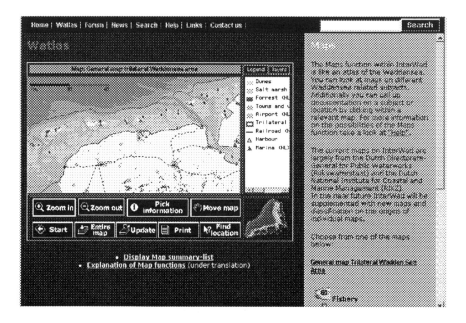

Fig. 11. Interwad Online GIS (courtesy of Interwad: www.waddenzee.nl)

Management, through decision-making and planning in the coastal zone requires ready access to data and information. At present, much of this data and information is scattered in many disparate forms and is held by many related and unrelated organisations (academic, commercial, private, and government). Much of this data is not catalogued or documented, and very little information about information (Metadata) is held.

Realistically there is a need to bring this data and information together under 'one roof'. But, there is also a need to go further than this and to standardise the collection and storage of further data and information in the future, so that new datasets become more widely applicable.

In practice, it is neither practical nor economically feasible to physically bring all the available data and information together in one place. However, in theory the rapid development of the Internet communications and networking technology offers both a practical and economic solution to the provision of rapid and efficient access to geospatial data and information. Instead of creating a 'new central building' approach to the problem, individual organisations would be linked via the Internet. This does, however, mean that there would need to be a national framework, established by a national co-ordinating body to look after the implementation of this proposed Internet-based data and information resource, that this organisation should be government in origin, with the power to implement such co-ordination, the required data quality and standards (Metadata), and to provide what must undoubtedly be seen as an essential source of data and information for the UK.

Progress to Date

What progress has been made to date with respect to the development of such an 'ideal' - a UK geospatial information resource for coastal zone management?

In the UK we are 'blessed' with a wide range of data and information sources, the national mapping agencies, the Ordnance Survey (OS) and the UK Hydrographic Office providing some of the best resources of land and marine data in the World. In recent years their products have become transformed to a digital format. Remote sensing data and imagery catalogues have also been produced e.g. SPOT Image. Relatively few organisations (government agencies interested in coastal management) have to date put GIS data onto their websites except English Nature. These go someway towards the ideal, by providing access to free downloadable ArcView datasets, although there is little documentation that is currently in a useful form.

The Ordnance Survey - under Geoffrey Robinson - was seriously looking at the pricing structure for key datasets, licensing, and the potential of the Internet (particularly e-commerce and on-line supply of data, and integration of other people's information) earlier in the year (Robinson, 1999, pp. 17-18). With his departure it remains to be seen what will now happen.

However, it is really only recently that real progress towards the potential for greater provision of access to such data and information has seriously been considered. To date this has been relatively slow, and whilst advances have been made, with all sorts of initiatives being undertaken e.g. AVID - UK Hydrographic Office and European equivalents:

Fig. 12. The ScotCoast, Forth Estuary Forum and Solway Firth Partnership websites

Added Value Information Dissemination (AVID) aims to gather, interpret and disseminate Hydrographic data held by the major European Hydrographic Offices.....work to define and develop efficient data distribution services that will make the government information widely available in a commercial format. The project is part of an initiative to assess how government information can be opened up to business and private citizens.... European industry is at a disadvantage compared with competitors in the US, where the Freedom of Information Act ensures that American public bodies are granted access to information systems free-of-charge, or for a small fee.

Despite the establishment of the National Geospatial Data Framework (NGDF), and the work of the Association for Geographic Information (AGI) on standards etc., there is seemingly still more talk than action, and in this respect we are some way behind the USA, Australia and Canada on the development of Internet-based geospatial data and information systems and provision, as well as national spatial data frameworks and Metadata standards.

In 1999 there were many events in the UK and Europe (e.g. The Netherlands) which have, not surprisingly considered the problem of access to geospatial data and information about the coastal zone. We have held two AGI Marine and Coastal Zone Management GIS Special Interest Group (AGI MCZM GIS SIG - http://www.abdn.ac.uk/geospatial/agi) one-day seminars in Carlisle (February 1999) and Cambridge at the Moller Centre (July 1999). All have highlighted the interest in the Internet for integrated coastal zone management, but all have drawn similar conclusions relating to concerns on: access to data and information for coastal zone management, including comparisons with the current state of play in the US, Australia and Canada.

InfoCoast'99 - February 1999 (Bridge, 1999)

- *European coastal zone data, information and knowledge networks (notably governmental/intergovernmental networks) generally deliver inadequately between users and providers*
- *Technologies should be designed to be fit for purpose and meet needs and capabilities of end users*
- *Information availability should be improved through better user awareness, making data and information available for uses other than their original purpose, free or low-cost accessibility of publicly-funded work and provision of standard maps and charts of the European coastal zone*
- *Information and knowledge flows should be facilitated by consistent and widespread use of simple Metadata records accessible through Internet-based and other gateways*
- *Using the knowledge base. Develop enabling mechanisms for networks that improve access to, and validation of, local knowledge by linking individuals, and local research by local, regional and NGO institutions*

Busby (1999, p.9) states:

The decision-maker needs to get access to all this information from a minimal number of sources, preferably through a single access point into a network of such resources. Why is a network necessary to achieve this?

- *Decision makers are seeking integrated solutions to complex issues*
- *No single organisation has all the answers*
- *Organised networks maximise value and minimise cost of information resources*

The way forward for the coastal zone involves building a distributed environmental information infrastructure.

Payne (1999, p. 13) from NOAA states:

To successfully address coastal issues, managers need information that is current, accurate, and reliable, in other words - useful.

A true decision support system refers to an information system or GIS tool that not only provides and extracts information from various data layers, but goes to the next step of providing options or answers to coastal resource management questions. Such systems can improve the ability of decision-makers to weigh alternative scenarios by including the impacts of changes to environmental, economic and other variables.

Scholten et al. (1999, p. 27) state:

Since environmental systems are usually multidimensional, a rich set of data must be considered when representing and modelling environmental systems.

In Coastal Zone Management (CZM) geographical information is invaluable because many of the numerous functions of the coast occupy, or make use of, space.

Integrated Mapping of the UK Marine and Coastal Zone - The Way Forward, CEFAS (June 1999) (Franklin and Hurrell, 1999)

- *Provide better access to information by: promoting awareness of what is available and who holds it; allowing better access to data held by public and commercial organisations; enabling users to access the information they need in the form in which they need it.*
- *Establish core datasets*
- *Carry out appropriate research, taking account of: identified data gaps; the dynamic nature of the marine and coastal zone; needs of different users for information on different scales.*

NORCOAST - 1999

Better information and understanding on coastal issues is needed in all areas. Both national, regional and local authorities have limited knowledge of the coastal marine area, and decisions are therefore sometimes based on limited or erroneous data. This also gives problems in achieving consensus on which are the most important problems to be solved (NORCOAST, 1999, p. 32).

In general more attention should be focused on information and development of knowledge and understanding on coastal issues, and on improving the exchange of information between authorities and agencies, to facilitate better decision making. Registering data and planning issues in co-operation with other agencies could ease the problem, as suggested in Hordaland e.g. in a common GIS database for the coastal zone (NORCOAST, 1999, p. 37).

There is now a need to develop a top-down approach to data collection, archiving, standards, co-ordination and dissemination so as to ensure that a European-wide approach to integrated coastal and marine resource management can be undertaken in the future to facilitate integrated coastal management (Green and King, 1999b, p.13).

NORCOAST, 1999. Statement that *National governments need to do more on co-ordinating data collection and centralising the results* (p. 15).

Two problems: one is to be able to get a general view of the enormous amount of information in reports and data files and how to get access to it. The other is how to use the information properly. As Burbridge (1999) states: *We have the knowledge but do we have the competence to use it?* (NORCOAST, 1999, p.15.)

Green Government (Green and King, 1999a)

Green and King (1999a, p. 19) stated:

Moreover, the continuing general lack of widespread use of information technologies such as remotely sensed data, geographic information systems (GIS) and the Internet for data collection, information and knowledge generation or dissemination respectively is currently preventing access to the very information that is needed by people who manage the coast. This is in turn blocking co-ordination and integration between the relevant responsible bodies.

We have the means to collect data, to store, process and analyse and output information and above all the required computer-based tools to achieve this. What we lack is government leadership to ensure the co-ordination of data and information collection, its maintenance, access and delivery to the end-user (Green and King, 1999a, p.20).

The worrying thing is that many of these statements continue to reiterate statements made many years ago; for example, comments and statements from Riddell (1992) and Green (1995b, 1996), Jeffries-Harris (1992), and Atkins (1995, p. 8-10) from Scottish Natural Heritage (SNH) who noted in relation to data and information for coastal zone management that:

The high level requirements are:

- **Ease of access to information**
- **Capability to do spatial analysis**
- **Integration for information from different sources**
- **Present information in different formats**
- **Know what information is available**

Project managers require to be able to undertake:

- **Spatial analysis**
- **Query different data at the same time**
- **View graphically different resources at the same time**
- **Have the ability to overlay resources spatially**
- **Isolate areas and look at the relationship between all activities**
- **Be able to generate basic statistics**
- **Be able to generate basic business graphics**
- **Be more responsive and timely to queries**
- **Query on an *ad hoc* basis**
- **Know where information is held and how to access it**
- **Be able to produce mapped information as well as reports**

Whilst Australia, the US and Canada have gone on to make considerable progress in addressing the potential and the problems that need to be overcome in order to develop access to data and information, the UK has in many respects stagnated, preferring to continue with a largely 'piecemeal approach' to virtually everything that concerns the provision of practical access to data and information for Integrated Coastal Zone Management (ICZM). This is not to say that good work has not been undertaken e.g. the OS and the UK Hydrographic Office; the numerous fora and voluntary groups e.g. Forth Estuary Forum (FEF), Moray Firth Partnership (MFP), Solway Firth Partnership (SFP), Firth of Clyde Forum, and the work of English Nature (EN), Scottish Natural Heritage (SNH), and the Joint Nature Conservation Committee (JNCC). However, what is lacking is the existence of an equivalent 'federal' body, and a 'federal-level' initiative, to ensure that all the spatial data and information that we currently hold about the UK, and that is continuing to be collected, is properly documented, catalogued, and its collection and dissemination (where possible) co-ordinated for the benefit of the UK end-user whether they be researchers, environmental consultancies, or educators.

Summary and Conclusions

It is abundantly clear that there is now a major and growing demand for access to both data and information - perhaps more so than ever before. The Internet has given us a 'good taste' of what is currently available and possible. Considerable progress has been made in recent years with communications and networking technology, hardware and software, the Freedom of Information Act, such that increasingly we are becoming an Information Technology (IT) aware and user-based society, both at work and at home. There are still problems of a technical nature which have been discussed, but moreover there also appears to be a number of institutional problems that are currently limiting the more widespread use of this technology in the workplace for a wide range of applications, one of which is Integrated Coastal Zone Management. Recently Hewson (1999, p. 5) stated:

Referring to DSL....This is what the Internet is supposed to be about... Why do the Americans have DSL racing through their nation when we don't?.. Quite why the Americans seem able to overcome these (problems) when we cannot remains a mystery... If we sit on the sidelines for long... we will see France and Germany begin to offer cheap, reliable broadband data access to individuals and businesses while we struggle to cope with obsolete or over-priced old technology. The business community's growing shift to network applications will be encouraged on the Continent and stifled here... The real culprits lie in a government that lacks the vision and the courage to move beyond hype and hot air to the hard work of policy and implementation.... Britain is stumbling into the new millennium with a telecommunications network designed for the 1970s - not because we cannot afford the future but because no politician has the guts to make it happen

      In the UK we now need to 'take the bull by the horns' and to set about effectively mirroring the work of the US, Canada and Australia as far as: setting up and developing a National Geospatial Data Framework (NGDF) is concerned, and attending to spatial data and Metadata standards, amongst other things. Clearly there will still be a need for setting up of committees and further discussion etc., but we also need far more concerted action rather than words. Once again, the attitude to something that must be seen as essential - a national information resource - is apparently different in the US to the UK:

- **Quality of life in a free society is determined by the collective decisions of its individual citizens acting in the home, the workplace and together as members of the community**
- **Responsible stewardship of our natural and cultural resources depends on making current, accurate and complete information available**
- **What strategies would pool more public and private financing for standardised data development and stewardship?**
- **How can we make data sharing easier, to avoid wasteful redundant data collection?**

- How can we remove the technical and institutional barriers to widespread use of geospatial data?
- How can new Internet tools help?
- What kinds of public-private partnerships should be formed to address these issues?
- Using common solutions and sharing data will dramatically improve the way we make decisions
- Millions of Web sites and a wide range of technologies offer access to maps, earth images and other geospatial data
- But it is hard to find the data you really need, and hard to use it in your GIS
- As these geodata discovery and geoprocessing interoperability problems get solved, one result will be that online geoprocessing services, not only online geodata, will become an integral part of your information environment
- Barriers to data sharing slow our progress in many endeavours and, ultimately in building liveable communities
- Geodata serves as the information base underlying liveable communities
- However, the kinds of data needed, who creates and maintains what information, and how effective co-ordination is achieved are among the challenges of this complex, multifaceted issue
- Framework data provide a base on which to collect, register, integrate and analyze other information... they are an essential element of NSDI

(after Conference flyer for: **Making Living Communities a Reality - 1999 National Geodata Forum**)

In the past we have had (and still do have) major national organisations who have provided essential map data and information which has formed and still is a major national resource worth 'billions' of pounds. But, despite the attempts to 'commercialise' the national resources, and to break up many organisations into smaller economic units, at the end of the day some things just can not operate like this; and one of these is a national resource of geospatial data and information, one that is co-ordinated and managed by a government body for the good of the nation.

The problem in the UK at present is that we are still far too good at convening lots of committees, 'talking shop', producing glossy brochures, developing our image at the expense of content and substance, and saying a lot of things that do not ever see the light of day. We need to get past this stage and to focus on providing low-cost data and information that will form the basis for a national resource, and to make the best possible use of the technology that we now have at our disposal to make this a reality. We need to address and resolve all the issues that are currently major stumbling blocks to maximising our use of data and information in the workplace.

To conclude, in summary, we need to:

- **Raise awareness that data and information is fundamental to our knowledge base, understanding, planning, management and decision-making in the workplace**

- **Educate people in the current and future technologies**
- **Develop a new attitude and perspective on technology, data and information needs and availability (different world with different technology)**
- **Develop an holistic view of the coast which requires integration of data and information held by many disparate organisations**
- **Recognise that 'piecemeal solutions' do not work**
- **Provide practical and operational solutions to access to, and use and communication of both data and information in the workplace**

Moreover we need to:

- **Reach the people in the UK Government who have the vision, the influence, and the decision-making power to make things happen, and not those who do not**

Perhaps the new millennium will see the electronic information age become a reality for Integrated Coastal Zone Management (ICZM).

References

Atkins, S. (1995). *User Requirements for Information Management in the 'Focus on Firths' Project*, 15p.
Bridge, L. (ed). (1999). *Info-Coast'99 Symposium Report*, Coastlink, Coastal and Marine Observatory at Dover, UK and EUCC-UK, Brampton, UK.
Burbridge, P. (1999). Sea-Land Integration: What Enabling Mechanisms do we need to make Integrated CZM a Reality? *NORCOAST: Integrated Coastal Zone Planning and Management Seminar, Aalborg, Denmark, 31st May - 1st June 1999, pp. 10-11.*
Busby, J. R. (1999). Principles, Practices and Current Directions in Information Management: An Overview. In, Bridge, L. (ed)., 1999. *Info-Coast'99 Symposium Report*, Coastlink, Coastal and Marine Observatory at Dover, UK and EUCC-UK, Brampton, UK, pp. 7-9.
Franklin, F.L., and Hurrell, V.L. (1999). Integrated Mapping of the UK Marine and Coastal Zone - The Way Forward. *Report* of a MAFF sponsored workshop held at CEFAS Lowestoft Laboratory on 17th-18th June 1999.
Green, D.R. (1996). Between the Desktop and the Deep Blue Sea, *Mapping Awareness*, 10(6):19-22.
Green, D.R. (1995a). Internet, the WWW and Browsers: The Basis for a Network-Based Geographic Information System (GIS) for Coastal Zone Management, *AGI'95 Conference Proceedings*, pp. 5.1.1-5.1.12.
Green, D.R. (1995b). Preserving a Fragile Marine Environment: Integrating Technology to Study the Ythan Estuary, *Mapping Awareness*, 9(3):28-30.
Green, D.R. and King, S.D. (1999a). Coastal Disintegration - The Legacy of Poor Co-ordination. *Green Government*, February 1999, pp. 19-20.
Green, D.R. and King, S.D. (1999b). Integrated Information for Integrated Coastal Planning: Reconciling Onshore and Offshore Data Sets in Countries Around

the North Sea. NORCOAST, Integrated Coastal Zone Planning and Management Seminar in Aalborg, 31st May - 1st June 1999. *Summary*, pp. 12-13.

Hewson, R. (1999). Innovation Section. *Sunday Times.* p.5

Jefferies-Harris, T. (1992). GIS Makes Inroads into UK Marine Environment, *GIS Europe*, 1(8):36-38.

National Geodata Forum (1999). Making Living Communities a Reality. *Conference Flyer*.

NORCOAST (1999a). *NORCOAST: Integrated Coastal Zone Planning and Management Seminar*, Aalborg, Denmark, 31st May - 1st June 1999. 28p. + Annex.

NORCOAST (1999b). *NORCOAST: Review of National and Regional Planning Processes and Instruments in the North Sea Regions - Summary.* NORCOAST Project Secretariat, Aalborg, Denmark, 53p.

Payne, J.L. (1999). Coastal Decision-Making: The Role of Information, Technology, and Capacity Building, In, Bridge, L. (ed.), 1999. *Info-Coast'99 Symposium Report*, Coastlink, Coastal and Marine Observatory at Dover, UK and EUCC-UK, Brampton, UK, pp. 11-14.

Riddell, K.J. (1992). Geographical Information and GIS: Keys to Coastal Zone Management, *GIS Europe*, 1(5):22-25.

Robinson, G. (1999). New Business Model, Pricing, and Licensing Announced for O.S., *Mapping Awareness*, October 1999, pp. 16-19.

Scholten, H., Fabbri, K., Uran, O., and Romao, T. (1999). Understanding the Role of Geo Information Communication Technology in Building a Desirable Future for CZM, In, Bridge, L. (ed.), 1999. *Info-Coast'99 Symposium Report*, Coastlink, Coastal and Marine Observatory at Dover, UK and EUCC-UK, Brampton, UK, pp. 27-33.

---

[i] Graphic image from maps.esri.com with Data from the City of Ontario, California. Copyright © *ESRI. Al rights reserved.*

List of Contributors

A. Alesheikh
Department of Geomatics Engineering,
The University of Calgary, 2500
University Drive N.W., Calgary,
Alberta, Canada
T2N 1N4

R. Andersen
Mapping & GIS Solutions
Intergraph, 35 Cotham Rd
Kew 3101, Melbourne, Australia

J. Aston
19 Murray Street
Townsville, Queensland
Australia, 4810.

P.J. Bacon
CEH Banchory, Hill of Brathens
Banchory, Aberdeenshire
AB31 4BY, Scotland, UK

P.S. Balson
Geophysics & Marine Geoscience
British Geological Survey
Keyworth, Nottingham, NG12 5GG
England, UK

M. Barnsley
Department of Geography
University of Wales Swansea
Singleton Park
Swansea, SA2 8PP
Wales, UK

L. Borgniet
Cemagref, Group d'Aix-en-Provence,
Le Tholonet, F-AIX-EN-PROVENCE,
France

C. Bristow
School of Earth Sciences
Birkbeck College
University of London
Malet St, London, WC1E 7HX
England, UK

N. J. Brown
CEH Monks Wood
Abbots Ripton, Huntingdon
Cambridgeshire, PE28 2LS
England, UK

R. Canessa
Axys Environmental Consulting,
PO Box 2219, 2045 Mills Road West
Sidney, British Columbia
Canada, V8L 3S8

M.A. Chapman
Department of Civil Engineering
Ryerson University
350 Victoria Street
Toronto, Ontario
Canada, M5B 2K3

H. Coccossis
University of the Aegean, 17 Karantoni
Street, GR 81100
Mytilini, Greece

R. Cox
CEH Monks Wood
Abbots Ripton, Huntingdon
Cambridgeshire, PE28 2LS
England, UK

B. Crawley
Project Component Manager
World Bank/Government of Samoa-
Infrastructure Asset Management
Project, Department of Lands Surveys
and Environment
Samoa

M. A. Damoiseaux
Hoofd ICT
Waterschap Peel en Maasvallei
Postbus 3390, 5902 RJ Venlo
The Netherlands.

A. Davison
WWF Scotland
8 The Square, Aberfeldy
Perthshire, PH15 2DD
Scotland, UK

A.M. Denniss
Geospatial Information Systems British
Geological Survey
Keyworth, Nottingham, NG12 5GG
England, UK

K. Dimitriou
University of the Aegean, 17 Karantoni
Street, GR 81100
Mytilini, Greece

A. Fernández-Palacios
Servicio de Evaluación de Recursos
Naturales. Dirección General de
Planificación. Consejería de Medio
Ambiente, Junta de Andalucía.
Pabellón de Nueva Zelanda, Avenida
de las Acacias, s/n. Isla de Cartuja.
41092 Sevilla, España/Spain

S. Fletcher
School of Maritime and Coastal
Studies, Southampton Institute
East Park Terrace, Southampton Hants.
SO14 0RD, England, UK

D. R. Green
Centre for Marine and Coastal Zone
Management, Department of
Geography and Environment
University of Aberdeen, Elphinstone
Road, Aberdeen, AB24 3UF Scotland,
UK

H. Goodwin
Norwegian Polar Institute, Polar
Environmental Centre, N-9296
Tromsø, Norway

M.P. Harris
CEH Banchory, Hill of Brathens
Banchory, Aberdeenshire
AB31 4BY, Scotland, UK

T. Horsman
Geological Survey of Canada Atlantic,
P.O. Box 1006, Dartmouth Nova
Scotia, Canada, B2Y 4A2.

L. P. Humphries
University of Sunderland
School of Computing and Technology,
The David Goldman Informatics
Centre, St Peter's
Sunderland, SR6 0DD
Tyne and Wear
England, UK

K.A. Jenner
Geological Survey of Canada
(Atlantic), 1 Challenger Drive
P.O. Box 1006, Dartmouth
Nova Scotia, Canada, B2Y 4A2

C. Johnston
Joint Nature Conservation Committee
Monkstone House
Peterborough
Cambridgeshire
PE1 1JY, England, UK

List of Contributors

G. Jones
Department of Geography
Graham Hills Building, 50 Richmond
Street, Glasgow, G1 1XN
Scotland, UK

H. Karimi
Director, Geoinformatics
Dept of Information Science and
Telecommunications
University of Pittsburgh
Pittsburgh, PA 15260, USA

C.P. Keller
Department of Geography
University of Victoria
PO BOX 3050 STN CSC, Victoria
British Columbia, Canada, V8W 3P5

A. Sh. Khabidov
Institute for Water and Environmental
Problems,
Russian Academy of Sciences
105, Papanintsev Street, 656099
Barnaul, Russia

M. Kerdreux
Ifremer - DEL/AO
BP 70, 29290 PLOUZANE
France

S.D. King
Centre for Marine & Coastal Zone
Management, Department of
Geography and Environment
University of Aberdeen, Elphinstone
Road, Aberdeen, AB24 3UF Scotland,
UK

I. Leaver
The Crown Estate
10 Charlotte Square
Edinburgh, EH2 4DR
Scotland, UK

O. Lemoine
Ifremer
BP 133, 17390 LA TREMBLADE
France

C.N. Ligdas
Web Development Project
Robert Gordon University
St Andrews Building
Aberdeen, AB25 1HG
Scotland, UK

D. Livingstone
School of Earth Sciences and
Geography, Penrhyn Road
Kingston-upon-Thames
Surrey, KT1 2EE, England, UK

Ana Lloret
Centro de Estudios de Puertos y Costas
del CEDEX 81,
Antonio Lopez - 28026 Madrid Spain

J. Loder
Department of Geography and
Environment, University of Aberdeen.
Elphinstone Road, Old Aberdeen,
Aberdeen, AB24 3UF Scotland, UK

U. Lohr
TopoSys Topographische Systemdaten
GmbH
Wilhelm-Hauff-Straße 41, D-88214
Ravensburg, Germany

R.A. Longhorn
Director, IDG (UK) Ltd,
EC Projects Office
Neihaff, L-9161 Ingeldorf,
Luxembourg

L. Loubersac
Ifremer - DEL/AO
BP 70]
29290 PLOUZANE, France

T. McCarthy
Airborne Videography Ltd.
22 Heathfield Gardens, Chiswick
London, W4 4JY, England, UK

J. McGlade
Dept of Mathematics
University College, London
England, UK

R. Newsham
British Geological Survey
Kingsley Dunham Centre
Keyworth, Nottingham
NG12 5GG, England, UK

J. Ojeda Zújar
Departamento de Geografía física y
AGR; Facultad de Geografía e
Historia, Universidad de Sevilla. C/M^a
de Padilla s/n, 41004 Sevilla
España/Spain.

H. Okayama
Center for Environmental Remote
Sensing, Chiba University
1-33 Yayoi, Inage, Chiba 263-8522,
Japan

R. Pakeman,
Macaulay Land Use Research Institute,
Craigiebuckler
Aberdeen, AB15 8QH
Scotland, UK

R. Palerud
Akvaplan-niva AS
9296 Tromsø, Norway

E. Parrilla
Departamento de Geografía física y
AGR; Facultad de Geografía e
Historia, Universidad de Sevilla. C/M^a
de Padilla s/n, 41004 Sevilla
España/Spain.

C.I.S. Pater
English Nature, Northminster House
Peterborough, PE1 1UA
England, UK

J.M. de la Peña
Centro de Estudios de Puertos y Costas
del CEDEX 81,
Antonio Lopez - 28026 Madrid Spain

A.M. Pérez Romero
Departamento de Ingenieria Gráfica.
Universidad de Sevilla. Centro de
Enseñanzas Integradas (Antigua Univ.
Laboral). Sevilla, Spain

J.M. Pérez
Departamento de Geografía física y
AGR; Facultad de Geografía e
Historia, Universidad de Sevilla. C/M^a
de Padilla s/n, 41004 Sevilla
España/Spain.

J. Populus
Ifremer - DEL/AO
BP 70, 29290 PLOUZANE
France

J. Prou
Ifremer
BP 7004
TARAVAO
Tahiti, Polynésie Française

J. Raper
Department of Information Science
City University, Northampton Square
London, EC1V OHB, England, UK

S.T. Ray
1 Chantry Hill
Slapton, Kingsbridge
Devon, TQ7 2QY
England, UK

List of Contributors

J. Robertson
Department of Land Economy
University of Aberdeen
Elphinstone Road
Aberdeen, AB24 3UF
Scotland, UK

I. Rodríguez
Universidad Rey Juan Carlos
Escuela Superior de Ciencias
Experimentales y Tecnologia
C/ tulipan s/n; 28933 - Mostoles
(Madrid), Spain

E. Sanchez Rodriguez
Departamento de Geografía física y
AGR; Facultad de Geografía e
Historia, Universidad de Sevilla. C/M^a
de Padilla s/n, 41004 Sevilla
España/Spain.

S. Shanmugam
Centre for Geo-science and
Engineering
Anna University
Madras 600 025, India.

A.G. Sherin
Geological Survey of Canada
(Atlantic), 1 Challenger Drive
P.O. Box 1006, Dartmouth, Nova
Scotia, Canada, B2Y 4A2

A. Simms
Department of Geography, Memorial
University of Newfoundland, St.
John's, Newfoundland, Canada
A1B 3X9

É.L. Simms
Department of Geography, Memorial
University of Newfoundland, St.
John's, Newfoundland, Canada
A1B 3X9

J.S. Smith
Department of Geography and
Environment, University of Aberdeen,
Elphinstone Road Aberdeen, AB24
3UF, Scotland, UK

J. Sun
272-0813 Ichikawa City, Nakayama 4-
7-15, Nikkou Mansion 203
Japan.

A.G. Thomson
CEH Monks Wood
Abbots Ripton
Huntingdon
Cambridgeshire
PE28 2LS, England, UK

D.G. Tragheim
Information Systems Group, British
Geological Survey, Keyworth,
Nottingham, NG12 5GG
England, UK

R.A. Wadsworth
CEH Monks Wood
Abbots Ripton , Huntingdon
Cambridgeshire
PE28 2LS, England, UK

S. Wanless
CEH Banchory , Hill of Brathens
Banchory, Aberdeenshire
AB31 4BY, Scotland, UK

P.A.G. Watts
Ordnance Survey of Great Britain
Romsey Road, Maybush Southampton,
SO16 4GU
England, UK

A.D. Webb
English Nature
Northminster House
Peterborough
PE1 1UA, England, UK

P. Wright
The United Kingdom Hydrographic
Office, Admiralty Way, Taunton
Somerset, TA1 2DN
England, UK

M.Yates
CEH Monks Wood
Abbots Ripton, Huntingdon
Cambridgeshire
PE28 2LS, England

Index

Abiotic, 425, 428
Accretive, 210, 211
Accuracy, 80, 81, 125, 139, 141, 165, 193, 194, 244, 264, 265, 268, 278, 339, 349, 351, 352, 353, 357, 358, 359, 367, 402, 409, 430, 440, 444, 492, 501, 507, 518, 524, 534,
Acoustic Remote Sensing, 487, 488, 490
Administrative Boundaries, 9, 421
Admiralty Hydrographic Surveys, 26
Advanced Very High Resolution Radiometer (AVHRR), 386
Aeolian Deflation, 511, 514, 515, 522
AEPS (Arctic Environmental Protection Strategy), 52
Aerial Photography, 64, 94, 193, 218, 248, 280, 402, 403, 404, 474, 477, 478, 480, 485, 519
Aerial Videography, 402, 404, 412
Aerotriangulation Techniques, 404
Agenda 21, 49, 86, 104, 422, 424
Airborne Sensors, 474, 477
Algal Growth, 502
Altimetric Data, 520
Amphipods, 301
Analog (Analogue), , 111, 113, 114, 115, 403, 405, 517
Analytical Models, 437
Andalucian System of Environmental Information (SinambA), 526
Andalusian Cartographic Institute, 517
Animated GIF, 558
Animation, 451, 457, 460
Anthropic Pressure, 512, 519, 520
Anti-Submarine Warfare, 80
Aquaculture, 38, 43, 275, 276, 277, 278, 280, 281, 284, 287, 290, 291, 293, 420, 439
Arc Macro Language (AML), 430
Archival Datasets, 113, 129
Arctic Ocean, 47, 52
Artificial Reefs, 115, 116, 117
ASCII, 126, 151, 196, 268

Asian Development Bank (ADB), 88
Asset Recording, 402
Association for Geographic Information (AGI), 571
Atlantic Institute (AI), 544
Atlantic Provinces of Canada, 371, 372, 384
Atmospheric Correction, 534
Australian Coastal Atlas, 564, 567
Autocad, 262, 264
Autodesk, 452, 456
Azimuth Angle, 387, 388, 389, 391, 392, 395, 397
Azores, 41

Babtie Group, 399, 400
Backshore, 28, 174, 179, 203, 372, 375, 376, 378, 379, 381, 517, 518, 519
Baltic Coastal and Marine Protected Areas (BSPAs), 52
Baltic Sea Study, 41
Baltic Sea, 37, 41, 47, 49, 50, 51, 52, 343, 548, 550
Barents Sea, 45, 163
Barometric Pressure, 26
Barrages, 79
Barrier Islands, 201, 202, 407, 512
Base Station, 264
Baseline Data, 90, 498
Baseline Information, 488
Bathymetric Chart of the Oceans, 546
Bathymetric Data, 124, 134
Bathymetric Surveys, 198, 467, 469
Bayhead Barrier Beaches, 372, 374
Baymouth Bars, 211
Beach Cannibalism, 27
Beach Nourishment, 4, 13, 14, 17
Beach Profiles, 235, 236, 242, 433
Bedrock Geology, 22
Benchmarks, 141, 409
Benthic Communities, 165, 166, 170
Benthic Feeder, 222
Benthivore, 305
Berm, 210

BIDS (Bath Information and Data Services), 481
Binary Classification., 177
Biodeposition, 261
Biodiversity, 3, 5, 16, 18, , 52, 247, 248, 250, 260, 436
Bioenergetics, 304, 343
Biological Communities, 490, 491
Biological Processes, 248
Biomass, 254, 265, 271, 284, 285, 286, 287, 293, 299, 300, 303, 304, 317, 318, 319, 324, 333, 334, 346, 501, 502, 503, 504, 505, 506, 507, 508, 509, 547
Biomass Consumption, 346
BIOTA, 233, 244, 425, 426, 428, 430
Biotope, 418, 487, 491, 492, 493, 494, 495
Bird Habitat, 158
Biscay Bay, 380, 382
Black Sea, 37, 47, 48, 49, 51, 549
Blowouts, 514
Blueprint, 87
Botanical Survey, 430
Bottom Type, 279, 503, 504, 505
Brachiopods, 493
Breakline, 144, 194
Breakwaters, 13, 15, 19, 519
Breeding Colony, 221, 222
Breeding Cycle, 221, 226
Breeding Season, 221, 229
British Admiralty Charts, 165
British Coal, 174, 177, 183
British Geological Survey (BGS), 129, 193
Bromley Model, 304
Brooding Stock, 290
Buffer Area, 155, 168, 169, 359, 360, 422
Buffer Zone, 494
Building Regulations, 416, 421

Cadastral, 65, 263
Cadmium, 527
Calendar Year, 225
Calibration, 141, 338, 359

Canopy Coverage, 502, 504, 507
Capacity Building, 87, 88, 89
Carrying Capacity, 261, 423, 442
CD-ROM, 45, 428, 453, 456, 464, 549
CEFAS, 572, 577
Centre for Earth Observation (CEO), 40
Centroid, 279
CGI Scripts, 455, 460, 558
Change Detection Technique, 535
Chart Differencing, 202, 517, 522
Climate Change, 46, 425, 430, 544
Clyde Estuary, 149, 151, 152, 154, 155, 157, 161
Coal Mining, 174
Coast Protection Act, 7, 8
Coastal Authorities, 10
Coastal Cell, 65
Coastal Conservation, 3, 91
Coastal Defence and Protection, 1, 2, 4, 5, 17, 18
Coastal Defence Schemes, 83
Coastal Dune System, 247, 248, 259, 512
Coastal Dynamics, 21, 25
Coastal Ecosystems, 420, 421, 426
Coastal Engineering Works, 420
Coastal Flood Protection, 158
Coastal Geomorphology, 372, 373, 378, 410
Coastal Information System (CIS), 371
Coastal Management, 1, 2, 3, 4, 5, 6, 7, 10, 16, 17, 18, 32, 48, 52, 65, 84, 86, 87, 90, 91, 92, 99, 105, 106, 173, 191, 213, 220, 273, 371, 384, 416, 417, 421, 430, 437, 438, 439, 440, 441, 442, 443, 444, 445, 447, 448, 449, 561, 569, 573
Coastal Morphology, 189, 210, 410
Coastal Planning, 9, 11, 19, 146
Coastal Processes, 2, 3, 8, 9, 11, 16, 19, 28, 188, 189, 206, 400, 420
Coastal Protection Authorities (CPAs), 3, 8, 11, 17
Coastal Protection, 7, 78, 87, 184
Coastal Strandplains, 23

Coastal Zone Colour Scanner (CZCS), 44
Coastal Zone Management (CZM), 572
Coastal Zone Mapping Project, 76, 80, 83, 84
Coastal Zone, 2, 3, 11, 16, 17, 19, 21, 23, 24, 31, 32, 36, 38, 40, 41, 42, 43, 44, 45, 49, 50, 51, 52, 53, 55, 56, 57, 58, , 62, 63, 67, 68, 69, 71, 72, 73, 75, 76, 77, 78, 79, 81, 83, 85, 86, 91, 95, 97114, 116, 129, 144, 205, 211, 212, 225, 248, 261, 262, 273, 274, 371, 382, 383, 402, 412, 420, 425, 426, 428, 449, 451, 452, 454, 468, 470, 473, 474, 477, 478, 483, 485, 525, 527, 531, 541, 544, 548, 549, 553, 562, 563, 564, 569, 571, 572, 573, 574
Colliery Spoil, 173, 174, 189
Commercial Fishery, 301, 317
Commercial Fishing Fleet, 297
Common Fisheries Policy (CFP), 298
Community Stakeholders, 440
COMPAS, 455, 464
Conglomerate, 22
Conical Projection, 262
Consumption Levels, 297, 300
Contaminated Area, 282
Continental Shelf, 79, 98
Contingency Planning, 70
Copepods, 301, 305
Copyright, 72, 111, 403, 555, 559, 561
Core Sediments, 163
Corine, 418
Cost-Benefit Analysis, 442
Covariance Matrix, 362, 363, 364, 367
Critical Zone, 421
Cross-Border Programmes, 36
Crude Oil, 164
Crustaceans, 301
CSIRO, 88

Danish Hydraulic Institute (DHI), 467
Data Accuracy, 357
Data Exchange Format, 107
Data Lineage, 357, 358

Data Protection Act, 461
Data Quality, 47, 113, 114, 115, 358, 425, 569
Data Transfer Standards, 357
Datum, 82, 138, 139, 141, 143, 150, 218, 266, 267, 271, 407, 410, 470, 489
Decision Support System (DSS), 425, 428, 485, 561
DEM, 349, 350, 351, 352, 353, 354
Demersal, 298, 301, 305, 306, 308, 339, 345
Density Slicing, 503, 530
Department of Fisheries and Oceans, 282
Department of the Environment (DoE), 3, 7, 16
Depositional Coasts, 399, 400
Depositional Sinks, 191
Descriptive Statistics, 438, 487
DGPS, 139
DHTML, 556
DIGEST, 546
Digital Geographic Information Working Group (DGIWG), 546
Digital Image Processing System, 404
Digital Photogrammetry, 191, 193, 198, 202
Digital Photography, 404, 412
Digital Terrain Model (DTM), 150, 269
Digitiser, 141
Directorate of the North-Netherlands, 107
Discriminant Analysis, 503
Disparate Data, 70, 117, 133, 145, 444, 564
Distance Meter, 518
Distance Referencing System, 372, 375
Distributed Database, 410, 425, 428
Drainage Basin, 50, 151
Dredging, 79, 262, 270, 271
Drilling Operations, 79
Drowned Delta Fans, 26
Dune Landforms, 25
DXF (Data eXchange Format), 81

Dynamic Segmentation, 371, 374, 375, 383, 384

Echo Sounder, 490, 491
Ecological Balance, 100
Ecological Diversity, 251
ECS (Electronic Chart Systems), 78
Edge Matching, 82
EEA (European Economic Area), 36, 41
EEA (European Environment Agency), 36, 37, 38, 40, 41, 45, 46, 53, 59
EEZ, 97, 98, 99, 100, 101, 102, 103
EGII (European Geographic Information Infrastructure), 35, 38, 44, 543, 553
EIONET (European Environment Information and Observation Network), 45
Electronic Chart Display and Information System (ECDIS), 78, 546
Electronic Navigation Charts (ENC), 546
ELOISE (European Land-Ocean Interaction StudiEs), 42
Embayment, 30, 133, 134, 136, 137, 138, 142, 143, 144, 146, 158
EMMA (European Maritime Multimedia Data Agency), 40
English Nature (EN), 2, 16, 19, 452, 464, 480, 499, 559, 569, 574
ENRICH (the European Network for Research into Global Change), 42
Environment Agency, 15, 58, , 64, 88, 134, 137, 140, 141, 144, 146, 244, 435, 480
Environment Assessment Program for Asia and Pacific, 89
Environmental Assessment, 87, 89, 165, 168, 442
Environmental Change, 42, 91, 167, 221, 298
Environmental Database, 113, 115, 116, 127, 129, 163, 165, 171

Environmental Impact Assessment, 163, 167, 170, 171, 442
Environmental Impact Statements, 416, 421
Environmental Resource Information Systems, 88
Epsilon Band Model, 359, 360, 364
ERGIS (European Marine Resource Geographical Information Service), 40
Error Band Model, 359, 360
Error Matrix, 362
Error Models, 355, 356, 368
EUREKA, 38, 45, 47, 53
EUROGI (the European Umbrella Organisation for Geographic Information), 35, 55, 56, 58, 543, 544, 546, 547, 551, 553
EuroGOOS, 550
EUROMAR, 45, 47
European Bank for Reconstruction and Development, 58
European Commission, 38, 42, 43, 45, 52, 58, 59, 231, 544, 550, 551, 553
European Community, 39, 51, 334, 488
European Economic Area (EEA), 36, 41
European Environment Agency (EEA), 40, 45, 58, 59
European Geographic Information Infrastructure, 35, 543, 553
European Investment Bank, 58
European Sea Level Observatory System (EOSS), 43
European Union (EU) , 42, 45, 231, 298, 344, 486
Eustatic Change, 23
Evaluative Criteria, 442
Exclusive Economic Zones (EEZs), 97, 99
Expert System, 428, 448
Exxon Valdez, 164, 171, 228, 232

Faecal Coliform, 279

FAO (Food and Agriculture Organisation), 51, 297, 342, 344, 548
Federal Geographic Data Committee (FGDC), 544
Feeding Behaviour, 221, 231
Feeding Sites, 221
Feeding Strategies, 297
Fen Orchid, 247, 250, 260
Field Checking, 359
Field Cruises, 163
Fifth Framework Programme, 35, 53, 57, 58
File on Forth, 457
Financial Liability, 149
Firth of Clyde Forum, 574
Firths Initiative, 452, 564
Fish Stocks, 52, 79, 297, 298, 300, 339, 550
Fisheries Agency, 94
Fisheries Data, 297, 340
Fishing Grounds, 79, 222
Fitness of Data, 357
Fixed Assets, 402
Flight Path, 350, 351
Flood Control, 544
Flood Defences, 134, 136, 480
Flood Index Values, 155
Flood Management, 70, 155
Flood Maps, 154
Flood Warning Procedures, 15
Flood Zone, 152, 154, 156
Flora and Fauna, 38, 89, 218, 491
Flow of Biomass, 299
Fluvial Processes, 209, 211
Fluvioglacial Landforms, 24
Food and Agriculture Organisation (FAO), 51, 297, 342, 344, 548
Food Supply, 222, 230
Foraging Efficiency, 228
Foraging Model, 221
Foredune Retreat, 511, 514, 518, 520
Foreshore, 12, 14, 64, 179, 184, 191, 203, 204, 372, 375, 376, 378, 379, 381, 524

Forth Estuary Forum (FEF), 451, 564, 574
Fragile Environment, 164, 170
Framegrabbed, 405
Framework for Action, 87
Freighter Traffic, 134
Fuzzy Logic, 423, 443

Gastric Evacuation, 304, 343
Generalization Error, 360
Geodetic Control Networks, 359
Geographical Information System (GIS), 134, 173, 221, 248, 425, 451
Geography, 62, 64, 250, 475, 552
Geology, 179, 194, 207, 231, 439
Geomatics, 261, 439
Geomorphological Unit, 201
Geomorphology, 21, 23, 144, 173, 174, 184, 399, 400, 402
Geo-Referenced, 138, 218, 251, 261, 264, 279
Geo-Referencing, 44, 218, 244, 259, 264
Geospatial, 63, 68, 70, 71, 79, 83, 129, 359, 452, 547, 548, 555, 556, 561, 562, 563, 564, 569, 571, 576
Geostatistics, 270, 343
German Umbrella Organisation for Geoinformation (DDGI), 544
GIS-Based Modelling Approach, 225
GISDATA, 35, 55, 56, 58
Glacial Deposits, 22
Glacial Till, 23, 191
Global Climate Change, 35, 38, 544
Global Emergency Management and Maritime Information Systems, 551
Global Ocean Observing System (GOOS), 41, 45, 51, 548, 549, 550
Global Positioning System (GPS), 139, 404, 452, 490, 524
Global Spatial Data Infrastructure (GSDI), 55, 59, 543, 544, 551, 552, 553
Global Warming, 62, 70, 149, 155, 160

Government Select Committees (House of Lords Select Committee on Science and Technology Remote Sensing and Digital Mapping 1983), 75
Grab Sampling, 487, 488, 490, 491
Graphical Editing Tools, 265
Greenhouse Gases, 156
Grid Format, 202
Gridded Bathymetric Data, 83
Ground Control Survey, 193
Ground Surveying Equipment, 140
Ground Swath, 502
Group Decision Support System (GDSS), 445
Groyne, 12, 215
GUI (Graphic User Interface), 430, 433

Harbour Maritime Service, 266, 271
Harvest Areas, 293
Harvest Plan Map, 292, 293
Harvesting Tracks, 293
Hazard Management, 4, 17
HCI (Human Computer Interface), 455, 459, 460, 484, 485
Helsinki Commission – Baltic Marine Environment Protection Commission (HELCOM), 50
Helsinki Convention (HELCOM), 52
Her Majesty's Stationery Office (HMSO), 84, 231
Heritage Coast, 5, 18
Hierarchical Database, 474
High Resolution Imagery, 233
High Resolution Visible (HRV), 386
Human Computer Interface (HCI), 453, 459, 460, 465
Human Dimension of Global Environmental Change Programme (HDP), 41
Hydrodynamic Model, 236, 244, 262
Hydrogeologic Conditions, 417
Hydrographic Data, 75, 76, 80, 81, 82, 83, 470, 547
Hydrographic Office, 80, 83
Hydrographic Surveys, 81, 138

Hydrology, 38, 251, 276, 278, 280, 552
Hydrophytes, 247
HyperText Markup Language (HTML), 455, 460

Ice-Fronts, 23
IDRISI, 153, 154, 202, 203, 523
IFREMER, 261, 274
Image Analysis, 194
Image Processing, 203, 247, 351, 454, 485, 535, 556, 558
Imagemap, 404
IMPACT (Information Market ACTions), 40
Indicatrices, 385, 387, 388, 393
Infaunal, 491, 492
Information Society, 54, 551
Information Technology (IT), 50, 62, 453, 575
In-Situ Data, 501
INSROP (International Northern Sea Route Program), 165, 167, 171
Institut Géographique National (IGN), 262
Institute for Land Information and its Land Information Assembly (ILI.LIA), 544
Integrated Coastal Area Management (ICAM), 48, 415, 416, 421
Integrated Coastal Management (ICM), 86
Integrated Coastal Planning, 10, 206
Integrated Coastal Zone Management (ICZM), 86, 87, 96, 415, 553, 564, 574, 575, 577
Intelligent Vector Data, 80, 81
Interactive Cartography, 558
Intergovernmental Oceanographic Commission (IOC), 41
Intergraph, 193, 194, 198
International Commission for the Scientific Exploration of the Mediterranean (ICSEM), 41
International Coral Reef Initiative (ICRI), 87, 548

International Council for the Exploration of the Sea (ICES), 41, 297, 341, 344, 345
International Geosphere-Biosphere Programme (IGBP), 41, 547
International Maritime Organisation (IMO), 548
Internet, 71, 72, 73, 105, 111, 114, 147, 433, 451, 452, 453, 454, 455, 456, 457, 458, 459, 460, 461, 462, 465, 481, 484, 550, 552, 555, 556, 558, 559, 560, 561, 562, 563, 564, 569, 571, 573, 575, 576, 577
Interpolation, 138, 141, 142, 175, 237, 266, 280, 287, 289, 407, 408, 409, 412, 524, 534
Intertidal, 133, 134, 136, 138, 139, 141, 142, 143, 144, 145, 146, 147, 203, 245, 377, 401, 402, 403, 408, 428, 488, 499, 502
Interwad Online GIS, 568
Invasive Species, 247
Inventory of Biotopes, 491
Invertebrate Beds, 428
Isolines, 138, 142, 262
Isostatic Rebound, 23

Jacobian Matrix, 362
JAVA, 451, 456, 460, 462
JavaScript, 556
Joint Nature Conservation Committee (JNCC), 222, 231, 465, 499, 574
JPEG, 193

Kara Sea, 163
Kelp Forests, 491
Kenfig NNR, 247, 249, 250, 251, 253, 256, 258, 259, 260
King Canute, 61
Kriging, 342, 492

Lagoon, 93, 541
Land Drainage Act, 6, 7, 8
Land Ocean Interaction Study (LOIS), 233, 244, 425, 435, 436
Land Reclamation, 61, 101

Land Use Planning Policy, 154
Land-Ocean Interaction Study (LOIS), 193
Land-Ocean Interactions in the Coastal Zone (LOICZ), 545, 547
Landsat MSS, 386
Landsat TM, 479, 529
Landscape Mapping, 65
Land-Use Planning, 416, 421
Large Format Cameras, 403
Laser Distance Measuring Device, 406
Law of the Sea Convention, 98
Layer-Based Maps, 562
Layered Information Search and Retrieval System (LISaRS), 484
Leica, 518
Levees, 214, 215
Lithological Boundary, 26
Lithology, 21, 22, 191, 197, 198, 218
Littoral Drift, 11, 201, 202, 203, 511, 514, 528
Longitudinal Progradation, 201, 202, 203
Longshore Drift, 14, 198
Louisiana Coastal Geographical Information System Network (LCGISN), 455

Macro-Ecology, 250
Magnetic Tape, 68, 404
Management Strategies, 4, 8, 18, 63, 91, 167, 343
Man-Made Lakes, 205, 206, 207, 210, 211, 212
Mann-Whitney Test, 431
Map Projection, 259
Mariculture, 101
Marine Resource Management, 448
Maritime Boundaries, 97, 100, 102
Maritime Climate, 518
Marram Grass, 250
Mathematical Model, 234
Matrix Inversion, 266
Mean High Water, 134, 143, 152, 156, 158
Mean Low Water, 62

Mean Sea Level, 137
MEDCOAST, 549
Mercator Projection, 262
Mesh Size, 266, 300
Metadata, 71, 107, 113, 115, 125, 127, 165, 166, 169, 357, 544, 546, 548, 552
Metaphor, 455, 476
MIKE 21, 467, 468, 469, 470, 472
MIKE INFO, 467, 468, 469, 470, 471, 472
Ministry of Defence, 78
Ministry of Public Works, 517
Ministry of Transport, 160, 262, 264
Minnaert Constants, 385, 388, 393, 394, 395
Mobile Rover Receiver, 518
Model Aeroplanes, 402
Moray Firth Partnership (MFP), 564, 574
Morphodynamic Analysis, 401
Morphodynamic Equilibrium, 512
MPEG, 405
Multi-Sectoral Data, 89
Murmansk Marine Biological Institute (MMBI), 163

Nadir, 385, 386, 387, 393, 395, 396
National Control Frameworks, 406
National Environmental Management Strategies (NEMS), 87
National Geospatial Data Framework (NGDF), 63, 71, 571, 575
National Grid, 126, 406
National Height Database, 70
National Institute for Coastal and Marine Management (RIKZ), 111
National Nature Reserve (NNR), 225, 249, 250
National Oceanic and Atmospheric Administration (NOAA), 102
National Storm Tide Warning Service (STWS), 15
National Topographic Database of Great Britain (NRD), 66
National Trust (NT), 78, 155, 157

Natural Environment Research Council (NERC), 199, 425, 435, 436
Nature Conservancy Council (NCC), 147, 151, 499
Navtech Systems, 405
Nearest Neighbour, 534
Near-Infrared, 475
Nearshore, 15, 28, 98, 174, 178, 179, 191, 192, 194, 198, 206, 210, 211, 372, 399, 400, 401, 406, 419
Networking, 102, 114, 451, 452, 453, 457, 462, 564, 569, 575
Nikon, 194
Non-Lambertian, 385, 386
NORCOAST, 573, 577, 578
Normal Distribution, 360, 503
NTF (National Transfer Format – British Standard 7367), 81

ODBC Drivers, 167
Offset Printing, 109
Offshore, 13, 26, 489, 577
Oil Dispersal Model, 169
Oil Rig Fabrication, 30
Open GIS Consortium (OGC), 544
Oracle, 429
Ordnance Datum, 82, 137, 141, 143, 194
Ordnance Survey Northern Ireland, 65
Ordnance Survey of Great Britain (OS), 75, 139
Orthogonal Projection, 359
Orthophoto, 518
Overlay Mapping, 126
Overlay Procedure, 442
Oxygen, 527
Oyster Biomass, 261, 263, 271
Oyster Culture Production, 261

Pacific Commission, 94
Pacific Environmental Natural Resource Information Centre (PENRIC), 87
Pacific Regional Strategy, 87
Parabolic Dunes, 24
Pechora Sea, 163, 164, 167

Pelagic, 232, 298, 301, 305, 306, 308, 345
Photic Zone, 118
Photogrammetry, 193, 194, 359
Photomultiplier, 386
Pioneer Species, 247, 250, 259
Piscivore, 305
Pixel Resolution, 402
Plane Coordinates, 406, 409
Plane Table Survey, 94
Planimetry, 518
Planktivore, 305
Plankton, 14, 305, 306, 308, 550
Planning Zone, 151
Plant Ecology, 251
Point-in-Polygon, 287, 356, 357, 366, 367
Pollution, 5, 32, 78, 163, 225, 248, 279, 282, 283, 298, 349, 419, 420, 526, 550
Polychaete Worms, 301
Polynomial, 264, 534
Pop-Up Windows, 556
Position Logging, 405
Postcode Unit, 70
Post-Glacial Ledges, 25
Preglacial Coastline, 23
Preglacial River Valleys, 23
Primary Data Collection, 477
Principal Components Analysis (PCA), 525, 535
Pristine Areas, 164
Progradation, 202
Proprietary Format, 81, 262
Proximal Mapping, 280
Public Interest Groups, 440

Quadrant, 279, 505, 507
Quadrat, 430, 505, 507
Quality Assurance, 141, 165
Quality Control, 141, 165, 351, 353, 376, 433
Quantification, 91, 193, 202, 254, 270, 522
Quantitative Analysis, 114, 128, 145, 146, 184, 342

Quartile, 504
Quintic Interpretation, 138

Radar, 78, 94, 236, 352, 413
Radial Distortion, 518
Radiance Values, 502
Radio Telemetry Techniques, 221
Raised Shoreline, 28, 30, 33
Raster Data, 69, 82, 125, 151, 153, 175, 278
Raster DEM, 349, 351
Real Time Modelling, 102
Rectification, 259, 518, 524
Recurrent Glaciation, 21, 23
Reforestation, 512
Regional Electronic Navigation Chart Centre (RENC), 546
Regressive Coast, 519
Regulatory Framework, 98
Remotely Sensed Imagery, 64, 406, 408, 412, 475, 476, 477, 478, 480, 481, 483, 556, 564
Renewable Energy, 97
Residual, 358
Resource Conflicts, 91, 97, 101
River Boards Act, 6
Rock Revetment, 13
Rock-Girt Coastline, 25
RoxAnn, 490, 491, 492
Rutherford Appleton Laboratories, 556

Salinity, 283, 284
Saltmarsh Communities, 233
Saltwater Flooding, 523
Sampling Method, 166
Sampling Strategy, 113, 115, 128, 507
Satellite Imagery, 94, 125, 169, 250, 456, 474, 477, 547, 552
Satellite Interferometry, 402
Scale, 10, 25, 26, 31, 65, 66, 69, 79, 81, 82, 83, 88, 93, 102, 107, 109, 113, 114, 115, 117, 125, 127, 128, 133, 136, 137, 138, 142, 144, 146, 150, 151, 153, 154, 160, 165, 177, 193, 194, 202, 203, 214, 218, 231, 244, 251, 259, 266, 276, 360, 383,

400, 423, 453, 475, 476, 485, 488, 490, 501, 515, 522, 545, 546, 552, 555, 556
Scanned Aerial Photographs, 278
Scenic Heritage, 2, 16
Scientific Cruises, 163, 166
Scottish Executive, 72
Scottish Natural Heritage (SNH), 116, 131, 151, 452, 463, 464, 487, 488, 490, 574
Scuba, 501
Seabed Sediments, 11, 221, 222, 224
Seabird Colony, 221
Search Engine, 481, 483, 559
Sectoral Management, 99
Sediment Budget, 191, 193, 198, 517
Sediment Cells, 3, 4, 5, 9, 10, 11, 16, 17, 18
Sediment Transport, 11, 65, 78, 83, 202, 210, 211, 514
Semi-Automatic Digitising, 251
Sensitivity Index, 167, 168
Service Level Agreement (SLA), 72
Shelf-Ocean Boundary, 426
Shellfish Leases, 276
Shingle Spits, 24
Shipping Lanes, 80, 118, 407
Shoaling, 301
Shore-Based Fisheries, 79
Shoreline Displacement, 23, 24
Shoreline Equilibrium, 24
Shoreline Evolution, 66
Shoreline Management Plans (SMPs), 3, 8, 11, 17, 19, 61, 133, 134, 136, 145, 146, 147
Sieve Mapping, 116
Sink Holes, 179
Site of Special Scientific Interest (SSSIs), 250
Slope Maps, 179
Small-Format Aerial Photography, 94
Softcopy Photogrammetry, 511, 523
Soft-Shell Clam Beds, 275, 277
Solid Geology, 22, 188
Solway Firth Partnership (SFP), 6, 8, 20, 564, 570, 574

Sonar, 499
Spatial Analysis, 126, 144, 146, 163, 202, 218, 425, 442, 574
Spatial Averaging, 289
Spatial Decision Support System (SDSS), 114, 130, 416, 425, 437, 448
Spatial Pattern, 173, 188, 225, 248
Spatial Resolution, 188, 475, 476, 477, 478, 483, 485, 501
Spatial Sampling, 114
Spatial Statistics, 278
Spatially Referenced, 119, 371, 543
Spatio-Temporal, 231, 407, 410
Spawning Stock, 298, 318, 344
Special Areas of Conservation (SAC), 225, 487, 488
Spectral Properties, 385
Spheroid, 82
Spline Interpolation, 287
SPOT Satellite, 386
SQL Connectivity, 167
SQL Database, 167, 168
State of the Environment Report, 89
Statistical Analyses, 166, 504
Statutory Regulations, 106
Stereocorrelator, 521
Stereomodels, 193
Stereo-Pairs, 251
Stereoplotting, 202, 517, 518
Stock Assessment, 271, 317, 339
Stock Breeding, 422
Stock Recruitment, 324
Stomach Sampling, 300, 338, 342, 343
Storm Beaches, 28
Storm Waves, 118
Stratified Random Samples, 287
Submerged Kelp Beds, 501, 502
Subsidence, 14, 173, 174, 175, 177, 178, 179, 184, 185, 188, 207, 209
Subtidal, 133, 134, 136, 138, 141, 142, 145, 146, 147, 488
Supervised Clustering Technique, 236
Surface Analysis, 175
Surface Modelling, 406
Surface Profiling, 402, 406

Survey Reduction Program, 406
Surveying Techniques, 248
Sustainable Development, 3, 5, 16, 18, 85, 86, 90, 99, 100, 261, 415, 417, 423, 424, 474, 547, 548, 549, 552

TEMA (Training Education and Mutual Assistance), 549
Temporal Accuracy, 357
Temporal Change, 115, 133, 260, 517
Temporal Patterns, 400
Terrain Analysis, 138
Terrain Modelling, 146, 412
Terrestrial Habitats, 248
Territorial Waters, 98
Tessellated Data, 83
Thematic Layers, 218
Tidal Channel, 202, 203
Tidal Inundation, 134
TIN (Triangulated Irregular Network), 140, 142, 153, 202, 521
Tombolos, 211
Topography, 26, 28, 66, 67, 89, 93, 107, 118, 133, 134, 136, 138, 142, 144, 179, 194, 237, 426
Topology, 262
TopoSys, 349, 350, 351, 352
Total Allowable Catch (TAC), 301
Total Coliforms, 527
Total Station, 175, 406, 511
Transects, 140, 241, 266
Transfer Efficiency, 297, 346
Transformed Coordinates, 193
Transition Matrices, 248
Transition Zone, 210
Triangulated Irregular Networks (TINs), 138
Turbidity, 14, 503, 526, 530, 531, 534, 535, 536, 538

UK Digital Marine Atlas, 455
UK Hydrographic Office (HO), 75, 129, 133, 224, 231, 569, 574
UK Meteorological Office, 407
UK National Grid, 82

Uncertainty Model, 359, 361, 364, 365, 367
Unconsolidated Materials, 23, 400
UNEP (UN ENVIRONMENTAL PROGRAMME), 85, 88, 89, 415, 424, 548, 549, 552, 553, 554
UNESCO, 424, 548, 549, 550, 554
United Arab Emirates, 468
United Nations (UN), 85, 88, 95, 98, 99, 104, 294, 297, 342, 344, 347, 424, 548
UNIX Shell Programmes, 430
UNIX Workstations, 351, 430
UNIX, 351, 430
URLs, 556
US Defense Mapping Agency, 165
US Spatial Data Transfer Standard, 357
Usability, 128, 146, 457
User Communities, 72, 552
User Friendly, 454
User-Group, 118
Utilisation, 32
Utilities, 155
Utility Lines, 349

Value-Based Judgements, 154
Valued Ecosystem Components (VECs), 167
Vanuatu Resource Information System (VANRIS), 89
Variance-Covariance Matrix, 361, 362
Variogram, 263
Vector Coverage(s), 138, 251, 254, 256
Vector Data, 69, 81, 83, 278, 280
Vector Format, 518, 561
Vegetated Slacks, 247, 252, 254, 257, 258
Vegetation Coefficient, 234
Vegetation Cover, 234, 257, 282, 351
Vertical Aerial Photographs, 193, 372
Vertical Exaggeration, 194
Vertical Photograph, 403
VHS Tape, 405
Video Camera, 349, 350, 405, 558
Videography, 404, 405, 477

Virtual Reality Modelling Language (VRML), 456
Visualisation, 70, 71, 105, 107, 111, 128, 138, 142, 194, 196, 279, 340, 351, 406, 409, 425

W3C (WWW Consortium), 556
Wadden Sea, 105, 106, 107, 109, 110, 111
Waste Disposal, 97, 101, 420, 425
Water Circulation, 78, 208
Water Column Profiles, 163
Water Level Changes, 208
Water Level Fluctuation, 206, 207, 208, 211
Water Masses, 525, 526, 528, 531, 534, 535, 536, 538, 539
Water Quality, 78, 81, 275, 276, 277, 278, 280, 282, 283, 284, 293, 526, 528, 529, 538, 541
Wave Action, 12, 23, 28, 174, 203
Wave Buoys, 406
Wave Climate, 27, 207, 208
Wave Energy, 12, 13, 15, 184, 210, 400, 401
Wave Exposed, 490, 493
Wave Height, 29, 207, 208, 209, 406, 470
Wave Hindcasting System, 407
Wave Refraction, 403
Wave Shoaling Models, 406
Wave-Induced Currents, 11, 210
Waverley Committee, 6, 7
Webcams, 558
Weighted Scores, 214
Whitstable Judgement, 7
Wind-Generated Waves, 27
Windows NT, 193
World Commission on the Environment and Development, 99
World Geodetic System, 139
World Meteorological Office (WMO), 548
World Wide Web (WWW), 68, 71, 72, 114, 451, 456, 549, 558, 562

Zone of Influence, 422
Zoning Plan, 442